Food *Allergy*

Food *Allergy*

Edited by Soheila J. Maleki, A. Wesley Burks, and Ricki M. Helm

ASM PRESS

Washington, D.C.

Cover design molecule: Model of the major peanut allergen, Ara h 1, produced with the MPACK suite. (C. H. Schein and coworkers, 2006.)

Address editorial correspondence to ASM Press, 1752 N St., N.W., Washington, DC 20036-2904, USA

Send orders to ASM Press, P.O. Box 605, Herndon, VA 20172, USA
Phone: 800-546-2416; 703-661-1593
Fax: 703-661-1501
E-mail: books@asmusa.org
Online: http://estore.asm.org

Library of Congress Cataloging-in-Publication Data

Food allergy / edited by Soheila J. Maleki, A. Wesley Burks,
and Ricki M. Helm.
p. ; cm.
Includes bibliographical references and index.
ISBN-13: 978-1-55581-375-8 (alk. paper)
ISBN-10: 1-55581-375-5 (alk. paper)
1. Food allergy. 2. Food additives—Health aspects. I. Maleki, Soheila J.
II. Burks, A. Wesley. III. Helm, Ricki M.
[DNLM: 1. Food Hypersensitivity. WD 310 F68563 2007]

RC596.F6544 2007
363.19′2—dc22 2006013174

10 9 8 7 6 5 4 3 2 1

CONTENTS

CONTRIBUTORS

Kirsten Beyer • Department of Pediatric Pneumology and Immunology, Children's Hospital Charité, Humboldt University, Augustenburger Platz 1, 13353 Berlin, Germany

Werner Braun • Sealy Center for Structural Biology and Biophysics, Department of Biochemistry and Molecular Biology, University of Texas Medical Branch, 301 University Blvd., Galveston, TX 77555–0857

Ariana Buchanan • Pediatric Allergy and Immunology, Duke University Medical Center, Durham, N.C. 27710

Wesley Burks • Pediatric Allergy and Immunology, Duke University Medical Center, Durham, N.C. 27710

Ye Chu • Department of Horticulture, The University of Georgia Tifton Campus, Tifton, GA 31793–0748

Sarah Comstock • Division of Rheumatology, Allergy, & Clinical Immunology, University of California, Davis, School of Medicine, 451 E. Health Sciences Dr., Suite 6510, Davis, CA 95616

Steven M. Gendel • Food and Drug Administration, 6502 S. Archer Rd., Summit-Argo, IL 60501

Richard E. Goodman • Food Allergy Research and Resource Program, Department of Food Science and Technology, University of Nebraska, Lincoln, NE 68583–0955

Ricki M. Helm • Department of Microbiology/Immunology, University of Arkansas for Medical Sciences, Arkansas Children's Hospital Research Institute, Arkansas Children's Nutrition Center, 1120 Marshall St., Little Rock, AR 72202

Ovidiu Ivanciuc • Sealy Center for Structural Biology and Biophysics, Department of Biochemistry and Molecular Biology, University of Texas Medical Branch, 301 University Blvd., Galveston, TX 77555–0857

Stacie M. Jones • University of Arkansas for Medical Sciences, Division of Allergy and Immunology, Department of Pediatrics, Arkansas Children's Hospital, Little Rock, AR 72202

Gideon Lack • King's College London, Paediatric Allergy Research Department, St. Thomas' Hospital, Lambeth Palace Road, London SE1 7EH, United Kingdom

Samuel J. Landry • Department of Biochemistry SL43, Tulane University Health Sciences Center, 1430 Tulane Ave., New Orleans, LA 70112

Soheila J. Maleki • USDA, ARS, Southern Regional Research Center, 1100 Robert E. Lee Blvd., New Orleans, LA 70124

Adriano Mari • Allergy Unit, National Health Service, Viale di Tor di Quinto, 33/A, I-00191 Rome, Italy

Peggy Ozias-Akins • Department of Horticulture, The University of Georgia Tifton Campus, Tifton, GA 31793–0748

Tamara T. Perry • University of Arkansas for Medical Sciences, Department of Pediatrics, Division of Allergy and Immunology, Arkansas Children's Hospital, Little Rock, AR 72202

Laurent Pons • Pediatric Allergy and Immunology, Duke University Medical Center, Durham, N.C. 27710

Maria Laura Ramos • Department of Horticulture, The University of Georgia Tifton Campus, Tifton, GA 31793–0748

Shridhar K. Sathe • Department of Nutrition, Food, & Exercise Sciences, 402 Sandels Building, Florida State University, Tallahassee, FL 32306–1493

Enrico Scala • Istituto Dermopatico dell'Immacolata, IDI-IRCCS, Via dei Monti di Creta 104, I-00167 Rome, Italy

Catherine H. Schein • Sealy Center for Structural Biology and Biophysics, Department of Biochemistry and Molecular Biology, University of Texas Medical Branch, 301 University Blvd., Galveston, TX 77555–0857

Amy M. Scurlock • University of Arkansas for Medical Sciences, Division of Allergy and Immunology, Department of Pediatrics, Arkansas Children's Hospital, Little Rock, AR 72202

Suzanne S. Teuber • Division of Rheumatology, Allergy, & Clinical Immunology, University of California, Davis, School of Medicine, 451 E. Health Sciences Dr., Suite 6510, Davis, CA 95616

Victor Turcanu • King's College London, Department of Asthma, Allergy & Respiratory Science, 5th Floor, Thomas Guy House, Guy's Hospital, London SE1 9RT, United Kingdom

Mikhael Wallowitz • Division of Rheumatology, Allergy, & Clinical Immunology, University of California, Davis, School of Medicine, 451 E. Health Sciences Dr., Suite 6510, Davis, CA 95616

Carmen D. Westphal • Center for Food Safety and Applied Nutrition, U.S. Food and Drug Administration, Laurel, MD 20708

John Wise • Food Allergy Research and Resource Program, Department of Food Science and Technology, University of Nebraska, Lincoln, NE 68583–0955

PREFACE

The incidence of food allergy appears to be increasing, as is our understanding of the underlying mechanisms, treatment options, and allergenic proteins within food sources. The aim of this book is to translate how this vast array of information may fit into the development of novel therapies and treatments, better diagnosis and detection of allergens, prediction tools for determining potential allergenicity of novel proteins, and reduction in allergenicity of the food supply. We would like to provide clinicians, scientists, educators, and regulators with an authoritative account of the state of knowledge of current topics in food allergy from a multifaceted perspective of well-known experts in the fields of medicine, biochemistry, immunology, bioinformatics, and food science. Particular emphasis has been placed on (i) clinical aspects of food allergy and advances in diagnosis, (ii) cutting-edge immunochemical theories and molecular mechanisms, (iii) immunotherapy and the role of animal models in advancing food allergy research, (iv) detection of and response to existing and novel allergens in the food supply, (v) the use of the most important new bioinformatics tools and characterization and identification of potential allergens with these tools, and (vi) food processing and allergenicity and methods for preparing hypoallergenic food allergen sources. Relatively few foods are commonly involved in food allergy, despite the variety of foods in a normal diet. With true food allergy, the immune system responds to an offending substance that is normally considered to be safe. Proteins in foods can cause both a true food allergy—an immediate hyperimmune response that results in the production of immunoglobulin E (IgE)—and food intolerance, which is usually a metabolically mediated abnormal response to ingested proteins.

In Chapter 1, Perry et al., pediatric clinician-scientists, present the clinical manifestations of true food allergy as opposed to intolerance and IgE-mediated (type I) versus non-IgE-mediated food reactions. Food hypersensitivity (allergy) is synonymous with reactions that involve the IgE mechanism; however, food hypersensitivity disorders can involve cell-mediated immune responses that are mixed with IgE-mediated mechanisms or that may be unrelated to IgE-mediated responses. Food intolerance is a general term describing an abnormal physiologic response as a result of toxic contaminants, pharmacologic properties of food, metabolic disorders, and idiosyncratic responses.

IgE-mediated hypersensitivity accounts for the majority of well-characterized food allergic reactions, although non-IgE-mediated immune mechanisms are believed to be responsible for a variety of hypersensitivity disorders. Acute IgE-mediated allergic reactions can involve one or more body systems with manifestations presenting as cutaneous, gastrointestinal, respiratory, and cardiovascular symptoms. In each case, signs and symptoms of body systems responding to the protein source are discussed with introductory statements discussing therapy, prevention, and patient education.

In Chapter 2, Mari and Scala explore current and novel methods for the detection and diagnosis of food allergy. The clinical approach involves various procedures for the diagnosis of food allergy, taking the reader beyond the scope of performing a complete medical history and physical examination. Highlighted are biological, functional, and laboratory assessments for diagnosing IgE-mediated food allergy, with a discussion of diagnostic reagents and the relevance of allergenic molecules and epitopes of common food allergens. The skin test (prick, prick-prick, or intradermal) is generally considered apart from other true provocation and/or challenge tests (oral, conjunctival, nasal, and bronchial), as it is used in the allergy diagnosis regardless of the organ involved in the allergic disease. The purpose of an in vivo test is to cause an allergic reaction in a tested organ by the direct application of an allergen on a body surface (i.e., skin during skin testing). Differing from the in vivo diagnostic test methods, in vitro analysis detects the presence of specific IgE by means of functional or nonfunctional assays. Both analytical methods contribute significantly to the rapid detection and diagnosis of food allergy.

Teuber et al. introduce the so-called big eight food allergen sources in Chapter 3, discussing immunochemical aspects and clinical impact of food allergy and how cultural and geographical influences could potentially alter the significance of the traditional big eight food categories. The traditional big eight foods—cow's milk, crustaceans, fish, chicken eggs, peanuts, soy, tree nuts, and wheat—and the increasing incidence of sesame food allergy are all discussed. Relevant and most recent information on allergenic food sources, prevalence, cross-reactivity, age of onset, natural history, threshold levels, severity of reaction, and the relevance and the risk assessment of pollen allergens and their association with food allergy are described in detail. Ultimately, allergens will continue to be identified in foods; however, there is a lack of investigations regarding immune decision making (tolerogenic versus allergenic) by the gastrointestinal tract mucosal immune system.

In Chapter 4, Turcanu and Lack discuss the contribution of molecular and immunological responses to food proteins. An outline of the immune structures that interact with the intestinal contents and their specialization to identify and respond to potentially dangerous molecules while maintaining tolerance towards nonpathogenic compounds are presented. In this respect, the ontogenesis of gut immune responses to foods is characterized with further comparisons to these responses induced by commensal pathogenic microorganisms. An identification of the immunological components of such responses that may be amenable to immunomodulation and could thus be used for therapeutic interventions in food allergies follows. A summary includes the most relevant questions that must be answered to design an efficient approach that could lead to the induction and long-term establishment of immune tolerance to foods. Thus, by characterizing the responding immune structures and the processes involved, a link between the information regarding the sequence and structure of food antigens and the clinical characteristics of food allergies is characterized.

As an example, in Chapter 5, Landry provides a profound discussion of the relationship of allergen structure and T-cell epitopes as they relate to the emerging complex interactions between food proteins and allergen processing. Direct evidence suggests that antigen-allergen structure has an influence on epitope dominance. This chapter presents an emerging picture of the complex interplay among allergen processing, T-cell activation, and the T-cell repertoire that shapes the T-cell response to the allergen. Beginning with immune responses to human immunodeficiency virus envelope protein and continuing with responses to hen egg and mouse lysozyme, ovalbumin, bovine lactoglobulin, and birch pollen Bet v 1, structural and epitope dominance that shapes the T-cell response to allergens is discussed. There is a review of structural data and how such data can be used, with implications for the development of allergy, evolution of the immune system, and allergy immunotherapy.

Recognition and identification of food allergy necessarily lead to various types of treatments of food allergy and animal models under investigation, with thoughts toward future objectives. In Chapter 6, Burks et al. highlight traditional and novel therapeutic options for the treatment of food allergy. Included in these options are current peptide immunotherapy, traditional Chinese medicine, fusion and mutated protein immunotherapy, and anti-IgE therapy. New therapies currently under investigation in the research setting should help the physician greatly improve care for patients with food-induced allergic reactions while reducing the risk of anaphylaxis in these patients. Within in the research setting, various animal models and human clinical trials are put forth to alleviate the symptoms and begin to cure specific food allergies.

In Chapter 7, Helm presents an overview of the laboratory and domestic animal models that are currently under investigation as mechanistic, immunotherapeutic, and risk assessment-allergen prediction models. An outline for an ideal food allergy animal model is presented with the advantages and disadvantages of using small and large animal models. Immunomodulatory responses to food proteins are summarized, and the murine models and the neonatal swine model that are being used to address mechanistic studies for food allergic sensitization and different immunotherapeutic strategies are presented. In addition, animal models for determining protein allergenicity detection with risk assessment-allergenicity prediction are discussed. As knowledge about immunological mechanisms, sensitization by food proteins, treatment options, and animal models used for predicting protein allergenicity is advanced, a continuous need for standardization prior to application is required.

In Chapter 8, Westphal identifies the problems and limitations in the detection of food allergens that enter the food chain. Emphasis is placed upon the protection of allergic consumers, which depends greatly on the ability of the food industry to establish and implement plans to avoid cross-contact and ensure the accuracy of food labels. To do so, the food industry has to rely on detection methods to evaluate the presence of food allergens in either food ingredients or final products, as well as to evaluate the adequacy of cleaning procedures. Immunoassays in both qualitative and quantitative formats are currently the leading tests commercially available. However, the application of additional technologies such as PCR is expanding. In spite of significant progress, the detection of food allergens is not free of challenges. For example, the simultaneous detection of food allergens can be complicated due to modification of allergenic proteins during food processing. In addition, new approaches focusing on the detection of single proteins as markers are beginning to emerge.

With the considerable amount of information recently accumulated on the sequences and structures of the proteins that cause food allergies, approaches using proteomics and genomics to interpret this data can predict the potential allergenicity of novel proteins. In Chapter 9, Goodman and Wise discuss the assessment of novel protein allergenicity in genetically modified organisms. The chapter is devoted primarily to evaluating whether the introduced protein is an allergen or is sufficiently similar to suspect potential cross-reactivity, with strong emphasis on the use of computer sequence comparisons between the introduced protein and known allergens. A perspective for allergenic assessment in genetically modified organisms is presented with practical details; evaluations of the criteria established by the U.S. Department of Agriculture, the International Life Sciences Institute, and jointly by the Food and Agriculture Organization of the United Nations and the World Health Organization are compared. Examples are presented that suggest that predictive allergenicity is within the realm of 35% amino acid sequence identity in 80 amino acids; however, the authors clearly demonstrate that we do not have sufficient data to suggest that identities as low as 35% in 80 amino acids are predictive of cross-reactivity. Stressed is the importance of using appropriate scoring matrices and gap and mismatch penalties with criteria that are robust enough to identify potentially cross-reactive proteins without a high likelihood of false-positive predictions.

In Chapter 10, Gendel provides his utilization of this concept. He suggests that the diversity among allergen-related bioinformatic resource databases provides descriptive information that could lead to principles of good database practices. A separation of analytical resources and data is recommended, as well as utilization of data resources toward a more user-friendly simple data exchange. Recommended is a complete characterization that includes a description of the contents of the database, the criteria used to determine which information is included, and information on how the database is updated.

In Chapter 11, Schein et al. then thoroughly describe how the Structural Database of Allergenic Proteins (SDAP) can be used to identify allergens and define structurally related allergenic proteins from different sources. This user-friendly database contains the sequences of <800 allergens. SDAP-Food is a separate directory of food allergens that is cross-referenced to all other known allergenic proteins. This Web resource is designed to be user friendly and to provide clinicians, patients, food scientists, and industrial engineers with an easy way to predict cross-reactive food sources and to analyze common properties of allergenic proteins. The chapter also contains a list of Web sites that contain complimentary information about allergy and allergens. Many clear examples are given of practical uses of SDAP. Theses include rapid FASTA searches for determining the relatedness of a test protein to known allergens, determining appropriate allergen nomenclature, and viewing the structure of the allergens and their epitopes. The chapter also summarizes how the incorporated tools of SDAP, including a novel property distance scale, are being used to analyze IgE epitopes. This information will allow intelligent choices about which new proteins could be introduced into the food chain.

Using the combined information on allergen identification, proteomics, and genomics, various methods to reduce the allergenicity of known food allergens are presented by Ozias-Akins et al. in Chapter 12. Hypoallergenic sources with respect to individual allergens within a food source that down-regulate allergens using gene knockdown, gene knockouts, and RNA silencing are highlighted. Mutation strategies that lead to hypoallergenic proteins are included in the final segment of this chapter, posing the question of how realistic a strategy for allergen gene replacement is.

Regardless of the likely availability of food sources, industrial processing of foods and how differing geographic and cooking methods uniquely influence allergenicity of common foods are addressed in Chapter 13 by Maleki and Sathe. To understand what makes a food allergenic, it is important to study the allergen within the context of that food and to define the interaction of the food components within that matrix. The authors describe some of the structural and molecular components that contribute to allergenicity of proteins and foods and the food-processing methods involved in altering the allergenicity of proteins. They also address possible options for a reduction in the allergenicity of foods.

The book summarizes the clinical and molecular immunology of food allergy, along with computational prediction tools, processing and genetic modification of food proteins for the reduction of allergenic potential, and the development of immunotherapeutic tools. We have compiled current information that demonstrates that a multifaceted approach will allow more rapid advances towards solving and/or treating allergic disease than any single independent approach.

We express our thanks to the clinical and basic scientists who contributed material to this book. We also thank Kenneth H. Roux, Sue Hefle, Barry K. Hurlburt, Anupurna Kewalramani, and Martin C. Cannon for their comments and suggestions.

Ricki M. Helm
Wesley Burks
Soheila J. Maleki

Section I

CLINICAL ASPECTS

Food Allergy
Edited by S. J. Maleki et al.

Chapter 1

Clinical Manifestations of Food Allergic Disease

Tamara T. Perry, Amy M. Scurlock, and Stacie M. Jones

An adverse food reaction is a general term that can be applied to a clinically abnormal response to an ingested food or food additive. Adverse reactions to foods are classified as either food hypersensitivity (allergy) or food intolerance (Anderson and Sogn, 1984; Sampson and Burks, 1996). The utilization of these terms has allowed better communication regarding various reactions to food components.

Food hypersensitivity (allergy) is due to an immunologic reaction resulting from the ingestion of a food or food additive (Table 1). This reaction occurs only in some patients, may occur after only a small amount of the substance is ingested, and is unrelated to any physiologic effect of the food or food additive. To most physicians, the term is synonymous with reactions that involve the immunoglobulin E (IgE) mechanism, of which anaphylaxis is the classic example. Several other food hypersensitivity disorders involve cell-mediated immune responses that are mixed with IgE-mediated mechanisms or may be unrelated to IgE-mediated responses.

Food intolerance (Table 2) is a general term describing an abnormal physiologic response to an ingested food or food additive. This reaction has not been proven to be immunologic in nature and may be caused by many factors including toxic contaminants (e.g., histamine in scromboid fish poisoning, toxins secreted by *Salmonella*, *Shigella*, and *Campylobacter*), pharmacologic properties of food (e.g., caffeine in coffee, tyramine in aged cheeses, and sulfites in red wine), characteristics of the host such as metabolic disorders (e.g., lactase deficiency), and idiosyncratic responses. The term food intolerance has often been overused and, like the term food allergy, has been applied incorrectly to all adverse reactions to foods.

IgE-mediated (type I) hypersensitivity accounts for the majority of well-characterized food allergic reactions, although non-IgE-mediated immune mechanisms are believed to be responsible for a variety of hypersensitivity disorders. In this chapter, we will examine adverse food reactions that are IgE mediated and non-IgE mediated and those entities that have characteristics of both.

Tamara T. Perry, Amy M. Scurlock, and Stacie M. Jones • University of Arkansas for Medical Sciences, Division of Allergy and Immunology, Department of Pediatrics, Arkansas Children's Hospital, Little Rock, AR 72202.

Table 1. Food hypersensitivity disorders[a]

IgE-mediated	Mixed IgE and non-IgE mediated	Non-IgE mediated
Generalized		
Anaphylaxis		
Cutaneous		
Utricaria	Atopic dermatitis	Contact dermatitis
Angioedema		Dermatitis herpetiformis
Morbilliform rash		
Flushing		
Gastrointestinal		
Pollen-food allergy syndrome	Eosinophillic esophagitis	Food protein-induced enterocolitis
Immediate gastrointestinal hypersensitivity	Eosinophilic gastroenteritis	Food protein-induced enteropathy
		Food protein-induced proctitis
		Celiac disease
Respiratory		
Acute rhinoconjunctivitis	Asthma	Food-induced pulmonary hemosiderosis
Laryngospasm		(Heiner's syndrome)
Bronchospasm		

[a]Reproduced with permission from Sampson (2005).

Table 2. Food intolerance: nonimmunologic adverse reactions

Toxic, pharmacologic	Nontoxic, intolerance
Bacterial food poisoning	Lactase deficiency
Heavy metal poisoning	Galactosemia
Scromboid fish poisoning	Pancreatic insufficiency
Caffeine	Gallbladder-liver disease
Tyramine	Gastroesophageal reflux
Histamine	Hiatal hernia
Monosodium glutamate	Gustatory rhinitis
	Anorexia nervosa

PREVALENCE

The true prevalence of adverse food reactions is still unknown. Up to 25% of the general population believe that they may be allergic to some food (Sloan and Powers, 1986). The actual prevalence of food allergy appears to be 1.5 to 2% of the adult population (Sampson, 1998), and the prevalence of adverse food reactions in young children is estimated to be approximately 6 to 8% (Bock, 1987; Sampson, 2004). Evidence suggests that the prevalence of food allergy is increasing in industrialized societies; studies in the United States and the United Kingdom have reported a twofold increase in reported peanut allergy over the last decade (Grundy et al., 2002; Sicherer et al., 2003) In fact, Grundy et al. (2002) reported a threefold increase in the number of children sensitized to peanuts when birth cohorts from 1989 and 1994 were compared.

The vast majority of food allergic reactions are secondary to a limited number of foods. The most common foods causing allergic reactions include milk, egg, peanuts, tree nuts, and fish in children and peanuts, tree nuts, fish, and shellfish in adults (Sampson, 1999). Burks and colleagues (1998) described the utility of a limited panel of food skin tests to screen for food allergy in 165 children with atopic dermatitis (AD). Eighty-nine percent of children with food allergies were appropriately identified by skin prick tests (SPTs) of seven foods (milk, egg, peanut, fish, cashew, soy, and wheat); most had allergies to only one or two foods.

IMMUNOLOGIC RESPONSES TO FOOD ALLERGENS

IgE-Mediated Reactions

Oral tolerance to food allergens occurs as the result of an appropriate suppression of the immune system when the gut mucosa comes in contact with dietary proteins. A variety of hypersensitivity responses to an ingested food antigen result from the lack of development of oral tolerance or a breakdown of oral tolerance in the gastrointestinal tract of genetically predisposed individuals (Mayer, 2003). Either a failure to develop or a breakdown in oral tolerance results in excessive production of food-specific IgE antibodies. These food-specific antibodies bind high-affinity IgE receptors (FcεRI) on mast cells and basophils and low-affinity IgE receptors (FcεRII) on macrophages, monocytes, lymphocytes, eosinophils, and platelets (Sampson and Burks, 1996). Upon reexposure to the offending food, the food allergen binds to the food-specific antibodies on mast cells or basophils, and mediators such as histamine, prostaglandins (PGs), and leukotrienes are released. These mediators then promote vasodilatation, smooth muscle contraction, and mucus secretion, resulting in the symptoms of immediate hypersensitivity. The activated mast cells also release various cytokines important in inflammatory cell recruitment responsible for late-phase allergic responses. With repeated ingestion of a specific food allergen, mononuclear cells are stimulated to secrete histamine-releasing factors. The spontaneous generation of histamine-releasing factors by the activated mononuclear cells in vitro has been associated with increased cutaneous irritability in children with AD. A rise in plasma histamine levels has been associated with IgE-mediated allergic symptoms after physician-supervised food challenges (Ohtsuka et al., 1993; Sampson and Jolie, 1984). In IgE-mediated gastrointestinal reactions, endoscopic observation has revealed local vasodilatation, edema, mucus secretion, and petechial hemorrhaging. Increased stool and serum PGE_2 and PGF_2 have been seen after food challenges, causing diarrhea.

Non-IgE-Mediated Reactions

Although a variety of reports have discussed other immune mechanisms causing food allergic reactions, the scientific evidence supporting these mechanisms is limited. Type III (antigen-antibody complex-mediated) hypersensitivity reactions have been examined in several studies. While IgE-food antigen complexes are seen more commonly in patients with food hypersensitivity, there is little support for food antigen immune complex-mediated disease. Type IV (cell-mediated) hypersensitivity has been discussed for several disorders where the clinical symptoms do not appear until several hours after the ingestion of the suspected food. This type of immune response may contribute to some adverse food reactions (i.e., enterocolitis), but evidence for a specific cell-mediated hypersensitivity disorder is lacking.

CLINICAL MANIFESTATIONS OF FOOD HYPERSENSITIVITY

IgE-Mediated Hypersensitivity

Generalized Anaphylaxis

Generalized anaphylaxis is defined by the American Academy of Allergy, Asthma, and Immunology (1998) as the "immediate systemic reaction caused by rapid, IgE-mediated immune release of potent mediators from tissue mast cells and peripheral basophils." Food-induced generalized anaphylaxis involves multiple organ systems as a result of food ingestion in a previously sensitized individual. The onset of symptoms is abrupt, and most often occurs within minutes of ingestion. Any organ system or groups of systems can be involved in an anaphylactic reaction, and symptoms are due to the effects of potent intracellular mediators such as histamine, tryptase, and leukotrienes that are released from mast cells and basophils during an allergic reaction.

Acute IgE-mediated allergic reactions can involve one or more body systems. Acute cutaneous manifestations include symptoms such as erythema, urticaria, or angioedema. Gastrointestinal symptoms of food allergy can include cramping, abdominal pain, nausea, vomiting, and diarrhea. The respiratory tract is often involved in acute allergic reactions and can result in mild symptoms of the upper respiratory tract such as rhinorhea, sneezing, or congestion, while lower respiratory involvement is related to more serious symptoms such as wheezing, difficulty breathing, or respiratory failure. The cardiovascular system, if involved, often represents a serious manifestation of allergic reactions to foods and includes symptoms such as tachycardia, hypotension, syncope, and cardiac failure. Severe or life-threatening generalized anaphylactic reactions involving the respiratory and cardiovascular systems can culminate in death if left untreated.

Fatal and near-fatal reactions to foods have been previously described (Bock et al., 2001; Sampson et al., 1992; Yunginger et al., 1991). The most common allergens reported to cause fatal or near-fatal reactions are peanuts and tree nuts. Fatality risk factors include a history of asthma, history of previous serious reaction, and delayed administration of epinephrine. In theses studies, all individuals had known allergies to the causal food and had unintentionally ingested the food allergen. The results of these studies convey the importance of education on allergen avoidance, strict label-reading practices to avoid allergic reactions, and appropriate treatment strategies to employ during a reaction for food-allergic individuals.

Cutaneous

The skin is the most common target organ in IgE-mediated food hypersensitivity reactions, and cutaneous symptoms occur in >80% of allergic reactions to foods (Perry et al., 2004). The ingestion of food allergens can either lead to immediate cutaneous symptoms or exacerbate chronic conditions such as AD. Acute urticaria and angioedema are the most common cutaneous manifestation of food hypersensitivity reactions, generally appearing within minutes of ingestion of the food allergen (Color Plate 1). Flushing, pruritis, and morbilliform rash are other acute cutaneous manifestations that commonly occur during allergic reactions to foods.

Respiratory and Ocular

Upper respiratory symptoms such as rhinorhea, sneezing, nasal congestion, and pruritis are frequently experienced during allergic reactions to foods. Nasal symptoms typically occur in conjunction with other organ system involvement, and isolated nasal symptoms are considered rare (Bock and Atkins, 1990). Ocular symptoms commonly occur concurrently with respiratory manifestations of IgE-mediated reactions to foods (Sampson, 1998; Sampson and Burks, 1996). Symptoms may include periocular erythema, pruritus, conjunctival erythema, and tearing (Color Plate 2).

Lower respiratory symptoms are potentially life-threatening manifestations of IgE-mediated reactions to foods (Sampson, 1998; Sampson and Burks, 1996). Symptoms can include wheezing, coughing, and laryngospasm and require prompt medical intervention. In a retrospective chart review of 253 failed oral food challenges, Perry and colleagues (2004) found that 26% of participants experienced lower respiratory symptoms; each of the foods tested carried a similar risk for eliciting lower respiratory symptoms. Although lower respiratory symptoms can occur in any person experiencing food anaphylaxis, patients with an underlying lung disease such as asthma are at increased risk of severe symptoms. Lower respiratory symptoms due to food allergy are temporally related to ingestion and are typically accompanied by other organ system involvement. It is not likely that chronic lower respiratory symptoms or poorly controlled asthma are sole manifestations of food allergy (James, 2003a, 2003b).

Gastrointestinal

The signs and symptoms of food-induced IgE-mediated gastrointestinal allergy may be secondary to a variety of syndromes, including the pollen-food allergy syndrome and immediate gastrointestinal hypersensitivity (Bock, 1982). The pollen-food allergy syndrome (Table 3) is considered a form of contact urticaria that is confined almost exclusively to the oropharynx and rarely involves other target organs. The symptoms include rapid onset of pruritus and angioedema of the lips, tongue, palate, and throat. The symptoms generally resolve quite rapidly without progression to systemic involvement. This syndrome is most commonly associated with the ingestion of fresh fruits and vegetables, not processed or cooked foods. Interestingly, patients with allergic rhinitis secondary to certain

Table 3. Pollen-food allergy syndrome (oral allergy syndrome)

Pollen	Fruit and/or vegetable
Birch	Apple, cherry, apricot, carrot, potato, kiwi, hazelnut, celery, pear, peanut, soybean
Ragweed	Melon (cantaloupe, honeydew, etc.), banana
Grass	Kiwi, tomato, watermelon, potato
Mugwort	Celery, fennel, carrot, parsley
Latex	Banana, avocado, chestnut, kiwi, fig, apple, cherry

airborne pollens (especially ragweed, birch, and mugwort pollens) are frequently afflicted with this syndrome. Patients with ragweed may experience these symptoms following contact with certain melons (watermelons, cantaloupe, honeydew, etc.) and bananas. Those patients with birch sensitivity often have symptoms following the ingestion of raw potatoes, carrots, celery, apples, and hazelnuts. The diagnosis of this syndrome is made after a suggestive history and positive SPTs with the implicated fresh fruits or vegetables (Bock, 1985). The caveat in this syndrome is that the commercially available allergen extracts for fruits and vegetables may be heat labile and often do not have the reliability of an allergen from the fresh food. It may be necessary to use the prick-to-prick method, where the device used for introducing the allergen into the skin is initially pricked into the fresh food.

Immediate gastrointestinal hypersensitivity (Table 4) is a form of IgE-mediated gastrointestinal hypersensitivity, which may accompany allergic manifestations in other target organs (Bock and Atkins, 1990; Sampson, 1998). The symptoms vary but may include

Table 4. Gastrointestinal food allergic disorders

Disorder	Key features	Causal food[a]	Natural history[a]
IgE mediated			
Pollen-food allergy syndrome	Oral pruritus, mild angioedema of oral cavity	Fresh (uncooked) fruits and vegetables	Unknown
Immediate gastrointestinal hypersensitivity	Acute nausea, vomiting, pain, diarrhea	M, E, S, W, PN, TN, F, SF	80% of cases resolve after 1–3 yr of protein elimination diet except PN, TN, F, SF, which are likely lifelong
Mixed IgE and non-IgE mediated			
Eosinophilic esophagitis	Dysphagia, postprandial nausea and vomiting, epigastric pain	Multiple	1–2 yr to protracted
Esoinophilic gastroenteritis	Vomiting, abdominal pain, diarrhea, malabsorption, failure to thrive	Multiple	1–2 yr to protracted
Non-IgE mediated			
Food protein-induced enterocolitis	Vomiting, diarrhea, poor growth, lethargy, dehydration	M, S, grains	50% of cases resolve by 18 mo; 90% resolve by 36 mo with avoidance
Food protein-induced enteropathy	Malabsorption, emesis, poor growth, diarrhea	M, S	Prolonged
Food protein-induced proctocolitis	Bloody diarrhea, mucus in stools, normal growth	M, S	Typically resolves after 6–24 mo of avoidance
Celiac disease	Malabsorption, failure to thrive, diarrhea	W, barley, rye	Requires lifelong avoidance

[a]M, milk; E, egg; S, soy; PN, peanut; W, wheat; TN, tree nut; F, fish; SF, shellfish.

nausea, abdominal pain, abdominal cramping, vomiting, and/or diarrhea. In studies of children with AD and food allergy, the frequent ingestion of a food allergen appears to induce partial desensitization of gastrointestinal mast cells, resulting in less-pronounced symptoms. The diagnosis of these symptoms is made by a suggestive clinical history, positive SPTs, complete elimination of the suspected food allergen for up to 2 weeks with resolution of symptoms, and oral food challenges. After avoidance of a particular food for 10 to 14 days, it is not unusual for symptoms of vomiting to occur during a challenge, although the patient previously ingested the food without having vomiting each time it was eaten.

Mixed IgE-Mediated and Non-IgE-Mediated Hypersensitivity

Cutaneous

AD is a chronic skin disorder that generally begins in early infancy and is characterized by typical distribution, extreme pruritus, chronically relapsing course, and association with asthma and allergic rhinitis (Color Plate 3) (Eigenmann et al., 1998). Food allergy has been strongly correlated with the development and persistence of AD, especially during infancy and early childhood. Approximately 35 to 40% of children who are <5 years old with moderate to severe AD will be allergic to at least one food (Burks, 2003; Eigenmann et al., 1998). These patients typically fail to respond to conventional medical therapy or may have frequent exacerbations of underlying skin disease if causal foods are not strictly avoided. The foods most commonly related to chronic cutaneous symptoms include milk, eggs, peanuts, soy, wheat, fish, and tree nuts. Due to the chronic nature of symptoms, it may be difficult to identify causal foods based on patient and/or parental report, and supervised oral food challenges may be warranted to aid in the diagnosis of food allergy. Dietary elimination of relevant food allergens may result in clearing of the skin. Some patients continue to have ongoing skin disease, due to sensitization to aeroallergens or due to non-allergic triggers (Color Plate 4).

Gastrointestinal

Allergic eosinophilic gastroenteropathies are a group of disorders characterized by eosinophilic infiltration of at least one layer of the gastrointestinal tract, absence of vasculitis, and peripheral eosinophilia (Bock, 1987; Sampson and Anderson, 2000). These disorders are defined by the site(s) of involvement and include eosinophilic esophagitis (EE) and eosinophilic gastroenteritis. Symptoms for each of these syndromes are generally related to the specific anatomical site of involvement; however, failure to thrive, irritability, abdominal pain, and dysphagia can be common manifestations (Rothenberg, 2004).

EE is characterized by severe gastroesophageal reflux-like symptoms, dysphagia, epigastric pain, and postprandial nausea and vomiting (Sicherer, 2003). Eosinophils are not normally found in the esophageal mucosa, and symptoms are due to infiltration of eosinophils and associated inflammatory mediators. EE is clinically distinguishable from gastroesophageal reflux disease (GERD) because it is refractory to aggressive management with acid blockers, proton pump inhibitors, and promotility medications that are typically effective in the treatment of GERD. Other distinguishing characteristics include normal pH probe results, patient or family history of atopy, and peripheral eosinophilia. On endoscopy, EE patients can have visually normal-appearing esophageal mucosa, although esophageal furrowing and rings have been previously reported (Liacouras and Ruchelli, 2004). On histologic examination, esophageal biopsies from patients with

EE typically contain >20 eosinophils per high-powered field (HPF) compared with <5 eosinophils per HPF in patients with GERD (Color Plate 5). Proximal and midesophageal lesions are common in EE, while reactive eosinophilic infiltrates due to GERD are mainly limited to the distal esophagus (Fox et al., 2002).

Treatment of EE involves strict elimination of the offending food(s); however, identification of a causal food(s) may prove to be difficult. Positive food-specific IgE levels and SPTs may offer clues to the identification of food allergens; however, these diagnostic tools are often negative in EE. Patch skin testing may prove to be a useful diagnostic aid for some patients with EE (Andrews et al., 2004); however, further studies are needed to better define the utility of this diagnostic tool. When the causal food cannot be readily identified, a hypoallergenic diet with an elemental formula and a few low-risk foods (such as non-citrus fruits and vegetables) should be initiated. Symptoms will typically improve within 10 days, and food groups can be slowly reintroduced into the diet at approximately 2- to 3-week intervals. The offending food(s) is identified by the recurrence of symptoms upon reintroduction. Repeat and frequent endoscopy may be necessary to document remission or exacerbations. Other therapies have proven clinical benefit but do not have long-term clinical efficacy if the offending food remains in the diet. These therapies include systemic and topical steroids, cromolyn sodium, and leukotriene receptor antagonists. EE typically has a prolonged course, and reintroduction of foods can be considered after a minimum of 2 years of avoidance and significant improvement of esophageal histologic findings.

Eosinophilic gastroenteritis is characterized by eosinophilic infiltration of the stomach, small intestine, or both, with variable involvement of the large intestine (Rothenberg, 2004). Symptoms are related to the area and severity of involvement and can include vomiting, abdominal pain, diarrhea, malabsorption, and failure to thrive. Severe symptoms can mimic pyloric stenosis or other forms of gastric outlet obstruction when duodenal involvement is present. Because eosinophils are normally found in the stomach and intestine, endoscopic findings are more difficult to interpret than with EE. In addition, multiple sites (up to eight) may need to be biopsied to effectively exclude eosinophilic gastroenteritis because the eosinophilic infiltrates may be quite patchy. Typically, biopsies to establish a diagnosis of eosinophilic gastroenteritis will have 20 to 40 eosinophils per HPF. Treatment involves elimination of the offending food(s). If food-specific IgE levels or SPTs identify a specific food(s), that food or foods should be eliminated first. If there are no clues to the offending agent based on history, serum levels, or skin testing, then an elemental diet, followed by the slow reintroduction of foods, should be carried out. Similar to EE, eosinophilic gastroenteritis usually follows a prolonged course; foods can be reintroduced after several months to years of avoidance and according to endoscopic findings.

Non-IgE-Mediated Food Hypersensitivity

Cutaneous

Food-induced contact dermatitis has been reported in individuals without IgE antibodies to the causal food (Hjorth and Roed-Petersen, 1976). This reaction typically occurs in food handlers and can be confirmed by patch testing. Implicated foods include raw fish, shellfish, meats, and eggs (Sampson, 2003).

Dermatitis herpetiformis (DH) is a skin manifestation of celiac disease (gluten-sensitive enteropathy). DH is a chronic blistering skin rash characterized by chronic, pruritic, papulovesicular lesions that are symmetrically distributed over the extensor surfaces and buttocks (Hall, 1992). Histologic examination of skin lesions demonstrates granular IgA deposition at the dermal-epidermal junction (Egan et al., 1997). Gastrointestinal clinical manifestations are generally milder, and histologic examination of the gut mucosa shows less-extensive changes than with those patients presenting with primary gastrointestinal disease. Elimination of gluten from the diet typically results in resolution of skin and gut lesions.

Gastrointestinal

Food protein-induced enterocolitis syndrome (FPIES) is a disorder that presents most commonly in early infancy. Acute symptoms are typically isolated in the gastrointestinal tract and consist of profuse, recurrent vomiting and/or diarrhea. Symptoms are often severe enough to cause dehydration or be erroneously diagnosed as sepsis. Chronic symptoms due to FPIES include failure to thrive, anemia, and hypoalbuminemia. Cow's milk and/or soy protein in infant formulas or maternal breast milk is most often responsible for this syndrome, although FPIES due to solid food has been previously reported (Sicherer, 2005). Objective findings on stool examination consist of the presence of gross or occult blood, polymorphonuclear neutrophils, eosinophils, Charcot-Leyden crystals, and positive reducing substances (indicating malabsorbed sugars). SPTs for the putative food protein are characteristically negative. Jejunal biopsies classically reveal flattened villi, edema, and increased numbers of lymphocytes, eosinophils, and mast cells. A food challenge with the responsible protein generally results in vomiting and/or diarrhea within minutes to several hours, occasionally leading to shock (Goldman et al., 1963; Sampson and Anderson, 2000).

It is not uncommon to find children who are intolerant of both cow's milk and soy protein. Approximately 50% of children with non-IgE-mediated cow's milk allergy will have concomitant soy allergy; therefore, it is recommended that infants with FPIES due to cow's milk avoid soy products. Elimination of the offending allergen generally results in improvement or resolution of the symptoms within 72 h, although secondary disaccharidase deficiency may persist longer. This disorder tends to subside by 18 to 24 months of age. Oral food challenges should be done in a medical setting because they can induce severe vomiting, diarrhea, dehydration, or hypotension. Symptoms should be treated primarily with an intravenous fluid bolus and aggressive maintenance of adequate hydration. Patients with FPIES typically do not respond to antihistamines and epinephrine.

Dietary protein enteropathy is characterized by diarrhea, vomiting, malabsorption, and poor weight gain. It is clinically distinguishable from FPIES because stools are typically nonbloody, vomiting is less prominent, and reexposure does not elicit acute symptoms after a period of avoidance. Other clinical features include abdominal pain and distension, hypoproteinemia, and edema. Patients usually present in the first year of life; the most common causal food is cow's milk, although other foods such as soy, eggs, and grains have been associated with food protein-induced enteropathy. Histologic examination reveals patchy villous atrophy, mononuclear cell infiltrates, and few eosinophils. Symptoms typically resolve within 72 h after dietary elimination. Dietary protein enteropathy is usually outgrown within 6 to 24 months after allergen avoidance.

Dietary protein proctocolitis generally presents in the first few months of life and is often secondary to cow's milk or soy protein hypersensitivity (Crowe and Perdue, 1992).

Infants with this disorder often do not appear ill; the condition is generally discovered because of the presence of blood (gross or occult) in patients' stools. Other distinguishing features include normal growth and an absence of vomiting. Gastrointestinal lesions are usually confined to the rectum but can involve the entire large bowel and consist of eosinophilic infiltrates (5 to 20 infiltrates per HPF) or eosinophilic abscesses in the epithelium and lumina propria. If lesions are severe with crypt destruction, polymorphonuclear leukocytes are also prominent in this disorder (Jenkins et al., 1984). It is thought, but without well-controlled studies, that cow's milk- and soy protein-induced proctocolitis resolves after 6 months to 2 years of allergen avoidance. Elimination of the offending food allergen leads to resolution of hematochezia within 72 h, but the mucosal lesions may take up to 1 month to disappear and range from patchy mucosal injection to severe friability with small aphthoid ulcerations and bleeding.

Celiac disease (or gluten-sensitive enteropathy) is an extensive enteropathy leading to malabsorption. Total villous atrophy and extensive cellular infiltrates are associated with sensitivity to gliadin, the alcohol-soluble portion of gluten found in wheat, rye, and barley. Celiac disease is almost exclusively limited to genetically predisposed individuals who express the HLA-DQ2 and/or HLA-DQ8 heterodimer (Papadopoulos et al., 2001; Sollid and Thorsby, 1993). Patients often have presenting symptoms of diarrhea or frank steatorrhea, abdominal distention, flatulence, failure to thrive, and occasionally nausea and vomiting. Oral ulcers and other extraintestinal symptoms secondary to malabsorption are not uncommon. Serologic testing aids in the diagnosis, and the most reliable diagnostic test is measurement of IgA to human recombinant tissue transglutaminase (Hill et al., 2005). Anti-endomysial IgA antibodies are accurate; however, measurement of endomysial IgA is observer dependent and therefore more subject to human error. One caveat to the diagnosis of celiac disease lies with those patients with concomitant IgA deficiency. IgA antibody levels will be low in these individuals; therefore, serologic tests to detect IgA are unreliable with this population. Regardless of the results of serologic testing, confirmation with endoscopic examination with biopsies should be performed for all individuals with suspected celiac disease. Histologic examination of the small bowel reveals total villous atrophy and inflammatory infiltrates. Findings on endoscopy and clinical symptoms resolve with strict elimination of gluten from the diet. Patients must remain on a gluten-free diet for their entire lifetimes.

Respiratory

Food-induced pulmonary hemosiderosis (also known as Heiner's syndrome) is characterized by recurrent pulmonary infiltrates, hemosiderosis, gastrointestinal blood loss, iron deficiency anemia, and failure to thrive (Heiner et al., 1962). Symptoms are associated with non-IgE-mediated hypersensitivity to cow's milk with peripheral eosinophilia and multiple cow's milk precipitins as diagnostic clues. Strict dietary elimination of milk reverses symptoms.

MEDICAL THERAPY FOR FOOD ALLERGIES

Once the diagnosis of food allergy is established, the only proven therapy is the strict elimination of the food from the patient's diet. Elimination diets may lead to malnutrition and/or eating disorders, especially if these diets include a large number of foods and/or are utilized for extended periods of time. Studies have shown that symptomatic food sensitivity generally is lost over time except for sensitivity to peanuts, tree nuts, fish, and shellfish (Wood, 2003).

Symptomatic food sensitivity is usually very specific, so patients rarely react to more than one member of a botanical family or animal species. Certain factors place some individuals at increased risk for more severe anaphylactic reactions: (i) history of a previous anaphylactic reaction; (ii) history of asthma, especially if poorly controlled; (iii) allergy to peanuts, nuts, fish, and shellfish; (iv) use of β-blockers or angiotensin-converting enzyme inhibitors; and (v) (possibly) being female.

Medications

Several medications have been used in an attempt to protect patients with food hypersensitivity, including oral cromolyn, H_1 and H_2 antihistamines, corticosteroids, and PG synthetase inhibitors. Some of these medications may modify food allergy symptoms, but overall they have minimal efficacy or unacceptable side effects. The use of epinephrine is vitally important in acute anaphylaxis. The importance of prompt epinephrine administration when symptoms of systemic reactions to foods develop cannot be overemphasized. Autoinjectable epinephrine (trade names include EpiPen, EpiPen Jr, and Twin Ject) can be given intramuscularly or subcutaneously, with most recent studies suggesting that intramuscularly administration is more effective (Simons et al., 2001). If an autoinjector is unavailable, the recommended weight-based dose is 0.01 mg/kg of body weight (maximum initial dose, 0.3 mg).

Other Therapies

Recent blinded, placebo-controlled studies of rush immunotherapy for the treatment of peanut hypersensitivity demonstrated efficacy with a small number of patients (Oppenheimer et al., 1992). The adverse reaction rates were significant and preclude general clinical application at this time.

Newer types of vaccines for immunotherapy specifically for food-induced anaphylaxis being developed (Burks et al., 2001) include (i) humanized anti-IgE monoclonal antibody therapy, (ii) plasmid-DNA immunotherapy, (ii) peptide fragments using so-called overlapping peptides, (iv) cytokine-modulated immunotherapy, (v) immunostimulatory sequence-modulated immunotherapy, (6) bacterial-encapsulated allergen immunotherapy, and (vi) engineered recombinant protein immunotherapy.

Additionally, recent studies with humanized, monoclonal antibody anti-IgE have been utilized in phase I trials for patients with peanut allergy (Leung et al., 2003). This type of therapy appears to be a promising option for the future for patients with a history of food-induced anaphylaxis or with a food allergy where the patient is at risk for a future systemic, anaphylactic reaction.

NATURAL HISTORY AND PROGNOSIS

Most food allergies develop in the first 1 to 2 years of life (Color Plate 6), and the majority of individuals develop oral tolerance over time (Bock, 1982; Bock, 1987; Bock and Atkins, 1990). Studies suggest that nearly 80% of children with milk allergy will outgrow the allergy by the age of 3 years (Host and Halken, 1990); egg allergy is typically outgrown by the school-age years (Dannaeus and Inganas, 1981; Ford and Taylor, 1982). Children who develop their food sensitivity after 3 years of age are less likely to lose their food reactions over a period of several years. Patients who develop very mild reactions (skin symptoms only) to peanuts early in life (i.e., during the first 12 to 24 months) may outgrow their symptoms (Hourihane et al., 1998; Skolnick et al., 2001). Allergies to foods such as peanuts, tree nuts, fish, and shellfish are generally not outgrown, no matter at what age they develop. These individuals appear likely to retain their allergic sensitivity for a lifetime, although recent reports have suggested that approximately 20% of peanut-allergic and 9% of tree nut-allergic individuals may outgrow their allergy (Fleischer et al., 2003; Fleischer et al., 2005). Consequently, several groups are evaluating new strategies to desensitize patients to these foods.

PRIMARY PREVENTION OF FOOD ALLERGIES

Primary prevention of food allergies relates to blocking immunologic sensitization to foods. Recommendations have been published by the Nutritional Committee of the American Academy of Pediatrics (2000), and updates to these guidelines are expected in 2006. Primary prevention measures are aimed at individuals at high risk of developing food allergies and include infants with a biparental history (or a history involving one sibling plus one parent) of atopic disease such as asthma, allergic rhinitis, or food allergies. Prevention strategies include supplementation with hypoallergenic formula if the infant is not breast-feeding or if supplementation is needed during breast-feeding. Supplementation with soy formula is not recommended as an alternative to cow's milk for patients with non-IgE-mediated food allergy, due to the high risk of concomitant soy allergy in these individuals. Other primary prevention strategies include delayed introduction of solids until 6 months of age, with continued restriction of cow's milk until 6 to 12 months of age; restriction of eggs until 24 months of age; and restriction of peanuts, tree nuts, and fish until 36 months of age. Similar recommendations have been published by the European Society for Pediatric Gastroenterology, Hepatology, and Nutrition (Host et al., 1999).

PATIENT EDUCATION

Patient education and support are essential for food-allergic patients. In particular, adults and older children prone to anaphylaxis (and their parents) must be informed in a direct but sympathetic way that these reactions are potentially fatal.

When eating away from home, food-sensitive individuals should feel comfortable requesting information about the contents of prepared foods. For the school-aged child, the American Academy of Pediatrics Committee on School Health has recommended that schools be equipped to treat anaphylaxis in allergic students. Children over the age of 7 years can usually be taught to inject themselves with epinephrine. The physician must be willing to explain and, with the parents, help instruct school personnel about these issues. In the home, parents should consider the need to eliminate the incriminated allergen, or if this is not practical, to place warning stickers on foods with the offending antigens.

A variety of groups can provide support, advocacy, and education, including The Food Allergy & Anaphylaxis Network (10400 Easton Pl., Suite 107, Fairfax, VA 22030-5647;

www.foodallergy.org), the Allergy & Asthma Network Mothers of Asthmatics (2751 Prosperity Ave., Suite 150, Fairfax, VA 22031; www.aanma.org), and the Asthma and Allergy Foundation of America (1125 15th St., NW, Suite 502, Washington, DC 20005; www.aafa.org).

REFERENCES

American Academy of Allergy, Asthma, and Immunology. 1998. Anaphylaxis in schools and other childcare settings. AAAAI Board of Directors. *J. Allergy Clin. Immunol.* **102:**173–176.

American Academy of Pediatrics. 2000. Committee on Nutrition. Hypoallergenic infant formulas. *Pediatrics* **106:**346–349.

Anderson, J., and D. Sogn. 1984. *Adverse Reactions to Foods.* American Academy of Allergy and Immunology Committee on Adverse Reactions to Foods and the National Institute of Allergy and Infectious Disease, National Institute of Allergy and Infectious Disease. National Institutes of Health publication no. 84-2442. National Institutes of Health, Bethesda, Md.

Andrews, T., J. M. Spergel, and C. A. Liacouras. 2004. Treatment of eosinophilic esophagitis with specific food elimination diet directed by a combination of prick skin test and patch tests. *J. Allergy Clin. Immunol.* **113:**S486.

Bock, S. A. 1982. The natural history of food sensitivity. *J. Allergy Clin. Immunol.* **69:**173–177.

Bock, S. A. 1985. Natural history of severe reactions to foods in young children. *J. Pediatr.* **107:**676–680.

Bock, S. A. 1987. Prospective appraisal of complaints of adverse reactions to foods in children during the first 3 years of life. *Pediatrics* **79:**683–688.

Bock, S. A., and F. M. Atkins. 1990. Patterns of food hypersensitivity during sixteen years of double-blind, placebo-controlled food challenges. *J. Pediatr.* **117:**561–567.

Bock, S. A., A. Munoz-Furlong, and H. A. Sampson. 2001. Fatalities due to anaphylactic reactions to foods. *J. Allergy Clin. Immunol.* **107:**191–193.

Burks, A. W., J. M. James, A. Hiegel, G. Wilson, J. G. Wheeler, S. M. Jones, and N. Zuerlein. 1998. Atopic dermatitis and food hypersensitivity reactions. *J. Pediatr.* **132:**132–136.

Burks, W. 2003. Skin manifestations of food allergy. *Pediatrics* **111:**1617–1624.

Burks, W., G. Bannon, and S. B. Lehrer. 2001. Classic specific immunotherapy and new perspectives in specific immunotherapy for food allergy. *Allergy* **56**(Suppl.) **67:**121–124.

Crowe, S. E., and M. H. Perdue. 1992. Gastrointestinal food hypersensitivity: basic mechanisms of pathophysiology. *Gastroenterology* **103:**1075–1095.

Dannaeus, A., and M. Inganas. 1981. A follow-up study of children with food allergy. Clinical course in relation to serum IgE- and IgG-antibody levels to milk, egg and fish. *Clin. Allergy* **11:**533–539.

Egan, C. A., S. O'Loughlin, S. Gormally, and F. C. Powell. 1997. Dermatitis herpetiformis: a review of fifty-four patients. *Ir. J. Med. Sci.* **166:**241–244.

Eigenmann, P. A., S. H. Sicherer, T. A. Borkowski, B. A. Cohen, and H. A. Sampson. 1998. Prevalence of IgE-mediated food allergy among children with atopic dermatitis. *Pediatrics* **101:**E8.

Fleischer, D. M., M. K. Conover-Walker, L. Christie, A. W. Burks, and R. A. Wood. 2003. The natural progression of peanut allergy: resolution and the possibility of recurrence. *J. Allergy Clin. Immunol.* **112:** 183–189.

Fleischer, D. M., M. K. Conover-Walker, E. C. Matsui, and R. A. Wood. 2005. The natural history of tree nut allergy. *J. Allergy Clin. Immunol.* **116:**1087–1093.

Ford, R. P., and B. Taylor. 1982. Natural history of egg hypersensitivity. *Arch. Dis. Child* **57:**649–652.

Fox, V. L., S. Nurko, and G. T. Furuta. 2002. Eosinophilic esophagitis: it's not just kid's stuff. *Gastrointest. Endosc.* **56:**260–270.

Goldman, A. S., D. W. Anderson, Jr., W. A. Sellers, S. Saperstein, W. T. Kniker, and S. R. Halpern. 1963. Milk allergy. I. Oral challenge with milk and isolated milk proteins in allergic children. *Pediatrics* **32:**425–443.

Grundy, J., S. Matthews, B. Bateman, T. Dean, and S. H. Arshad. 2002. Rising prevalence of allergy to peanut in children: data from 2 sequential cohorts. *J. Allergy Clin. Immunol.* **110:**784–789.

Hall, R. P., III. 1992. Dermatitis herpetiformis. *J. Investig. Dermatol.* **99:**873–881.

Heiner, D. C., J. W. Sears, and W. T. Kniker. 1962. Multiple precipitins to cow's milk in chronic respiratory disease. A syndrome including poor growth, gastrointestinal symptoms, evidence of allergy, iron deficiency anemia, and pulmonary hemosiderosis. *Am. J. Dis. Child* **103:**634–654.

Hill, I. D., M. H. Dirks, G. S. Liptak, R. B. Colletti, A. Fasano, S. Guandalini, E. J. Hoffenberg, K. Horvath, J. A. Murray, M. Pivor, and E. G. Seidman. 2005. Guideline for the diagnosis and treatment of celiac disease in children: recommendations of the North American Society for Pediatric Gastroenterology, Hepatology and Nutrition. *J. Pediatr. Gastroenterol. Nutr.* **40:**1–19.

Hjorth, N., and J. Roed-Petersen. 1976. Occupational protein contact dermatitis in food handlers. *Contact Derm.* **2:**28–42.

Host, A., and S. Halken. 1990. A prospective study of cow milk allergy in Danish infants during the first 3 years of life. Clinical course in relation to clinical and immunological type of hypersensitivity reaction. *Allergy* **45:**587–596.

Host, A., B. Koletzko, S. Dreborg, A. Muraro, U. Wahn, P. Aggett, J. L. Bresson, O. Hernell, H. Lafeber, K. F. Michaelsen, J. L. Micheli, J. Rigo, L. Weaver, H. Heymans, S. Strobel, and Y. Vandenplas. 1999. Dietary products used in infants for treatment and prevention of food allergy. Joint Statement of the European Society for Paediatric Allergology and Clinical Immunology (ESPACI) Committee on Hypoallergenic Formulas and the European Society for Paediatric Gastroenterology, Hepatology and Nutrition (ESPGHAN) Committee on Nutrition. *Arch. Dis. Child* **81:**80–84.

Hourihane, J. O., S. A. Roberts, and J. O. Warner. 1998. Resolution of peanut allergy: case-control study. *BMJ* **316:**1271–1275.

James, J. M. 2003a. Food allergy, respiratory disease, and anaphylaxis, p. 529–37. *In* D. Y. M. Leung, H. A. Sampson, R. S. Geha, S. J. Szefler (ed.), *Pediatric Allergy: Principles and Practice.* Mosby, St. Louis, Mo.

James, J. M. 2003b. Respiratory manifestations of food allergy. *Pediatrics* **111:**1625–1630.

Jenkins, H. R., J. R. Pincott, J. F. Soothill, P. J. Milla, and J. T. Harries. 1984. Food allergy: the major cause of infantile colitis. *Arch. Dis. Child* **59:**326–329.

Leung, D. Y., H. A. Sampson, J. W. Yunginger, A. W. Burks, Jr., L. C. Schneider, C. H. Wortel, F. M. Davis, J. D. Hyun, and W. R. Shanahan, Jr. 2003. Effect of anti-IgE therapy in patients with peanut allergy. *N. Engl. J. Med.* **348:**986–993.

Liacouras, C. A., and E. Ruchelli. 2004. Eosinophilic esophagitis. *Curr. Opin. Pediatr.* **16:**560–566.

Mayer, L. 2003. Mucosal immunity. *Pediatrics* **111:**1595–1600.

Ohtsuka, T., S. Matsumaru, K. Uchida, M. Onobori, T. Matsumoto, K. Kuwahata, and M. Arita. 1993. Time course of plasma histamine and tryptase following food challenges in children with suspected food allergy. *Ann. Allergy* **71:**139–146.

Oppenheimer, J. J., H. S. Nelson, S. A. Bock, F. Christensen, and D. Y. Leung. 1992. Treatment of peanut allergy with rush immunotherapy. *J. Allergy Clin. Immunol.* **90:**256–262.

Papadopoulos, G. K., C. Wijmenga, and F. Koning. 2001. Interplay between genetics and the environment in the development of celiac disease: perspectives for a healthy life. *J. Clin. Investig.* **108:**1261–1266.

Perry, T. T., E. C. Matsui, M. K. Conover-Walker, and R. A. Wood. 2004. Risk of oral food challenges. *J. Allergy Clin. Immunol.* **114:**1164–1168.

Rothenberg, M. E. 2004. Eosinophilic gastrointestinal disorders (EGID). *J. Allergy Clin. Immunol.* **113:**11–28.

Sampson, H. A. 1998. Adverse reactions to foods, p. 1162–1182. *In* E. Middleton, C. Reed, E. Ellis, N. Adkinson, J. W. Yunginger, and W. Busse (ed.), *Allergy: Principles and Practice*, 5th ed., vol. II. Mosby-Year Book, Inc., St. Louis, Mo.

Sampson, H. A. 1999. Food allergy. Part 1: immunopathogenesis and clinical disorders. *J. Allergy Clin. Immunol.* **103:**717–728.

Sampson, H. A. 2003. Adverse reactions to foods, p. 1619–1643. *In* N. F. Adkinson, B. S. Bochner, J. W. Yunginger, S. T. Holgate, W. W. Busse, and F. E. R. Simons (ed.), *Middleston's Allergy: Principles and Practice,* 6th ed. Mosby, Philadelphia, Pa.

Sampson, H. A. 2004. Update on food allergy. *J. Allergy Clin. Immunol.* **113:**805–819.

Sampson, H. A. 2005. Food allergy: when mucosal immunity goes wrong. *J. Allergy Clin. Immunol.* **115:**139–141.

Sampson, H. A., and J. A. Anderson. 2000. Summary and recommendations: classification of gastrointestinal manifestations due to immunologic reactions to foods in infants and young children. *J. Pediatr. Gastroenterol. Nutr.* **30**(Suppl.)**:**S87–S94.

Sampson, H. A., and A. W. Burks. 1996. Mechanisms of food allergy. *Annu. Rev. Nutr.* **16:**161–177.

Sampson, H. A., and P. L. Jolie. 1984. Increased plasma histamine concentrations after food challenges in children with atopic dermatitis. *N. Engl. J. Med.* **311:**372–376.

Sampson, H. A., L. Mendelson, and J. P. Rosen. 1992. Fatal and near-fatal anaphylactic reactions to food in children and adolescents. *N. Engl. J. Med.* **327:**380–384.

Sicherer, S. H. 2003. Clinical aspects of gastrointestinal food allergy in childhood. *Pediatrics* **111:**1609–1616.

Sicherer, S. H. 2005. Food protein-induced enterocolitis syndrome: case presentations and management lessons. *J. Allergy Clin. Immunol.* **115:**149–156.

Sicherer, S. H., A. Munoz-Furlong, and H. A. Sampson. 2003. Prevalence of peanut and tree nut allergy in the United States determined by means of a random digit dial telephone survey: a 5-year follow-up study. *J. Allergy Clin. Immunol.* **112:**1203–1207.

Simons, F. E., X. Gu, and K. J. Simons. 2001. Epinephrine absorption in adults: intramuscular versus subcutaneous injection. *J. Allergy Clin. Immunol.* **108:**871–873.

Skolnick, H. S., M. K. Conover-Walker, C. B. Koerner, H. A. Sampson, W. Burks, and R. A. Wood. 2001. The natural history of peanut allergy. *J. Allergy Clin. Immunol.* **107:**367–374.

Sloan, A. E., and M. E. Powers. 1986. A perspective on popular perceptions of adverse reactions to foods. *J. Allergy Clin. Immunol.* **78:**127–133.

Sollid, L. M., and E. Thorsby. 1993. HLA susceptibility genes in celiac disease: genetic mapping and role in pathogenesis. *Gastroenterology* **105:**910–922.

Wood, R. A. 2003. The natural history of food allergy. *Pediatrics* **111:**1631–1637.

Yunginger, J. W., D. R. Nelson, D. L. Squillace, R. T. Jones, K. E. Holley, B. A. Hyma, L. Biedrzycki, K. G. Sweeney, W. Q. Sturner, and L. B. Schwartz. 1991. Laboratory investigation of deaths due to anaphylaxis. *J. Forensic Sci.* **36:**857–865.

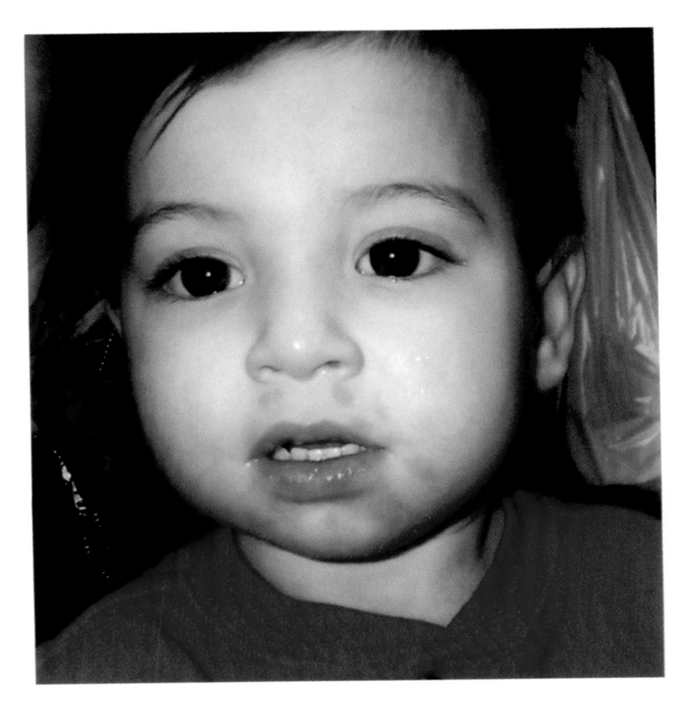

Color Plate 1. Perioral urticaria and angioedema of the lower lip in a 3-year-old child after a positive food challenge to milk.

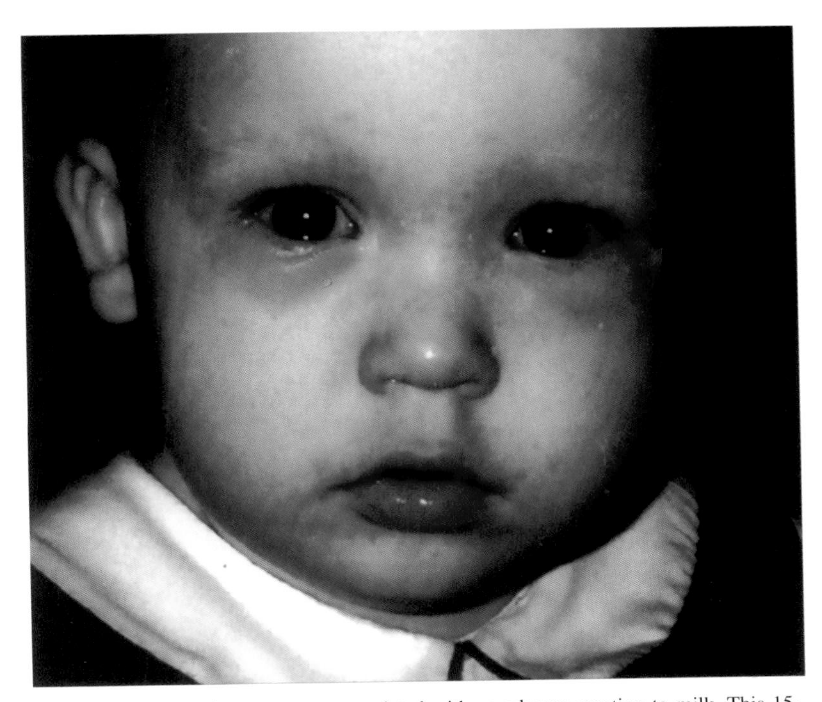

Color Plate 2. Ocular symptoms associated with an adverse reaction to milk. This 15-month-old child developed immediate periorbital and perioral angioedema with associated urticarial lesions within 2 to 3 min of milk ingestion. Other manifestations included intense conjunctival injection, rhinorrhea, and hoarseness.

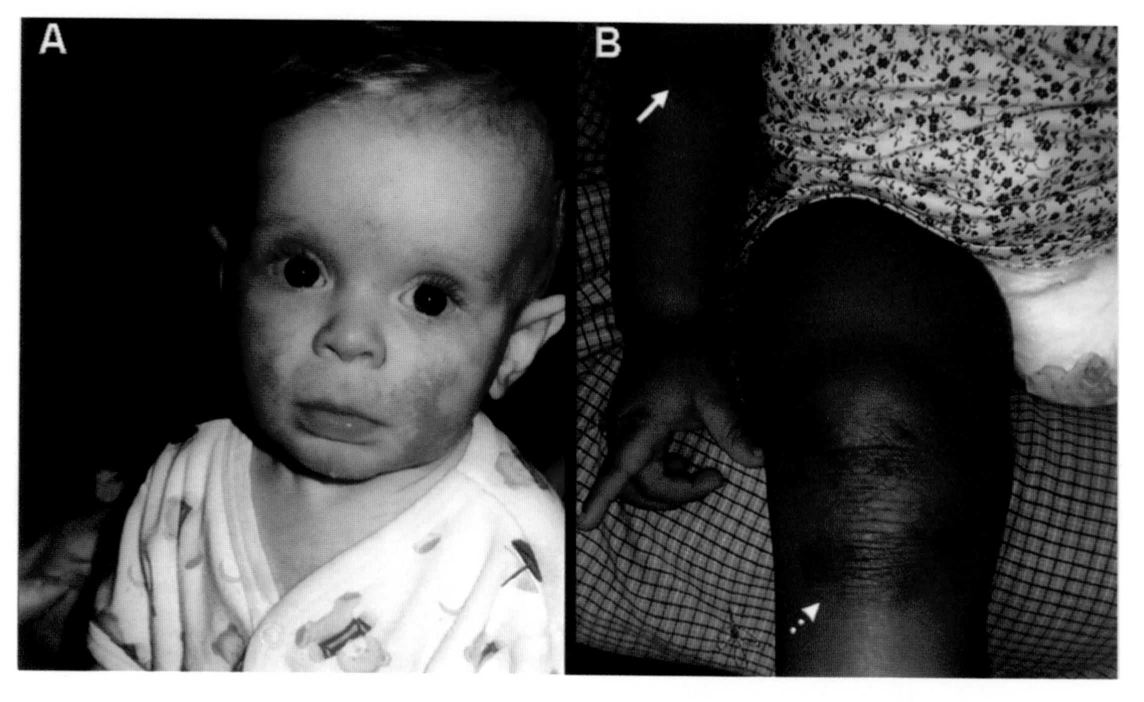

Color Plate 3. AD in early childhood. (A) An 8-month-old child with known egg allergy and the typical maculopapular, erythematous rash commonly seen on the face of infants and children < 2 years old. (B) A 2-year-old child with peanut allergy. The acute skin changes in the antecubital fossa (solid arrow) were noted after an accidental exposure to peanut butter. The hyperpigmented, lichenified lesions on the extensor surface of the knee (dashed arrow) represent long-standing AD, likely due to other triggers.

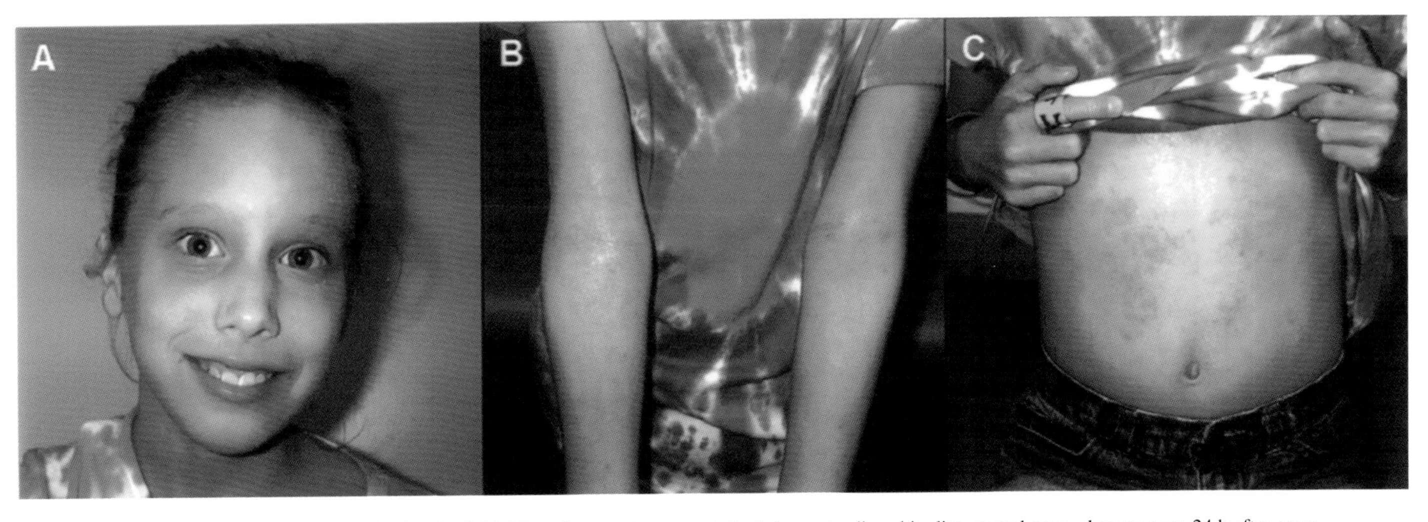

Color Plate 4. AD in an older food-allergic child. These images demonstrate both long-standing skin disease and acute changes seen 24 h after an accidental exposure to milk in this 8-year-old child with milk allergy and multiple aeroallergen sensitivities. (A) Signs of allergic disease with allergic shiners, chelitis, and AD rash on forehead, eyelids, and perioral region. (B) Chronic antecubital fossa lesions typical of older children with AD. (C) Acute flare of AD with intense erythema and rash on the abdomen.

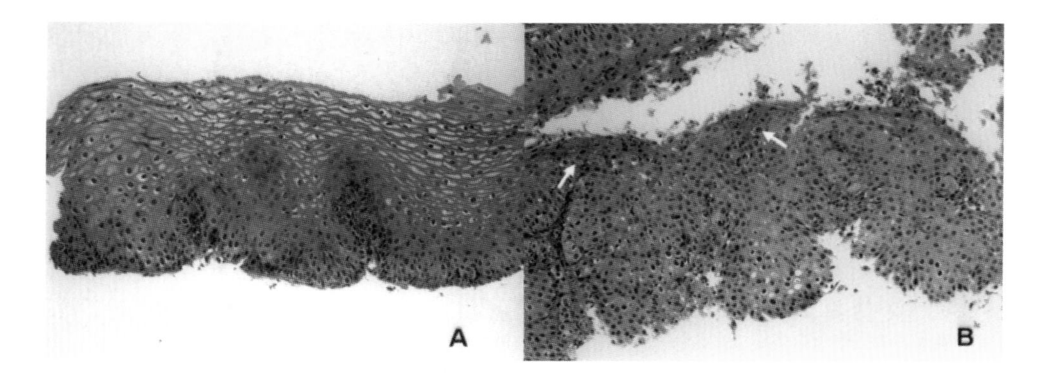

Color Plate 5. Endoscopic punch biopsy specimens from normal esophageal mucosa (A) and from a patient with EE (B). (A) Normal esophagus without evidence of eosinophilic infiltrate from a patient with GERD. (B) Esophageal biopsy obtained from a patient who presented with feeding intolerance, failure to thrive, and recurrent vomiting. Note the intraepithelial eosinophilic infiltrates (arrows).

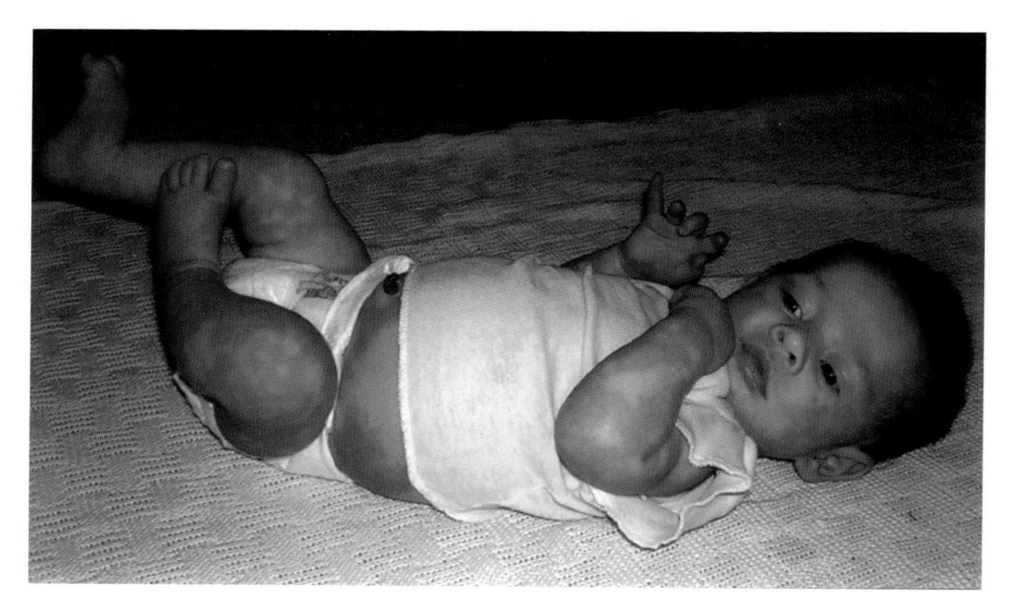

Color Plate 6. Three-week-old, previously healthy infant who presented with a 2-day history of a diffuse, erythematous, raised, maculopapular rash. The rash was most prominent after breast-feeding. An allergy evaluation was positive for milk allergy with a CAP-RAST specific to milk (>100 kU_A/liter). The rash resolved with maternal dietary restriction of milk protein and dietary supplement with a protein hydrolysate formula.

Food Allergy
Edited by S. J. Maleki et al.

Chapter 2

Exploring Current and Novel Methods for the Detection and Diagnosis of Food Allergy: the Clinical Approach

Adriano Mari and Enrico Scala

Food allergy constitutes a deviation from the physiologic immune response normally leading to tolerance to food proteins. An immune mechanism, mainly the production of allergen-specific immunoglobulin E (IgE), plays an important role in food allergy-related disorders. Although the lack of objective measures of real prevalence of food allergy may lead to an overestimation of the problem (Woods et al., 2001; Woods et al., 2002), recent interview-based epidemiological surveys indicate that food allergy affects about 6% of infants and/or children and is present in about 3% of the adult population (Kanny et al., 2001; Eigenmann et al., 2001; Sicherer et al., 2003; Mattila et al., 2003; Moneret-Vautrin et al., 2004; Zuberbier et al., 2004; Roehr et al., 2004).

The recently revised nomenclature for allergic diseases (Johansson et al., 2004) includes several manifestations of food allergy such as asthma, urticaria and/or angioedema, anaphylaxis, eczema, and rhinitis and briefly defines when the term "food allergy" should be used. A useful division in allergic and nonallergic manifestation is introduced for each disease. Nonallergic manifestations include toxic or pharmacologically induced allergy-like manifestations. Allergic diseases are further divided into IgE-mediated and non-IgE-mediated forms. This classification and nomenclature can be applied to other well-characterized food-induced manifestations such as oral allergy syndrome (OAS), food-dependent exercise-induced anaphylaxis, gastrointestinal (GI) anaphylaxis, eosinophilic esophagitis-gastroenteritis, and food protein-induced enteropathy (Sampson, 1999a; Sicherer and Teuber, 2004). Figure 1 shows a very general flow chart adapted from the report of the Nomenclature Review Committee of the World Allergy Organization (Johansson et al., 2004), which proceeds through the symptom evaluation process of food allergy.

The aim of a correct diagnosis of food allergy is to establish a causal relationship between food ingestion and the clinical symptoms reported by the patient and to identify the

Adriano Mari • Allergy Unit, National Health Service, Viale di Tor di Quinto, 33/A, I-00191 Rome, Italy.
Enrico Scala • Istituto Dermopatico dell'Immacolata, IDI-IRCCS, Via dei Monti di Creta 104, I-00167 Rome, Italy.

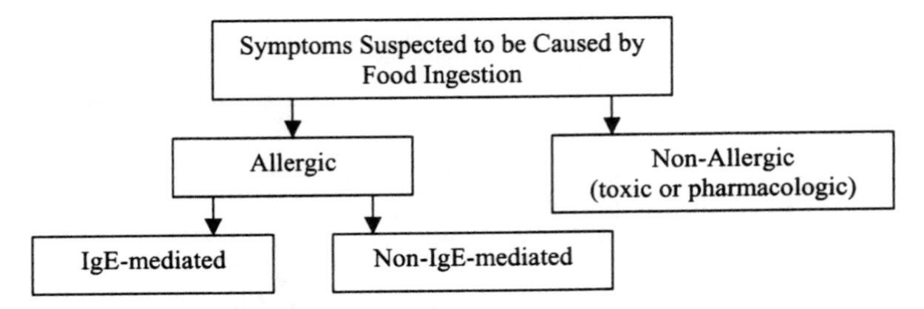

Figure 1. General flow chart for a possible food allergy mechanism: from suspicion to immune mechanism (adapted from Johansson et al., 2004).

triggering allergen and the immune mechanism determining the reaction. The current approach to food allergy diagnosis using well-established tools including medical history, skin testing, IgE detection, and oral challenge is depicted in Fig. 2. On the basis of future availability of immunological reagents (allergenic molecules) and in vitro assays from advanced biotechnology (microarray), this chart could be modified. Advantages and limitations of current and future resources will be analyzed here. Many unproven in vivo or in vitro methods, including IgG detection, have currently been proposed for the diagnosis of food allergy. Neither the demonstration of underlining biologic mechanisms nor any clinical validation studies have ever been reported for any of these alternative methods. Their uselessness, limitations, and risks have been discussed in several authoritative documents and reported previously (Bindslev-Jensen and Poulsen, 1996; Fox et al., 1999; Terr, 2000; Lewith et al., 2001; Semizzi et al., 2002; Teuber and Porch-Curren, 2003; Senna et al., 2004; Niggemann and Gruber, 2004).

MEDICAL HISTORY AND PHYSICAL EXAMINATION

The first approach for evaluating allergic reactions to food is an accurate recording of symptoms, which are spontaneously reported by patients or caregivers in the case of children (Sampson, 1999b; Niggemann, 2004). As patients frequently misinterpret or overlook true allergy symptoms, a full questionnaire containing simple and easy-to-understand questions should be administered to the patient. By means of the medical history, the physician may determine whether the symptoms under evaluation are consistent with an adverse reaction to food, and a toxic non-immune-mediated reaction can be ruled out.

The clinical history should be extremely thorough. In addition to the symptom description and questions on the suspected food causing the reaction, the following aspects should be carefully explored: (i) the amount ingested, (ii) the food processing procedure (handling, preservation, peeling, boiling, roasting, or unprocessed), (iii) timing of symptom onset and the relation of symptoms to foods ingested in the last 24 h, (iv) previous and successive safe exposures to suspected foods, (v) coexisting facilitating factors (e.g., physical exercise, alcohol ingestion, medication use other than antiallergic drugs, acute or chronic gastroenteric diseases), and (vi) food-induced reactions occurring through other routes of exposure (e.g., inhalation of cooking vapors or skin contact during food handling and manipulation) (Lee et al., 2001). Records or reports from the emergency department or from doctors who assisted the patient during the allergic reaction are useful

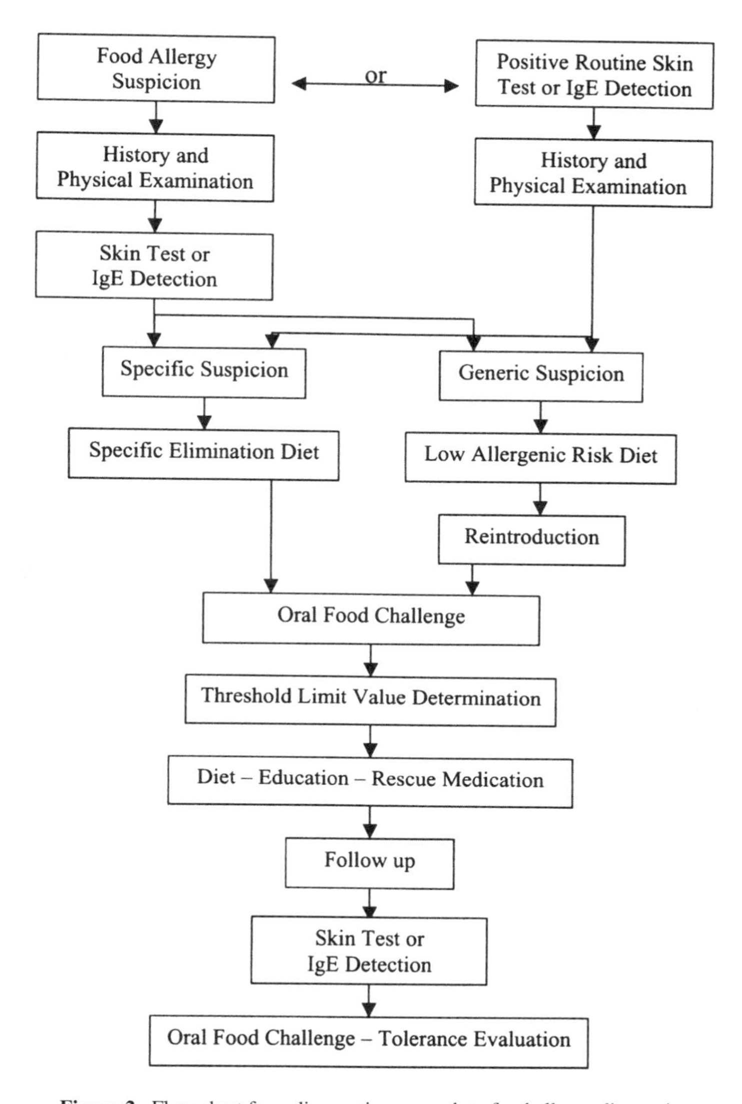

Figure 2. Flow chart for a diagnostic approach to food allergy diagnosis.

for a more precise definition of the clinical picture severity. A careful description of the symptom course may be useful for patient presenting severe reactions (e.g., lower respiratory tract symptoms or anaphylaxis) to focus the earliest alarming symptoms (e.g., OAS or sneezing). GI manifestations with oropharyngeal pruritus and local angioedema, acute gastric and abdominal pain, nausea, vomiting, and diarrhea may occur as isolated symptoms in adverse reactions to foods. Skin symptoms include itching, whealing, angioedema, and eczema. Airway symptoms are bronchospasm, nasal itching, sneezing, and nasal watery secretion. A conjunctival reaction, with redness, itching, and chemosis, is frequently associated. Hypotension may take place during severe reactions, sometimes leading to a fatal outcome. Multiorgan symptoms may occur in a puzzling and changing combination within the same patient and between reactions to the same triggering food. Mueller grading, based on the severity of reactions to Hymenoptera stings (Mueller, 1966; Muller et al., 2004), has been usefully adopted for food-induced allergic reactions. As for hymenoptera venom allergy, it helps to determine the degree of therapeutic intervention such as diet limitations and pharmacological therapy. In chronic disorders related to food allergies (e.g., atopic dermatitis, nonacute asthma, and allergic eosinophilic gastroenteritis), the medical history may have a poor predictive accuracy. A much higher predictive value can be obtained from a careful anamnesis when reactions with a sudden onset (e.g., anaphylaxis, urticaria, and acute asthma) occur after the ingestion of single or few identifiable foods. In obtaining the history, it is important to know all ingested foods, regardless of whether they represent the largest or a minor part of the meal or if they are common in a patient's diet or are rarely ingested (e.g., exotic fruits and spices). The presence of a so-called hidden allergen should always be suspected when all the information from the patient leads to an allergenic source to which the patient is not sensitized (Koppelman et al., 1999; Moneret-Vautrin et al., 1994; Jones et al., 1992). That is also the case for reactions induced by contamination with non-food allergens ingested during meals (Morisset et al., 2003b; Edston and Van Hage-Hamsten, 2003; Guillet et al., 2003; Bernardini et al., 2002; Matsumoto et al., 2001; Blanco et al., 1997; Sanchez-Borges et al., 1997; Erben et al., 1993).

The physical examination is also important for defining the presence of any other nonallergic disease affecting the patient that could either increase risk during an allergic reaction (e.g., ischemic cardiac disease) or could interfere with the pharmacological treatment of the food-induced allergic disease (e.g., diabetes). Furthermore, a careful evaluation should be carried out to estimate other associated allergic diseases (e.g., inhalant allergy or atopic dermatitis), whose presence could increase both the severity of an allergic reaction to foods and the outcome of some in vivo diagnostic procedures (e.g., food challenge). It should also be considered that the use of drugs to treat respiratory allergic diseases such as rhinitis or asthma could mask mild earlier symptoms triggered by an allergenic food, leading the patient to increase the ingested amount. This could lead to an overexposure to the offending food, eliminating the attempt to define safe threshold levels of allergenic contaminants (Mills et al., 2004).

A particular aspect of food allergy is the occupational and nonoccupational exposure to nebulized allergens. This kind of exposure should be labeled as a noningestive food allergy (Roberts and Lack, 2003; Schuller et al., 2005). The acquired sensitization during work-related food handling or processing may cause allergic reactions outside the workplace because of the ingestion of the offending inhaled food allergen (Tabar et al., 2004). As for ingested food, this may be caused by true food allergens or by nonfood allergen-contaminated foodstuffs (Scala et al., 2001; Armentia et al., 2004).

DIAGNOSTIC METHODS—IN VIVO TESTS

The purpose of an in vivo test is to cause an allergic reaction in a tested organ through the direct application of an allergen on a body surface (i.e., on the skin during skin testing). Therefore, in vivo tests should be considered as provocation tests, able to reveal the presence of sensitizing IgE and its functional activity. The skin test (ST) (prick, prick-prick, or intradermal) is generally considered apart from other true provocation and/or challenge tests (oral, conjunctival, nasal, or bronchial), as it is used in the allergy diagnosis regardless of the organ involved in the allergic disease. The main features of most important in vivo diagnostic test are reported in Table 1. Features of in vitro ones are reported for comparative purposes.

Skin Test

The prick (or puncture) ST constitutes a quick and useful method of demonstrating an IgE-mediated response to specific foods. Several disposable devices (common needle, bifurcated needle, lancet, or commercial single- or multiple-use devices such as DermaPIK, Duotip-Test, and MultiTest) made from different materials (steel or plastic) may be used. Several studies have compared different devices for performing STs (Adinoff et al., 1989; Nelson et al., 1998; Corder et al., 1996; Corder and Wilson, 1995), showing that each device varies greatly in several characteristics, including the size of reactions at both positive and negative test sites. The authors suggest that the operator performing STs should be tested in the use of the device to establish criteria for positive and negative tests. The device is generally used on the skin with a drop of a commercially glycerinated food extract, along with appropriate positive (histamine) and negative (saline) controls. In the case of a positive test, the outcome is the appearance of erythema and whealing of different sizes, depending on several individual and objective factors (Niggemann and Beyer, 2005). Recording reactivity may be done by means of a visual score comparing the tested allergen reactivity to the histamine reactivity or by measuring the wheal size by its diameter or area, the latter representing the most precise measure of ST reactivity (Poulsen et al., 1993). ST is generally reported as safe, with very rare reports of generalized or fatal reactions (Liccardi et al., 2003; Vanin et al., 2002; Turkeltaub and Gergen, 1989).

The ST still represents the first diagnostic approach for the food-allergic subject, as it allows screening of a large number of potential allergenic sources in a relatively inexpensive way. The negative predictive value of the ST is generally high (<90%) (Norgaard et al., 1992; Sampson 1988; Pucar et al., 2001), while the positive predictive value is generally lower (≈60%) (Sampson 1988; Pucar et al., 2001). In fact, while a positive ST response may be confirmatory in patients with a recent and clear-cut history of a food-induced allergic reaction, when the ST is used for screening, many subjects show positive skin reactions but not all of them show related symptoms on exposure (Bodtger, 2004). For suspected foods, any therapeutic decision about withdrawing the foods from the diet should be taken after a food challenge, currently considered the "gold standard" for decision making.

Prick-Prick Skin Test

Trying to overcome the problem of labile allergenic molecules present in plant-derived foods and extracts, the prick-prick test has become routine (Rosen et al., 1994). It consists of pricking the freshly supplied allergenic source, followed by a prick on the subject's skin. It is generally performed with standardized steel or plastic lancets, having the same readout of the commercial extract ST. Safety of the prick-prick test is generally considered high, although more attention should be paid when the test is performed under certain specific conditions (i.e., with patients reporting anaphylactic reactions) (Devenney and

Table 1. Features of in vivo and in vitro tests for food allergy diagnosis

Test	Allergenic preparation(s) in a single run	Reagent		Functional	Blood sampling	Serum Amt.	Operator	Standardization	Automation	Cost
		Extracts[b]	Molecules[c]							
In vivo										
Skin test	Multiple[d]	Yes	Yes[e]	Yes	No		Experienced	Very difficult	Impossible	Low
DBPCFC	Single	Yes	No[f]	Yes	No		Highly experienced	Very difficult	Impossible	High
In vitro										
IgE (direct)	Single to multiple	Yes	Yes	No	Yes	Variable[a]	Technician	Possible	Yes	Variable[a]
IgE (microarray)	Multiple[g]	Yes	Yes	No	Yes	Low	Technician	Possible	Yes	Medium-low
IgE (immunoblot)	Single to multiple	Yes	Yes	No	Yes	Variable[a]	Experienced	Very difficult	No	Variable[a]
Basophil activation test	Single to multiple	Yes	Yes	Yes	Yes	Variable[a]	Highly experienced	Difficult	No	Variable[a]

[a]Variable, the amount of serum needed and the cost of the whole test increase as the number of tests performed increases.
[b]Standardized extract preparations (in terms of their allergenic components) cannot easily be obtained.
[c]Allergenic molecule reagents can be quantified by weight.
[d]An upper limit of 10 to 30 allergenic preparations should be considered.
[e]Applied only to research studies to determine allergenicity.
[f]Provocation tests have been carried out on conjunctival, nasal, and bronchial mucosa with inhalant allergens.
[g]Virtually hundreds of tests could be performed.

Falth-Magnusson 2000; Novembre et al., 1995). The sensitivity of the prick-prick test is reported to be higher than the classical ST with commercial extracts, mainly for plant-derived foods (Ortolani et al., 1989; Norgaard et al., 1992) because of the differences in the reagent quality rather than in the technique per se. A possible pitfall in the use of the prick-prick test is the unknown concentration of allergens in the source. If a strong suspicion is present and a negative test is recorded, retesting the subject with another freshly supplied foodstuff could be required.

Intradermal Test

The intradermal injection of sterile allergenic preparations at a lower concentration than those used for ST is still in use. This technique allows careful establishment of the amount of the preparation to be tested. Major drawbacks include (i) how to determine the cutoff point for a negative-positive test, (ii) that it is not as painless a procedure as the ST, (iii) that it needs a more experienced operator, and (iv) that it has a greater risk of generalized adverse reactions (Sampson et al., 1996). Intradermal ST is no longer recommended for the evaluation of food allergies.

Labial Test

Although it is not performed using the skin as a testing organ, the labial food challenge has been proposed as an alternative to ST and oral challenge for the routine diagnosis of food allergy (Rance and Dutau, 1997). Studies performed to compare ST to labial food challenge report a higher sensitivity and the chance to replace the ingestion food challenge by the labial test (Rance and Dutau, 1997). Unless it is called "challenge," the test exposes the labial surface, not protected by the epidermis, to the allergenic extract or to the fresh food. This resembles the way the ST is performed more closely than a food challenge. Unlike with a real food challenge, no ingestion occurs, thus missing the true simulation of a food intake (digestion and adsorption) by the subject. Recording of a local reaction (swelling of the lips and contiguous urticaria) is unique, and the readout is difficult to standardize. Furthermore, it is not devoid of adverse generalized reactions (4.5% of the positive tests), possibly due to rapid local adsorption or to the ingestion of the food preparation (Rance and Dutau, 1997). It would be of interest to evaluate whether the labial test was more sensitive than the ST and if it was helpful in evaluating the allergenicity of native and recombinant allergenic molecules.

Atopy Patch Test

The identification of a delayed component in IgE-mediated food-induced allergic diseases, mainly atopic dermatitis (Sutas et al., 2000; Breuer et al., 2004) and GI tract eosinophilic diseases (Heine, 2004), and the presence of a cell-mediated immunopathogenic mechanism in these conditions (Sutas et al., 2000; Kerschenlohr et al., 2003) have led to the use of the patch test in the diagnosis of food allergy (Niggemann et al., 2000; Spergel and Brown-Whitehorn, 2005). The patch test is performed by the same testing technique used with allergic contact dermatitis, and the same reactivity grading is adopted, as an eczematous reaction is expected (Darsow et al., 2004). In a large European multicenter study, an interesting evaluation of this test was carried out by testing both inhalant and food allergens (Darsow et al., 2004). The authors define the atopy patch test to be used in addition to ST and IgE as a tool to prove the clinical relevance of a given sensitization. The atopy patch test was not completely devoid of adverse reactions, as 7% of the subjects experienced relevant local reactions and (in one case) respiratory symptoms were recorded during the test. The conclusions from the study were that, unless 7% of the subjects otherwise labeled as having an intrinsic eczema were recorded positive to the atopy patch test, nothing can be said about its clinical impact, as no intervention studies have been performed to define the benefit of allergen avoidance in such cases.

The atopy patch test cannot be used as a routine resource for food allergy diagnosis until extensive multicenter validation studies are published.

Diagnostic Diets

With a suspicion of food allergy, determined by the clinical history and/or by a positive ST or IgE detection, a diagnostic elimination diet should be carried out before an oral challenge test (Sampson, 1999b; Niggemann, 2004).

When there is a specific suspicion of a single food or a few foods, the elimination diet should be based on strict avoidance of the specific foods. Depending on the frequency of symptom appearance, the diet should be carried out for a reasonable time period. For instance, if symptoms appear several times in a week, an elimination diet lasting 2 to 3 weeks would be enough to determine the relevance of the suspected food, whereas in the case of two to three suspected food reactions in a year, a 6- to 12-month elimination diet period should be considered. The elimination diet should be stopped in the case of symptom recurrence, while in the case of a successful diet a supervised oral food challenge should be performed before the decision to definitely withdraw the food from the diet is made.

When there is a general suspicion of a food allergy-mediated disease based mostly not on a clear-cut clinical history but on a positive ST or IgE, the elimination diet should be based on the exclusion of all the suspected foods and the use of an oligoallergenic food diet. An oligoallergenic food diet generally includes those foods that have rarely, if ever, been described as allergenic. Its composition varies from country to country, depending on eating habits. Even in the case of an oligoallergic approach, the diet should be stopped if no changes occur in the appearance of the disease, and an open careful reintroduction of excluded foods can be done exclusively when no severe reactions are under evaluation (e.g., with atopic dermatitis). The reintroduction should be performed by allowing the introduction of a single food at a time into the diet. In the case of recurrence of symptoms after the introduction of a food, the food should be withdrawn again and the reintroduction continued with the remaining foods. If severe reactions to foods are under evaluation, supervised oral food challenges should be carried out before reintroduction.

Oral Challenge Tests

The gold standard of any diagnostic procedure in food allergy remains the positive outcome of a standardized food challenge, usually in the form of a double-blind placebo-controlled food challenge (DBPCFC) (Bindslev-Jensen et al., 2004). As with other mucosal provocation tests, the oral food challenge is an attempt to reproduce the clinical manifestations of food-induced allergic diseases by administering one or more suspected allergenic food sources to the subject under controlled conditions. Several documents have been published addressing the need to define when a food challenge should be performed (Bock et al., 1988; Sampson, 1999b; Bindslev-Jensen et al., 2004). The aim of a food challenge is either to rule out or to define the role of a food suspected of causing an allergic disease. On the basis of the challenge result, it can be decided whether the specific food must be withdrawn from the patient's diet, leading to symptom clearing, or if it can be ingested without problem or reintroduced in the diet, in the case where a tolerance condition has been reached after a previous food allergy diagnosis. Either an open food challenge or a DBPCFC requires a careful evaluation of the subject for general health status and a more specific evaluation of the allergic diseases other than the food allergy problem under evaluation (Niggemann, 2004; Bindslev-Jensen et al., 2004). Inclusion criteria and eligibility are different for the open challenge and the DBPCFC. The former can be performed mainly with children who are younger than 4 years of age or when a negative result is expected, rendering the DBPCFC unnecessary (Niggemann, 2004; Bindslev-Jensen et al., 2004). The DBPCFC is always needed for older children and for adults when the diagnosis of food allergy must be unequivocally established, either for routine diagnostic purposes or for research needs (Sampson, 1999b; Niggemann, 2004; Bindslev-Jensen et al., 2004).

A number of studies have addressed the use of the DBPCFC for different allergenic food sources, clearly demonstrating how the DBPCFC is the reference in vivo test with the highest sensitivity and specificity compared to the ST and IgE detection (Norgaard et al., 1992; Fiocchi et al., 2002; Pastorello et al., 2002; Morisset et al., 2003a; Gray et al., 2004).

Recently, the usefulness of the DBPCFC in determining threshold levels of allergenic foods causing symptoms in allergic subjects has been explored (Bindslev-Jensen et al., 2002; Taylor et al., 2004; Morisset et al., 2003a; Osterballe and Bindslev-Jensen, 2003). The definition of threshold levels is primarily useful for those patients with severe or potentially fatal or near-fatal reactions who must manage their allergic reactivity (Moneret-Vautrin and Kanny, 2004). For many patients, knowing about their threshold level of reactivity may be helpful in improving their quality of life by controlling exposure to the triggering food from a certain level upward. Threshold level data are now available for milk, egg, peanuts, wheat flour, and sesame (Bindslev-Jensen et al., 2002), and the recent publication of a consensus protocol will help to estimate threshold doses for other allergenic foods and to define which is the lowest amount that elicits mild, objective symptoms in highly sensitive individuals (Taylor et al., 2004). In fact, studies carried out by means of the DBPCFC with patients allergic to peanuts showed an increase in the eliciting dose threshold causing symptoms in subjects treated with humanized monoclonal anti-IgE (Leung et al., 2003). Moreover, the definition of the lowest threshold levels in a population of patients with severe food allergy reactions may help the industry in choosing detection tests with the highest level of sensitivity and specificity, capable of detecting food allergens accidentally contaminating industrial products (Moneret-Vautrin and Kanny, 2004).

The need to evaluate potential allergenicity of food components present in transgenic organisms and the evaluation of the exposure risk to foods newly introduced on the market are the most recent applications of the DBPCFC (Poulsen, 2004). Among other tests used to evaluate allergenicity of new proteins, such as ST, basophil activation, and IgE binding, the DBPCFC is considered closest to the physiopathological response.

As for the prick-prick test, the natural nonstandardized source material is used in the DBPCFC, since neither food-derived testing material, standardized in terms of its allergenic molecule content, nor purified allergenic molecules have ever been used in the oral challenge test. A further limitation of the correct evaluation of a negative test is the capability to reproduce the physiopathological condition present during a food allergy reaction, mainly for intestinal permeability. Both the last two issues should be carefully addressed to further reinforce the usefulness of the oral challenge as the gold standard for food allergy diagnosis.

GI Tract Endoscopic Evaluation and Testing

Some procedures, presently not widely accepted, have recently been proposed for studying food allergies. This has mainly been done through the endoscopic evaluation of the GI tract mucosa after a food challenge. It helps to understand local morphological and immunological changes (Ulanova et al., 2000; Lin et al., 2002) in either food-induced systemic symptoms or, better, in isolated allergic GI tract diseases. This approach may be useful when negative ST and IgE detection are opposed to a positive DBPCFC result, probably due to local GI tract mucosa IgE production (Lin et al., 2002). A single study reported performing GI tract mucosa testing by local application of an allergen preparation. In this colonoscopic allergen provocation test, extracts were injected into the mucosa of the large bowel. The response was similar to that assessed by ST, with a wheal-and-flare reaction of the provoked areas measured and compared with unprovoked and histamine-challenged tissue (Bischoff et al., 1997). Due to the invasiveness of the test, it cannot be used for routine evaluation of the subjects, and its use should be limited to research studies.

Conjunctival, Nasal, and Bronchial Provocation Tests

Provocation tests not performed by ingestion of food are mainly used to evaluate allergy symptoms caused by respiratory mucosal contact with aerodispersed allergens (Roberts and Lack, 2003). Symptoms related to inhalation of food allergens are present in highly sensitized subjects (Sampson et al., 1992) in either domestic or occupational settings. Studies have been performed by using a food allergen bronchial provocation test showing both early and late-phase reactions (Roberts et al., 2002). Specific attention should be paid to occupational exposure to food allergens (Jeebhay et al., 2001; Taylor et al., 2000). As for the ingestion food challenge test, the respiratory mucosa testing can be a valuable method for defining the triggering allergen, threshold values of reactivity, and the underlying immune mechanism. Once validated, the use of other mucosal surfaces, such as the nasal tissues or conjunctiva, for testing nebulized food allergens may replace the bronchial provocation test. It could be helpful in reducing time and costs for bronchial challenges in research studies.

DIAGNOSTIC METHODS—IN VITRO TESTS

Differing from the in vivo diagnostic test methods, in vitro analysis detects the presence of specific IgE by functional or nonfunctional assays. The latter are represented by those tests, aiming to show IgE binding to an allergenic preparation by an enzyme-linked immunosorbent assay (ELISA), radioallergosorbent assay (RAST), and immunoblotting, whereas the former explore the capability of IgE to cause mediator release or detect cell activation markers from isolated effector cells (basophils and mast cells). The main features of in vitro tests are reported in Table 1 for comparative purposes.

Nonfunctional IgE Detection

Direct specific IgE determination and the ST represent the diagnostic tests used most routinely for defining a food IgE sensitization. Direct IgE in vitro detection has moved from experimental and time-consuming tests to almost fully automated third-generation systems for routine laboratory testing (Li et al., 2004; Johansson, 2004). In all the systems, the binding of IgE to the allergen preparation is detected by labeled anti-IgE. Several implementations of the reagents (e.g., the amount of available allergen preparation or the use of a combination of monoclonal anti-IgE recognizing different epitopes on the IgE molecule instead of or added to a rabbit polyclonal antibody) (Sander et al., 2005) have increased both sensitivity and specificity of most IgE in vitro assays; for several food allergen preparations, they have almost the same sensitivity as the ST. Like the ST, the IgE determination defines the sensitization status of a subject. Clinical history and a subsequent oral challenge are needed to define the clinical relevance of the positive in vitro result. Either solid-phase or liquid-phase systems test discrete allergen preparations as single reagents. It represents a major limitation to a wider use of laboratory tests for IgE, as it increases the cost of the test severalfold when an increasing number of allergens are used. In addition to a particular patient's condition (e.g., severe and generalized eczematous lesions) when the ST cannot be performed, it is suggested that the IgE determination be used to confirm a positive ST. Several studies have recently shown that IgE concentrations with semiquantitative systems (e.g., ImmunoCAP and Immulite) can be used to avoid the need for a DBPCFC for a percentage of patients (Eigenmann, 2004). Cutoff limits have been determined for some foods, including peanuts, tree nuts, eggs, and milk (Boyano-Martinez et al., 2001; Rance et al., 2002; Hill et al., 2001; Clark and Ewan, 2003; Hill et al., 2004; Shek et al., 2004). This adds value to IgE determination in both the first diagnostic approach and in the follow-up phase (Fig. 2) (Sampson, 2001; Sampson, 2002; Sicherer, 2003). An additional advantage of the use of IgE determination is in relation to the development of tolerance. Recent studies have shown that lower IgE levels may be predictive of tolerance (Clark and Ewan, 2003; Shek et al., 2004).

A new kind of in vitro test for preliminary screening of the food-allergic subject is available (Phadiatop; AlaTOP). It consists of a mixture of predefined allergen preparations that are considered the most relevant from statistical and geographical perspectives and in a nonstandard selection among different producers. These reagents are not useful for the specific diagnosis of food allergy but can be used when a large number of subjects have to be screened for epidemiological or other public health survey purposes. Once positive, the multiple food allergen test needs to be followed by specific IgE determination for each of the food preparations present in the test. It should be always considered that less-common food allergies are missed by this kind of test. Conversely, when clinical indications for the presence of atopic disease are lacking, this single multiallergen screening assay for IgE antibody might exclude an allergic disease (Hamilton and Adkinson, 2003).

IgE Immunoblotting

An allergen extract may be subjected to a primary separation to obtain a qualitative description of the components. Such assays are dependent on molecular weight (sodium dodecyl sulfate-polyacrylamide gel electrophoresis [SDS-PAGE]), isoelectric point (isoelectric focusing [IEF]), or both (two-dimensional gel electrophoresis). The immunoblotting technique differs from the direct IgE determination performed by the ELISA or RAST techniques, as the components of the allergenic preparation, previously separated by SDS-PAGE or IEF, are electrotransferred onto nitrocellulose membranes, where they are probed with the subject's serum. IgE binding is shown by the addition of a labeled anti-IgE. The sensitivity is equal but sometimes lower than that of the direct ELISA or RAST assays. Sera with a low IgE concentration and recognizing several components in the allergenic preparation may have their IgE binding spread on several bands, at levels that are not high enough to be detected by the immunoblotting assay (Gruber et al., 2000). The IgE immunoblotting assay is mainly considered a research tool, as it needs a good level of expertise to be performed. The identification of the components is often not possible on the basis of a visual evaluation. The identification of the components can be done by the inhibition of the IgE binding with previously purified components. Thus, the IgE immunoblot is only a step toward allergenic component identification, but it does not represent a better routine solution for diagnosis at the molecular level.

Microarray

The use of biochips or microarray in genetic studies has been transferred to the proteomic field, allowing the quantitative definition of multiple analytes in a single run using very small amounts of all reagents (Hamilton and Adkinson, 2004; Bacarese-Hamilton et al., 2004). In allergy diagnoses, the methodological bases are similar to that of a solid-phase direct IgE determination but differ greatly in the way the allergenic preparations are presented and coupled to the solid phase (Harwanegg and Hiller, 2004). A very small amount of allergenic preparation is spotted in a 200-µm area (average diameter). Hundreds of these spots can be arrayed onto a Teflon-modified glass surface, with up to 50 to 100 allergenic preparations tested by a single assay on a squared surface that is no more than 10 mm in size. The amount of serum needed for testing is very small (20 to 50 µl). A fluorescent anti-IgE antibody detects the bound IgE. Fluorescence intensity is analyzed by a laser-based reader. Specific software guides the image analysis. A semiquantitative result is expressed in arbitrary units toward a reference standard. The solid-phase microarray is only one example, as several other systems, currently used for other analyte determinations, are suitable for IgE detection (Abassi et al., 2004; Kricka et al., 2003; Kingsmore and Patel, 2003; Bacarese-Hamilton et al., 2004). The application of microarrays for IgE determination is in its early infancy. Studies have been published showing the potential application of this new methodology in the diagnosis of allergic diseases (Musu et al., 1997; Su et al., 1999; Wiltshire et al., 2000; Kim et al., 2002a; Suck et al., 2002; Deinhofer et al., 2004). Hiller et al. (2002), pioneering in this field with a large international collaborative study, showed how ST, IgE dot blot, and IgE detection by microarray give almost the same results in terms of identification of allergenic components involved in the sensitization. Although this new biotechnology has to be fully validated against other well-established in vivo and in vitro assays (Hamilton and Adkinson, 2004), it certainly deserves great attention. The chance to test up to 100 or more allergens could mean that IgE testing replaces the ST as the first approach in the food allergy testing. An entire new generation of products that could be stronger, lighter, more precise, and cheaper than previous products could represent the real novelty in the next 5 to 10 years (Harwanegg et al., 2003). More than for ST or for classic IgE detection systems, a microarray determination with a huge output in potential positive results should be carefully evaluated from a clinical point of view; sensitization should be confirmed by appropriate additional challenges (Hamilton and Adkinson, 2004).

Functional IgE Detection

Basophil- and mast cell-based tests aim to detect serum IgE by their ability to activate these effector cells. After these effector cells, having IgE bound on their membranes, are challenged with specific allergenic preparations, the cells release mediators, thus showing both the presence of IgE and their biological activity. Although tissue mast cells are implicated in the large majority of food allergy reactions (Church and Levi-Schaffer, 1997), they are present in very small amounts in the target organs (Berger et al., 2002), and it is very difficult to isolated them. For this reason, human mast cells are rarely used for this in vitro test. Conversely, human basophils used for functional tests can be easily obtained by peripheral blood sampling of the subjects under evaluation (direct test) or from a donor (indirect test) (Kleine Budde et al., 2001). Transfected rat basophilic leukemia cells expressing the human high-affinity IgE receptor on their surface are also used. They can be used as an alternative to human mast cells (Dibbern et al., 2003). In the direct basophil degranulation-activation test, cells from the food-allergic subject are supposed to have specific IgE on their surface that will be cross bridged after the addition of specific allergenic preparations to which the subject is sensitized. In the indirect test, basophil are taken from a donor, IgE on the donor's basophils is stripped from the membrane by a mild lactic acid treatment, and the patient's serum is added, followed by the specific allergenic preparation as in the direct test (Kleine Budde et al., 2001). The same procedure is used with rat basophilic leukemia-transfected cells. Markers of effector cell activation are the release of preformed mediators, histamine (Wedi et al., 2000), newly formed leukotriene C4 (Moneret-Vautrin et al., 1999), or the expression of cell surface activation markers such as CD63 (Lambert et al., 2003; Erdmann et al., 2003) or CD203 (Hauswirth et al., 2002; Buhring et al., 2004; Boumiza et al., 2003). The use of flow cytometry has increased the use of basophil activation testing to confirm the diagnosis of food allergy. It remains a second-level approach to confirm, by a functional test, suspicions with results from the preliminary ST and IgE detection tests (Ebo et al., 2005; Ebo et al., 2004). With all the above-reported tests, confirmation of the diagnosis would require an oral challenge test. Extensive validation studies are needed before what is now a research tool now can become a routine tool.

DIAGNOSTIC REAGENTS—FRESH FOOD PREPARATIONS

Fresh food preparations are the basic reagents for oral food challenges and for the prick-prick ST. With oral food challenges, the major issue is to replicate the so-called normal exposure to the suspected food. This requires the selection of a foodstuff as it is normally consumed. When the food is consumed normally, either raw or processed, both preparations should be available for testing to define a possible tolerance to the processed food. When the food is normally consumed as processed, it should always be tested in the same form in which it is available to the consumer. It should also be considered that different cooking processes and time might lead to different allergen structure modifications

(Maleki, 2004). The evaluation of several cooked allergenic food preparations could also be relevant to increasing the quality of life for the allergic patient.

A major problem in food challenges is masking some typical food flavors and tastes. Several recipes are available, and pretesting should be considered for any new food to be tested. As other foodstuffs are used for masking the food tested, a careful preliminary evaluation of these other foodstuffs should be carried out (Vlieg-Boerstra et al., 2004). A major problem in fresh food preparation is how to get a testing material containing the relevant food allergen and how to get a standard concentration of each allergenic molecule. The stability and the relative constant concentration of allergenic molecules within animal-derived foods do not require particular attention in the preparation of the challenging material, whereas the majority of plant-derived food allergens are easily degraded during preparation and vary greatly in their concentration in natural starting material. Natural or induced ripening, storage, or freezing may influence the allergen content of the material. Consensus documents produced to define and to standardize the DBPCFC have considered the need to adjust the procedure of food challenge preparation on the basis of the current knowledge of allergen stability (Niggemann and Beyer, 2005). In the case of subjective symptoms (OAS) or a first negative food challenge, the test should be repeated with a new food preparation (Niggemann and Beyer, 2005). It would be useful to evaluate allergenic molecule concentrations before the testing material was used.

The same rules apply to fresh foods used in the prick-prick test. As this test is mostly used for testing fresh fruits, it requires the use of unpeeled samples, as much of the allergen content is located in the outer surface of the fruit (i.e., lipid transfer proteins). Skin testing with fresh food preparations allows us to obtain better results than with allergenic extract preparations tested either in vivo or in vitro (Ortolani et al., 1989; Norgaard et al., 1992; Rance et al., 1997; Kim et al., 2002b).

DIAGNOSTIC REAGENTS—ALLERGENIC EXTRACTS

Food allergen extracts represent the most common reagents used for both in vivo and in vitro diagnoses. Allergen extracts are heterogeneous and contain a complex combination of a number of different proteins, glycoproteins, and polysaccharides. Only a few of these components are recognized as allergenic by testing for IgE binding. Furthermore, these components are not present in extracts in constant amounts; their concentration can change, depending on the origin of the source material and extraction and preservation conditions. In many cases, specific conditions, such as low temperatures or low levels of enzyme inhibitors, are applied during the extraction procedure to guarantee a reliable allergenic content (Vieths et al., 1998). Furthermore, processing of the food source and the nature of the matrix in which the allergen is processed must also be evaluated (Maleki, 2004).

In recent decades, great efforts have been expended to attempt to standardize allergenic extract preparations. Biochemical (overall protein content, SDS-PAGE, and IEF), immunochemical (ELISA inhibition and immunoblotting), and biological (ST and basophil activation) methods have been used (van Ree, 1997). As they do not distinguish the reactivity caused by single components, these methods define the overall allergenicity of the extract preparations, which is expressed in arbitrary units differing for each manufacturer. Using appropriate tools, it has been shown that these preparations may vary greatly in their composition, leading to the conclusion that the current reliability of allergenic extracts from food sources is low (Cuesta-Herranz et al., 1998; Skamstrup Hansen et al., 2001; Akkerdaas et al., 2003). Recently, the European Union funded a project specifically addressing questions on the use of allergenic molecules and allergen-specific monoclonal antibodies as useful tools in determining the specific allergen contents of allergenic extracts and on the creation of international reference standards to harmonize the allergenic extract preparations among different manufacturers (van Ree, 2004). No food allergenic extracts have been considered in the project, as it exclusively focuses on inhalant allergenic sources (mites, Der f 1 and Der f 2 and Der p 1 and Der p 2; grass pollen, Phl p 1 and Phl p 5; birch pollen, Bet v 1; olive pollen, Ole e 1) to be used in specific immunotherapy. By the way, the results obtained seem to be promising for the application of these new tools to allergen standardization of food extracts as well.

DIAGNOSTIC REAGENTS—ALLERGENIC MOLECULES AND EPITOPES

Beginning in the late 1970s, there has been a rising interest in the identification and characterization of allergenic components within allergenic extracts. The application of more sophisticated biochemical and immunochemical tools increased this knowledge during the 1980s. The greatest impulse in identifying and characterizing the nature of allergenic molecules has come from the association of previously available laboratory tools with novel molecular biology methodology. Figure 3A reports the number of allergenic molecule-dedicated studies showing these trends (data are taken from the Allergome reference database). The overall result of these studies has been the identification of 329 food allergens and 83 isoforms of some of the identified allergens (data are taken from allergenic molecule Web databases and from the literature as of 15 February 2005) (Table 2). Data are also available from several recently published review papers (Bush and Hefle, 1996; Breiteneder and Ebner, 2000; Breiteneder and Radauer, 2004; Breiteneder and Mills, 2005). Many other recognized allergenic sources still require allergenic molecule identification (Allergome database, allergenic source section). Globalization of food products and changes in eating habit are leading to identification of new sources and molecules as allergenic (Sten et al., 2002; Bolhaar et al., 2004) in previously unexposed populations.

The large majority of the published papers have been mainly dedicated to basic bench studies, whereas applications in the clinical field have been more recent (Fig. 3B). As in the case of respiratory allergens, the application of allergenic molecules to the diagnostic workup and to epidemiological studies should lead to a better knowledge of the impact of each IgE-recognized structure in affected populations. Allergenic molecules can be used as the most objective tool for the best diagnosis and unequivocal identification of affected subjects. For example, the use of Bet v 1 as a marker of a condition associating Fagales pollen allergy with OAS to apple and hazelnut in a population not exposed to birch pollen but to other Fagales pollens has been reported. Unless the situation is different than in northern European countries, Bet v 1-sensitized subjects in a southern European area generally considered free of birch allergy problems were clearly identified (Mari et al., 2003).

Studying food allergenic molecules instead of extracts means that the distinction between inhalant and food allergens in terms of primary sensitizers is being superceded. Specific IgE production, induced by molecule sensitization via the respiratory mucosa, corecognize the same epitope on food allergens (Ferreira et al., 2004). Examples are

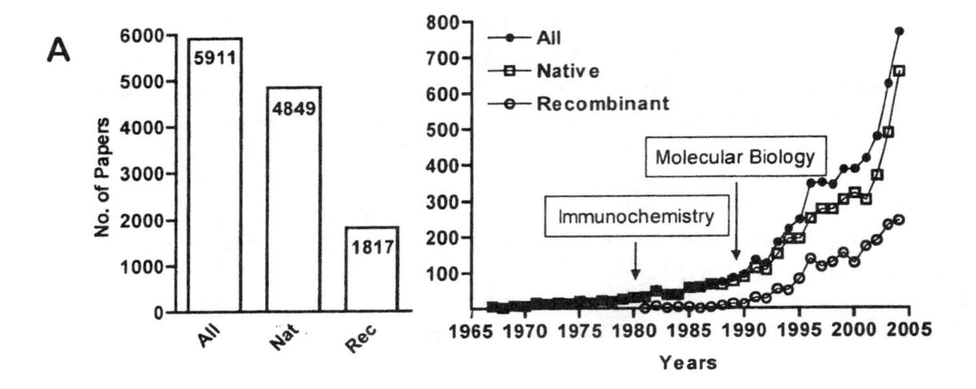

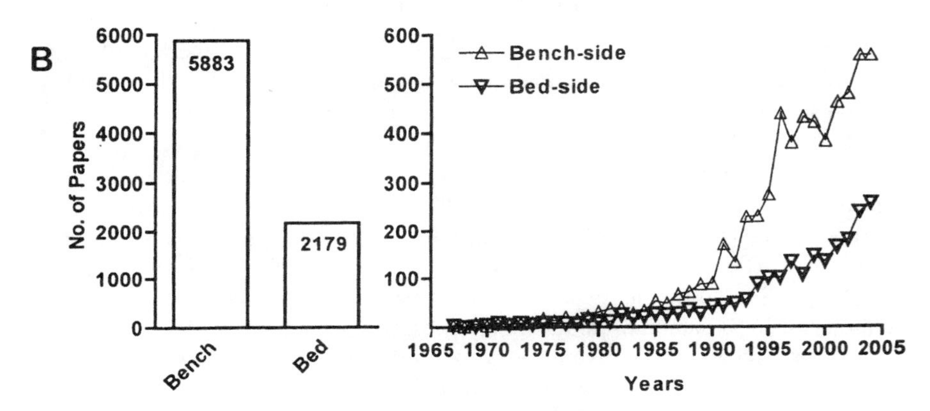

Figure 3. Scientific publications related to allergenic molecules. (A) Overall number of papers, dedicated to native or to recombinant molecules, are indicated on the left. (Right) Three-group distribution during the last 40 years. (B) Papers dedicated to basic research (benchside, biochemistry, molecular biology, immunochemistry, immune mechanisms, and genetics) or to clinical research (environmental detection, diagnosis, epidemiology, and immunotherapy) are reported as either overall numbers or distribution during the last 40 years.

Bet v 1-, hevein-, and profilin-related allergens (Breiteneder and Radauer, 2004). The inverse process can be recorded as well. Tropomyosin from crustaceans, which mostly induce severe food allergy reactions after primary sensitization by the ingestion route, leads to sensitization to mite and other arthropod tropomyosins to which patients are exposed in the indoor environment, and it is capable of causing respiratory symptoms by inhalation (Ayuso et al., 2002). In addition, the tropomyosin model is interesting because this molecule, released in the environment by mites and other arthropods, can sensitize subjects not otherwise exposed to the ingested allergen by the inhalation route (Fernandes et al., 2003). Recently, an interesting natural model was reported for kiwifruit allergy (Bublin et al., 2004). Subjects from three European countries showed different patterns of molecular reactivity; Italian subjects never exposed to gold kiwifruit were reactive to its allergens. Interestingly, the gold kiwifruit species seemed to lack any homologous molecule to the Act d 1 allergen identified in green kiwifruit. This would allow subjects with isolated Act d 1 reactivity to consume gold kiwifruit. From the above example, it should be assumed that the IgE corecognition concept could be applied to all the allergenic molecules where a single reagent could be identified as the marker of the primary sensitizer and the clinical picture was determined by different exposure modalities to IgE-corecognized allergen. The final clinical picture will depend on how the single subject is exposed to corecognized allergens that are highly influenced by personal social, geographical, and cultural variables. The application of allergenic molecules in routine diagnosis may qualify them as markers of a condition (Mari, 2001). They will also allow a true mapping of the sensitization and a thorough objective evaluation of population sensitivity (Mari et al., 2003).

The previous technical limitations of allergen purification by biochemical-immunochemical methods to obtain reasonable amounts of allergen to be used in the diagnosis have now largely been overcome by cloning. The production of allergens by molecular biology techniques may provide purified, reproducible, theoretically unlimited quantities of allergenic proteins for diagnosis and immunotherapy (Wallner et al., 2004). The diagnostic application of these purified recombinant food allergens seems to lead to better results than with extracts applied either in vivo or in vitro (Beyer 2003; Bohle and Vieths, 2004). Several expression systems can be used for recombinant molecule production and are available. For instance, to obtain a recombinant molecule lacking the disturbing effect of glycan side chains, prokaryotic and non-plant-derived eukaryotic systems may be suitable, whereas tobacco plants are not. At the moment, the availability of all the relevant food-derived molecules to be used as a complete panel for the diagnostic process is a major limitation. To overcome these limitations, broad collaborative agreements should be reached among research laboratories worldwide (Hiller et al., 2002).

A further step toward a more-refined diagnosis with relevant prognostic implications is the use of IgE-recognized epitopes. Dissection of allergenic molecules by IgE epitope mapping is producing new information on these topics. Several studies have shown different prognostic outcomes, depending on whether a subject is sensitized or not to linear or conformational epitopes (Reese et al., 2001; Chatchatee et al., 2001; Vila et al., 2001; Busse et al., 2002; Järvinen et al., 2002; Robotham et al., 2002; Beyer et al., 2003; Shreffler et al., 2004). Allergenic epitopes of many food allergens have been mapped, and the exact sites of IgE binding have been determined. Both conformational and linear epitopes may be responsible for allergic reactions. IgE antibodies directed to sequential epitopes are able to react to any form of food (e.g., cooked or hydrolyzed). On the other hand, patients with IgE antibodies that recognize conformational epitopes can tolerate small amounts of the food when the tertiary structure of the protein is modified and the conformational epitopes are destroyed. Patients affected by egg and milk allergies with IgE antibodies directed toward linear epitopes tend to have persistent allergy, whereas in the presence of IgE antibodies directed to conformational epitopes, such patients tend to acquire clinical tolerance (Chatchatee et al., 2001; Järvinen et al., 2002). Determining epitope-specific binding correlates with clinical reactivity better than does simple quantitative IgE values for the whole protein. Higher levels of IgE antibodies to specific linear epitopes from casein were found in cases of persistent cow's milk allergy than in patients who developed tolerance (Urisu et al., 1999). Similarly, it was demonstrated that the presence of IgE antibodies to linear allergenic epitopes of α-lactalbumin and β-lactoglobulin might be considered a marker of persistent cow's milk allergy (Järvinen et al., 2001). Similar

Table 2. Features of allergenic molecule databases on the Web

Website	URL	Data source	Allergenic source(s)	Allergenic molecule(s)	Allergen sequence(s)	Biochemistry or molecular biology	Clinical and epidemiological data	Internal search engine	Internal data cross-linking	Computational tools[b]	User alert	References[b]	Last update (mo day, yr)
Official Allergen Nomenclature[a]	http://www.allergen.org	Submission	No	All	No	Yes	No	No	No	No	No	+	Feb. 02, 2005
Food Allergy and Resource Program, (Farrp) allergen database	http://www.allergenonline.com	Protein databases	No	All	Links	Yes	No	Yes	No	+	Yes	No	Oct. 1, 2004
Allergome	http://www.allergome.org/	Literature and protein databases	All	All	Links	Yes	Yes	Yes	Yes	No	Yes	+++	Feb. 15, 2005
Allermatch	http://www.allermatch.org/	Protein databases	No	All	Links	Yes	No	No	No	+++	No	No	Dec. 10, 2004
Aller-Predict	http://sdmc.i2r.a-star.edu.sg/Templar/DB/Allergen/Predict.Predict.html	Protein databases	No	All	Yes	No	No	Yes	Yes	+++	No	No	Given for each allergen

Biotechnology Information for Food Safety	http://www.iit.edu/~sgendel/fa.htm	Protein databases	No	All	Links	No	No	No	No	No	No	+	Not given
Central Science Laboratory (DAB) Allergen Database	http://www.csl.gov.uk/allergen	Protein databases	No	All	Links	Yes	No	Yes	No	No	No	+	Given for each allergen
InformAll (Protall)	http://foodallergens.ifr.ac.uk/	Literature	Food	Food	Links	Yes	Yes	Yes	No	No	No	++	Jan. 18, 2005
Structural Database of Allergenic Proteins (SDAP)	http://fermi.utmb.edu/SDAP/	Protein databases	No	All	Yes	Yes	No	Yes	No	+++	No	No	Given for each allergen

[a]Official Web site of the IUIS/WHO Allergen Nomenclature Sub-Committee.
[b]An arbitrary relative grading of the supported computational tools and of the linked references is indicated by plus signs.

data have been obtained with allergenic molecules from peanuts, where the application of microarray analyses of IgE-reactive epitopes leads to novel prognostic information (Shreffler et al., 2004).

A further interesting example of how the biochemical definition of the nature of an epitope may lead to a relevant clinical outcome is given by cross-reactive carbohydrate determinants (CCD). These epitopes, mainly represented by O-glycan side chains of plant and arthropod glycoproteins, are capable of binding IgE, but they have been reported to show a negligible level of biological activity (Van Der Veen et al., 1997; Mari et al., 1999). IgE reactivity to CCD seems to be quite common among pollen allergic subjects, creating much specific but irrelevant in vitro positive results (Mari et al., 1999; Mari, 2002). The presence of any glycoprotein in extracts, allergenic or not, allows the binding of specific IgE to CCD (Iacovacci et al., 2001). The spreading of IgE-positive results does not match any clinical reactivity, as these subjects continue the intake of foods containing large amounts of glycoproteins (Mari, 2002).

The allergenic molecule-based approach is leading to a reconsideration of the immunotherapy approach to food allergy. The major drawback of injection-specific immunotherapy with food allergenic extracts is the frequent occurrence of severe adverse reactions following administration (Nelson et al., 1997). Several studies have approached this topic with animal models by using hypoallergenic variants of relevant food allergens. Promising results are linked to both less-frequent adverse reactions that could be obtained with the lower IgE-binding capacity of hypoallergens and the enhancement of the protective immune response obtained by allergen conjugation with specific adjuvants (Li and Sampson, 2004).

MANAGING THE KNOWLEDGE OF ALLERGENIC MOLECULES

As the amount of available data on allergens increases, the organization of this knowledge has been attempted. Much work has been done by several organizations and institutions to classify already-described allergenic molecule and to have lists available from the internet (Table 2). By supplying criteria for the assignment of names to molecules, the International Union of Immunological Societies-World Health Organization (IUIS/WHO) Allergen Nomenclature Sub-Committee is keeping confusion in this field at the lowest possible level. Other initiatives have been taken to complete the list of identified allergens and to supply additional biochemical, immunological, and clinical information for each allergen.

As soon as a large number of allergens have been identified and characterized, sequences obtained from these lists have been used for computational purposes. The need to recognize the allergenic nature of novel proteins has led to the application of available algorithms (BLAST, FASTA) for comparative analysis and to the development of completely new ones for the more specific purpose of recognition of allergenic structures. Table 2 lists Web-based databases, which allow the performance of such comparative evaluations by the algorithms originally available from BLAST or FASTA or algorithms available on the Web. At the moment, no single criteria or algorithm seems to satisfy the need for the certain identification of a new molecule as allergenic, as no single unifying feature has been found. Unfortunately, results obtained with each of the original algorithms are not completely comparable, as they are based on different allergen databases created by different allergenic molecule selection criteria. Unless rigid criteria are defined for the nomenclature of allergens, there is no general agreement on scientific criteria for the classification of a protein as allergenic. Allergens listed in the official IUIS/WHO Web site are included on the basis of documented in vitro IgE reactivity (Larsen and Lowenstein, 1996).

Moreover, the concept of IgE corecognition of allergenic molecules should lead to a simplification of the number of unique molecules needed for diagnosis and therapy. On the basis of their reciprocal relationship, allergens cluster in groups (e.g., Bet v 1 like, profilins, or nonspecific lipid transfer proteins). For each group, it will be useful to define the best representative molecule as a marker for a defined IgE reactivity. Specific expert systems should help to interpret such data.

Data from high-throughput microtechnologies (e.g., allergen microarrays) will generate huge amounts of data that need to be managed to extract meanings from them (data mining). This has already occurred in other research fields such as genetics (Schulze and Downward, 2001). In a circular fashion, all of these tools will generate an increasing number of scientific publications. Published data and information will need to be extracted and organized (literature mining) (Chaussabel, 2004).

Bioinformatics applied to allergy must face the rapid evolution of biologic research from a local to a global scale, which will lead to the management of data produced by these high-throughput yet experimental technologies. Doubtless, all the forthcoming tools from bioinformatics will further assist in managing future food allergy diagnostics and therapy.

CONCLUSIONS

A combination of in vivo and in vitro tools are needed to reach a good level of diagnostic reliability in food allergy diseases. The advanced measurements of specific food allergen-serum IgE concentrations may enable us to determine more-specific diagnostic decision points for predicting the outcome of oral food challenge tests and for predicting the natural history of the food allergic disorder. IgE quantification could be enhanced by testing the largest panel of discrete molecules available for each relevant food source, as has been demonstrated for grass and birch pollens (Mari, 2003; Mari et al., 2003). The relevance of specific IgE dosages can be further increased by excluding IgE reactivity to CCD by nonglycosylated recombinant molecules produced in prokaryotic expression systems. The use of the best standardized DBPCFC will help clear the scenario of food allergy diseases from unnecessary epidemiological overestimation and will also aid in carefully defining the therapeutic approach that will provide the best quality of life to affected subjects (Niggemann and Beyer, 2005). Classical diagnostic tools such as ST, IgE, and DBPCFC based on freshly prepared or commercially available extracts may seem difficult to standardize. Great expectations arise from the combined use of allergens obtained by molecular biology techniques, exactly as they are seen and processed by the immune system, sophisticated assays that allow suitable multiple-IgE detection of hundreds of allergenic molecules and global sharing of a huge mass of data. This novel approach will certainly represent developments that are not too far in the future for food allergy diseases.

REFERENCES

Abassi, Y. A., J. A. Jackson, J. Zhu, J. O'Connell, X. Wang, and X. Xu. 2004. Label-free, real-time monitoring of IgE-mediated mast cell activation on microelectronic cell sensor arrays. *J. Immunol. Methods* **292:** 195–205.

Adinoff, A. D., D. M. Rosloniec, L. L. McCall, and H. S. Nelson. 1989. A comparison of six epicutaneous devices in the performance of immediate hypersensitivity skin testing. *J. Allergy Clin. Immunol.* **84:**168–174.

Akkerdaas, J. H., M. Wensing, A. C. Knulst, M. Krebitz, H. Breiteneder, S. De Vries, A. H. Penninks, R. C. Aalberse, S. L. Hefle, and R. van Ree. 2003. How accurate and safe is the diagnosis of hazelnut allergy by means of commercial skin prick test reagents? *Int. Arch. Allergy Immunol.* **132:**132–140.

Armentia, A., M. Lombardero, C. Martinez, D. Barber, J. M. Vega, and A. Callejo. 2004. Occupational asthma due to grain pests Eurygaster and Ephestia. *J. Asthma* **41:**99–107.

Ayuso, R., G. Reese, S. Leong-Kee, M. Plante, and S. B. Lehrer. 2002. Molecular basis of arthropod cross-reactivity: IgE-binding cross-reactive epitopes of shrimp, house dust mite and cockroach tropomyosins. *Int. Arch. Allergy Immunol.* **129:**38–48.

Bacarese-Hamilton, T., A. Ardizzoni, J. Gray, and A. Crisanti. 2004. Protein arrays for serodiagnosis of disease. *Methods Mol. Biol.* **278:**271–284.

Berger, P., C. N'guyen, M. Buckley, E. Scotto-Gomez, R. Marthan, and J. M. Tunon-De-Lara. 2002. Passive sensitization of human airways induces mast cell degranulation and release of tryptase. *Allergy* **57:**592–599.

Bernardini, R., E. Novembre, E. Lombardi, N. Pucci, F. Marcucci, and A. Vierucci. 2002. Anaphylaxis to latex after ingestion of a cream-filled doughnut contaminated with latex. *J. Allergy Clin. Immunol.* **110:** 534–535.

Beyer, K. 2003. Characterization of allergenic food proteins for improved diagnostic methods. *Curr. Opin. Allergy Clin. Immunol.* **3:**189–197.

Beyer, K., L. Ellman-Grunther, K. M. Järvinen, R. A. Wood, J. Hourihane, and H. A. Sampson. 2003. Measurement of peptide-specific IgE as an additional tool in identifying patients with clinical reactivity to peanuts. *J. Allergy Clin. Immunol.* **112:**202–207.

Bindslev-Jensen, C., B. K. Ballmer-Weber, U. Bengtsson, C. Blanco, C. Ebner, J. Hourihane, A. C. Knulst, D. A. Moneret-Vautrin, K. Nekam, B. Niggemann, M. Osterballe, C. Ortolani, J. Ring, C. Schnopp, and T. Werfel. 2004. Standardization of food challenges in patients with immediate reactions to foods—position paper from the European Academy of Allergology and Clinical Immunology. *Allergy* **59:**690–697.

Bindslev-Jensen, C., D. Briggs, and M. Osterballe. 2002. Can we determine a threshold level for allergenic foods by statistical analysis of published data in the literature? *Allergy* **57:**741–746.

Bindslev-Jensen, C., and L. K. Poulsen. 1996. What do we at present know about the ALCAT test and what is lacking? *Monogr. Allergy* **32:**228–232.

Bischoff, S. C., J. Mayer, J. Wedemeyer, P. N. Meier, G. Zeck-Kapp, B. Wedi, A. Kapp, Y. Cetin, M. Gebel, and M. P. Manns. 1997. Colonoscopic allergen provocation (COLAP): a new diagnostic approach for gastrointestinal food allergy. *Gut* **40:**745–753.

Blanco, C., J. Quiralte, R. Castillo, J. Delgado, C. Arteaga, D. Barber, and T. Carrillo. 1997. Anaphylaxis after ingestion of wheat flour contaminated with mites. *J. Allergy Clin. Immunol.* **99:**308–313.

Bock, S. A., H. A. Sampson, F. M. Atkins, R. S. Zeiger, S. Lehrer, M. Sachs, R. K. Bush, and D. D. Metcalfe. 1988. Double-blind, placebo-controlled food challenge (DBPCFC) as an office procedure: a manual. *J. Allergy Clin. Immunol.* **82:**986–997.

Bodtger, U. 2004. Prognostic value of asymptomatic skin sensitization to aeroallergens. *Curr. Opin. Allergy Clin. Immunol.* **4:**5–10.

Bohle, B., and S. Vieths. 2004. Improving diagnostic tests for food allergy with recombinant allergens. *Methods* **32:**292–299.

Bolhaar, S. T., R. van Ree, C. A. Bruijnzeel-Koomen, A. C. Knulst, and L. Zuidmeer. 2004. Allergy to jackfruit: a novel example of Bet v 1-related food allergy. *Allergy* **59:**1187–1192.

Boumiza, R., G. Monneret, M. F. Forissier, J. Savoye, M. C. Gutowski, S. Powell, and J. Bienvenu. 2003. Marked improvement of the basophil activation test by detecting CD203c instead of CD63. *Clin. Exp. Allergy* **33:**259–265.

Boyano-Martinez, T., C. Garcia-Ara, J. M. Diaz-Pena, F. M. Munoz, S. G. Garcia, and M. M. Esteban. 2001. Validity of specific IgE antibodies in children with egg allergy. *Clin. Exp. Allergy* **31:**1464–1469.

Breiteneder, H., and C. Ebner. 2000. Molecular and biochemical classification of plant-derived food allergens. *J. Allergy Clin. Immunol.* **106:**27–36.

Breiteneder, H., and E. N. Mills. 2005. Molecular properties of food allergens. *J. Allergy Clin. Immunol.* **115:**14–23.

Breiteneder, H., and C. Radauer. 2004. A classification of plant food allergens. *J. Allergy Clin. Immunol.* **113:**821–830.

Breuer, K., A. Heratizadeh, A. Wulf, U. Baumann, A. Constien, D. Tetau, A. Kapp, and T. Werfel. 2004. Late eczematous reactions to food in children with atopic dermatitis. *Clin. Exp. Allergy* **34:**817–824.

Bublin, M., A. Mari, C. Ebner, A. Knulst, O. Scheiner, K. Hoffmann-Sommergruber, H. Breiteneder, and C. Radauer. 2004. IgE sensitization profiles toward green and gold kiwifruits differ among patients allergic to kiwifruit from 3 European countries. *J. Allergy Clin. Immunol.* **114:**1169–1175.

Buhring, H. J., A. Streble, and P. Valent. 2004. The basophil-specific ectoenzyme E-NPP3 (CD203c) as a marker for cell activation and allergy diagnosis. *Int. Arch. Allergy Immunol.* **133:**317–329.

Bush, R. K., and S. L. Hefle. 1996. Food allergens. *Crit. Rev. Food Sci. Nutr.* **36**(Suppl.)**:**S119–S163.

Busse, P. J., K. M. Järvinen, L. Vila, K. Beyer, and H. A. Sampson. 2002. Identification of sequential IgE-binding epitopes on bovine αs2-casein in cow's milk allergic patients. *Int. Arch. Allergy Immunol.* **129:** 93–96.

Chatchatee, P., K. M. Järvinen, L. Bardina, K. Beyer, and H. A. Sampson. 2001. Identification of IgE- and IgG-binding epitopes on αs1-casein: differences in patients with persistent and transient cow's milk allergy. *J. Allergy Clin. Immunol.* **107:**379–383.

Chaussabel, D. 2004. Biomedical literature mining: challenges and solutions in the 'omics' era. *Am. J. Pharmacogenomics* **4:**383–393.

Church, M. K., and F. Levi-Schaffer. 1997. The human mast cell. *J. Allergy Clin. Immunol.* **99:**155–160.

Clark, A. T., and P. W. Ewan. 2003. Interpretation of tests for nut allergy in one thousand patients, in relation to allergy or tolerance. *Clin. Exp. Allergy* **33:**1041–1045.

Corder, W. T., M. B. Hogan, and N. W. Wilson. 1996. Comparison of two disposable plastic skin test devices with the bifurcated needle for epicutaneous allergy testing. *Ann. Allergy Asthma Immunol.* **77:**222–226.

Corder, W. T., and N. W. Wilson. 1995. Comparison of three methods of using the DermaPIK with the standard prick method for epicutaneous skin testing. *Ann. Allergy Asthma Immunol.* **75:**434–438.

Cuesta-Herranz, J., M. Lazaro, A. Martinez, E. Alvarez-Cuesta, E. Figueredo, J. Martinez, C. Cuesta, and M. Las-Heras. 1998. A method for quantitation of food biologic activity: results with peach allergen extracts. *J. Allergy Clin. Immunol.* **102:**275–280.

Darsow, U., J. Laifaoui, K. Kerschenlohr, A. Wollenberg, B. Przybilla, B. Wuthrich, S. Borelli Jr., F. Giusti, S. Seidenari, K. Drzimalla, D. Simon, R. Disch, S. Borelli, A. C. Devillers, A. P. Oranje, L. De Raeve, J. P. Hachem, C. Dangoisse, A. Blondeel, M. Song, K. Breuer, A. Wulf, T. Werfel, S. Roul, A. Taieb, S. Bolhaar, C. Bruijnzeel-Koomen, M. Bronnimann, L. R. Braathen, A. Didierlaurent, C. Andre, and J. Ring. 2004. The prevalence of positive reactions in the atopy patch test with aeroallergens and food allergens in subjects with atopic eczema: a European multicenter study. *Allergy* **59:**1318–1325.

Deinhofer, K., H. Sevcik, N. Balic, C. Harwanegg, R. Hiller, H. Rumpold, M. W. Mueller, and S. Spitzauer. 2004. Microarrayed allergens for IgE profiling. *Methods* **32:**249–254.

Devenney, I., and K. Falth-Magnusson. 2000. Skin prick tests may give generalized allergic reactions in infants. *Ann. Allergy Asthma Immunol.* **85:**457–460.

Dibbern, D. A., G. W. Palmer, P. B. Williams, S. A. Bock, and S. C. Dreskin. 2003. RBL cells expressing human Fc varε RI are a sensitive tool for exploring functional IgE-allergen interactions: studies with sera from peanut-sensitive patients. *J. Immunol. Methods* **274:**37–45.

Ebo, D., M. Hagendorens, C. Bridts, A. Schuerwegh, L. De Clerck, and W. Stevens. 2004. In vitro allergy diagnosis: should we follow the flow? *Clin. Exp. Allergy* **34:**332–339.

Ebo, D. G., M. M. Hagendorens, C. H. Bridts, A. J. Schuerwegh, L. S. De Clerck, and W. J. Stevens. 2005. Flow cytometric analysis of in vitro activated basophils, specific IgE and skin tests in the diagnosis of pollen-associated food allergy. *Cytometry B Clin. Cytom.* **64:**28–33.

Edston, E., and M. Van Hage-Hamsten. 2003. Death in anaphylaxis in a man with house dust mite allergy. *Int. J. Legal Med.* **117:**299–301.

Eigenmann, P. A. 2004. Do we have suitable in-vitro diagnostic tests for the diagnosis of food allergy? *Curr. Opin. Allergy Clin. Immunol.* **4:**211–213.

Eigenmann, P. A., F. D. Pastore, and S. A. Zamora. 2001. An internet-based survey of anaphylactic reactions to foods. *Allergy* **56:**540–543.

Erben, A. M., J. L. Rodriguez, J. McCullough, and D. R. Ownby. 1993. Anaphylaxis after ingestion of beignets contaminated with Dermatophagoides farinae. *J. Allergy Clin. Immunol.* **92:**846–849.

Erdmann, S. M., N. Heussen, S. Moll-Slodowy, H. F. Merk, and B. Sachs. 2003. CD63 expression on basophils as a tool for the diagnosis of pollen-associated food allergy: sensitivity and specificity. *Clin. Exp. Allergy* **33:**607–614.

Fernandes, J., A. Reshef, L. Patton, R. Ayuso, G. Reese, and S. B. Lehrer. 2003. Immunoglobulin E antibody reactivity to the major shrimp allergen, tropomyosin, in unexposed Orthodox Jews. *Clin. Exp. Allergy* **33:**956–961.

Ferreira, F., T. Hawranek, P. Gruber, N. Wopfner, and A. Mari. 2004. Allergic cross-reactivity: from gene to the clinic. *Allergy* **59:**243–267.

Fiocchi, A., G. R. Bouygue, P. Restani, G. Bonvini, R. Startari, and L. Terracciano. 2002. Accuracy of skin prick tests in IgE-mediated adverse reactions to bovine proteins. *Ann. Allergy Asthma Immunol.* **89:**26–32.

Fox, R. A., B. M. Sabo, T. P. Williams, and M. R. Joffres. 1999. Intradermal testing for food and chemical sensitivities: a double-blind controlled study. *J. Allergy Clin. Immunol.* **103:**907–911.

Gray, H. C., T. M. Foy, B. A. Becker, and A. P. Knutsen. 2004. Rice-induced enterocolitis in an infant: TH1/TH2 cellular hypersensitivity and absent IgE reactivity. *Ann. Allergy Asthma Immunol.* **93:**601–605.

Gruber, C., D. Buck, U. Wahn, and B. Niggemann. 2000. Is there a role for immunoblots in the diagnosis of latex allergy? Intermethod comparison of in vitro and in vivo IgE assays in spina bifida patients. *Allergy* **55:**476–483.

Guillet, M. H., C. Kauffmann-Lacroix, F. Dromer, C. Larsen, and G. Guillet. 2003. Urticaria and anaphylactic shock due to food allergy to *Penicillium italicum*. *Rev. Fr. Allergol. Immunol. Clin.* **43:**520–523.

Hamilton, R. G., and N. F. Adkinson, Jr. 2003. Clinical laboratory assessment of IgE-dependent hypersensitivity. *J. Allergy Clin. Immunol.* **111:**687–701.

Hamilton, R. G., and N. F. Adkinson, Jr. 2004. In vitro assays for the diagnosis of IgE-mediated disorders. *J. Allergy Clin. Immunol.* **114:**213–225.

Harwanegg, C., and R. Hiller. 2004. Protein microarrays in diagnosing IgE-mediated diseases: spotting allergy at the molecular level. *Expert Rev. Mol. Diagn.* **4:**539–548.

Harwanegg, C., S. Laffer, R. Hiller, M. W. Mueller, D. Kraft, S. Spitzauer, and R. Valenta. 2003. Microarrayed recombinant allergens for diagnosis of allergy. *Clin. Exp. Allergy* **33:**7–13.

Hauswirth, A. W., S. Natter, M. Ghannadan, Y. Majlesi, G. H. Schernthaner, W. R. Sperr, H. J. Buhring, R. Valenta, and P. Valent. 2002. Recombinant allergens promote expression of CD203c on basophils in sensitized individuals. *J. Allergy Clin. Immunol.* **110:**102–109.

Heine, R. G. 2004. Pathophysiology, diagnosis and treatment of food protein-induced gastrointestinal diseases. *Curr. Opin. Allergy Clin. Immunol.* **4:**221–229.

Hill, D. J., R. G. Heine, and C. S. Hosking. 2004. The diagnostic value of skin prick testing in children with food allergy. *Pediatr. Allergy Immunol.* **15:**435–441.

Hill, D. J., C. S. Hosking, and L. V. Reyes-Benito. 2001. Reducing the need for food allergen challenges in young children: a comparison of in vitro with in vivo tests. *Clin. Exp. Allergy* **31:**1031–1035.

Hiller, R., S. Laffer, C. Harwanegg, M. Huber, W. M. Schmidt, A. Twardosz, B. Barletta, W. M. Becker, K. Blaser, H. Breiteneder, M. Chapman, R. Crameri, M. Duchene, F. Ferreira, H. Fiebig, K. Hoffmann-Sommergruber, T. P. King, T. Kleber-Janke, V. P. Kurup, S. B. Lehrer, J. Lidholm, U. Muller, C. Pini, G. Reese, O. Scheiner, A. Scheynius, H. D. Shen, S. Spitzauer, R. Suck, I. Swoboda, W. Thomas, R. Tinghino, M. Van Hage-Hamsten, T. Virtanen, D. Kraft, M. W. Muller, and R. Valenta. 2002. Microarrayed allergen molecules: diagnostic gatekeepers for allergy treatment. *FASEB J.* **16:**414–416.

Iacovacci, P., C. Pini, C. Afferni, B. Barletta, R. Tinghino, E. Schinina, R. Federico, A. Mari, and G. Di Felice. 2001. A monoclonal antibody specific for a carbohydrate epitope recognizes an IgE-binding determinant shared by taxonomically unrelated allergenic pollens. *Clin. Exp. Allergy* **31:**458–465.

Järvinen, K. M., K. Beyer, L. Vila, P. Chatchatee, P. J. Busse, and H. A. Sampson. 2002. B-cell epitopes as a screening instrument for persistent cow's milk allergy. *J. Allergy Clin. Immunol.* **110:**293–297.

Järvinen, K. M., P. Chatchatee, L. Bardina, K. Beyer, and H. A. Sampson. 2001. IgE and IgG binding epitopes on alpha-lactalbumin and beta-lactoglobulin in cow's milk allergy. *Int. Arch. Allergy Immunol.* **126:**111–118.

Jeebhay, M. F., T. G. Robins, S. B. Lehrer, and A. L. Lopata. 2001. Occupational seafood allergy: a review. *Occup. Environ. Med.* **58:**553–562.

Johansson, S. 2004. ImmunoCAP specific IgE test: an objective tool for research and routine allergy diagnosis. *Expert Rev. Mol. Diagn.* **4:**273–279.

Johansson, S. G., T. Bieber, R. Dahl, P. S. Friedmann, B. Q. Lanier, R. F. Lockey, C. Motala, J. A. Ortega Martell, T. A. Platts-Mills, J. Ring, F. Thien, P. Van Cauwenberge, and H. C. Williams. 2004. Revised nomenclature for allergy for global use: report of the Nomenclature Review Committee of the World Allergy Organization, October 2003. *J. Allergy Clin. Immunol.* **113:**832–836.

Jones, R. T., D. L. Squillace, and J. W. Yunginger. 1992. Anaphylaxis in a milk-allergic child after ingestion of milk-contaminated kosher-pareve-labeled "dairy-free" dessert. *Ann. Allergy* **68:**223–227.

Kanny, G., D. A. Moneret-Vautrin, J. Flabbee, E. Beaudouin, M. Morisset, and F. Thevenin. 2001. Population study of food allergy in France. *J. Allergy Clin. Immunol.* **108:**133–140.

Kerschenlohr, K., S. Decard, B. Przybilla, and A. Wollenberg. 2003. Atopy patch test reactions show a rapid influx of inflammatory dendritic epidermal cells in patients with extrinsic atopic dermatitis and patients with intrinsic atopic dermatitis. *J. Allergy Clin. Immunol.* **111:**869–874.

Kim, T. E., S. W. Park, N. Y. Cho, S. Y. Choi, T. S. Yong, B. H. Nahm, S. Lee, and G. Noh. 2002a. Quantitative measurement of serum allergen-specific IgE on protein chip. *Exp. Mol. Med.* **34:**152–158.

Kim, T. E., S. W. Park, G. W. Noh, and S. S. Lee. 2002b. Comparison of skin prick test results between crude allergen extracts from foods and commercial allergen extracts in atopic dermatitis by double-blind placebo-controlled food challenge for milk, egg, and soybean. *Yonsei Med. J.* **43:**613–620.

Kingsmore, S. F., and D. D. Patel. 2003. Multiplexed protein profiling on antibody-based microarrays by rolling circle amplification. *Curr. Opin. Biotechnol.* **14:**74–81.

Kleine Budde, I., P. G. de Heer, J. S. Der Zee, and R. C. Aalberse. 2001. The stripped basophil histamine release bioassay as a tool for the detection of allergen-specific IgE in serum. *Int. Arch. Allergy Immunol.* **126:**277–285.

Koppelman, S. J., M. Wensing, G. A. de Jong, and A. C. Knulst. 1999. Anaphylaxis caused by the unexpected presence of casein in salmon. *Lancet* **354:**2136

Kricka, L. J., T. Joos, and P. Fortina. 2003. Protein microarrays: a literature survey. *Clin. Chem.* **49:**2109.

Lambert, C., L. Guilloux, C. Dzviga, C. Gourgaud-Massias, and C. Genin. 2003. Flow cytometry versus histamine release analysis of in vitro basophil degranulation in allergy to Hymenoptera venom. *Cytometry B Clin. Cytom.* **52:**13–19.

Larsen, J. N., and H. Lowenstein. 1996. Allergen nomenclature. *J. Allergy Clin. Immunol.* **97:**577–578.

Lee, S. Y., K. S. Lee, C. H. Hong, and K. Y. Lee. 2001. Three cases of childhood nocturnal asthma due to buckwheat allergy. *Allergy* **56:**763–766.

Leung, D. Y., H. A. Sampson, J. W. Yunginger, A. W. Burks, Jr., L. C. Schneider, C. H. Wortel, F. M. Davis, J. D. Hyun, and W. R. Shanahan, Jr. 2003. Effect of anti-IgE therapy in patients with peanut allergy. *N Engl. J. Med.* **348:**986–993.

Lewith, G. T., J. N. Kenyon, J. Broomfield, P. Prescott, J. Goddard, and S. T. Holgate. 2001. Is electrodermal testing as effective as skin prick tests for diagnosing allergies? A double blind, randomised block design study. *BMJ* **322:**131–134.

Li, T. M., T. Chuang, S. Tse, D. Hovanec-Burns, and A. S. El Shami. 2004. Development and validation of a third generation allergen-specific IgE assay on the continuous random access IMMULITE 2000 analyzer. *Ann. Clin. Lab. Sci.* **34:**67–74.

Li, X. M., and H. A. Sampson. 2004. Novel approaches to immunotherapy for food allergy. *Clin. Allergy Immunol.* **18:**663–679.

Liccardi, G., A. Salzillo, G. Spadaro, G. Senna, G. W. Canonica, G. D'Amato, and G. Passalacqua. 2003. Anaphylaxis caused by skin prick testing with aeroallergens: case report and evaluation of the risk in Italian allergy services. *J. Allergy Clin. Immunol.* **111:**1410–1412.

Lin, X. P., J. Magnusson, S. Ahlstedt, A. Dahlman-Hoglund, L. A. Hanson, O. Magnusson, U. Bengtsson, and E. Telemo. 2002. Local allergic reaction in food-hypersensitive adults despite a lack of systemic food-specific IgE. *J. Allergy Clin. Immunol.* **109:**879–887.

Maleki, S. J. 2004. Food processing: effects on allergenicity. *Curr. Opin. Allergy Clin. Immunol.* **4:**241–245.

Mari, A. 2003. Skin test with a timothy grass (*Phleum pratense*) pollen extract vs IgE to a timothy extract vs IgE to rPhl p 1, rPhl p 2, nPhl p 4, rPhl p 5, rPhl p 6, rPhl p 7, rPhl p 11, and rPhl p 12: epidemiological and diagnostic data. *Clin. Exp. Allergy* **33:**43–51.

Mari, A. 2002. IgE to cross-reactive carbohydrate determinants: analysis of the distribution and appraisal of the in vivo and in vitro reactivity. *Int. Arch. Allergy Immunol.* **129:**286–295.

Mari, A. 2001. Multiple pollen sensitization: a molecular approach to the diagnosis. *Int. Arch. Allergy Immunol.* **125:**57–65.

Mari, A., P. Iacovacci, C. Afferni, B. Barletta, R. Tinghino, G. Di Felice, and C. Pini. 1999. Specific IgE to cross-reactive carbohydrate determinants strongly affect the in vitro diagnosis of allergic diseases. *J. Allergy Clin. Immunol.* **103:**1005–1011.

Mari, A., M. Wallner, and F. Ferreira. 2003. Fagales pollen sensitization in a birch-free area: a respiratory cohort survey using Fagales pollen extracts and birch recombinant allergens (rBet v 1, rBet v 2, rBet v 4). *Clin. Exp. Allergy* **33:**1419–1428.

Matsumoto, T., Y. Goto, and T. Mike. 2001. Anaphylaxis to mite-contaminated flour. *Allergy* **56:**247

Mattila, L., M. Kilpelainen, E. O. Terho, M. Koskenvuo, H. Helenius, and K. Kalimo. 2003. Food hypersensitivity among Finnish university students: association with atopic diseases. *Clin. Exp. Allergy* **33:** 600–606.

Mills, E. N., E. Valovirta, C. Madsen, S. L. Taylor, S. Vieths, E. Anklam, S. Baumgartner, P. Koch, R. W. Crevel, and L. Frewer. 2004. Information provision for allergic consumers—where are we going with food allergen labelling? *Allergy* **59:**1262–1268.

Moneret-Vautrin, D. A., and G. Kanny. 2004. Update on threshold doses of food allergens: implications for patients and the food industry. *Curr. Opin. Allergy Clin. Immunol.* **4:**215–219.

Moneret-Vautrin, D. A., G. Kanny, M. Morisset, F. Rance, M. F. Fardeau, and E. Beaudouin. 2004. Severe food anaphylaxis: 107 cases registered in 2002 by the Allergy Vigilance Network. *Allerg. Immunol.* (Paris) **36:**46–51.

Moneret-Vautrin, D. A., J. Sainte-Laudy, G. Kanny, and S. Fremont. 1999. Human basophil activation measured by CD63 expression and LTC4 release in IgE-mediated food allergy. *Ann. Allergy Asthma Immunol.* **82:**33–40.

Moneret-Vautrin, D. A., R. Hatahet, and G. Kanny. 1994. Risks of milk formulas containing peanut oil contaminated with peanut allergens in infants with atopic dermatitis. *Pediatr. Allergy Immunol.* **5:**184–188.

Morisset, M., D. A. Moneret-Vautrin, G. Kanny, L. Guenard, E. Beaudouin, J. Flabbee, and R. Hatahet. 2003a. Thresholds of clinical reactivity to milk, egg, peanut and sesame in immunoglobulin E-dependent allergies: evaluation by double-blind or single-blind placebo-controlled oral challenges. *Clin. Exp. Allergy* **33:** 1046–1051.

Morisset, M., L. Parisot, G. Kanny, and D. A. Moneret-Vautrin. 2003b. Food allergy to moulds: two cases observed after dry fermented sausage ingestion. *Allergy* **58:**1195–1216.

Mueller, H. L. 1966. Diagnosis and treatment of insect sensitivity. *J. Asthma Res.* **3:**331–333.

Muller, U. R., D. B. Golden, P. J. Demarco, and R. F. Lockey. 2004. Immunotherapy for hymenoptera venom and biting insect hypersensitivity. *Clin. Allergy Immunol.* **18:**541–559.

Musu, T., J. Rabillon, C. Pelletier, B. David, and J. P. Dandeu. 1997. Simultaneous quantitation of specific IgE against 20 purified allergens in allergic patients sera by checkerboard immunoblotting (CBIB). *J. Clin. Lab. Anal.* **11:**357–362.

Nelson, H. S., J. Lahr, A. Buchmeier, and D. McCormick. 1998. Evaluation of devices for skin prick testing. *J. Allergy Clin. Immunol.* **101:**153–156.

Nelson, H. S., J. Lahr, R. Rule, A. Bock, and D. Leung. 1997. Treatment of anaphylactic sensitivity to peanuts by immunotherapy with injections of aqueous peanut extract. *J. Allergy Clin. Immunol.* **99:**744–751.

Niggemann, B. 2004. Role of oral food challenges in the diagnostic work-up of food allergy in atopic eczema dermatitis syndrome. *Allergy* **59**(Suppl. 78)**:**32–34.

Niggemann, B., and K. Beyer. 2005. Diagnostic pitfalls in food allergy in children. *Allergy* **60:**104–107.

Niggemann, B., and C. Gruber. 2004. Unproven diagnostic procedures in IgE-mediated allergic diseases. *Allergy* **59:**806–808.

Niggemann, B., S. Reibel, and U. Wahn. 2000. The atopy patch test (APT)—a useful tool for the diagnosis of food allergy in children with atopic dermatitis. *Allergy* **55:**281–285.

Norgaard, A., P. S. Skov, and C. Bindslev-Jensen. 1992. Egg and milk allergy in adults: comparison between fresh foods and commercial allergen extracts in skin prick test and histamine release from basophils. *Clin. Exp. Allergy* **22:**940–947.

Novembre, E., R. Bernardini, G. Bertini, G. Massai, and A. Vierucci. 1995. Skin-prick-test-induced anaphylaxis. *Allergy* **50:**511–513.

Ortolani, C., M. Ispano, E. A. Pastorello, R. Ansaloni, and G. C. Magri. 1989. Comparison of results of skin prick tests (with fresh foods and commercial food extracts) and RAST in 100 patients with oral allergy syndrome. *J. Allergy Clin. Immunol.* **83:**683–690.

Osterballe, M., and C. Bindslev-Jensen. 2003. Threshold levels in food challenge and specific IgE in patients with egg allergy: is there a relationship? *J. Allergy Clin. Immunol.* **112:**196–201.

Pastorello, E. A., S. Vieths, V. Pravettoni, L. Farioli, C. Trambaioli, D. Fortunato, D. Luttkopf, M. Calamari, R. Ansaloni, J. Scibilia, B. K. Ballmer-Weber, L. K. Poulsen, B. Wuthrich, K. S. Hansen, A. M. Robino, C. Ortolani, and A. Conti. 2002. Identification of hazelnut major allergens in sensitive patients with positive double-blind, placebo-controlled food challenge results. *J. Allergy Clin. Immunol.* **109:**563–570.

Poulsen, L. K. 2004. Allergy assessment of foods or ingredients derived from biotechnology, gene-modified organisms, or novel foods. *Mol. Nutr. Food Res.* **48:**413–423.

Poulsen, L. K., C. Liisberg, C. Bindslev-Jensen, and H. J. Malling. 1993. Precise area determination of skin-prick tests: validation of a scanning device and software for a personal computer. *Clin. Exp. Allergy* **23:**61–68.

Pucar, F., R. Kagan, H. Lim, and A. E. Clarke. 2001. Peanut challenge: a retrospective study of 140 patients. *Clin. Exp. Allergy* **31:**40–46.

Rance, F., M. Abbal, and V. Lauwers-Cances. 2002. Improved screening for peanut allergy by the combined use of skin prick tests and specific IgE assays. *J. Allergy Clin. Immunol.* **109:**1027–1033.

Rance, F., and G. Dutau. 1997. Labial food challenge in children with food allergy. *Pediatr. Allergy Immunol.* **8:**41–44.

Rance, F., A. Juchet, F. Bremont, and G. Dutau. 1997. Correlations between skin prick tests using commercial extracts and fresh foods, specific IgE, and food challenges. *Allergy* **52:**1031–1035.

Reese, G., R. Ayuso, S. M. Leong-Kee, M. J. Plante, and S. B. Lehrer. 2001. Characterization and identification of allergen epitopes: recombinant peptide libraries and synthetic, overlapping peptides. *J. Chromatogr. B Biomed. Sci. Appl.* **756:**157–163.

Roberts, G., N. Golder, and G. Lack. 2002. Bronchial challenges with aerosolized food in asthmatic, food-allergic children. *Allergy* **57:**713–717.

Roberts, G., and G. Lack. 2003. Relevance of inhalational exposure to food allergens. *Curr. Opin. Allergy Clin. Immunol.* **3:**211–215.

Robotham, J. M., S. S. Teuber, S. K. Sathe, and K. H. Roux. 2002. Linear IgE epitope mapping of the English walnut (Juglans regia) major food allergen, Jug r 1. *J. Allergy Clin. Immunol.* **109:**143–149.

Roehr, C. C., G. Edenharter, S. Reimann, I. Ehlers, M. Worm, T. Zuberbier, and B. Niggemann. 2004. Food allergy and non-allergic food hypersensitivity in children and adolescents. *Clin. Exp. Allergy* **34:**1534–1541.

Rosen, J. P., J. E. Selcow, L. M. Mendelson, M. P. Grodofsky, J. M. Factor, and H. A. Sampson. 1994. Skin testing with natural foods in patients suspected of having food allergies: is it a necessity? *J. Allergy Clin. Immunol.* **93:**1068–1070.

Sampson, H. A. 1988. Comparative study of commercial food antigen extracts for the diagnosis of food hypersensitivity. *J. Allergy Clin. Immunol.* **82:**718–726.

Sampson, H. A. 1999a. Food allergy. Part 1: immunopathogenesis and clinical disorders. *J. Allergy Clin. Immunol.* **103:**717–728.

Sampson, H. A. 1999b. Food allergy. Part 2: diagnosis and management. *J. Allergy Clin. Immunol.* **103:**981–989.

Sampson, H. A. 2001. Utility of food-specific IgE concentrations in predicting symptomatic food allergy. *J. Allergy Clin. Immunol.* **107:**891–896.

Sampson, H. A. 2002. Improving in-vitro tests for the diagnosis of food hypersensitivity. *Curr. Opin. Allergy Clin. Immunol.* **2:**257–261.

Sampson, H. A., L. Mendelson, and J. P. Rosen. 1992. Fatal and near-fatal anaphylactic reactions to food in children and adolescents. *N Engl. J. Med.* **327:**380–384.

Sampson, H. A., J. P. Rosen, J. E. Selcow, L. Mendelson, M. P. Grodofsky, J. M. Factor, S. A. Bock, A. W. Burks, J. M. James, R. Zeiger, and J. W. Yunginger. 1996. Intradermal skin tests in the diagnostic evaluation of food allergy. *J. Allergy Clin. Immunol.* **98:**714–715.

Sanchez-Borges, M., A. Capriles-Hulett, E. Fernandez-Caldas, R. Suarez-Chacon, F. Caballero, S. Castillo, and E. Sotillo. 1997. Mite-contaminated foods as a cause of anaphylaxis. *J. Allergy Clin. Immunol.* **99:**738–743.

Sander, I., S. Kespohl, R. Merget, N. Goldscheid, P. O. Degens, T. Bruning, and M. Raulf-Heimsoth. 2005. A new method to bind allergens for the measurement of specific IgE antibodies. *Int. Arch. Allergy Immunol.* **136:**39–44.

Scala, E., M. Giani, L. Pirrotta, E. C. Guerra, S. Cadoni, C. R. Girardelli, O. D. Pita, and P. Puddu. 2001. Occupational generalised urticaria and allergic airborne asthma due to *Anisakis simplex*. *Eur. J. Dermatol.* **11:**249–250.

Schuller, A., M. Morisset, F. Maadi, M. N. Kolopp Sarda, S. Fremont, L. Parisot, G. Kanny, and D. A. Moneret-Vautrin. 2005. Occupational asthma due to allergy to spinach powder in a pasta factory. *Allergy* **60:**408–409.

Schulze, A., and J. Downward. 2001. Navigating gene expression using microarrays—a technology review. *Nat. Cell Biol* **3:**E190–E195.

Semizzi, M., G. Senna, M. Crivellaro, G. Rapacioli, G. Passalacqua, W. G. Canonica, and P. Bellavite. 2002. A double-blind, placebo-controlled study on the diagnostic accuracy of an electrodermal test in allergic subjects. *Clin. Exp. Allergy* **32:**928–932.

Senna, G., G. Passalacqua, C. Lombardi, and L. Antonicelli. 2004. Position paper: controversial and unproven diagnostic procedures for food allergy. *Allerg. Immunol.* (Paris) **36:**139–145.

Shek, L. P., L. Soderstrom, S. Ahlstedt, K. Beyer, and H. A. Sampson. 2004. Determination of food specific IgE levels over time can predict the development of tolerance in cow's milk and hen's egg allergy. *J. Allergy Clin. Immunol.* **114:**387–391.

Shreffler, W. G., K. Beyer, T. H. Chu, A. W. Burks, and H. A. Sampson. 2004. Microarray immunoassay: association of clinical history, in vitro IgE function, and heterogeneity of allergenic peanut epitopes. *J. Allergy Clin. Immunol.* **113:**776–782.

Sicherer, S. H. 2003. Beyond oral food challenges: improved modalities to diagnose food hypersensitivity disorders. *Curr. Opin. Allergy Clin. Immunol.* **3:**185–188.

Sicherer, S. H., A. Munoz-Furlong, and H. A. Sampson. 2003. Prevalence of peanut and tree nut allergy in the United States determined by means of a random digit dial telephone survey: a 5-year follow-up study. *J. Allergy Clin. Immunol.* **112:**1203–1207.

Sicherer, S. H., and S. Teuber. 2004. Current approach to the diagnosis and management of adverse reactions to foods. *J. Allergy Clin. Immunol.* **114:**1146–1150.

Skamstrup Hansen, K., C. Bindslev-Jensen, P. S. Skov, S. H. Sparholt, H. G. Nordskov, N. R. Niemeijer, H. J. Malling, and L. K. Poulsen. 2001. Standardization of food allergen extracts for skin prick test. *J. Chromatogr. B Biomed. Sci. Appl.* **756:**57–69.

Spergel, J. M., and T. Brown-Whitehorn. 2005. The use of patch testing in the diagnosis of food allergy. *Curr. Allergy Asthma Rep.* **5:**86–90.

Sten, E., P. Stahl Skov, S. B. Andersen, A. M. Torp, A. Olesen, U. Bindslev-Jensen, L. K. Poulsen, and C. Bindslev-Jensen. 2002. Allergenic components of a novel food, Micronesian nut Nangai (Canarium indicum), shows IgE cross-reactivity in pollen allergic patients. *Allergy* **57:**398–404.

Su, X., F. T. Chew, and S. F. Li. 1999. Self-assembled monolayer-based piezoelectric crystal immunosensor for the quantification of total human immunoglobulin E. *Anal. Biochem.* **273:**66–72.

Suck, R., A. Nandy, B. Weber, M. Stock, H. Fiebig, and O. Cromwell. 2002. Rapid method for arrayed investigation of IgE-reactivity profiles using natural and recombinant allergens. *Allergy* **57:**821–824.

Sutas, Y., O. Kekki, and E. Isolauri. 2000. Late onset reactions to oral food challenge are linked to low serum interleukin-10 concentrations in patients with atopic dermatitis and food allergy. *Clin. Exp. Allergy* **30:** 1121.

Tabar, A., M. Alvarez-Puebla, B. Gomez, R. Sanchez-Monge, B. Garcia, S. Echechipia, J. Olaguibel, and G. Salcedo. 2004. Diversity of asparagus allergy: clinical and immunological features. *Clin. Exp. Allergy* **34:**131–136.

Taylor, A. V., M. C. Swanson, R. T. Jones, R. Vives, J. Rodriguez, J. W. Yunginger, and J. F. Crespo. 2000. Detection and quantitation of raw fish aeroallergens from an open-air fish market. *J. Allergy Clin. Immunol.* **105:**166–169.

Taylor, S. L., S. L. Hefle, C. Bindslev-Jensen, F. M. Atkins, C. Andre, C. Bruijnzeel-Koomen, A. W. Burks, R. K. Bush, M. Ebisawa, P. A. Eigenmann, A. Host, J. O. Hourihane, E. Isolauri, D. J. Hill, A. Knulst, G. Lack, H. A. Sampson, D. A. Moneret-Vautrin, F. Rance, P. A. Vadas, J. W. Yunginger, R. S. Zeiger, J. W. Salminen, C. Madsen, and P. Abbott. 2004. A consensus protocol for the determination of the threshold doses for allergenic foods: how much is too much? *Clin. Exp. Allergy* **34:** 689–695.

Terr, A. I. 2000. Controversial and unproven diagnostic tests for allergic and immunologic diseases. *Clin. Allergy Immunol.* **15:**307–320.

Teuber, S. S., and C. Porch-Curren. 2003. Unproved diagnostic and therapeutic approaches to food allergy and intolerance. *Curr. Opin. Allergy Clin. Immunol.* **3:**217–221.

Turkeltaub, P. C., and P. J. Gergen. 1989. The risk of adverse reactions from percutaneous prick-puncture allergen skin testing, venipuncture, and body measurements: data from the second National Health and Nutrition Examination Survey 1976–80 (NHANES II). *J. Allergy Clin. Immunol.* **84:**886–890.

Ulanova, M., M. Torebring, S. A. Porcelli, U. Bengtsson, J. Magnusson, O. Magnusson, X. P. Lin, L. A. Hanson, and E. Telemo. 2000. Expression of CD1d in the duodenum of patients with cow's milk hypersensitivity. *Scand. J. Immunol.* **52:**609–617.

Urisu, A., K. Yamada, R. Tokuda, H. Ando, E. Wada, Y. Kondo, and Y. Morita. 1999. Clinical significance of IgE-binding activity to enzymatic digests of ovomucoid in the diagnosis and the prediction of the outgrowing of egg white hypersensitivity. *Int. Arch. Allergy Immunol.* **120:**192–198.

Van Der Veen, M. J., R. van Ree, R. C. Aalberse, J. Akkerdaas, S. J. Koppelman, H. M. Jansen, and J. S. Van Der Zee. 1997. Poor biologic activity of cross-reactive IgE directed to carbohydrate determinants of glycoproteins. *J. Allergy Clin. Immunol.* **100:**327–334.

Vanin, E., S. Zanconato, E. Baraldi, and L. Marcazzo. 2002. Anaphylactic reaction after skin-prick testing in an 8-year-old boy. *Pediatr. Allergy Immunol.* **13:**227–228.

van Ree, R. 2004. The CREATE project: EU support for the improvement of allergen standardization in Europe. *Allergy* **59:**571–574.

van Ree, R. 1997. Analytic aspects of the standardization of allergenic extracts. *Allergy* **52:**795–805.

Vieths, S., A. Hoffmann, T. Holzhauser, U. Muller, J. Reindl, and D. Haustein. 1998. Factors influencing the quality of food extracts for in vitro and in vivo diagnosis. *Allergy* **53:**65–71.

Vila, L., K. Beyer, K. M. Järvinen, P. Chatchatee, L. Bardina, and H. A. Sampson. 2001. Role of conformational and linear epitopes in the achievement of tolerance in cow's milk allergy. *Clin. Exp. Allergy* **31:** 1599–1606.

Vlieg-Boerstra, B. J., C. M. Bijleveld, S. van der Heide, B. J. Beusekamp, S. A. Wolt–Plompen, J. Kukler, J. Brinkman, E. J. Duiverman, and A. E. Dubois. 2004. Development and validation of challenge materials for double–blind, placebo-controlled food challenges in children. *J. Allergy Clin. Immunol.* **113:**341–346.

Wallner, M., P. Gruber, C. Radauer, B. Maderegger, M. Susani, K. Hoffmann–Sommergruber, and F. Ferreira. 2004. Lab scale and medium scale production of recombinant allergens in Escherichia coli. *Methods* **32:**219–226.

Wedi, B., V. Novacovic, M. Koerner, and A. Kapp. 2000. Chronic urticaria serum induces histamine release, leukotriene production, and basophil CD63 surface expression—Inhibitory effects of anti-inflammatory drugs. *J. Allergy Clin. Immunol.* **105:**552–560.

Wiltshire, S., S. O'Malley, J. Lambert, K. Kukanskis, D. Edgar, S. F. Kingsmore, and B. Schweitzer. 2000. Detection of multiple allergen-specific IgEs on microarrays by immunoassay with rolling circle amplification. *Clin. Chem.* **46:**1990–1993.

Woods, R. K., M. Abramson, M. Bailey, and E. H. Walters. 2001. International prevalences of reported food allergies and intolerances. Comparisons arising from the European Community Respiratory Health Survey (ECRHS) 1991–1994. *Eur. J. Clin. Nutr.* **55:**298–304.

Woods, R. K., R. M. Stoney, J. Raven, E. H. Walters, M. Abramson, and F. C. Thien. 2002. Reported adverse food reactions overestimate true food allergy in the community. *Eur. J. Clin. Nutr.* **56:**31–36.

Zuberbier, T., G. Edenharter, M. Worm, I. Ehlers, S. Reimann, T. Hantke, C. C. Roehr, K. E. Bergmann, and B. Niggemann. 2004. Prevalence of adverse reactions to food in Germany—a population study. *Allergy* **59:**338–345.

Food Allergy
Edited by S. J. Maleki et al.

Chapter 3

The Big Eight Foods: Clinical and Epidemiological Overview

Suzanne S. Teuber, Kirsten Beyer, Sarah Comstock, and Mikhael Wallowitz

IDENTIFICATION OF THE "BIG EIGHT"

Peanuts, tree nuts (e.g., walnuts, cashews, and Brazil nuts), cow's milk, soy, wheat, hen's egg, fish, and crustaceans are considered the "Big Eight" foods in discussions of immunoglobulin E (IgE)-mediated or -associated food allergy both internationally and in the United States. These foods account for approximately 90% of food allergies in the best-studied group in the United States: children with atopic dermatitis (Sampson and McCaskill, 1985; Burks et al., 1998; Ellman et al., 2002). Importantly, these foods, particularly peanuts and tree nuts, account for reported fatal and near-fatal reactions to foods in the United States and Europe (Yunginger et al., 1988; Sampson et al., 1992; Bock et al., 2001). Internationally, the Codex Alimentarius Commission (1999) recommended in 1999 that member countries adopt this list of eight common foods and take steps to ensure that manufacturers within member nations list these foods or ingredients derived from these foods on labels. Thus, the term "Big Eight" was coined; these eight foods are the focus of subsequent U.S. federal legislation, the Food Allergen Labeling and Consumer Protection Act (PL 108-282), that took effect in 2006. Under this legislation, all products containing Big Eight foods require clear, so-called plain English labeling designed to help consumers navigate the confusing world of ingredient lists. For example, "nondairy cheese" made from soybeans but containing added casein had been allowed to merely list "casein" as an ingredient without identifying that this is a cow's milk protein. To an uninformed parent of a child diagnosed with cow's milk allergy, this may appear the perfect substitute for regular cheese—with horrific consequences. Exceptions

Suzanne S. Teuber • Division of Rheumatology, Allergy & Clinical Immunology, University of California, Davis, School of Medicine, 451 E. Health Sciences Dr., Suite 6510, Davis, CA 95616. ***Kirsten Beyer*** • Department of Pediatric Pneumology and Immunology, University Hospital Charité, Humboldt University, Augustenburger Platz 1, 13353 Berlin, Germany. ***Sarah Comstock*** • Division of Rheumatology, Allergy & Clinical Immunology, University of California, Davis, School of Medicine, 451 E. Health Sciences Dr., Suite 6510, Davis, CA 95616. ***Mikhael Wallowitz*** • Division of Rheumatology, Allergy & Clinical Immunology, University of California, Davis, School of Medicine, 451 E. Health Sciences Dr., Suite 6510, Davis, CA 95616.

to the United States labeling rule will be allowed by petition and notification processes, similar to exceptions approved within the evolving framework of the recent European Union (EU) directive (2003/89/EC) on allergen labeling that took effect at the end of November 2005 (European Commission, 2003). For example, in the EU, casein, fish gelatin, and fish isinglass (a product derived from fish collagen) will continue to be used as clarifying agents in beer, cider, or wine without being labeled; industry has satisfactorily demonstrated to a scientific panel that no allergen capable of eliciting an allergic reaction is present in the finished products (European Commission, 2005).

Of note, sesame is considered to be an emerging important allergen associated with severe allergic reactions in the United States and in much of the developed world (Dalal et al., 2002). It is included in the 12-item list of EU Directive 2003/89/EC, which regulates the labeling of ingredients in food (cereals containing gluten, fish, crustaceans, hen's eggs, peanuts, nuts, soy, cow's milk, celery, mustard, sesame, and sulfites) (European Commission, 2003). Sesame will thus be included in the following overview.

Although the Big Eight foods account for about 90% of clinical reactions in children with atopic dermatitis in referral centers, the pattern of foods causing reactions prompting emergency department visits is somewhat different in the United States. In a recent multicenter study of emergency department visits for food allergy, which gives a national perspective, peanuts and tree nuts accounted for 21% of visits, crustaceans accounted for 19%, and fish accounted for 10%, but fruits as a group were significant and represented 12% of allergic reactions (Clark et al., 2004). However, the Big Eight foods remain highly useful for labeling rules because the vast majority of individuals with food allergy will be helped, and specific fruits or vegetables are easier to avoid once the allergy is identified and the patient receives appropriate education on avoidance. Fruits and vegetables are not as commonly found as hidden ingredients in processed foods but rather pose problems in restaurants or when the patient is dining away from home. In contrast, edible seeds (e.g., sesame, mustard, coriander, or sunflower seeds), which are associated with potentially life-threatening allergies, are sometimes found in the United States as hidden ingredients in spices and flavorings and will not require disclosure (Dalal et al., 2002; Menendez-Arias et al., 1988; Manzanedo et al., 2004; Kelly and Hefle, 2000).

CULTURAL AND GEOGRAPHIC INFLUENCES

As pointed out in Chapter 2, approximately 6 to 8% of infants and young children have true food allergy (in North America, Europe, Australia, and New Zealand, cow's milk, hen's eggs, peanuts, wheat, and soy dominate), while about 2 to 4% of older children and adults have true food allergy (peanuts, tree nuts, and seafood dominate). However, this oft-reported estimated prevalence in adults represents an underestimate if the pollen-food syndrome is also counted in the prevalence estimates; such food allergy is usually mild. Nor do these numbers include celiac disease, which affects about 1% of the population (Treem, 2004). It should also be noted that many of the epidemiologic reports of food allergy are, in fact, surveys and not clinical studies with food allergy proven by challenge. Despite this shortcoming, definite trends emerge along geographic and cultural lines.

The key food allergens differ from country to country, based on local dietary habits in infancy, childhood, or adulthood. For example, in Japan, a survey of pediatric emergency department visits for food allergy showed reactions to hen's egg, cow's milk, wheat, buckwheat, and fish to be most common (Imai and Iikura, 2003). The importance of buckwheat allergy in Japan was supported by a previous survey of elementary school nurses in Yokohama, Japan, which revealed a 1.3% prevalence of food allergy among the approximately 90,000 school children, with buckwheat allergy in 0.22% of children; such an allergy is very rare in the United States. It was also noted that reactions to buckwheat involved wheezing or anaphylactic shock in 30.4% of affected children and that some students had accidental ingestions due to buckwheat noodles in school lunches (Takahashi, 1998). Moreover, edible bird's nest, derived from the saliva of a cave-dwelling swift in Asia, was reported to be the most important food allergen in a Singapore pediatric population associated with systemic allergic reactions. In 124 emergency department cases of food allergy, 34 were due to edible bird's nest (27%), 30 cases were due to crustaceans (24%), and 14 were due to egg or cow's milk combined (13%). The other cases were due to Chinese herbs and miscellaneous foods, but none were due to peanuts or tree nuts (Goh et al., 1999).

Regional dietary habits and methods of food preparation that are culturally defined clearly play a role in the prevalence of specific food allergies in various countries. At the same time that peanut allergy is a disease of developed countries, peanuts are a staple food in many developing countries, with volumes of consumption similar to those in North America (Beyer et al., 2001). Dietary practice in the United States is predominantly to eat roasted peanut products, not boiled or fried peanuts, as in African and Asian populations where peanut allergy is very uncommon (Hourihane, 2002; Goh et al., 1999; Hill et al., 1999). The per capita consumption of peanuts in China and the United States is essentially the same, but there is virtually no peanut allergy in China (Hill et al., 1999). The high heat of dry roasting at 180°C (typical of preparation methods in the United States) and the process of maturation and curing have been shown to increase the allergenicity of peanut proteins, as discussed in detail in Chapter 13 (Maleki et al., 2000; Chung et al., 2000; Beyer et al., 2001; Maleki et al., 2003). Clearly, consumption is not the only factor related to development of peanut allergy, since young Israeli children commonly receive a peanut-containing snack, and yet sesame seed allergy appears to be the major food allergy in this population (Dalal et al., 2002).

Pollinosis can also be a determinant of the pattern of food allergies seen. This is particularly notable with birch pollen exposure and subsequent food allergy to hazelnut. Where birch trees (and other members of the family *Betulaceae*) are present, cross-reactivity between the major birch pollen allergen, Bet v 1, and a Bet v 1-like protein in hazelnut can lead to a significant prevalence of hazelnut allergy, i.e., the pollen-food syndrome. For this reason, allergy to hazelnut is more common in northern Europe than in southern Europe (Hirschwehr et al., 1992; Ortolani et al., 2000). For example, hazelnut was reported as the most common cause of IgE-mediated food allergy in Switzerland (Etesamifar and Wuthrich, 1998); upon double-blind, placebo-controlled food challenge (DBPCFC) of patients in Zurich and Copenhagen, many subjects exhibited reactions localized to the oral mucosa that were thought to be associated with birch pollen allergy (Ortolani et al., 2000). Patients without birch pollen allergy who had hazelnut allergy tended to have more severe reactions and showed IgE against non-pollen-related proteins.

SPECIFIC FOODS: THE BIG EIGHT PLUS SESAME

In the following section, the eight dominant foods associated with food allergic reactions are reviewed, along with sesame. The reader is also directed to Table 1 for a summary.

Cow's Milk

Cow's milk allergy is the most common food allergy in infants reported from developed Western nations. It can occur in breast-fed infants but is most common in cow's milk formula-fed infants. The spectrum of food allergic disorders associated with cow's milk is broad and most commonly involves the skin and/or the gastrointestinal tract. IgE-mediated mechanisms account for about 60% of milk allergic disorders in young children, with the majority of IgE-mediated reactions involving the skin and the majority of non-IgE-mediated reactions involving the gastrointestinal tract. Acute urticaria and atopic dermatitis are two common forms of IgE-mediated skin reactions associated with milk allergy. Gastrointestinal syndromes associated with milk allergy are milk-protein induced enterocolitis, enteropathy with protein loss, eosinophilic gastroenteropathy, proctocolitis, constipation, infantile colic, gastroesophageal reflux, and esophagitis (Heine et al., 2002; Ravelli et al., 2001). The immune mechanisms underlying some of these disorders are yet to be fully elucidated. Most infants with cow's milk allergy will do well on either soy formula, extensively hydrolyzed formulas based on either casein or whey in which the proteins have been hydrolyzed to such a degree that the polypeptides are short, nonantigenic fragments (<1,200 Da) (Moro et al., 2002) or amino acid formulas. However, a proportion with IgE-mediated cow's milk allergy will also develop allergy to soy (about 10 to 14%) (Zeiger et al., 1999). Some children, estimated at about 10% of all children with cow's milk allergy, display a symptom complex known as multiple food protein intolerance. This syndrome often starts with symptoms while breast-feeding and will continue when extensively hydrolyzed formulas or soy formulas are tried; the symptom complex extends to other food proteins, such as hen's egg and wheat. These children often have failure to thrive and require an amino acid-based formula (Heine et al., 2002).

Prevalence

Several prospective studies have determined that the prevalence of cow's milk allergy, proven by challenges, is approximately 2% in infants. However, the rates vary in different selected populations from 0.5% to 7.5% (Bahna, 2002). Host and Halken (1990) followed a cohort of 1,749 infants in Odense, Denmark, to their third birthday. In this group, 117 infants developed possible symptoms of cow's milk allergy and underwent further evaluation. Thirty-nine infants had positive challenges (2.2%). Of these, 64% had cutaneous reactions, 59% had gastrointestinal reactions, and 33% had respiratory reactions.

Age of Onset and Natural History

Cow's milk allergy is a disorder with onset in infancy and, less commonly, in early childhood. Rare cases of adult onset cow's milk allergy are also seen (Bahna, 2002). Host and Halken (1990) rechallenged cow's milk-allergic children at 6- to 12-month intervals up to the third birthday. At 1 year of age, 56% (22 of 39) of the infants became tolerant to cow's milk; at 2 years, 77% were tolerant; and at 3 years, 87% were tolerant. Those with severe systemic IgE-mediated reactions were less likely to become tolerant (Host and Halken, 1990).

Table 1. The big eight foods plus sesame

Food	Estimated prevalence	Most common allergic disorder(s)	Common age (yr) of onset	Outgrown?	Deaths?
Peanut	0.8% of children, 0.6% of adults	IgE-mediated acute reaction, atopic dermatitis	2	20%	Yes
Tree nuts	0.2% of children, 0.5% of adults	IgE-mediated acute reaction, atopic dermatitis	3–5	Unknown	Yes
Crustaceans	0.1% of children <5 yr, 2% of adults	IgE-mediated acute reaction, atopic dermatitis	5–adult	Unknown	Yes
		Occupational asthma	Adult		
		Occupational contact urticaria	Adult		
Cow's milk	2% of children, 0.3% of adults	IgE-mediated acute reaction, non-IgE gastrointestinal reaction, atopic dermatitis, food-induced pulmonary hemosiderosis	<1	85%	Yes
Hen's egg	1.3% of children, 0.2% of adults	Atopic dermatitis, IgE-mediated acute reaction	<1	55–80%	Yes
		Occupational asthma	Adult		
Fish	0.2% of children, 0.5% of adults	IgE-mediated acute reaction, atopic dermatitis	2–adult	Unknown	Yes
		Occupational contact urticaria	Adult		
Soy	0.4% of infants, 0.04% of adults	Atopic dermatitis, IgE-mediated acute reaction, non-IgE gastrointestinal reaction	<1	85%	Yes
Wheat	1%, celiac disease patients	Atopic dermatitis, celiac disease, IgE-mediated acute reaction	<1–childhood	85% IgE associated	Unknown
	Unknown for other disorders	Occupational asthma; food-dependent, exercise-associated anaphylaxis	Adult	0% Gluten enteropathy	
Sesame	Unknown	IgE-mediated acute reaction	Both young children and adults	20% of young children	Unknown

Cross-Reactivity

Although soy formulas and extensively hydrolyzed cow's milk formulas are good substitutes for most infants with IgE-associated or -mediated cow's milk allergy, Bellioni-Businco et al. (1999) report that in Italy, a goat's milk-based formula is sometimes recommended for infants with cow's milk allergy. The authors studied 26 children with proven IgE-mediated cow's milk allergy. They performed skin prick tests with cow's milk and goat's milk and DBPCFC with both types of milk, along with soy as a control. Extensive cross-reactivity was found: all children had positive skin tests to both types of milk, all had positive oral challenge results with cow's milk, and 24 of 26 also had positive challenge results with goat's milk. Therefore, goat's milk is not a safe substitute for children with cow's milk allergy. Despite this report, Pessler et al. (2004) reviewed how goat's milk continued to be promoted in lay sources as a substitute for cow's milk in allergic children; they presented a case of near-fatal anaphylaxis to goat's milk in a 4-month-old child who had presented with only postprandial emesis to cow's milk formula as a neonate. The child had been doing well on soy formula (Pessler et al., 2004). Mare's milk has also been marketed as a substitute, although it is not as widely available. Mare's milk was also studied by Businco et al. (2000) with more-favorable results. Twenty-five children with IgE-mediated cow's milk allergy were challenged by DBPCFC, as well as skin testing, to both cow's milk and mare's milk. All were positive by skin testing and oral challenge to cow's milk, but only 2 of 26 children had a positive skin test to mare's milk, and only 1 child reacted upon challenge with mare's milk. Therefore, mare's milk appears to be a good substitute for most children, but the authors caution that a titrated oral challenge should be performed under the supervision of a physician first.

Caseins, lactoglobulins, and bovine serum albumin are the main allergens in cow's milk, with caseins being the most allergenic and antigenic proteins (Shek et al., 2005). Bovine serum albumin and bovine gamma globulin are also present in beef but are heat labile (Werfel et al., 1997). Some children with cow's milk allergy will also react in DBPCFC to beef, but it appears that this is uncommon, as is isolated beef allergy (Werfel et al., 1997).

Threshold for Clinical Reactions

Some patients are exceedingly sensitive to cow's milk. Morisset et al. (2003) analyzed 59 positive DBPCFCs for cow's milk and reported a threshold dose of 3 mg of milk protein (equals 0.1 ml of milk). However, only one of these 59 patients reacted at this very low level, but 30% of the patients also showed great sensitivity, reacting to <6.8 ml. The lowest provoking dose reported in the literature by double-blind challenge is 1.5 mg of cow's milk protein in an infant formula and, by open challenge, just 0.6 mg (Taylor et al., 2002).

Risk of Severe Reactions

Cow's milk allergy covers many different clinical syndromes, with most presentations at no risk of causing a near-fatal or fatal reaction. However, children or adults with IgE-mediated acute reactions are at risk of developing severe systemic reactions, particularly those patients with IgE directed towards linear epitopes on the milk proteins, high titers of IgE to cow's milk, and a history of previous systemic reactions (Sampson, 2004). Pumphrey reported two deaths from milk allergy, including one in which a 12-year-old boy properly received two doses of self-injected epinephrine but still could not be revived

(Pumphrey, 2000). Clark et al. (2004) reported 6% of emergency department visits were for cow's milk allergy in a United States study. Among children in Italy, cow's milk was the trigger of anaphylaxis in 12 of 54 episodes (22%) (Novembre et al., 1998).

Crustaceans

Commonly ingested invertebrate crustaceans include shrimp, lobster, crab, and crayfish. Shrimp leads consumption in the United States, with approximately 2.7 lb consumed annually per capita, while crab lags behind at 0.4 lb annually per capita, and rates of lobster and crayfish consumption are lower still (Lehrer et al., 2003). Crustacean allergy is an important allergy, mainly reported in adults associated with immediate IgE-mediated systemic reactions upon ingestion but also with contact urticaria on handling shellfish and occupational asthma. Food-dependent, exercise-induced anaphylaxis has also been seen; in a review of cases in Japan, shrimp was the second-leading cause of this syndrome after wheat (Harada et al., 2000). Interestingly, up to one-third of 303 snow crab-processing plant workers reported asthma, and one-quarter reported a skin rash when working with crabs (Cartier et al., 1984). Patients with crustacean food allergy are at risk for severe asthmatic and/or systemic allergic reactions if they inhale cooking steam or aerosols containing food allergens, a possible occurrence in restaurants, processing plants, fisherman's wharves, and homes. Shrimp allergens can be easily recovered and measured in distillates of cooking steam from boiling shrimp (Goetz and Whisman, 2000).

Seafood allergy, either finned fish or crustacean, has nothing to do with reactions to iodinated contrast used for radiological procedures; however, people with allergies do have a higher rate of reaction than nonatopics, possibly due to a general state of increased histamine releasability from mast cells and basophils. However, asthmatics have the highest relative risk for a reaction to contrast dye (Morcos and Thomsen, 2001). Many years ago, patients were told that they were iodine allergic if they had allergy to seafood; this fallacy has continued in an oral tradition to this day.

Prevalence

The Food Allergy and Anaphylaxis Network (FAAN) sponsored a random digit dial population survey of crustacean, fish, and mollusk food allergies in the United States. In a sample population of 14,948 individuals, 2.0% (303 persons) reported shellfish allergy. Over half of these shellfish-allergic individuals were age 41 or older. In contrast, the allergy was rare in children age 17 and younger. Only one child under the age of 5 was reported to be allergic, giving a prevalence in this young age group of 0.1%, which rose to 0.7% in the age group of 6- to 17-year-old children (Sicherer et al., 2004a). Data reflecting both food allergies and intolerances in Europe, Iceland, New Zealand, Australia, and the United States was obtained from the European Community Respiratory Health Survey. In this study, 2.8% reported shrimp as causing "illness or trouble" nearly always when eaten (Woods et al., 2001). A population survey in France using mailed questionnaires did not calculate prevalences for the individual food allergens, but crustaceans accounted for 8% of food allergies reported. In that report, it was also noted that 97% of cases of severe anaphylaxis occurred in adults over the age of 30, and seafood was the most frequent cause (Kanny et al., 2001). Of note, crustaceans accounted for 19% of 678 emergency department visits for food allergy in a multicenter United States study; therefore, it was even more frequent than peanut or tree nut allergy (12 and 9%, respectively) (Clark et al., 2004).

Age of Onset and Natural History

As seen from the prevalence studies, crustacean allergy is predominantly (in Western culture) an allergy of adults, although it can be seen rarely in children. The onset is usually in young adulthood. There are no data on how common it is to develop tolerance after a period of avoidance, but in a recent survey, 4% reported outgrowing the allergy (Sicherer et al., 2004a). It appears to be a long-lasting allergy. One of us has seen a 70-year-old male with a recent systemic reaction to shrimp prompting an emergency room visit; his last reaction had been approximately 45 years before (S. S. Teuber, unpublished data).

Cross-Reactivity

Almost all patients with crustacean allergy show IgE binding to the major allergen, tropomyosin. Tropomyosin is a pan-allergen among invertebrate species; thus, the risk of cross-reactivity to other crustaceans is very high, at about 75% in small challenge studies (Waring et al., 1985). In a recent population survey, 38% of those with crustacean allergy reported having reactions to more than one type (i.e., crab, lobster, and shrimp) (Sicherer et al., 2004a). Interestingly, some rare shrimp-allergic patients may clinically react to only one species of shrimp; sera demonstrated a lack of cross-reactivity between brown and white shrimp when investigated in vitro and by skin testing (Morgan et al., 1989). Crustacean allergy can also portend a risk, albeit lower, of mollusk allergy (clam, scallop, oyster, etc.) (Sicherer, 2001). Mollusk allergy is less well defined, but 14% of respondents in the FAAN survey with crustacean allergy also reported reacting to mollusks (Sicherer et al., 2004a).

There are reports of immunotherapy with dust mites inducing clinical allergy to certain mollusks and crustaceans on the basis of cross-reactive tropomyosin. It can be normal for specific IgE levels to initially increase during the initiation of immunotherapy vaccines for allergic rhinitis or asthma; in occasional unfortunate individuals, this may result in a food allergy. In a prospective study of 17 patients receiving dust mite immunotherapy, 2 patients developed cross-reactive IgE towards dust mite, snail, and shrimp tropomyosin. These two patients also developed new mild oral symptoms upon shrimp ingestion (van Ree et al., 1996). However, a follow-up study of 31 patients on dust mite immunotherapy who regularly ingested shellfish failed to show any who developed IgE to shrimp tropomyosin (Asero, 2005).

Threshold for Clinical Reactions

There are no clinical studies of oral challenges to determine the lowest observed adverse effect levels, nor are there well-documented case reports of hidden crustacean allergens eliciting a reaction that could help estimate the threshold dose for a response. Certainly, the dose is likely to be low if airborne allergens can induce a severe response and if kissing after eating shrimp can cause anaphylaxis (Steensma, 2003).

Risk of Severe Reactions

Crustaceans can cause severe systemic reactions and have been implicated in at least one fatality (Yunginger et al., 1988; Steensma, 2003). Although Pumphrey (2000) reported three fatalities from seafood in the United Kingdom, it was unclear if the triggers were fish or crustaceans. In a French registry of severe food anaphylaxis, shellfish was one of the main precipitants (Moneret-Vautrin et al., 2004). In the FAAN random digit dial seafood allergy survey, 40% of shellfish reactions resulted in physician evaluation or an emergency department visit (Sicherer et al., 2004a).

Fish

Allergy to finned fish clinically manifests as IgE-associated atopic dermatitis in sensitized infants and young children and as IgE-mediated acute reactions in both children and adults. Reactions have been reported from cutaneous exposure upon handling fish, ingestion, and inhalational exposure of aerosolized proteins, such as those encountered in commercial canning facilities or restaurant or home cooking. Sometimes a reaction to histamine in slightly spoiled fish, particularly of the scombroid type, can cause an allergy-like syndrome, termed scombroid poisoning (Lehane and Olley, 2000).

Prevalence

The FAAN random digit dial survey of 2002 obtained valuable information on fish allergy in the United States (Sicherer et al., 2004a). A total of 0.2% of children were reported to have fish allergy, while 0.5% of adults reported reactions. Of the 64 respondents represented by these percentages, 29 cases (45%) were reportedly diagnosed by a physician, and another 26 (41%) were highly convincing (Sicherer et al., 2004a).

Age of Onset and Natural History

Sensitization to fish in children has been reported to be significant in countries with high per capita fish consumption. In the United States FAAN population survey, fish allergy was most often seen in adults. The prevalence more than doubled from children to adults, and no child younger than 5 years was reported with the condition, supporting the clinical impression that fish allergy is usually acquired in young adulthood in the United States (Sicherer et al., 2004a). In contrast, in a series of children with food allergy from southern Europe and Scandinavia, fish is prominent. For example, in 54 Italian children, 19% of anaphylaxis episodes to foods were caused by fish (Novembre et al., 1998). In Finland, it was estimated that 3% of children were fish allergic, with onsets when the children were <3 years old (Saarinen and Kajosaari, 1980). From a series of 355 food-allergic children in Spain, it was reported that 30% of children had fish allergy; in most children, the onset was prior to 2 years of age (Crespo et al., 1995). Fish allergy is considered a long-lived allergy; in the FAAN seafood study, only 3.5% of fish-allergic individuals reported that they had outgrown their fish allergy (Sicherer et al., 2004a). Tolerance to fish in previously allergic patients in small numbers has been reported (Kajosaari, 1982; Solensky, 2003). Overall, it appears that fish allergy is usually persistent.

Cross-Reactivity

There is no cross-reactivity between finned fish and crustaceans or mollusks. The main allergen in fish is parvalbumin, which is present in the muscle tissue of all fish, while the main allergen in crustacean muscle is tropomyosin (Taylor et al., 2004). However, among fish species, the range of clinical cross-reactivity varies from patient to patient. Isolated fish allergy to one species definitely exists, but most patients and most challenge studies support the presence of significant cross-reactivity. For instance, in a rigorous study of eight patients with cod allergy, all showed clinical reactivity also to mackerel, herring, and plaice, which was supported by in vitro studies as well (Hansen et al., 1997). In contrast, in a different study of 10 confirmed fish allergic subjects, although skin tests were positive for 10 fish species in 8 patients, 7 patients reacted to only 1 species of fish upon open oral challenge, 1 patient reacted to 2 species, and only 1 patient reacted to 3 species. These patients were all challenged with four to six species each

(Bernhisel-Broadbent et al., 1992a). In the FAAN random digit dial survey, 67% of fish-allergic respondents reported allergy to more than one species of fish (Sicherer et al., 2004a). Overall, it appears that patients with fish allergy should avoid all fish unless they have undergone challenge with selected other species and can obtain it without risk of cross-contamination. The latter is a crucial point, since there is a risk of severe or fatal reactions to fish. Another food challenge that appears worth doing with fish-allergic patients is with canned tuna—it appears that most patients will tolerate fish in this form (Bernhisel-Broadbent et al., 1992b), but there is a case report of allergy to both fresh fish and canned tuna (Kelso et al., 2003).

Threshold for Clinical Reactions

In a review of threshold doses by Taylor et al. (2002), Bindslev-Jensen and Hansen reported doses as low as 5 mg of minced fish in double-blind challenges caused reactions in allergic persons. In children with predominantly atopic dermatitis, Sicherer et al. (2000a) reported that 17% reacted to the first dose of 400 or 500 mg.

Risk of Severe Reactions

Although many cases in young children are associated with atopic dermatitis flares, fish allergy has the potential to be life threatening. In the FAAN survey, 55% of reactions resulted in physician evaluation or an emergency department visit (Sicherer et al., 2004a). Several deaths have been documented in the literature due to fish allergy (Bock et al., 2001; Yunginger et al., 1988).

Hen's Eggs

Hen's egg is an important childhood allergy, with persistence into adulthood in a minority of patients. In some countries, such as Japan and Spain, it is the leading cause of food allergy in children <2 years of age, surpassing cow's milk (Crespo et al., 1995; Imai and Iikura, 2003; Han et al., 2004). In Korea, egg was recently shown to greatly outpace sensitization to other foods in children with atopic dermatitis; 87 of 266 children had specific IgE blood test values highly suggestive of clinical allergy (specific IgE by Pharmacia CAP FEIA of >7 kU/liter over age 2 or >2 kU/liter under age 2), compared to only 12 of 266 for cow's milk, 8 for peanut, and 3 for soy (Han et al., 2004). Hen's egg allergy is most often IgE mediated. Clinical reactions involve the skin (e.g., urticaria, angioedema, and atopic dermatitis), the respiratory system (e.g., wheezing), and/or the gastrointestinal system (e.g., vomiting, diarrhea) and can range from mild reactions to severe anaphylaxis. More rarely, food protein-induced enterocolitis, enteropathy, or eosinophilic gastroenteritis syndromes are caused by hen's egg. Exposure to aerosolized dried egg powder is also a cause of occupational asthma in adults and has even been associated with the subsequent development of food allergy upon ingestion in some cases of longstanding aerosolized egg exposure in such individuals (Leser et al., 2001; Escudero et al., 2003).

Another issue that arises with hen's egg sensitivity of the IgE-mediated type is the potential for anaphylaxis to viral vaccines that use eggs in the manufacturing process. Such vaccines include measles, mumps, and rubella (MMR), influenza, and yellow fever. The MMR vaccine has been shown to be safe in children with anaphylactic sensitivity to egg (James et al., 1995). The MMR vaccines are prepared from viral strains multiplied in chick embryo fibroblast cultures, rather than embryos themselves, and only contain miniscule amounts of hen egg protein (James et al., 1995). However, yellow fever and influenza vaccines are prepared from embryonated eggs and may contain residual egg proteins at higher levels. Indeed, James et al. (1998) reported log differences in ovalbumin content among influenza vaccine manufacturers. James et al. (1998) also demonstrated safe administration of the influenza vaccine to egg-allergic children but with the caveat that the particular vaccine lots that were used had low ovalbumin levels measured. Patients with a history of egg allergy who require vaccination with yellow fever or influenza should be referred to an allergist for management (Kelso et al., 1999; Zeiger, 2002).

Prevalence

The prevalence of hen's egg allergy is estimated to be approximately 1.3% in the United States, with most of these cases being young children; in adults, the estimate is 0.2% (Sampson, 2004). The prevalence among referral populations of young children with moderate to severe atopic dermatitis can be much higher (Sampson, 2004).

Age of Onset and Natural History

Hen's egg allergy usually develops within the first 2 years of life and in the majority resolves by school age. In a prospective study of 100 children with atopic dermatitis in Sweden, 58 children developed IgE sensitization to foods as measured in vitro (Gustafsson et al., 2003). Hen's egg was the most common sensitizer, with 46 of 58 children positive; oral challenges to prove clinical allergy were not performed. However, it was noted that no child developed sensitization after 3 years of age. At 7 years of age, only 12 of the 46 still had detectable IgE levels to hen's egg; thus, IgE sensitization had resolved in 74% of the children (Gustafsson et al., 2003). In a survey of French schoolchildren, 23 had a history of egg allergy, and 9 (39%) had become tolerant at an average age of 4.7 years (Rance et al., 2005). Sensitization to egg in early childhood appears to be associated with the later development of sensitization to aeroallergens (Nickel, 1997) or asthma (Tariq et al., 2000).

Cross-Reactivity

In the United States, eggs of other birds are infrequently consumed. Although in other countries, duck, quail, and goose eggs are more frequently eaten, there are no reports of cross-reactivity with hen's egg. Moreover, there is a report of a patient with isolated duck and goose egg allergy who was tolerant of hen's egg (Anibarro et al., 2000).

Threshold for Clinical Reactions

In a series of 124 oral food challenges performed for egg, 16% of subjects were reactive at <65 mg hen's egg (equals about 6.5 mg of hen's egg protein) (Morisset et al., 2003). The lowest observed threshold dose in food challenges for egg white was reported by Morisset et al., at <2 mg of crude egg white (equals about 0.2 mg of hen's egg protein). The lowest dose was observed in France, with 0.13 mg of hen's egg protein (Taylor et al., 2002).

Risk of Severe Reactions

In patients with acute IgE-mediated reactions to egg, there is a definite risk of severe anaphylaxis. In an Italian study of 54 episodes of anaphylaxis, 6 (11%) were caused by hen's egg (Novembre et al., 1998).

Peanuts

Peanut allergy affects both children and adults. The most common type of reaction is an IgE-mediated systemic reaction, but flares of atopic dermatitis in young children are also fairly common. Reactions are often severe, and hypersensitivity persists throughout life for most patients. In fact, in a recent report, peanuts accounted for over half of the food-induced anaphylaxis fatalities in a United States series (Bock et al., 2001). This is a serious concern, especially knowing that among participants of a study who were followed over a period of 5 years, >50% of peanut allergic individuals suffered from allergic reactions after accidental ingestion (Sicherer et al., 1998). The same study also recorded an average of two accidental ingestions per person over the same period. Peanuts or peanut flour are ubiquitous as food ingredients and are so widely consumed that avoidance is extremely difficult and requires a tremendous amount of self discipline. Since this allergy affects children, the burden extends to family members and other caretakers. Indeed, food allergy, not only peanut allergy, has been shown to have a significant negative impact on the quality of life of patients and families (Avery et al., 200; Sicherer et al., 2001a).

Prevalence

Peanut allergy affects approximately 0.8% of children in the United States and 0.6% of adults, based on a random digit dial telephone survey in 2002 of self-reported or parent-reported allergy (Sicherer et al., 2003). A similar survey had been performed in 1997, with a reported rate of peanut allergy in children of 0.4%. Therefore, a doubling of the rate in the 5-year interval can be concluded. A similar finding was reported from a population cohort study on the Isle of Wight, United Kingdom (Sicherer et al., 1999; Grundy et al., 2002).

Age of Onset and Natural History

In >70% of children with peanut allergy, symptoms develop upon the first known exposure, raising concerns for transplacental or neonatal sensitization via breast milk (Hourihane et al., 1997a; Sicherer et al., 1998; Sicherer et al., 2001b). Peanut allergen can be found in breast milk (Vadas et al., 2001). The first allergic reaction to peanuts develops in most children between 14 and 24 months of age, most commonly at home (Sicherer et al., 1998; Sicherer et al., 2001b). Detailed information from the FAAN voluntary peanut and tree nut registry (5,149 peanut- and/or tree nut-allergic individuals) revealed cutaneous reactions (urticaria, pruritus, erythema, and angioedema) to be the most frequently reported first known reaction to peanuts in 89% of patients. Other organ systems affected included the respiratory (42%), gastrointestinal (26%), and cardiovascular (4%) systems (Sicherer et al., 2001b). Just over half (54%) of all children were reported to have symptoms isolated to one organ system during their first reaction. Almost one-third (32%) developed symptoms in two systems, 10 to 15% had symptoms in three systems, and 1% had symptoms in four systems (Sicherer et al., 2001b).

Although peanut allergy that develops in childhood is unlikely to be outgrown, current reports suggest that clinical tolerance develops in approximately 20% of children who developed peanut allergy as infants or very young children (Hourihane et al., 1998; Skolnick et al., 2001). Children with no reactions for 2 years, a history of only mild reactions, and a peanut-specific IgE level of <5 kU/liter are good candidates for evaluation of possible clinical tolerance prior to entering school (Skolnick et al., 2001). However, children who experience an allergic reaction after the age of 5 years are unlikely to develop clinical tolerance (Sampson, 2004). Unfortunately, tolerance is not always maintained in this population. A small minority of patients with peanut allergy may redevelop clinical reactivity, even after having a negative peanut challenge result (Busse et al., 2002; Fleischer et al., 2003). One study determined that the rate of recurrent reactions in peanut-allergic individuals who had passed oral food challenges was 7.9% (Fleischer et al., 2004). Another estimated the recurrence rate at 14% (Busse et al., 2002). Both studies found that those patients who consumed concentrated forms of peanut frequently (i.e., ate peanut butter again rather than just accidental small exposures) had a significantly lower chance of having a recurrence of their allergy (Busse et al., 2002; Fleischer et al., 2004). A recommendation to regularly consume peanuts in this population may be difficult because patients who have outgrown their peanut allergy often dislike the taste of peanuts or are afraid to eat them and therefore limit the amount they consume (Fleischer et al., 2003).

Cross-Reactivity

The risk of clinical cross-reactivity (as opposed to clinically irrelevant IgE binding, which is quite common) with other legumes is quite low in general, around 5% (Bernhisel-Broadbent and Sampson, 1989; Bernhisel-Broadbent et al., 1989). However, an exception exists with lupine (*Lupinus albus*). Lupine seed flour and bran are sometimes added as flour enhancers in Australia and Europe. The exact prevalence of cross-reactivity to lupine flour in patients with peanut allergy is not known, due to the small numbers reported, but appears to be very high (Moneret-Vautrin et al., 1999). Patients with peanut allergy in geographic areas in which lupine seed flour is used must be educated to avoid this product.

There is a moderate occurrence of tree nut allergy in patients with peanut allergy, i.e., coallergy. In fact, about one-quarter to one-third of patients in self-selected or referral populations who have a peanut allergy will report clinical allergy to a tree nut (Sicherer et al., 2001b). For example, 34% of 102 children with acute reactions to peanut had subsequently experienced reactions to at least one tree nut (Sicherer et al., 1998). However, this prevalence of coallergy was not seen in the random digit dial survey of the general population in the United States on the prevalence of peanut and tree nut allergy, in which only four subjects (2.4%) were reported to be allergic to both peanuts and at least one tree nut (Sicherer et al., 1999a). Ewan (1996) also concluded from her reported series of 62 peanut- or tree nut-allergic patients that in patients with peanut allergy, tree nut allergies tended to continue to develop over time. Thus, it appears that patients with systemic allergic reactions to peanuts are at risk of developing a separate allergy to other seed storage proteins. These observations have led to the recommendation that patients with peanut allergy should, from the time that the peanut allergy is identified, avoid tree nuts unless they have already knowingly ingested a particular nut without any adverse reaction (Sicherer et al., 1998). The converse advice rarely applies to newly diagnosed tree nut-allergic patients, since most will have already eaten and tolerated peanuts when a tree nut allergy becomes apparent (usually later than for peanuts, at around age 3 to 4 years of age).

Although some patients with peanut allergy develop tree nut allergy, to date, this appears to be predominantly due to cosensitization and not classic cross-reactivity. For instance, when sera were used that had been obtained from several patients highly allergic clinically to both peanut and walnut and whose IgE bound the vicilin from both seeds, there was no evidence of high-affinity cross-reactive IgE between the vicilin from walnut (Jug r 2) and crude peanut extract containing vicilin (Ara h 1) (Teuber et al., 1999). Also, no IgE epitopes were found in common upon comparison of linear epitopes from cashew vicilin (Ana o 1) with published Ara h 1 linear epitopes, although sera from patients highly allergic to both peanut and cashew were included in the study (Wang et al., 2002; Astwood et al., 2002). At least from this example of IgE directed towards vicilin allergens in peanuts and tree nuts, it appears that patients undergo separate sensitization events even when allergy to peanuts and tree nuts coexists. However, de Leon et al. (2003) published a small study of four patients who did show some cross-reactivity in vitro between peanuts and tree nuts that correlated with clinical reactivity to peanuts, almonds, Brazil nuts, and hazelnuts. The specific proteins involved in the cross-reactivity were not identified, but there will likely be subsets of patients described and defined in more detail who do exhibit allergy due to classic cross-reactivity of IgE towards peanut with other seeds, tree nuts, or pollens as more investigations are performed.

Threshold for Clinical Reactions

Several well-designed threshold studies have been performed with defatted peanut flour. Considering that a single peanut contains about 200 mg of protein, the results were of concern. An initial study published in 1997 reported that doses as low as 100 μg of protein could precipitate subjective oral responses in two subjects and that objective responses were seen at a dose as low as 2 to 5 mg of peanut protein (Hourihane et al., 1997b). In a later study, subjective oral symptoms were included as a positive outcome but verified by repetitive challenge with active dose versus placebo to ensure that the reaction was reproducible. The lowest reactive dose (oral symptoms) was again 100 μg of protein, a quantity equal to 1/2,000 of a single peanut. With a dose of 3 mg of peanut protein, half of the 26 patients had reacted (Wensing et al., 2002a). Sensitivity can be so great in some individuals that mucosal contact from kissing can result in a reaction (Hallett et al., 2002). Aerosolized peanut dust may also result in upper respiratory symptoms on airplanes or other enclosed spaces. However, the odor of peanut butter, although it may trigger an intense emotional response that may mimic an allergic reaction, is extremely unlikely to induce an allergic response, based on a carefully performed study (Sicherer et al.1999b; Simonte et al., 2003).

Risk of Severe Reactions

Unfortunately, predicting who will progress from a mild cutaneous reaction to more serious reactions is not possible at this time. A recent disturbing finding was that 72% of a group of children in the United States with peanut allergy experienced allergic reactions as a result of accidental exposure after initial diagnosis, with the majority experiencing potentially life-threatening reactions, regardless of the nature of their initial reaction (Vander Leek et al., 2000). In the same study, the few children who did not have a reaction to accidental exposure were those who had low peanut-specific IgE levels and had actually become tolerant to peanuts. Such findings support the management recommendation that physicians should consider prescribing self-injected epinephrine to patients with systemic IgE-mediated reactions to foods, particularly in the case of peanut, tree nut, or seafood allergy, as part of an emergency plan with instructions on when to use it (Sampson, 2004; Sicherer et al., 2004). However, encouraging data on the risk of a subsequent severe reaction come from a study carried out in the United Kingdom by Ewan and Clark (2001). They reported on 567 unselected referrals of patients with confirmed peanut or tree nut allergy to their allergy clinic, in which participants were given a comprehensive management and avoidance plan based on their history of reaction and health status and taught to recognize mild, moderate, and severe reactions and how to use all medications prescribed. Only 15% of the patients had further reactions during follow-up, and only 0.5% (three patients) had a severe reaction. Nine of the 26 patients with moderate or severe reactions received an injection of epinephrine. The epinephrine was effective in all nine cases. Out of 172 patients who were not prescribed injectable epinephrine, only 1 patient had a reaction where epinephrine was needed. Although these data are quite encouraging and imply that education makes a difference, it was also in the United Kingdom that Pumphrey (2000) reported that 29 of 37 food allergy-related fatalities from 1992 to 1998 were in patients who initially only had a mild reaction to a food and were thus not given self-injected epinephrine.

Soy

Food allergy to soy has been reported in North and South America, Europe, Asia, and Australia (Ahn et al., 2003; Awazuhara et al., 1997; Magnolfi et al., 1996; Giampietro et al., 1992; Herian et al., 1990; Bock and Atkins 1990; Foucard and Malmheden-Yman, 1999; Bishop et al., 1990). In most cases, it is a disorder of infancy and early childhood. The spectrum of clinical reactions includes IgE-associated atopic dermatitis, non-IgE-mediated gastrointestinal reactions (mild to severe, including food protein-induced enterocolitis syndrome), and IgE-mediated acute reactions.

Soy protein is commonly used to enhance the protein content of health foods and beverages. It is marketed to women by supplement manufacturers as an alternative to regulate hormone fluctuations, and it provides inexpensive filler for many packaged foods. Thus, it is ubiquitous in food systems around the world and can be difficult to avoid. Fortunately, the majority of patients with soy allergy are very young and will eventually outgrow their allergy. The infant with food allergy to soy formula can be switched to a cow's milk formula, if she or he is not allergic to cow's milk, or to an extensively hydrolyzed formula. In rare cases, a baby may need to be switched to an amino acid formula. Most baby foods packaged in jars do not include soy protein, and dry cereals can be found that do not contain soy products. For those extremely rare patients with severe, IgE-mediated soy allergy, soy lecithin and soybean oil may contain variable concentrations of allergenic proteins. While no reactions to soybean oil have been reported, there are several case reports of adverse reactions to soy proteins contained in soy lecithin, which can be found as a health food supplement, as an emulsifier in foods, and in pharmaceutical products, normally in amounts far too low to cause a reaction (Palm et al., 1999; Renaud et al., 1996). The allergens in soy lecithin have been sequenced and partially identified (Gu et al., 2001). Awazuhara et al. (1997) have shown that IgE from soy-allergic patients could bind proteins extracted from soy lecithin, but no detectable IgE binding to highly processed soy oil extract was seen. In contrast, another group showed that IgE-binding proteins were indeed present in soybean oil aqueous extract from both a highly processed oil and a cold-pressed oil. However, challenge testing has not yet been performed, and the proteins recovered from the oil were in exceedingly low concentrations that may not be relevant (0.32 and 1.8 μg/ml) (Errhali et al., 2002).

Prevalence

Two studies give prospective data on the development of soy allergy in pediatric populations. In a birth cohort study done on the Isle of Wight with 1,218 children studied from birth to 4 years of age, none had clinical soy allergy (Tariq et al., 1996). A prospective study carried out in the United States of 480 children, followed from birth to 3 years of age, found that 0.4% of children developed soy allergy as infants (Young et al., 1994).

In the United Kingdom, soy was a reported cause of food allergy or intolerance in 0.04% of a large group of >16,000 individuals 15 years of age or older (Emmett et al., 1999). Among children with moderate to severe atopic dermatitis, generally, 1 to 10% reacted to soy upon oral challenge (Ellman et al., 2002; Burks et al., 1998; Bruno et al., 1997; Sampson et al., 1985; Bock and Atkins, 1990). A lower prevalence of reactivity was seen among atopic children with symptoms of food allergy (i.e., not a population restricted to moderate to severe atopic dermatitis) and among children with atopic parents, who were fed soy protein formula for the first 6 months to prevent cow's milk allergy. Only 1.2% of 505 children in the first group and 0.4% of 243 children in the second group had a positive skin test and positive food challenges to soy (Bruno et al., 1997). However, a much higher percentage of children showed positive skin prick tests to soy (6% in both groups), indicating the low positive predictive value of a positive prick skin test for soy in this population. Further highlighting the need for food challenges in most soy skin test-positive children, Magnolfi et al. (1996) screened a population of 704 atopic patients, ranging in age from 1 to 18 years, from a university allergy clinic in Milan for soy allergy. Positive skin tests were seen in 148 of 704 (21%) patients, but only 8 of 131 skin test-positive patients reacted on oral challenge. Thus, only 6% of patients with a positive skin prick test were actually soy allergic.

Age of Onset and Natural History

Allergy to soy in infant formulas can present as early as 1 month of age and is generally reported by 1 year of age (Bock and Atkins, 1990; Klemola et al., 2002). Soy allergy is generally a problem of infancy and early childhood. Some rare cases of adults with food allergy to soy have been reported (Bock, 1985; Herian et al., 1990), but most children outgrow their allergy to soy; this is less likely if the reaction is a severe systemic IgE-mediated reaction (Bock and Atkins, 1990; Sampson and McCaskill, 1985; Host and Halken, 1990; Perry et al., 2004; Giampietro et al., 1992).

Cross-Reactivity

Although there is extensive in vitro cross-reactivity, clinical cross-reactivity with peanuts or other legumes is actually uncommon. In patients with peanut allergy, about 5% may react with other legumes, including soy. In one study of adults (age range, 23 to 67 years) with soy allergy, 7 of 8 patients (88%) reported sensitivity to more than one legume, but other legume sensitivity was not confirmed by DBPCFC (Herian et al., 1990). Other studies, which included pediatric patients whose legume sensitivity was confirmed or refuted by DBPCFC, found that only 3 to 5% of patients had multiple legume sensitivities (Bernhisel-Broadbent et al., 1989a; Bock et al., 1989). Thus, it appears rare that any one patient will present with allergy to multiple legumes; avoidance of all legumes in soy-allergic patients is not currently recommended. However, a few deaths from anaphylaxis due to intake of soy by severely peanut allergic patients who had no prior knowledge of soy allergy have been reported (Foucard and Yman, 1999; Yunginger et al., 1991).

There has been one report that patients with birch allergy may be at risk for reactions to a soy protein isolate (Kleine-Tebbe et al., 2002). In this study, 20 patients who had not previously reported allergy to soy experienced severe reactions after consuming a dietary supplement containing a soy protein isolate. It was found that IgE specific for the major birch allergen, Bet v 1, was cross-reactive to proteins in this preparation of soy, specifically SAM22, a pathogenesis-related protein.

Coallergy to Cow's Milk

There is clinical evidence to support the concept that a percentage of patients with cow's milk allergy will also be allergic to soy, especially those patients with food protein-induced enterocolitis syndromes (Burks et al., 1994). Interestingly, there is also an in vitro study suggesting that in some IgE-mediated cow's milk allergy cases, reactions to soy are due to cross-reactivity between soy and bovine proteins. This was demonstrated by the use of monoclonal antibodies raised to casein, polyclonal antibodies towards cow's milk, and sera from cow's milk-allergic patients who had never knowingly eaten soy. These antibody preparations were able to bind a glycinin fraction from soy (Rozenfeld et al., 2002). However, the mere presence of IgE specific for a food does not indicate that a patient will have a clinical reaction upon ingestion of that food, and the clinical relevance of this novel finding is unknown.

The reported rates of coallergy to cow's milk and soy vary markedly, mainly depending on whether oral DBPCFCs were performed to confirm parental reports, excepting patients with a history of anaphylaxis and positive in vitro-specific IgE. For example, in one study, parents reported that 47 of 100 patients with proven cow's milk allergy had developed adverse reactions to soy (Bishop et al., 1990). In contrast, two recent studies of IgE-mediated cow's milk allergy where reactivity to soy was confirmed by DBPCFC showed that only 11 to 14% of a large cohort of proven cow's milk-allergic children were also allergic to soy (Zeiger et al., 1999; Klemola et al., 2002). Klemola et al. (2002) also included a group with non-IgE-associated cow's milk allergy; in this group, approximately 9% (but up to 28% of patients if more ambiguous delayed reactions were included as positives) were intolerant of soy formula. However, in a small study of 10 infants with the less-common condition of cow's milk protein-induced enterocolitis, 6 had reactions upon soy challenge (Burks et al., 1994). In summary, it has been found that only a minority of patients with IgE-associated or IgE-mediated cow's milk allergy were allergic to soy, but about half of children with cow's milk protein-induced enterocolitis syndrome had reactions to soy. Therefore, the Committee on Nutrition of the American Academy of Pediatrics (1998) concluded that most infants with IgE-mediated or IgE-associated allergy to cow's milk protein will do well on isolated soy protein-based formula and that there is no need to spend money on more expensive, less-palatable formulas. It has been noted, however, that these studies did not include many infants of <6 months of age, so these findings are of uncertain application to the very young infant (Klemola et al., 2002).

Threshold for Clinical Reactions

A study looking at dose-response in DBPCFC in atopic dermatitis found that 28% of soy-allergic patients reacted at the first dose (500 mg), suggesting that the true threshold was lower (Sicherer et al., 2000a). The lowest eliciting dose in a DBPCFC study that identified eight children with IgE-mediated soy allergy was 88 mg of soy protein, which was the first dose. A 4.6-year-old child with a radioallergosorbent class 6 IgE titer to soy began to react within 5 min with serious symptoms consisting of abdominal pain, vomiting, malaise, and collapse (Magnolfi et al., 1996). Since this was the initial dose, the actual threshold was probably lower. Zeiger et al. (1999) published titrated dose information on 13 patients with IgE-mediated soy allergy. The lowest provoking dose was 409 mg, elicited by 29 ml of a solution of 100 mg of soy formula powder/ml (powder containing 14.1% soy protein isolate).

Risk of Severe Reactions

Reactions to soy protein are commonly mild. These reactions generally consist of flares of atopic dermatitis or gastrointestinal symptoms. In the case of food protein-induced enterocolitis, a reaction can be life threatening. In IgE-mediated soy allergy, severe systemic allergic reactions can occur, but such reactions are not common (Herian et al., 1990). Although rare, life-threatening reactions to soy have been reported in the literature, including several deaths (Mortimer, 1961; Foucard and Yman, 1999; Yunginger et al., 1991). The deaths occurred in patients severely allergic to peanut who were not known to be allergic to soy (Foucard and Yman, 1999; Yunginger et al., 1991; for a review, see Sicherer et al., 2000b).

Tree Nuts

There are 12 major types of edible tree nuts produced in the world—almonds, Brazil nuts, cashews, chestnuts, coconuts, hazelnuts (filberts), macadamia nuts, pecans, pine nuts (pignolias or piñon nuts), pistachios, black walnuts, and English (or Persian) walnuts (Burket, 2000). Clinical reactions to tree nuts are IgE-mediated acute reactions and, in some cases, IgE-associated flares of atopic dermatitis. English walnuts, hazelnuts, almonds, cashews, and Brazil nuts are most frequently implicated in tree nut allergy, while pistachios, macadamia nuts, and pine nuts are less commonly reported as the cause of systemic allergic reactions (Teuber et al., 2003; Sicherer et al., 1999a; Ewan 1996; Sicherer et al., 2001b). Most discussions of tree nut allergy do not include chestnuts or coconut. Allergy to chestnut (*Castanea sativa*) is usually seen as a result of cross-reactivity with natural rubber latex allergens (Salcedo et al., 2001). Coconut (*Cocos nucifera* L.) food allergy is rarely reported. For example, only 4 of 1,667 tree nut-allergic patients in the FAAN peanut and tree nut registry wrote in "coconut" as an allergy (Sicherer et al., 2001b). Coconut food allergy has also been reported as a secondary phenomenon in patients with severe tree nut allergy to walnut. In that situation, walnut-directed IgE (the patients' primary allergy) was shown to be cross-reactive with coconut (Teuber and Peterson, 1999).

Prevalence

Several studies have examined tree nut allergy on a population basis, always in the context of surveys for peanut allergy as well. On the Isle of Wight in the United Kingdom, a cohort of 1,218 newborns was followed from birth to 4 years of age (Tariq et al., 1996). Two (0.16%) of the children were found to be tree nut sensitized by skin testing. Both children had experienced a clinical reaction upon ingestion of the nut, one reacting to a cashew and one reacting to a hazelnut. A second study, using at-home interviews of a nationally representative sample of 16,434 respondents 15 years of age and older, found the prevalence of self-reported tree nut allergy to be 0.38% (Emmett et al., 1999). In the United States, random digit dial telephone surveys for peanut and tree nut allergy were performed in 1997 and 2002 with nearly identical results for tree nut allergy (Sicherer et al., 1999; Sicherer et al., 2003). Tree nut allergy was reported in 0.2% of the children and 0.5% of the adults in the 2002 survey. From these surveys, it is estimated that about 0.5% of the United States population is allergic to tree nuts.

In the United States, the most commonly reported tree nut allergy in both the FAAN random digit dial survey and the FAAN peanut and tree nut registry was to English walnuts, by a large margin. In the random digit dial survey done in 1997, 24 patients reported allergy to walnuts, with <8 patients reporting allergy to each of the other nuts (Sicherer et al.,

1999a). In the FAAN peanut and tree nut registry, 34% of tree nut-allergic respondents reported allergy to walnuts, followed by cashews (20%), almonds (15%), pecans (9%), pistachios (7%), and other nuts (<5% each) (Sicherer et al., 2001b).

Age of Onset and Natural History

Although tree nut allergy usually presents later than peanut, milk, wheat, or soy allergy, the median age of onset is still early in life, typically around 3 to 4 years of age (Ewan and Clark, 2001; Sicherer et al., 1998; Sicherer et al., 2001b). In general, allergy to less-ubiquitous nuts in the culture, those which have yet to infiltrate processed foods and are often only consumed by adults, may present in the third decade of life or later. However, as mentioned previously, geographic location often determines how ubiquitous a particular tree nut will be in a population's diet. As with peanut allergy, patients who develop tree nut allergy tend to react upon the first-known contact with a nut, implying previous sensitizing exposure unknown to the patient and family (Sicherer et al., 2001b; Clark and Ewan, 2003a).

The age range of patients with life-threatening tree nut allergy in our database at the University of California, Davis, ranges from 2 to 73 years (Teuber, unpublished). Self-injected epinephrine has been successfully used to help reverse anaphylactic reactions in several patients >65 years old in our database with tree nut allergy after accidental ingestion. It is highly likely that a percentage of patients with tree nut allergy will become tolerant after some years of avoidance, based on the experience with peanut allergy, but there are no such studies to document this yet.

Cross-Reactivity

Patients with tree nut allergy are usually advised to avoid all tree nuts due to the risk of cross-reactivity among tree nuts, which exists for many patients but is not yet well characterized. For instance, pistachios and cashews are both in the Anacardaceae family and appear to have extensive cross-reactivity, although this is not always clinically correlated (Garcia et al., 2000). Walnuts and pecans are also in the same family (Juglandaceae). Bock and Atkins (1989) have performed the only study to date examining cross-reactivity among the tree nuts by double-blind food challenges. Fourteen children were challenged: 7 reacted to walnut, 6 reacted to cashew, 3 reacted to pecan, 2 reacted to pistachio, and 1 reacted to hazelnut. Only one patient reacted to two nuts, and one patient reacted to five nuts (Bock and Atkins, 1989). However, there are many self-reports of multiple-nut allergy, especially in those with severe, life-threatening allergy to at least one nut. Some cases of multiple-nut allergy may actually represent overreporting, due to mislabeling of foods or visual misidentification by patients of the implicated nut, but it would be very difficult to disprove cases of multiple-nut allergy when severe, life-threatening sensitivity is involved. Of the 1,667 tree nut-allergic participants in FAAN's peanut and tree nut allergy registry, 46% reported allergy to more than one tree nut (Sicherer et al., 2001b). Another study found 37% of 54 tree nut-reactive patients had convincing histories of reactions to more than one nut (Sicherer et al., 1998). Also, in a study of 1,000 children and adults conducted in the United Kingdom, 45% of those allergic to a tree nut were also allergic by history to another type of nut (Clark and Ewan, 2003b). These studies likely included more people with severe tree nut allergy than might be present in the general population, since many people with food allergy do not seek medical care. A more representative sample might be found in the random digit dial and follow-up surveys conducted in the United States for peanut and tree nut allergy. In the first study, only 5 of 59 (8.5%) tree nut-allergic participants reported allergy to more than one tree nut (Sicherer et al., 1999b). However, in the follow-up survey, 43 out of 82 (52%) of tree nut-allergic responders reported being allergic to more than one nut (Sicherer et al., 2003). Thus, the actual prevalence of allergy to more than one tree nut is unclear. It is advisable that tree nut-allergic patients avoid all tree nuts unless they have previously consumed the nut without adverse reaction, have passed an oral challenge, and importantly, can obtain the nut without any risk of cross-contamination (e.g., almonds harvested in the backyard).

Using sera from patients with severe, life-threatening reactions, the major allergens identified in all the tree nuts are generally seed storage proteins that tend to have significant homology. For example, it was recently shown that the cashew 2S albumin, Ana o 3, shares homology with the walnut 2S albumin (Jug r 1), including the sharing of an identical major IgE-binding epitope (Robotham et al., 2005). Walnut-directed IgE has been shown to have cross-reactivity against other tree nuts, including almonds (Teuber and Peterson, 1999) and hazelnuts and Brazil nuts (Asero et al., 2004). Lipid transfer protein was shown to be a major allergen in Italian patients with walnut allergy; this class of allergens is well known for exhibiting cross-reactivity that can be clinically relevant across species, but little is yet known about the cross-reactivity of lipid transfer proteins among tree nuts (Pastorello et al., 2004). These laboratory findings also support the recommendation to avoid other tree nuts unless a nut is known to be tolerated and can be obtained without cross-contamination.

Threshold for Clinical Reactions

Only hazelnuts have been subjected to a study specifically performed to determine the threshold dose required to elicit a clinical reaction. Thirty-one patients (almost all with oral symptoms only, consistent with the pollen-food syndrome, and only 1 patient with a history of anaphylaxis) underwent DBPCFCs, and 29 reacted to hazelnut. Almost all reactions were subjective and limited to itching of the oral cavity, but one patient had an objective reaction of lip swelling after 1 mg of hazelnut protein. One-half of patients reacted after eating 6 mg of hazelnut protein, and all patients reacted after 100 mg of hazelnut protein (Wensing et al., 2002b). No threshold dose studies have been done with the other tree nuts.

Risk of Severe Reactions

Many studies have been published verifying that allergic reactions to tree nuts are often systemic and can be life threatening or fatal (Yocum et al., 1999, Pumphrey, 2000; Sicherer et al., 2001b; Yunginger et al., 1988; Bock et al., 2001; Sampson et al., 1992; Asero et al., 2004; Sutherland et al., 1999; Ewan, 1996). Fatalities have been reported from ingestion of pecans, cashews, walnuts, Brazil nuts, pistachios, and hazelnuts (Yunginger et al., 1988; Sampson et al., 1992; Foucard and Yman, 1999; Pumphrey, 2000; Bock et al., 2001). Of food-induced fatalities, about one-eighth to one-third in different series are reported to be due to ingestion of a tree nut (Sampson et al., 1992; Pumphrey, 2000; Bock et al., 2001). It is obvious that tree nut allergy can be dangerous and needs to be treated as such by physicians and patients alike.

Wheat

Wheat is associated with several important, distinct disorders. The most important in terms of numbers and global impact is celiac disease, also known as gluten enteropathy or celiac sprue. This disorder represents an interaction of the immune system with certain epitopes on multiple, related gluten storage proteins. The disease only occurs in those persons with the correct genetic background, expressing the tissue type antigen HLA-DQ2 or HLA-DQ8. The result is a gluten-driven autoimmune disease with small bowel inflammation (an enteropathy with villous atrophy and malabsorption) and protean extraintestinal manifestations, including refractory iron deficiency anemia, infertility, osteoporosis, peripheral neuropathy, and dermatitis herpetiformis (an extremely pruritic vesicular rash). Unlike in previous decades, the disease is not only diagnosed in children with a classic malabsorption syndrome, but is now most commonly diagnosed in middle-aged adults, often after patients have had 8 to 10 years of poorly defined symptoms, with diarrhea seen in only half of patients (Treem, 2004). Other non-IgE-mediated wheat-induced disorders include food protein-induced enterocolitis and eosinophilic esophagitis or gastroenteritis. IgE-associated atopic dermatitis is seen in infancy and early childhood, IgE-mediated acute reactions are seen with patients of all ages, and food-associated, exercise-induced anaphylaxis is seen in some countries with a growing number of adults. Occupational asthma, known as baker's asthma, is yet another important manifestation of wheat allergy; such patients do not have associated food allergy, however. Wheat allergy is one of the common triggers of atopic dermatitis in young children and has recently been shown to be a common food allergy coexisting in refractory atopic dermatitis in children with cow's milk allergy (Jarvinen et al., 2003).

Prevalence

Celiac disease affects approximately 1% of the population in the United States, similar to European surveys (Treem, 2004). A prospective study followed a cohort of neonates in Colorado. Those who were DQ2 or DQ8 positive were followed prospectively. By age 5, 1 in 104 children overall had developed the autoimmune marker antibody IgA anti-tissue transglutaminase, which is seen in celiac disease (Hoffenberg et al., 2003). The prevalence of other wheat food allergy is less clear. In a population survey in Great Britain, 0.37% of 16,420 people reported allergy to wheat. However, this may also have included celiac patients; the types of reactions were not queried further, as the study was focused on peanut allergy (Emmett et al., 1999).

Age of Onset and Natural History

Wheat allergy usually has its onset in infancy or early childhood (Sampson, 2004). However, many of the reported cases of wheat-dependent, exercise-associated anaphylaxis due to wheat have been adult-onset cases (Palosuo et al., 1999). Wheat allergy associated with atopic dermatitis or non-IgE-mediated gastrointestinal disorders, besides celiac disease, is usually outgrown by school age. Celiac disease is a lifelong condition. It appears that children with severe, systemic IgE-mediated reactions are less likely to outgrow wheat allergy, but data are not available, nor are data available on the natural history of adult-onset wheat anaphylaxis.

Cross-Reactivity

In celiac disease, the gluten homolog proteins in barley (hordeins) and rye (secalins) are also disease inducing, but oats are tolerated by most people with the disorder (Treem, 2004; Janatuinen et al., 2002). There have been a few rare case reports documenting recurrence of villous atrophy with oats (Lundin et al., 2003). In IgE-associated or -mediated wheat allergy, cross-reactivity with other cereals is not as definite. In a series of 18 wheat-allergic children with immediate reactions (many with anaphylaxis), 10 (55.5%) reacted upon challenge with barley, while none reacted to rice or corn (Pourpak et al., 2005). Interestingly, 10 patients had positive corn skin tests (55.5%), and 9 had positive rice skin tests (50%), indicating a high degree of clinically irrelevant IgE binding (Pourpak et al., 2005). Another study indicated a lower rate of clinical cross-reactivity to other grains (Jones et al., 1995). Thirty-one challenge-proven cereal grain-allergic patients (most were allergic to wheat) were challenged with other grains; only 20% reacted to more than one grain (Jones et al., 1995).

Threshold for Clinical Reactions

Many studies indicate that DBPCFCs to wheat were performed, but the eliciting doses for immediate reactions have rarely been published. Sicherer et al. (2000a) published summarized results from 13 years of diagnostic food challenges in children. The starting doses were 400 mg or 500 mg; 25% of wheat-allergic children reacted to the initial dose, suggesting that the true threshold was lower. Almost 50% of positive wheat challenges resulted in a moderate to severe reaction (Sicherer et al., 2000a). A child from Bangkok reacted with a life-threatening anaphylactic reaction after 2.5 g of spaghetti (Daengsuwan et al., 2005). In celiac disease, complete disease remission has been documented by small intestinal biopsy when the ingested gluten was kept in the low-milligram range (<30 mg per day) by the use of gluten-free foods, which may still have traces of gluten (Collin et al., 2004).

Risk of Severe Reactions

Although the long-term risks of wheat or other gluten-containing grain in a patient with celiac disease can be significant in terms of the increased risk of gastrointestinal lymphoma, it is the IgE-mediated acute reactions that are immediately life threatening. Wheat-dependent, exercise-induced anaphylaxis is an important syndrome in adults (Palosuo, 2003). Fortunately, no deaths have been reported. This syndrome is likely to be underreported and may be unrecognized in some deaths due to exercise-induced anaphylaxis or other deaths while exercising. Anaphylaxis without exercise can also occur, particularly in children (Palosuo et al., 2001). Exact data on wheat-induced anaphylaxis are not available, but 6% of 229 severe allergic reactions reported to the French Allergo Vigilance Network from 2002 to 2004 were due to wheat (Moneret-Vautrin et al., 2005). However, no emergency department visits for wheat allergy out of 678 visits were recorded in the series by Clark et al. (2004) in the United States.

Sesame

Sesame allergy has been a growing concern in the last decade. The first case of sesame seed allergy was reported in 1950 (Rubenstein, 1950). Increasing use in North America and Europe appears to be paralleled by an increase in reported sesame-induced allergic reactions (Chiu and Haydik 1991; Moneret-Vautrin et al., 1997; Stern et al., 1998; Pajno et al., 2000). Although many of the studies of sesame allergy have involved small populations, there are a number of similarities compared with peanut and tree nut allergies. Reactions to sesame consist of immediate systemic allergic reactions, often life threatening, and some cases of atopic dermatitis flares. Unfortunately, similarly to peanut, sesame is difficult to avoid, due to its increasing use as an ingredient in processed foods and as an ingredient in cosmetics. Often, sesame seeds are found in a number of gluten-free foods (bread, cakes, pastries, and biscuits) used to treat celiac patients. (Pajno et al., 2000). At least one celiac patient has been reported to have developed a systemic reaction upon consumption of sesame (Pajno et al., 2000). In this case, the allergy developed in adulthood.

Prevalence

Accurate estimations of persons affected by sesame allergy are difficult, due to the recent emergence of sesame as a significant allergen and the relatively small groups previously examined. In 1992, a study of 422 adults with food allergy found that sesame seed allergy was extremely rare, being present in only 0.7% of all the patients with food allergies tested (Moneret-Vautrin and Kanny, 1992). A similar result came from another study, which found that only 1.2% of patients were allergic to sesame seed among 402 food-allergic patients seen between 1978 and 1991 (Kagi and Wuthrich, 1991). Despite the relatively low occurrence of sesame allergy, there is concern about the sudden rise in cases and the severity of the reactions.

Geographical differences exist in sesame allergy, as demonstrated by an Australian study that reported sesame seed sensitization was one-third the rate of sensitization to peanut and higher than the rate for any one tree nut (Sporik et al., 1996). The same study also reported that twice as many children were sensitized to both peanut and sesame as those who were sensitized to both tree nuts and sesame (Sporik et al., 1996). Although sensitization does not always correlate with clinical responsiveness, other countries have reported high rates of life-threatening reactions from sesame consumption. For example, sesame is a major cause of severe IgE-mediated food allergic reactions among infants and young children in Israel, second in frequency only to cow's milk as a cause of anaphylaxis (Dalal et al., 2002). Rance et al. (1999) reported that sesame seed allergy was the 15th-most-common pediatric food allergy in France (thus, not very common), while a second report from this group 2 years later indicated that sesame was the seventh-leading cause of severe food allergy among all age groups in France (Moneret-Vautrin et al., 2001).

Age of Onset and Natural History

Allergy to sesame, similar to allergy to peanuts and tree nuts, appears to develop mainly in young children; reactions can occur after the first known exposure. One study reported on an infant (11 months of age) who developed a severe reaction upon the first known exposure (Sporik et al., 1996). The same study found that 60% of children sensitized to sesame seed were <24 months of age (Sporik et al., 1996). In a more recent study of children in France, the mean age of onset was 5 years (ranging from 10 months to 16 years), and only five children in the study were <2 years of age (Agne et al., 2004). One 5-month-old infant in the study reacted to breast milk with facial edema after her mother had consumed sesame seed. The same infant experienced urticaria later at 12 months of age after eating sesame bread (Agne et al., 2004).

It appears that a proportion of children will outgrow sesame allergy, based on follow-up studies; 3 (13%) of the 23 participating children in Israel outgrew their allergy in 1 to 2 years (Dalal et al., 2003). Recently, a larger-scale study in France reported about 20% of the children in the study outgrew their allergy within 6 years (Agne et al., 2004). For the majority of patients afflicted with sesame allergy, however, it appears to persist into adulthood.

Cross-Reactivity

Multisensitization affects 60% of the sesame-allergic children in France, with nuts and peanuts being the most dominant (Agne et al., 2004). The dominant allergens in sesame are seed storage proteins (2S albumin, vicilin, and legumin group proteins) with homologs in peanuts, tree nuts, and other edible seeds (Beyer et al., 2002). In vitro cross-reactivity has not been explored yet to determine whether there is coallergy or cross-reactivity in such patients.

Threshold for Clinical Reactions

Clinically relevant reactions have been documented to occur in patients after exposure to as little as 30 mg of sesame seed powder (Morisset et al., 2003). Overall, 5 of 12 challenge-positive patients in this study had bronchospasm. Moreover, five of six subjects challenged with sesame oil developed reactions. Two of these five patients experienced anaphylactic shock after 1 ml and 5 ml of sesame seed oil, respectively. Interestingly, the doses of sesame oil that provoked reactions only contained a few milligrams of sesame protein, and yet the same patients required higher doses of sesame seed powder prior to an objective reaction (100 mg of powder up to 7 g) (Morisset et al., 2003). The proteins in the oil may be particularly available to interact with IgE.

Risk of Severe Reactions

Unfortunately, reactions tend to be very severe in nature, often resulting in an anaphylactic reaction (Chiu and Haydik, 1991; Kagi and Wuthrich, 1991; Kolopp-Sarda et al., 1997; Stern et al., 1998; Asero et al., 1999; Pajno et al., 2000). Occupational allergy to sesame resulting in asthma or severe reaction has also been previously documented (Alday et al., 1996; Keskinen et al., 1991).

REFERENCES

Agne, P. S., E. Bidat, P. S. Agne, F. Rance, and E. Paty. 2004. Sesame seed allergy in children. *Allerg. Immunol.* (Paris) **36:**300–305.

Ahn, K. M., Y. S. Han, S. Y. Nam, H. Y. Park, M. Y. Shin, and S. I. Lee. 2003. Prevalence of soy protein hypersensitivity in cow's milk protein-sensitive children in Korea. *J. Korean Med. Sci.* **18:**473–477.

Alday, E., G. Curiel, M. J. Lopez-Gil, D. Carreno, and I. Moneo. 1996. Occupational hypersensitivity to sesame seeds. *Allergy* **51:**69–70.

Anibarro, B., F. J. Seoane, C. Vila, and M. Lombardero. 2000. Allergy to eggs from duck and goose without sensitization to hen egg protein. *J. Allergy Clin. Immunol.* **105:**834–836.

Asero, R. 2005. Lack of de novo sensitization to tropomyosin in a group of mite-allergic patients treated by house dust mite-specific immunotherapy. *Int. Arch. Allergy Immunol.* **137:**62–65.

Asero, R., G. Mistrello, D. Roncarolo P. L. Antoniotti, and P. Falagiani. 1999. A case of sesame seed-induced anaphylaxis. *Allergy* **54:**526–527.

Asero, R., G. Mistrello, D. Roncarolo, and S. Amato. 2004. Walnut-induced anaphylaxis with cross-reactivity to hazelnut and Brazil nut. *J. Allergy Clin. Immunol.* **113:**358–360.

Astwood, J. D., A. Silvanovich, and G. A. Bannon. 2002. Vicilins: a case study in allergen pedigrees. *J. Allergy Clin. Immunol.* **110:**26–27.

Avery, N. J., R. M. King, S. Knight, and J. O. Hourihane. 2003. Assessment of quality of life in children with peanut allergy. *Pediatr. Allergy Immunol.* **14:**378–382.

Awazuhara, H., H. Kawai, and N. Maruchi. 1997. Major allergens in soybean and clinical significance of IgG4 antibodies investigated by IgE- and IgG4-immunoblotting with sera from soybean-sensitive patients. *Clin. Exp. Allergy* **27:**325–332.

Bahna, S. L. 2002. Cow's milk allergy versus cow milk intolerance. *Ann. Allergy Asthma Immunol.* **89**(Suppl.)**:** 56–60.

Bellioni-Businco, B., R. Paganelli, P. Lucenti, P. G. Giampietro, H. Perborn, and L. Businco. 1999. Allergenicity of goat's milk in children with cow's milk allergy. *J. Allergy Clin. Immunol.* **103:**1191–1194.

Bernhisel-Broadbent, J., and H. A. Sampson. 1989. Cross-allergenicity in the legume botanical family in children with food hypersensitivity. *J. Allergy Clin. Immunol.* **83:**435–440.

Bernhisel-Broadbent, J., S. Taylor, and H. A. Sampson. 1989a. Cross-allergenicity in the legume botanical family in children with food hypersensitivity. II. Laboratory correlates. *J. Allergy Clin. Immunol.* **84:** 701–709.

Bernhisel-Broadbent, J., S. M. Scanlon, and H. A. Sampson. 1992a. Fish hypersensitivity. I. In vitro and oral challenge results in fish-allergic patients. *J. Allergy Clin. Immunol.* **89:**730–737.

Bernhisel-Broadbent, J., D. Strause, and H. A. Sampson. 1992b. Fish hypersensitivity. II. Clinical relevance of altered fish allergenicity caused by various preparation methods. *J. Allergy Clin. Immunol.* **90:**622–629.

Beyer, K., L. Bardina, G. Grishina, and H. A. Sampson. 2002. Identification of sesame seed allergens by 2-dimensional proteomics and Edman sequencing: seed storage proteins as common food allergens. *J. Allergy Clin. Immunol.* **110:**154–159.

Beyer, K., E. Morrow, X. M. Li, L. Bardina, G. A. Bannon, A. W. Burks, and H. A. Sampson. 2001. Effects of cooking methods on peanut allergenicity. *J. Allergy Clin. Immunol.* **107:**1077–1081.

Bishop, J. M., D. J. Hill, and C. S. Hosking. 1990. Natural history of cow milk allergy: clinical outcome. *J. Pediatr.* **116:**862–867.

Bock, S. A. 1985. Natural history of severe reactions to foods in young children. *J. Pediatr.* **107:**676–680.

Bock, S. A., and F. M. Atkins. 1989. The natural history of peanut allergy. *J. Allergy Clin. Immunol.* **83:**900–904.

Bock, S. A., and F. M. Atkins. 1990. Patterns of food hypersensitivity during sixteen years of double-blind, placebo-controlled food challenges. *J. Pediatr.* **117:**561–567.

Bock, S. A., A. Munoz-Furlong, and H. A. Sampson. 2001. Fatalities due to anaphylactic reactions to foods. *J. Allergy Clin. Immunol.* **107:**191–193.

Bruno, G., P. G. Giampietro, M. J. Del Guercio, P. Gallia, L. Giovannini, C. Lovati, P. Paolucci, L. Quaglio, E. Zoratto, and L. Businco. 1997. Soy allergy is not common in atopic children: a multicenter study. *Pediatr. Allergy Immunol.* **8:**190–193.

Burket, S. 2000. *Industry & Trade Summary: Edible Nuts.* U.S. International Trade Commission, Washington, D.C.

Burks, A. W., H. B. Casteel, S. C. Fiedorek, L. W. Williams, and C. L. Pumphrey. 1994. Prospective oral food challenge study of two soybean protein isolates in patients with possible milk or soy protein enterocolitis. *Pediatr. Allergy Immunol.* **5:**40–45.

Burks, A. W., J. M. James, A. Hiegel, G. Wilson, J. G. Wheeler, S. M. Jones, and N. Zuelein. 1998. Atopic dermatitis and food hypersensitivity reactions. *J. Pediatr.* **132:**132–136.

Businco, L., P. G. Giampietro, P. Lucenti, F. Lucaroni, C. Pini, G. Di Felice, P. Iacovacci, C. Curadi, and M. Orlandi. 2000. Allergenicity of mare's milk in children with cow's milk allergy. *J. Allergy Clin. Immunol.* **105:**1031–1034.

Busse, P. J., A. H. Nowak-Wegrzyn, S. A. Noone, H. A. Sampson, and S. H. Sicherer. 2002. Recurrent peanut allergy. *N. Engl. J. Med.* **347:**1535–1536.

Cartier, A., J. L. Malo, F. Forest, M. Lafrance, L. Pineau, J. J. St. Aubin, and J. Y. Dubois. 1984. Occupational asthma in snow crab-processing workers. *J. Allergy Clin. Immunol.* **74:**261–269.

Chiu, J. T., and I. B. Haydik. 1991. Sesame seed oil anaphylaxis. *J. Allergy Clin. Immunol.* **88:**414–415.

Chung, S. Y., C. L. Butts, S. J. Maleki, and E. T. Champagne. 2000. Linking peanut allergenicity to the processes of maturation, curing, and roasting. *J. Agric. Food Chem.* **51:**4273–4277.

Clark, A. T., and P. W. Ewan. 2003a. Food allergy in childhood. *Arch. Dis. Child* **88:**79–81.

Clark, A. T., and P. W. Ewan. 2003b. Interpretation of tests for nut allergy in one thousand patients, in relation to allergy or tolerance. *Clin. Exp. Allergy* **33:**1041–1045.

Clark, S., S. A. Bock, T. J. Gaeta, B. E. Brenner, R. K. Cydulka, C. A. Camargo, and the MARC-8 Investigators. 2004. Multicenter study of emergency department visits for food allergies. *J. Allergy Clin. Immunol.* **113:**347–352.

Codex Alimentarius Commission. 1999. *Report of the Twenty-Third Session of the Codex Alimentarius Commission.* Alinorm 99/37. Codex Alimentarius Commission, Rome, Italy.

Collin, P., L. Thorell, K. Kaukinen, and M. Maki. 2004. The safe threshold for gluten contamination in gluten-free products. Can trace amounts be accepted in the treatment of celiac disease? *Aliment. Pharmacol. Ther.* **19:**1277–1283.

Committee on Nutrition, American Academy of Pediatrics. 1998. Soy protein-based formulas: recommendations for use in infant feeding. *Pediatrics* **101:**148–153.

Crespo, J. F., C. Pascual, A. W. Burks, R. M. Helm, and M. M. Esteban. 1995. Frequency of food allergy in a pediatric population from Spain. *Pediatr. Allergy Immunol.* **6:**39–43.

Dalal, I., I. Binson, A. Levine, E. Somekh, A. Ballin, and R. Reifen. 2003. The pattern of sesame sensitivity among infants and children. *Pediatr. Allergy Immunol.* **14:**312–316.

Dalal, I., I. Binson, R. Reifen, Z. Amitai, T. Shohat, S. Rahmani, A. Levine, A. Ballin, and E. Somekh. 2002. Food allergy is a matter of geography after all: sesame as a major cause of severe IgE-mediated food allergic reactions among infants and young children in Israel. *Allergy* **57:**362–365.

De Leon, M. P., I. N. Glaspole, A. C. Drew, J. M. Rolland, R. E. O'Hehir, and C. Suphioglu. 2003. Immunologic analysis of allergenic cross-reactivity between peanut and tree nuts. *Clin. Exp. Allergy* **33:**1273–1280.

Ellman, L.K., P. Chatchatee, S. H. Sicherer, and H. A. Sampson. 2002. Food hypersensitivity in two groups of children and young adults with atopic dermatitis evaluated a decade apart. *Pediatr. Allergy Immunol.* **13:** 295–298.

Emmett, S. E., F. J. Angus, J. S. Fry, and P. N. Lee. 1999. Perceived prevalence of peanut allergy in Great Britain and its association with other atopic conditions and with peanut allergy in other household members. *Allergy* **54:**380–385.

Errahali, Y., M. Morisset, D. A. Moneret-Vautrin, G. Kanny, M. Metche, J. P. Nicolas, and S. Fremont. 2002. Allergen in soy oils. *Allergy* **57:**648–649.

Escudero, C., S. Quirce, M. Fernandez-Nieto, J. de Miguel, J. Cuesta, and J. Sastre. 2003. Egg white proteins as inhalant allergens associated with baker's asthma. *Allergy* **58:**616–620.

Etesamifar, M., and B. Wüthrich. 1998. IgE-vermittelte Nahrungsmittelallergien bei 383 Patienten unter Berücksichtigung des oralen Allergie-Syndroms. *Allergologie* **21:**451–457.

European Commission. 2003. *Directive 2003/89/EC of the European Parliament and of the Council Amending Directive 2000/13/EC as Regards Indication of the Ingredients Present in Foodstuffs*, p. 15. Official Journal L 308. 25.11.2003. Publications Office, Luxembourg.

European Commission. 2005. *Directive 2005/26/EC* establishing a list of food ingredients or substances provisionally excluded from Annex IIIa of Directive 2000/13/EC of the European Parliament and of the Council, p. 0033–0034. Official Journal L 075. 22.03.2005. Publications Office, Luxembourg.

Ewan, P. W. 1996. Clinical study of peanut and nut allergy in 62 consecutive patients: new features and associations. *Brit. Med. J.* **312:**1074–1078.

Ewan, P. W., and A. T. Clark. 2001. Long-term prospective observational study of patients with peanut and nut allergy after participation in a management plan. *Lancet* **357:**111–115.

Fleischer, D. M., M. K. Conover-Walker, L. Christie, A. W. Burks, and R. A. Wood. 2003. The natural progression of peanut allergy: resolution and the possibility of recurrence. *J. Allergy Clin. Immunol.* **112:** 183–189.

Fleischer, D. M., M. K. Conover-Walker, L. Christie, A. W. Burks, and R. A. Wood. 2004. Peanut allergy: recurrence and its management. *J. Allergy Clin. Immunol.* **114:**1195–1201.

Foucard, T., and I. Malmheden Yman. 1999. A study on severe food reactions in Sweden—is soy protein an underestimated cause of food anaphylaxis? *Allergy* **54:** 261–265.

Garcia, F., I. Moneo, B. Fernandez, J. M. Garcia-Menaya, J. Blanco, S. Juste, and M. Gonzalo. 2000. Allergy to Anacardaceae: description of cashew and pistachio nut allergens. *J. Investig. Allergol. Clin. Immunol.* **10:**173–177.

Giampietro, P. G., V. Ragno, S. Daniele, A. Cantani, M. Ferrara, and L. Businco. 1992. Soy hypersensitivity in children with food allergy. *Ann. Allergy* **69:**143–146.

Goetz, D. W., and B. A. Whisman. 2000. Occupational asthma in a seafood restaurant worker: cross-reactivity of shrimp and scallops. *Ann. Allergy Asthma Immunol.* **85:**461–466.

Goh, D. L., Y. N. Lau, F. T. Chew, L. P. L. Shek, and B. W. Lee. 1999. Pattern of food-induced anaphylaxis in children of an Asian community. *Allergy* **54:**84–86.

Grundy, J., S. Matthews, B. Bateman, T. Dean, and S. H. Arshad. 2002. Rising prevalence of allergy to peanut in children: data from 2 sequential cohorts. *J. Allergy Clin. Immunol.* **110:**784–789.

Gu, X., T. Beardslee, M. Zeece, G. Sarath, and J. Markwell. 2001. Identification of IgE-binding proteins in soy lecithin. *Int. Arch. Allergy Immunol.* **126:**218–225.

Gustafsson, D., O. Sjoberg, and T. Foucard. 2003. Sensitization to food and airborne allergens in children with atopic dermatitis followed up to 7 years of age. *Pediatr. Allergy Immunol.* **14:**448–452.

Hallett, R., L. A. Haapanen, and S. S. Teuber. 2002. Food allergies and kissing. *N. Engl. J. Med.* **46:** 1833–1834.

Han, D. K., M. K. Kim, J. E. Yoo, S. Y. Choi, B. C. Kwon, M. H. Sohn, K.-E. Kim, and S. Y. Lee. 2004. Food sensitization in infants and young children with atopic dermatitis. *Yonsei Med. J.* **45:**803–809.

Hansen, T. K., C. Bindslev-Jensen, P. S. Skov, and L. K. Poulsen. 1997. Codfish allergy in adults: IgE cross-reactivity among fish species. *Ann. Allergy Asthma Immunol.* **78:**187–194.

Harada, S., T. Horikawa, and M. Icihashi. 2000. A study of food-dependent exercise-induced anaphylaxis by analyzing the Japanese cases reported in the literature. *Arerugi* **49:**1066–1073.

Heine, R. G., S. Elsayed, C. S. Hosking, and D. J. Hill. 2002. Cow's milk allergy in infancy. *Curr. Opin. Allergy Clin. Immunol.* **2:**217–225.

Herian, A. M., S. L. Taylor, and R. K. Bush. 1990. Identification of soybean allergens by immunoblotting with sera from soy-allergic adults. *Int. Arch. Allergy Appl. Immunol.* **92:**193–198.

Hill, D. J., C. S. Hosking, and R. G. Heine. 1999. Clinical spectrum of food allergy in children in Australia and South-East Asia: identification and targets for treatment. *Ann. Med.* **31:**272–281.

Hirschwehr, R., R. Valenta, C. Ebner, F. Ferreira, W. R. Sperr, P Valent, M. Robac, H. Rumpold, O. Scheiner, and D. Kraft. 1992. Identification of common allergenic structures in hazel pollen and hazelnuts: a possible explanation for sensitivity to hazelnuts in patients allergic to tree pollen. *J. Allergy Clin. Immunol.* **90:** 927–936.

Hoffenberg, E., T. Mackenzie, K. Barriga, G. S. Eisenbarth, F. Bao, J. E. Haas, H. Erlich, T. I. T. Bugawan, R. J. Sokol, J. M. Norris, and M. Rewers. 2003. A prospective study of the incidence of childhood celiac disease. *J. Pediatr.* **143:**308–314.

Host, A., and S. Halken. 1990. A prospective study of cow milk allergy in Danish infants during the first 3 years of life. Clinical course in relation to clinical and immunological type of hypersensitivity reaction. *Allergy* **45:**587–596.

Hourihane, J. O. 2002. Recent advances in peanut allergy. *Curr. Opin. Allergy Clin. Immunol.* **2:**227–231.

Hourihane, J. O., S. A. Kilburn, P. Dean, and J. O. Warner. 1997a. Clinical characteristics of peanut allergy. *Clin. Exp. Allergy* **27:**634–639.

Hourihane, J. O., S. A. Kilburn, J. A. Nordlee, S. L. Hefle, S. L. Taylor, and J. O. Warner. 1997b. An evaluation of the sensitivity of subjects with peanut allergy to very low doses of peanut protein: a randomized, double-blind, placebo-controlled food challenge study. *J. Allergy Clin. Immunol.* **100:**596–600.

Hourihane, J. O., S. A. Roberts, and J. O. Warner. 1998. Resolution of peanut allergy: case-control study. *Brit. Med. J.* **316:**1271–1275.

Imai, T., and Y. Iikura. 2003. The national survey of immediate type of food allergy. *Arerugi* **52:**1006–1013.

James, J. M., A. W. Burks, P. K. Roberson, and H. A. Sampson. 1995. Safe administration of the measles vaccine to children allergic to eggs. *New Engl. J. Med.* **332:**1262–1266.

James, J. M., R. S. Zeiger, M. R. Lester, M. B. Fasano, J. E. Gern, L. E. Mansfield, H. J. Schwartz, H. A. Sampson, H. H. Windom, S. B. Machtinger, and S. Lensing. 1998. Safe administration of influenza vaccine to patients with egg allergy. *J. Pediatr.* **133:**624–628.

Janatuinen, E. K., T. A. Kemppainen, R. J. Julkunen, V. M. Kosma, M. Maki, M. Heikkinen, and M. I. Uusitupa. 2002. No harm from five year ingestion of oats in celiac disease. *Gut* **50:**332–335.

Jarvinen, K. M., M. Turpeinen, and H. Suomalainen. 2003. Concurrent cereal allergy in children with cow's milk allergy manifested with atopic dermatitis. *Clin. Exp. Allergy* **33:**1060–1066.

Jones, S. M., C. F. Magnolfi, S. K. Cooke, and H. A. Sampson. 1995. Immunologic cross–reactivity among cereal grains and grasses in children with food hypersensitivity. *J. Allergy Clin. Immunol.* **96:**341–351.

Kagi, M., and B. Wuthrich. 1991. Falafel-burger anaphylaxis due to sesame seed allergy. *Lancet* **338:**582.

Kajosaari, M. 1982. Food allergy in Finnish children aged 1 to 6 years. *Acta Paediatr. Scand.* **71:**815–819.

Kanny, G., D. A. Moneret-Vautrin, J. Flabbee, E. Beaudouin, M. Morisset, and F. Thevenin. 2001. Population study of food allergy in France. *J. Allergy Clin. Immunol.* **108:**133–140.

Kanny, G., C. De Hauteclocque, and D. A. Moneret-Vautrin. 1996. Sesame seed and sesame seed oil contain masked allergens of growing importance. *Allergy* **51:**952–957.

Kelly, J. D., and S. L. Hefle. 2000. 2S methionine-rich protein (SSA) from sunflower seed is an IgE-binding protein. *Allergy* **55:**556–560.

Kelso, J. M., G. T. Mootrey, and T. F. Tsai. 1999. Anaphylaxis from yellow fever vaccine. *J. Allergy Clin. Immunol.* **103:**698–701.

Kelso, J. M., L. Bardina, and K. Beyer. 2003. Allergy to canned tuna. *J. Allergy Clin. Immunol.* **111:**901.

Keskinen, H., P. Ostman, et al. 1991. A case of occupational asthma, rhinitis and urticaria due to sesame seed. *Clin. Exp. Allergy* **21:**623–624.

Kleine-Tebbe, J., L. Vogel, et al. 2002. Severe oral allergy syndrome and anaphylactic reactions caused by a Bet v 1-related PR-10 protein in soybean, SAM22. *J. Allergy Clin. Immunol.* **110:**797–804.

Klemola, T., T. Vanto, K. Juntunen-Backman, K. Kalimo, R. Korpela, and E. Varjonen. 2002. Allergy to soy formula and to extensively hydrolyzed whey formula in infants with cow's milk allergy: a prospective, randomized study with a follow-up to the age of 2 years. *J. Pediatr.* **140:**219–224.

Kolopp-Sarda, M. N., D. A. Moneret-Vautrin, B. Gobert, G. Kanny, M. Brodschii, M. C. Bene, and G. C. Faure. 1997. Specific humoral immune responses in 12 cases of food sensitization to sesame seed. *Clin. Exp. Allergy* **27:**1285–1291.

Lehane, L., and J. Olley. 2000. Histamine fish poisoning revisited. *Int. J. Food Microbiol.* **58:**1–37.

Lehrer, S. B., R. Ayuso, and G. Reese. 2002. Current understanding of food allergens. *Ann. N. Y. Acad. Sci.* **964:**69–85.

Lehrer, S. B., R. Ayuso, and G. Reese. 2003. Seafood allergy and allergens: a review. *Mar. Biotechnol.* **5:**339–348.

Leser, C., A. L. Hartman, G. Pramal and B. Wuthrich. 2001. The "egg-egg" syndrome: occupational respiratory allergy to airborne egg proteins with consecutive ingestive egg allergy in the bakery and confectionery industry. *J. Investig. Allergol. Clin. Immunol.* **11:**89–93.

Lundin, K. E. A., E. M. Nilsen, H. G. Scott, E. M. Loberg, A. Gjoen, J. Bratlie, V. Skar, E. Mendez, A. Lovik, and K. Kett. 2003. Oats induced villous atrophy in celiac disease. *Gut* **52:**1649–1652.

Magnolfi, C. F., G. Zani, L. Lacava, M. F. Patria, and M. Bardare. 1996. Soy allergy in atopic children. *Ann. Allergy Asthma Immunol.* **77:**197–201.

Maleki, S. J., S. Y. Chung, E. T. Champagne, and J. P. Raufman. 2000. The effects of roasting on the allergenic properties of peanut proteins. *J. Allergy Clin. Immunol.* **106:**763–768.

Maleki, S. J., O. Viquez, T. Jacks, H. Dodo, E. T. Champagne, S. Y. Chung, and S. J. Landry. 2003. The major peanut allergen, Ara h 2, functions as a trypsin inhibitor, and roasting enhances this function. *J. Allergy Clin. Immunol.* **112:**190–195.

Manzanedo, L., J. Blanco, M. Fuentes, M. L. Caballero, and I. Moneo. 2004. Anaphylactic reaction in a patient sensitized to coriander seed. *Allergy* **59:**362–363.

Menendez-Arias, L., I. Moneo, J. Dominguez, and R. Rodriguez. 1988. Primary structure of the major allergen of yellow mustard (*Sinapsis alba* L.) seed, Sin a 1. *Eur. J. Biochem.* **177:**159–166.

Moneret-Vautrin, D. A., and G. Kanny. 1992. *Les allergies alimentaires.* Elsevier, Paris, France.

Moneret-Vautrin, D. A., C. Aaghassian, et al. 1997. Anaphylaxie "idiopathique" recidivante au sesame. Un allergene alimentaire en expansion. *Rev. Fr. Allergol.* **34:**487–489.

Moneret-Vautrin, D. A., L. Guerin, G. Kanny, J. Flabbee, S. Fremont, and M. Morisset. 1999. Cross-allergenicity of peanut and lupine: the risk of lupine allergy in patients allergic to peanuts. *J. Allergy Clin. Immunol.* **104:**883–888.

Moneret-Vautrin, D. A., G. Kanny, M. Morrisset, F. Rance, M. F. Fardeau, and E. Beaudouin. 2004. Severe food anaphylaxis: 107 cases registered in 2002 by the Allergy Vigilance Network. *Allergy Immunol.* (Paris) **36:**46–51.

Moneret-Vautrin, D. A., G. Kanny, and L. Parisot. 2001. Accidents graves par allergie alimentaire en France: frequence, caracteristiques cliniques et etiologiques. *Rev. Fr. Allergol. Immunol. Clin.* **41:**696–700.

Moneret-Vautrin, D. A., M. Morisset, J. Flabbee, E. Beaudouin, and G. Kanny. 2005. Epidemiology of life-threatening and lethal anaphylaxis: a review. *Allergy* **60:**443–451.

Morcos, S. K., and H. S. Thomsen. 2001. Adverse reactions to iodinated contrast media. *Eur. Radiol.* **11:** 1267–1275.

Morgan, J. E., C. E. O'Neil, C. B. Daul, and S. B. Lehrer. 1989. Species-specific shrimp allergens: RAST and RAST-inhibition studies. *J. Allergy Clin. Immunol.* **83:**1112–1117.

Morgan, J. E., C. B. Daul, and S. B. Lehrer. 1990. Characterization of important shrimp allergens by immunoblot analysis. *J. Allergy Clin. Immunol.* **85:**170.

Morisset, M., D. A. Moneret-Vautrin, G. Kanny, L. Guenard, E. Beaudouin, J. Flabbee, and R. Hatahet. 2003. Thresholds of clinical reactivity to milk, egg, peanut and sesame in immunoglobulin E-dependent allergies: evaluation by double-blind or single-blind placebo-controlled oral challenges. *Clin. Exp. Allergy* **33:**1046–1051.

Moro, G. E., A. Warm, S. Arslanoglu, and V. Miniello. 2002. Management of bovine protein allergy: new perspectives and nutritional aspects. *Ann. Allergy Asthma Immunol.* **89**(Suppl.)**:**91–96.

Mortimer, E. Z. 1961. Anaphylaxis following ingestion of soybean. *J. Pediatr.* **58:**90–92.

Nickel, R., M. Kulig, J. Forster, R. Bergmann, C. P. Bauer, S. Lau, I. Guggenmoos-Holzmann, and U. Wahn. 1997. Sensitization to hen's egg at the age of twelve months is predictive for allergic sensitization to common indoor and outdoor allergens at the age of three years. *J. Allergy Clin. Immunol.* **99:**613–617.

Novembre, E., A. Cianferoni, R. Bernardini, L. Mugnaini, C. Caffarelli, Gl Cavagni, A. Giovane, and A. Vierucci. 1998. Anaphylaxis in children: clinical and allergologic features. *Pediatrics* **101:**E8.

Ortolani, C., B. K. Ballmer-Weber, K. S. Hansen, M. Ispano, B. Wuthrich, C. Bindslev-Jensen, R. Ansaloni, L. Vannucci, V. Pravettoni, J. Scibilia, L. K. Poulsen, and E. A. Pastorello. 2000. Hazelnut allergy: a double-blind, placebo-controlled food challenge multicenter study. *J. Allergy Clin. Immunol.* **105:**577–581.

Osterballe, M., and C. Bindslev-Jensen. 2003. Threshold levels in food challenges and specific IgE in patients with egg allergy: is there a relationship? *J. Allergy Clin. Immunol.* **112:**196–201.

Pajno, G. B., G. Passalacqua, G. Magazzu, G. Barberio, D. Vita, and G. W. Canonica. 2000. Anaphylaxis to sesame. *Allergy* **55:**199–201.

Palm, M., D. A. Moneret-Vautrin, G. Kanny, S. Denery-Papini, and S. Fremont. 1999. Food allergy to egg and soy lecithins. *Allergy* **54:**114–115.

Palosuo, K. 2003. Update on wheat hypersensitivity. *Curr. Opin. Allergy Clin. Immunol.* **3:**205–209.

Palosuo, K., E. Varjonen, O.-M. Kekki, T. Klemola, N. Kalkkinen, H. Alenius, and T. Reunala. 2001. Wheat omega-5 gliadin is a major allergen in children with immediate allergy to ingested wheat. *J. Allergy Clin. Immunol.* **108:**634–638.

Pastorello, E. A., L. Farioli, V. Pravettoni, A. M. Robino, J. Scibilia, D. Fortunato, A. Conti, L. Borgonovo, A. Bengtsson, and C. Ortolani. 2004. Lipid transfer protein and vicilin are important walnut allergens in patients not allergic to pollen. *J. Allergy Clin. Immunol.* **114:**908–914.

Perry, T. T., E. C. Matsui, M. K. Conover-Walker, and R. A. Wood. 2004. Risk of oral food challenges. *J. Allergy Clin. Immunol.* **114:**1164–1168.

Pourpak, Z., M. Mesdaghi, M. Mansouri, A. Kazemnejad, S. B. Toosi, and A. Farhoudi. 2005. Which cereal is a suitable substitute for wheat in children with wheat allergy? *Pediatr. Allergy Immunol.* **16:**262–266.

Pumphrey, R. S. 2000. Lessons for management of anaphylaxis from a study of fatal reactions. *Clin. Exp. Allergy* **30:**1144–1150.

Rance, F., G. Kanny, G. Dutau, and D. A. Moneret-Vautrin. 1999. Food hypersensitivity in children: clinical aspects and distribution of allergens. *Pediatr. Allergy Immunol.* **10:**33–38.

Rance, F., X. Grandmottet, and H. Grandjean. 2005. Prevalence and main characteristics of schoolchildren diagnosed with food allergies in France. *Clin. Exp. Allergy* **35:**167–172.

Ravelli, A. M., P. Tobanelli, S. Volpi, and A. G. Ugazio. 2001. Vomiting and gastric motility in infants with cow's milk allergy. *J. Pediatr. Gastroenterol. Nutr.* **32:**59–64.

Renaud, C., C. Cardiet, and C. Dupont. 1996. Allergy to soy lecithin in a child. *J. Pediatr. Gastroenterol. Nutr.* **22:**328–329.

Robotham, J. M., F. Wang, V. Seamon, S. S. Teuber, S. K. Sathe, H. A. Sampson, K. Beyer, M. Seavy, and K. H. Roux. 2005. Ana o 3, an important cashew nut (Anacardium occidentale L.) allergen of the 2S albumin family. *J. Allergy Clin. Immunol.* **115:**1284–1290.

Rozenfeld, P., G. H. Docena, M. C. Anon, and C. A. Fossati. 2002. Detection and identification of a soy protein component that cross-reacts with caseins from cow's milk. *Clin. Exp. Immunol.* **130:**49–58.

Rubenstein, L. 1950. Sensitivity to sesame seed and sesame oil. *N. Y. State J. Med.* **50:**343.

Saarinen, U. M., and M. Kajosaari. 1980. Does dietary elimination in infancy prevent or only postpone a food allergy? A study of fish and citrus allergy in 375 children. *Lancet* **i:**166–167.

Salcedo, G., A. Diaz-Perales, and R. Sanchez-Monge. 2001. The role of plant panallergens in sensitization to natural rubber latex. *Curr. Opin. Allergy Immunol.* **1:**177–183.

Sampson, H. A. 2004. Update on food allergy. *J. Allergy Clin. Immunol.* **113:**805–819.

Sampson, H. A., and C. C. McCaskill. 1985. Food hypersensitivity and atopic dermatitis. Evaluation of 113 patients. *J. Pediatr.* **107:**669–675.

Sampson, H. A., L. Mendelson, and J. P. Rosen. 1992. Fatal and near-fatal anaphylactic reactions to food in children and adolescents. *N. Engl. J. Med.* **327:**380–384.

Shek, L. P., and B. W. Lee. 1999. Food allergy in children—the Singapore story. *Asian Pac. J. Allergy Immunol.* **17:**203–206.

Shek, L., L. Bardina, R. Castro, H. A. Sampson, and K. Beyer. 2005. Humoral and cellular responses to cow milk proteins in patients with milk-induced IgE-mediated and non-IgE-mediated disorders. *Allergy* **60:**912–919.

Sicherer, S. H. 2001. Clinical implications of cross-reactive food allergens. *J. Allergy Clin. Immunol.* **108:**881–890.

Sicherer, S. H., A. W. Burks, and H. A. Sampson. 1998. Clinical features of acute allergic reactions to peanut and tree nuts in children. *Pediatrics* **102:**e6.

Sicherer, S. H., A. Munoz-Furlong, A. W. Burks, and H. A. Sampson. 1999a. Prevalence of peanut and tree nut allergy in the US determined by a random digit dial telephone survey. *J. Allergy Clin. Immunol.* **103:**559–562.

Sicherer, S. H., T. J. Furlong, J. DeSimone, H. A. Sampson. 1999b. Self-reported allergic reactions to peanut on commercial airliners. *J. Allergy Clin. Immunol.* **103:**186–189.

Sicherer, S. H., E. H. Morrow, and H. A. Sampson. 2000a. Dose-response in double-blind, placebo-controlled oral food challenges in children with atopic dermatitis. *J. Allergy Clin. Immunol.* **105:**582–586.

Sicherer, S. H., H. A. Sampson, and A. W. Burks. 2000b. Peanut and soy allergy: a clinical and therapeutic dilemma. *Allergy.* **55:**515–521.

Sicherer, S. H., S. A. Noone, and A. Munoz-Furlong. 2001a. The impact of childhood food allergy on quality of life. *Ann. Allergy Asthma Immunol.* **87:**461–464.

Sicherer, S. H., T. J. Furlong, A. Munoz-Furlong, A. W. Burks, and H. A. Sampson. 2001b. A voluntary registry for peanut and tree nut allergy: characteristics of the first 5149 registrants. *J. Allergy Clin. Immunol.* **108:** 128–132.

Sicherer, S. H., A. Munoz-Furlong, and H. A. Sampson. 2003. Prevalence of peanut and tree nut allergy in the United States determined by means of a random digit dial telephone survey: a 5 year follow-up study. *J. Allergy Clin. Immunol.* **112:**1203–1207.

Sicherer, S. H., A. Munoz-Furlong, and H. A. Sampson. 2004a. Prevalence of seafood allergy in the United States determined by a random telephone survey. *J. Allergy Clin. Immunol.* **114:**159–165.

Sicherer, S. H., S. Teuber, and the Adverse Reactions to Foods Committee. 2004b. Food allergy: practice paper. *J. Allergy Clin. Immunol.* **114:**1146–1150.

Simonte, S. J., M. Songhui, S. Mofidi, and S. H. Sicherer. 2003. Relevance of casual contact with peanut butter in children with peanut allergy. *J. Allergy Clin. Immunol.* **112:**180–182.

Skolnick, H. S., M. K. Conover-Walker, C. B. Koerner, H. A. Sampson, W. Burks, and R. A. Wood. 2001. The natural history of peanut allergy. *J. Allergy Clin. Immunol.* **107:**367–374.

Solensky, R. 2003. Resolution of fish allergy: a case report. *Ann. Allergy Asthma Immunol.* **91:**411–412.

Sporik, R., and D. Hill. 1996. Allergy to peanut, nuts, and sesame in Australian children. Letter. *Brit. Med. J.* **313:**1477–1478.

Steensma, D. P. 2003. The kiss of death: a severe allergic reaction to a shellfish induced by a good-night kiss. *Mayo Clin. Proc.* **78:**221–222.

Stern, A., and B. Wuthrich. 1998. Non-IgE-mediated anaphylaxis to sesame. *Allergy* **53:**325–326.

Sutherland, M. F., R. E. O'Hehir, D. Czarny, and C. Suphioglu. 1999. Macadamia nut anaphylaxis: demonstration of specific IgE reactivity and partial cross-reactivity with hazelnut. *J. Allergy Clin. Immunol.* **104:**889–890.

Takahashi, Y., S. Ichikawa, Y. Aihara, and S. Yokota. 1998. Buckwheat allergy in 90,000 school children in Yokohama. *Arerugi* **47:**26–33.

Tariq, S. M., M. Stevens, S. Matthews, S. Ridout, R. Twiselton, and D. W. Hide. 1996. Cohort study of peanut and tree nut sensitisation by age of 4 years. *Brit. Med. J.* **313:**514–517.

Tariq, S. M., S. M. Matthews, E. A. Hakim, and S. H. Arshad. 2000. Egg allergy in infancy predicts respiratory allergic disease by 4 years of age. *Pediatr. Allergy Immunol.* **11:**162–167.

Taylor, S. L., J. L. Kabourek, and S. L. Hefle. 2004. Fish allergy: fish and products thereof. *J. Food Sci.* **69:**175–180.

Taylor, S. L., S. L. Hefle, C. Bindslev-Jensen, S. A. Bock, A. W. Burks, L. Christie, D. J. Hill, A. Host, J. O. Hourihane, G. Lack, D. D. Metcalfe, D. A. Moneret-Vautrin, P. A. Vadas, F. Rance, D. J. Skrypec, T. A. Trautman, I. Malmheden Yman, and R. S. Zeiger. 2002. Factors affecting the determination of threshold doses for allergenic foods: how much is too much? *J. Allergy Clin. Immunol.* **109:**24–30.

Teuber, S. S., and W. R. Peterson. 1999. Systemic allergic reaction to coconut (*Cocos nucifera*) in 2 subjects with hypersensitivity to tree nut and demonstration of cross-reactivity to legumin-like seed storage proteins: new coconut and walnut food allergens. *J. Allergy Clin. Immunol.* **103:**1180–1185.

Teuber, S. S., K. C. Jarvis, A. M. Dandekar, W. R. Peterson, and A. A. Ansari. 1999. Cloning and sequencing of a gene encoding a vicilin-like proprotein, Jug r 2, from English walnut kernel (*Juglans regia*): a major food allergen. *J. Allergy Clin. Immunol.* **104:**1311–1320

Teuber, S. S., S. S. Comstock, S. K. Roux, and K. H. Roux. 2003. Tree nut allergy. *Curr. Allergy Asthma Rep.* **3:**54–61.

Vadas, P., Y. Wai, W. Burks, and B. Perelman. 2001. Detection of peanut allergens in breast milk of lactating women. *JAMA* **285:**1746–1748.

Vander Leek, T. K., A. H. Liu, and S. A. Bock. 2000. The natural history of peanut allergy in young children and its association with serum peanut-specific IgE. *J. Pediatr.* **137:**749–755.

van Ree, R., L. Antonicelli, J. H. Akkerdaas, M. S. Garritani, R. C. Aalberse, and F. Bonifazi. 1996. Possible induction of food allergy during mite immunotherapy. *Allergy* **51:**108–113.

Wang, F., J. M. Robotham, S. S. Teuber, P. Tawde, S. K. Sathe, and K. H. Roux. 2002. Ana o 1, a cashew (*Anacardium occidental*) allergen of the vicilin seed storage protein family. *J. Allergy Clin. Immunol.* **110:** 160–166.

Waring, N. P., C. B. Daul, R. D. deShazo, M. L. McCants, and S. B. Lehrer. 1985. Hypersensitivity reactions to ingested crustacean: clinical evaluation and diagnostic studies in shrimp-sensitive individuals. *J. Allergy Clin. Immunol.* **76:**440–445.

Wensing, M., A. H. Penninks, S. L. Hefle, S. J. Koppelman, C. A. Bruijnzeel-Koomen, and A. C. Knulst. 2002a. The distribution of individual threshold doses eliciting allergic reactions in a population with peanut allergy. *J. Allergy Clin. Immunol.* **110:**915–920.

Wensing, M., A. H. Penninks, S. L. Hefle, J. H. Akkerdaas, R. van Ree, S. J. Koppelman, C. A. F. M. Bruijnzeel-Koomen, and A. C. Knulst. 2002b. The range of minimum provoking doses in hazelnut-allergic patients as determined by double-blind, placebo-controlled food challenges. *Clin. Exp. Allergy* **32:**1757–1762.

Werfel, S. J., S. K. Cooke, and H. A. Sampson. 1997. Clinical reactivity to beef in children allergic to cow's milk. *J. Allergy Clin. Immunol.* **99:**293–300.

Woods, R. K., M. Abramson, M. Bailey, and E. H. Walters on behalf of the European Community Respiratory Health Survey (ECRHS). 2001. International prevalences of reported food allergies and intolerances. Comparisons arising from the European Community Respiratory Health Survey (ECRHS) 1991–1994. *Eur. J. Clin. Nutr.* **55:**298–304.

Yocum, M., J. Butterfield, J. S. Klein, G. W. Volvheck, D. R. Schroeder, and M. D. Silverstein. 1999. Epidemiology of anaphylaxis in Olmsted County: a population-based study. *J. Allergy Clin. Immunol.* **104:** 452–456.

Young, E., M. D. Stoneham, A. Petruckevitch, J. Barton, and R. Rona. 1994. A population study of food intolerance. *Lancet* **343:**1127–1130.

Yunginger, J. W., D. R. Nelson, D.L. Squillace, R. T. Jones, K. E. Holley, B. A. Hyma, L. Biedrzycki, K. G. Sweeney, W. Q. Sturner, and L. B. Schwartz. 1991. Laboratory investigation of deaths due to anaphylaxis. *J. Forensic Sci.* **36:**857–865.

Yunginger, J. W., K. G. Sweeney, W. Q. Sturner, L. A. Giannandrea, J. D. Teigland, M. Bray, P. A. Benson, J. A. York, L. Biedrzycki, D. L. Squillace, et al. 1988. Fatal food-induced anaphylaxis. *JAMA* **260:** 1450–1452.

Zeiger, R. S. 2002. Current issues with influenza vaccination in egg allergy. *J. Allergy Clin. Immunol.* **110:** 834–840.

Zeiger, R. S., H. A. Sampson, S.A. Bock, A. W. Burks, K. Harden, S. Noone, D. Martin, S. Leung, and G. Wilson. 1999. Soy allergy in infants and children with IgE-associated cow's milk allergy. *J. Pediatr.* **134:**614–622.

Section II

IMMUNOCHEMICAL ASPECTS

Food Allergy
Edited by S. J. Maleki et al.

Chapter 4

Molecular and Immunological Responses to Food

Victor Turcanu and Gideon Lack

In the gut, the host's internal medium is separated only by a unicellular epithelial layer from a large and diverse microbial population of >400 species of bacteria (Kohler et al., 2003) that reaches up to 10^{12} organisms/g in the colon (Macpherson et al., 2002). The breakdown of this intestinal barrier has dramatic consequences, as seen in the case of necrotizing enterocolitis, in which an initial enterocyte lesion (presumably caused by pathogenic signaling triggered by bacterial compounds) leads to bacterial invasion and subsequently to necrosis of the intestinal tissue, massive influx of bacteria, coagulation pathology, and shock, correlating with a high risk of death (Hackam et al., 2005; Sharma et al., 2005).

However, the gut also encounters a large number of self and non-self antigens that are not dangerous (proteins from exfoliated gut epithelium, secreted antibodies, and, respectively, food antigens and nonpathogenic bacteria). Moreover, the constellation of non-self antigens will be permanently changing, depending upon a new environment, a change of diet, or travel.

Indeed, considering the contemporary increase in long-distance travel, it is easy to imagine an individual who takes a holiday to a faraway place and, once arriving, has a meal comprising mainly local foods. He will thus be suddenly exposed to amounts, in milligrams to grams, of several food proteins he has never encountered before. Moreover, by the end of the meal, he has also ingested a fair number of indigenous microorganisms he has never encountered before; most of these microorganisms will be nonpathogenic, but among them there may be a few pathogenic bacteria. Nevertheless, his gastrointestinal system (the digestive epithelia, the mucosal structures, the underlying gut-associated lymphoid tissue [GALT], and other components) will be able to generate a noninflammatory immune response toward the novel food and nonpathogenic bacterial antigens. Such a response will occur simultaneously with a protective immune response against the pathogens, but the overall outcome will not usually result in tissue damage. Should this outcome be different, remote holiday destinations and the exploration of exotic cuisines would be

Victor Turcanu • King's College London, Department of Asthma, Allergy & Respiratory Science, 5th Floor, Thomas Guy House, Guy's Hospital, London SE1 9RT, United Kingdom. ***Gideon Lack*** • King's College London, Paediatric Allergy Research Department, St. Thomas's Hospital, Lambeth Palace Road, London SE1 7EH, United Kingdom.

much less attractive than they actually are. Thus, the four main tasks that the gastrointestinal tract and its associated lymphoid tissue structures must simultaneously accomplish are the following.

1. Digestion and uptake of nutrients (mainly non-self proteins, sugars, lipids, etc.)
2. Avoidance of tissue-damaging immune responses to food antigens, despite their non-self immunogenic epitopes (which would normally elicit a response if injected, for example) and the microbial adjuvant-rich gut microenvironment (endotoxin, bacterial nucleic acids, etc.)
3. Avoidance of tissue-damaging (inflammatory) immune responses to commensal bacteria, despite their possession of non-self immunogenic epitopes and of adjuvants (danger signals, such as lipopolysaccharide [LPS]) that are normal components of the commensals
4. Generation of protective immune responses to pathogens to avoid being invaded by them

Excessive immune responses to commensals or to food antigens will result in autoimmunity or food allergy. Conversely, insufficient responses to pathogens might allow life-threatening infections. The complexity of the processes required to generate an appropriately balanced response is further compounded by the variability of food and microbial antigens.

Despite the progress in knowledge regarding the interactions between food antigens and the gut, unanswered questions remain, such as the following.

- How does the gut immune system maintain tolerance to food antigens that are mixed with commensal bacteria (containing LPS, cytosine-phosphate-guanosine [CpG], and other proinflammatory adjuvants) while remaining able to respond against pathogenic bacteria (containing similar molecules)?
- Why does an individual rarely become allergic to more than one food while maintaining oral tolerance of all other foods?
- What makes a particular food protein more (or less) allergenic than others, and why are some food allergies usually outgrown while other allergies persist?
- Could total dietary exclusion (or, conversely, early exposure) prevent food allergy?
- How can therapeutic oral tolerance be induced with an active, ongoing immune response (such as in allergy or autoimmune diseases)?

In the present review, we shall address these questions by bringing together recent findings on the ways in which different components of the gut tract respond to foods, commensals, and pathogens in a systematic overview. We propose that the outcome of an immune response directed to a non-self food antigen results from the integration of several responses achieved in multiple decision-making units that interact with each other.

Several cells that are localized in a circumscribed microenvironment and interact directly with a food can be grouped in such a unit. The outcome (pro- or noninflammatory response) resulting from the intraunit processes is further signaled to the adjacent units through multiple messengers (cytokines, chemokines, mobile cells, or cell fragments). We propose (Fig. 1) that the main such decision units are represented by the following structures.

1. The interface unit (Fig. 2). This unit consists of gastrointestinal epithelial cells (EC), some highly specialized epithelia (e.g., Paneth cells), and the associated, interacting intraepithelial lymphocytes (IEL; mainly T cells) above the basal membrane. EC bind microbes through specific receptors and react by producing pro- or antiinflammatory messengers and by expressing major histocompatibility complex class II (MHC-II) to interact with the IEL. In parallel, food antigen uptake is followed by digestion, eventual secretion of cytokines and chemokines, and/or presentation to IEL, which leads to further cellular interactions. IEL in their turn act upon the EC and, when activated, secrete cytokines or chemokines and may even induce EC apoptosis.
2. The lamina propria (LP) unit. The LP unit is located between the basal membrane and muscularis mucosae; it contains macrophages (Mf), associated LP lymphocytes (LPL), dendritic cells (DC), and other immune cells (mast cells, plasma cells, eosinopils, neutrophils, etc.). Mf (whose cytokine production was shown to follow M1/M2 polarization patterns analogous to those of Th1/Th2 (Mantovani et al., 2002) capture food antigens transferred to the LP through EC or M cells and present them to the T cells among LPL, leading to activation and cytokine-chemokine production. These messengers ensure both upstream and downstream integration as they act upon EC and, respectively, the immature DC (iDC) from the LP. The B cells from LP that predominantly secrete immunoglobulin A (IgA) will support the recognition and capture of antigens by Mf. The iDC from LP become mature DC (mDC) under the influence of the LP microenvironment and interact with T and B cells from LP while en route toward the secondary lymphoid structures.
3. Secondary lymphoid structures. These structures consist of Peyer's patches (PP), lymphoid follicles, and mesenteric lymph nodes (MLN). After the mDC that have captured antigens from the LP enter PP-MLN, they present the respective antigens to naïve and memory T cells that in turn activate B cells and lead to antibody production. Alternately, the highly differentiated gut epithelial M cells allow food antigens to reach the follicles and PP that the M cells directly overlay (Fig. 3). As a consequence of antigen presentation, naïve T cells are activated, subsequently mature, and provide help for antigen-specific B cells. Memory T and B cells leave the PP-MLN via efferent lymphatics and enter the circulation.

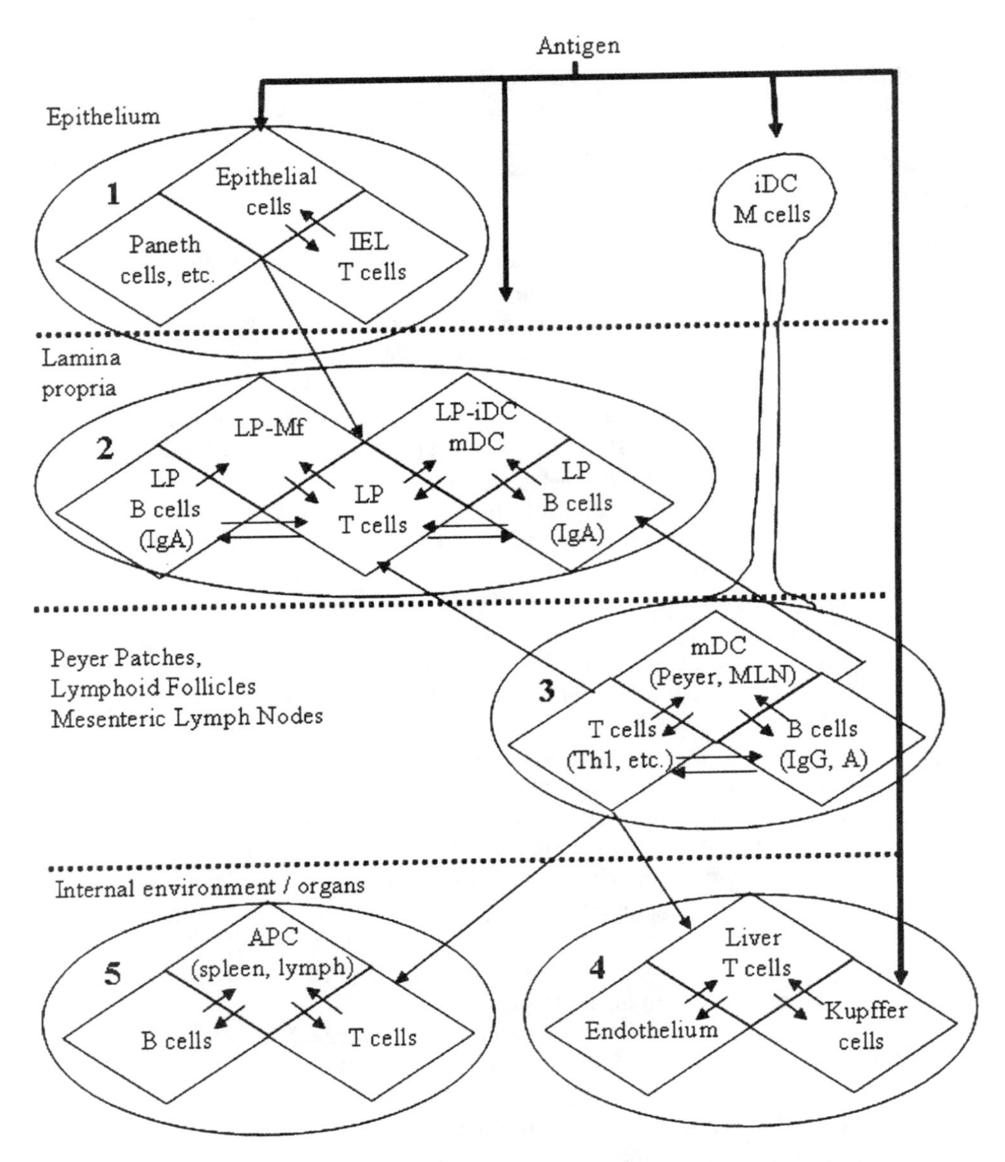

Figure 1. Gastrointestinal immunity decision-making systems. The spatial separation between these units allows cell-to-cell interactions and short-distance messengers to generate a microenvironment that allows independent immune decisions to be taken regarding inflammatory responses against one antigen or another. The integration of these autonomous units is ensured by long-distance messengers (cytokines and chemokines), migrating cells, cell-secreted exosomes, and cell fragments.

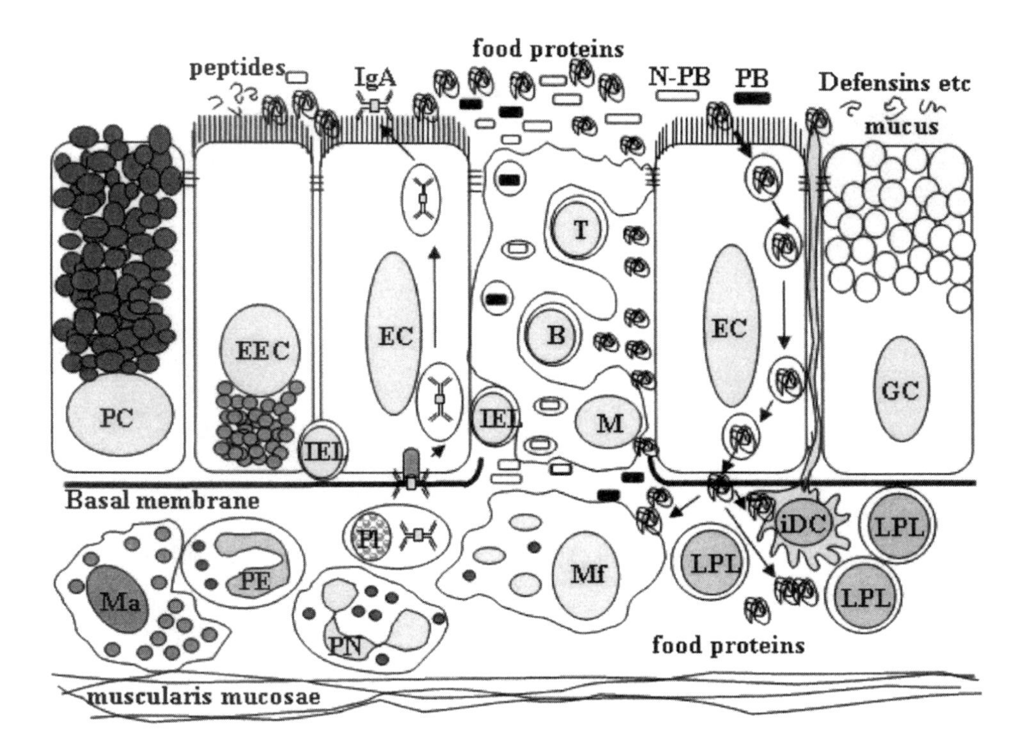

Figure 2. The intestinal mucosa comprises the cells and interstitial structures above the muscularis mucosa layer of smooth muscle cells. The intestinal epithelium comprises mainly EC, goblet cells (GC), M cells, enteroendocrine cells (EEC), Paneth cells (PC), IEL, and other immune cells. The LP is located between the basal membrane and the muscularis mucosae and normally contains Mf, LPL, DC, plasma cells (Pl), mast cells (Ma), eosinophils (PE), neutrophils (PN), etc.

4. The liver, through its immune tolerance-inducing structures, presumably involving the endothelia and the Kupffer cells, represents a major clearing site for IgA-containing antibody complexes (Rifai and Mannik, 1984). Moreover, due to its draining of the portal blood flow that brings intact food antigens, the liver has a key role in establishing immune tolerance to foods. Hence, portal injection of antigens induces tolerance (Wrenshall et al., 2001); transplanted livers from ovalbumin-fed mice transfer ovalbumin-specific tolerance (Li et al., 2004), while, conversely, liver transplantation from a peanut-allergic donor led to peanut allergy in the recipient (Legendre et al., 1997).
5. Other secondary lymphoid structures (the spleen) and other peripheral lymph nodes may also be involved in responses to food antigens, because such antigens can be detected in the circulation as soon as 5 min after ingestion. Local DC presumably capture and present these antigens in a tolerance-prone environment, since intravenous injection of deaggregated proteins induces tolerance (Braley-Mullen and Sharp, 1997).

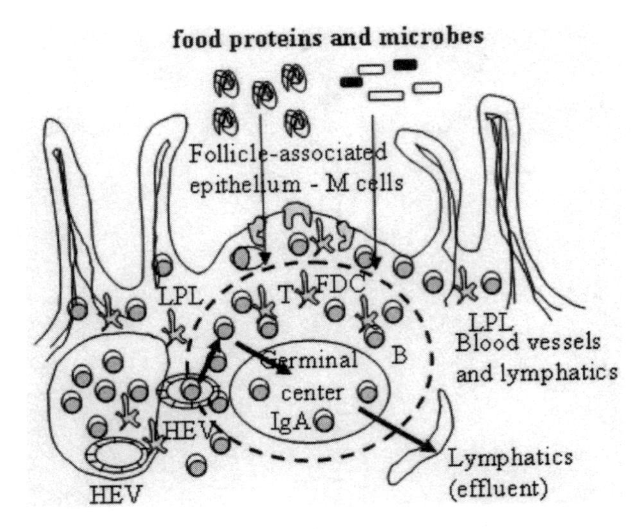

Figure 3. PP are secondary lymphoid organs formed by at least several primary and secondary lymphoid follicles. Food antigens and bacteria reach the PP either directly, through the follicle-associated M cells, or indirectly, carried by dendritic cells. Naïve T and B cells enter PP at the level of the high-endothelium venules and recognize specific antigens presented by the local APCs. Subsequent interaction between T and B cells leads to the formation of a secondary follicle with a germinal center. Memory T and B cells leave PP through the lymphatics.

Envisioning the theoretical concept of separate immune decision-making units (Fig. 4) would explain the paradoxical immune responses observed in the gut, for example, the spatial coexistence of inflammatory and suppressor responses. Thus, iDC are not known to have the capacity of antigen-specific uptake; so after maturation, DC present simultaneously an assortment of self, food, and pathogen-derived epitopes. Should the pathogenic epitopes among them drive an active response, this could lead to self antigen-directed inflammation due to the bystander adjuvant effect (Rudin et al., 2001). Conversely, regulatory T cells (Treg) that recognize self or tolerated antigens should in theory also suppress pathogen-specific responses due to bystander suppression (Weiner, 2001). If multiple immune decision-making units do really exist, T-cell tolerance could be established and maintained in one unit (e.g., the liver), so that other units (e.g., the PP) are allowed to generate active immune responses required for the defense against pathogens.

Another apparent paradox is the coexistence of (i) tolerance to non-self commensal bacteria that express however high levels of immune adjuvant molecules and (ii) immunity toward related pathogens. Moreover, conceiving the GALT and associated structures as a complex of interacting units would be helpful for designing therapeutic interventions by targeting specifically one or another of these units. We shall structure the present review as follows. First, a descriptive section will have a brief outline of the spatially distinct immune decision-making units and their integrative interactions, including discussion of the cellular and molecular structures involved in gastrointestinal responses to foods, commensal bacteria, and pathogens, as grouped in the proposed autonomous structures. This will be followed by an overview of the role of development in determining the

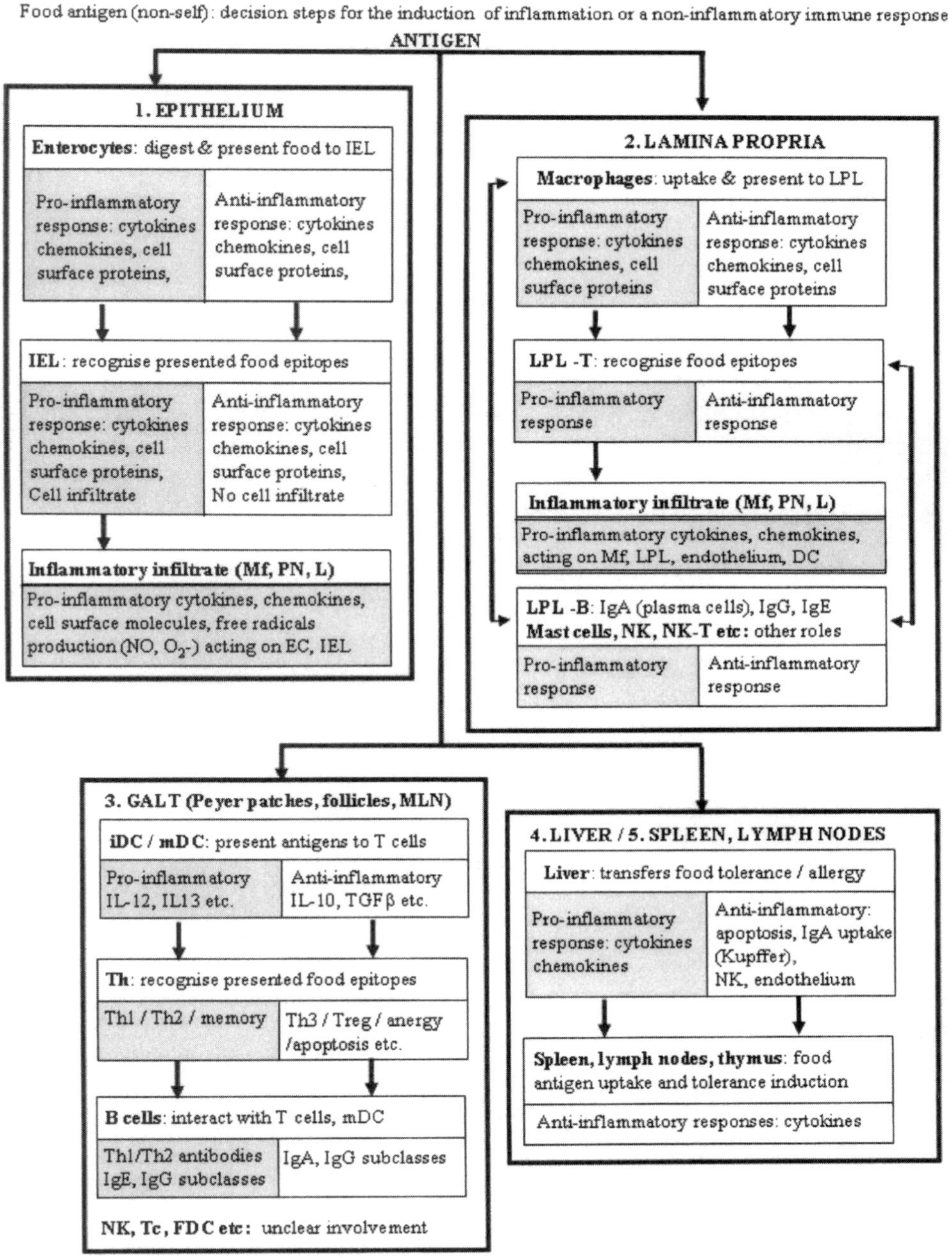

Food antigen (non-self): decision steps for the induction of inflammation or a non-inflammatory immune response

ANTIGEN

1. EPITHELIUM

Enterocytes: digest & present food to IEL

Pro-inflammatory response: cytokines chemokines, cell surface proteins,	Anti-inflammatory response: cytokines chemokines, cell surface proteins,

IEL: recognise presented food epitopes

Pro-inflammatory response: cytokines chemokines, cell surface proteins, Cell infiltrate	Anti-inflammatory response: cytokines chemokines, cell surface proteins, No cell infiltrate

Inflammatory infiltrate (Mf, PN, L)

Pro-inflammatory cytokines, chemokines, cell surface molecules, free radicals production (NO, O_2-) acting on EC, IEL

2. LAMINA PROPRIA

Macrophages: uptake & present to LPL

Pro-inflammatory response: cytokines chemokines, cell surface proteins	Anti-inflammatory response: cytokines chemokines, cell surface proteins

LPL -T: recognise food epitopes

Pro-inflammatory response	Anti-inflammatory response

Inflammatory infiltrate (Mf, PN, L)

Pro-inflammatory cytokines, chemokines, acting on Mf, LPL, endothelium, DC

LPL -B: IgA (plasma cells), IgG, IgE
Mast cells, NK, NK-T etc: other roles

Pro-inflammatory response	Anti-inflammatory response

3. GALT (Peyer patches, follicles, MLN)

iDC / mDC: present antigens to T cells

Pro-inflammatory IL-12, IL13 etc.	Anti-inflammatory IL-10, TGFβ etc.

Th: recognise presented food epitopes

Th1 / Th2 / memory	Th3 / Treg / anergy /apoptosis etc.

B cells: interact with T cells, mDC

Th1/Th2 antibodies IgE, IgG subclasses	IgA, IgG subclasses

NK, Tc, FDC etc: unclear involvement

4. LIVER / 5. SPLEEN, LYMPH NODES

Liver: transfers food tolerance / allergy

Pro-inflammatory response: cytokines chemokines	Anti-inflammatory: apoptosis, IgA uptake (Kupffer), NK, endothelium

Spleen, lymph nodes, thymus: food antigen uptake and tolerance induction

Anti-inflammatory responses: cytokines

Figure 4. Multiple levels of food antigen handling in the gut and associated tissues. Food antigen (non-self) decision steps for the induction of inflammation or a noninflammatory immune response are shown.

outcome of gut responses to foods, since such responses are also influenced by the history of previous exposures to food antigens in utero and after birth. The evolution of the mother-child integrated immunological unit from fetal life to the end of the breast-feeding period will be discussed, as well as characterization of the pathological immune responses to foods that occur in the gut and intestinal autoimmune diseases and food allergies.

Finally, we will describe the preventive and therapeutic proposed uses of oral tolerance and the perspectives and the questions that still remain unanswered.

INTERFACE UNIT

The interface unit comprises IEL and gut epithelia, differentiated in humans into four major cell types: enterocytes, goblet cells, enteroendocrine cells, and Paneth cells.

Enterocytes are continuously produced by the division of epithelial stem cells located at the base of the intestinal crypt, the average life of enterocytes being only 2 to 3 days on the crypt-villus path. Thereafter, enterocytes are shed into the gut lumen, providing a constant supply of self antigens. An elegant study designed to evaluate enzyme expression in the gut found that enterocytes represented 84.2% of the 56.23 million cells collected from 20-cm jejunal segments of healthy volunteers during 2 h (Glaeser et al., 2002). Interestingly, the shed enterocytes were not apoptotic (they could even be used to investigate enterocyte activity).

Morphologically, enterocytes are 20- to 26-μm-high columnar cells that are polarized so that their luminal surface has a striated brush border consisting of up to 3,000 closely packed, 1- to 1.4-μ-long microvilli per cell (Fawcett, 1994). Thin filaments (formed by a core polypeptide and oligosaccharide side chains) radiate from the tips of these microvilli, which intermingle and form a continuous 0.1- to 0.5-μm-thick glycocalyx. The glycocalyx is resistant to mucolytic and proteolytic agents and thus protects the cells; also, pancreatic and intestinal enzymes are adsorbed in its surface and ensure in situ protein digestion (Massey-Harroche, 2000). Thus, enterocyte luminal uptake is represented mainly by amino acids and short peptides; however, intact proteins are also captured and transferred to the basal pole, where they are released in the internal medium (Bland, 1998). As a consequence of this active process, although the enterocytes form an impermeable layer, due to the tight junctions that prevent paracellular transport of macromolecules (van der Wouden et al., 2003), food antigens such as lactoglobulin can be detected in the blood minutes after the ingestion of cow's milk (Husby et al., 1985). Unsurprisingly, food antigens that enter the circulation are excreted in the milk, providing the breast-feeding child with small amounts of the food antigen repertoire ingested by the mother (Hanson et al., 2001; Telemo et al., 1986).

The gut barrier is thus never completely impermeable; in fact, the intestine is permeable to macromolecules in the human fetus and undergoes a progressive decrease in permeability around birth. Gut closure occurs late after birth in ruminants and rodents (typically on the 20th day of life in the rat) because in these species placentas are impermeable to antibodies, and so IgG must be taken up from milk. In the case of humans, some authors consider that closure is complete at birth (Roberton et al., 1982), while others suggest that the small intestine is more permeable to food antigens at birth than at later time points (Eastham et al., 1978). The timing of gut closure in newborn children in relation to breast-feeding has been investigated by successive measurements of serum IgA concentration because the gut is impermeable to IgA only when fully closed (Vukavic, 1984). It was found that with neonates in whom breast-feeding was started in the first 24 h of life, serum IgA levels decreased until the third day after birth. Conversely, if breast-feeding was started after the first 24 h of life, serum IgA concentrations increased during the first days, suggesting that the gut remained more permeable. The pathological significance of increased gut permeability, especially with respect to food antigen sensitization, remains unclear. Gut permeability can be easily investigated using large-pore probes (lactullose and cellobiose), small-pore probes (<0.4 nm; D-mannitol and L-rhamnose), and specific probes (D-xylose, polyethylene glycol, etc.); it has been shown that gut permeability is increased in pathological conditions such as celiac sprue, infections, and inflammatory bowel diseases (IBDs) (DeMeo et al., 2002). The situation of food allergy associated or not with atopic eczema is, however, controversial; thus, intestinal permeability was found to be normal by some authors (Du Mont et al., 1984) but increased by others (Andre et al., 1987), specifically to polyethylene glycol (Jackson et al., 1981) or after a food challenge (Dupont et al., 1989). Given that increases in gastrointestinal permeability may be repeatedly induced by gut infections at any age and would not be selective for one food antigen or another, while food allergies are more prevalent only during the first years of life and are often limited to one food, it appears unlikely that sudden increases in the intestinal absorption of orally ingested antigens may trigger food allergies by themselves.

Enterocytes have important immunological functions: they express MHC-I and MHC-II, although their role in antigen presentation is unclear (Bland and Warren, 1986a, 1986b; Campbell et al., 1999). Moreover, under the regulation of stress molecules such as hsp110, enterocytes express CD1d, an invariable, MHC-I-related antigen whose ligation induces interleukin 10 (IL-10) secretion (Colgan et al., 1999) and which is involved in the presentation of certain conserved microbial antigens directly to the IEL (Colgan et al., 2003). Additionally, enterocytes express the MHC-like molecules MIC-A and MIC-B (Groh et al., 1996) and several Toll receptors (Bambou et al., 2004; Fusunyan et al., 2001), allowing them to recognize bacterial components. Enterocytes also express chemokine receptors (CXCR4, CCR5, CCR6, and CX_3CR1) and when activated produce a vast array of cytokines (such as IL-7, which acts as a growth factor for LP T cells) and chemokines (monocyte chemoattractant protein 1, RANTES, IL-8, ENA-78, interferon [IFN]-inducible protein 10, and macrophage inflammatory protein 3α) (Dwinell et al., 2003; Stadnyk, 2002; Watanabe et al., 1995).

A recently described function of enterocytes is the production of exosomes, small vesicles (30 to 90 nm) that carry adhesion molecules, membrane antigens, and MHC-I and -II involved in antigen presentation (Mallegol et al., 2005; van Niel et al., 2001). Exosomes are also produced by B cells (Raposo et al., 1996) and DC (Schartz et al., 2002). While their function is still unclear, it has been shown that exosomes can activate T cells to stimulate immune responses (Hwang et al., 2003). Moreover, exosome-like vesicles have also been found in human blood; hence, they may represent a pathway used for organ-to-organ communication (Caby et al., 2005). Thus, enterocytes not only can act as local nonprofessional antigen-presenting cells (APCs) but may also influence, through the exosomes they produce, immune responses that occur at remote sites, such as draining lymph nodes (Thery et al., 2002).

Another role of enterocytes is to limit the penetration of soluble antigens or microorganisms into the internal medium. Enterocytes use secretory IgA (sIgA) and, to a lesser extent, secretory IgM to capture such antigens, acting for that reason as an immune response gatekeeper (Brandtzaeg, 2002). Thus, enterocytes expressing sIgA receptors capture IgA (with bound antigens) on their basal side and then transport the IgA through intracellular vesicles and release it on the luminal side.

IEL are nearly all $\alpha\beta^+$ T cells in humans (MacDonald, 2003), in contrast with mice, where >70% of IEL are $\gamma\delta^+$ T cells (Hayday et al., 2001). IEL are the major cellular components that interact with enterocytes in the frame of the immune processes that occur in the interface unit; there are, on average, 10 to 20 IEL for 100 enterocytes in the human small bowel villi (Hayday et al., 2001). In humans, IEL from the small bowel are mainly $CD8^+$, with a few $CD4^+$ and even fewer $CD3^+$ $CD4^-$ $CD8^-$ double-negative cells (MacDonald, 2003). Their surface molecules comprise the gut-specific integrin $\alpha_E\beta_7$ (which binds epithelial E-cadherin) and the memory marker CD45RO (Brandtzaeg et al., 1998). The interaction between IEL and enterocytes has been further outlined by experiments showing that enterocytes are killed by ovalbumin-specific IEL if ovalbumin is expressed in enterocytes injected systemically and if ovalbumin-specific cytotoxic T cells are transferred into mice (Vezys et al., 2000). In this respect, IEL seem to belong to the effector arm of mucosal immunity; usually, naïve T cells encounter antigens and are activated in the PP and the MLN, where they differentiate and then finally migrate into the epithelium (and also into LP areas). Once at this level, IEL remain active; some of them spontaneously produce Th1-type cytokines (IFN-γ; IL-2), regardless of the food allergy status of the respective individual (Perez-Machado et al., 2004). Conversely, IEL survival itself depends upon their close interaction with enterocytes, as demonstrated by the fact that isolated IEL undergo massive apoptosis (Brunner et al., 2001).

The immunological functions of goblet cells, Paneth cells, and the enteroendocrine cells have only recently been discovered, but the multiplicity of their secretion products that are involved in multiple signaling and effector networks reveals their importance. Thus, goblet cells mainly produce the mucus that represents a physical barrier to microorganisms and also (due to its binding properties) forms a depot for secreted IgA, trapping commensals and contributing to the establishment of a protective bacterial microfilm in the gut (Kelly et al., 2004). Additionally, goblet cells secrete other proteins that might have a direct or an indirect effect on the immune responses of the gut; for example, they produce kallikrein, which is involved in intestinal inflammation (Colman et al., 1998; Stadnicki et al., 1998). The enteroendocrine cells form a very diverse population: at least 15 different cell types can be differentiated according to their morphology and secretion products (Hocker and Wiedenmann, 1998). Serotonin produced by these cells has been shown to act as a chemotactic factor for eosinophils (Boehme et al., 2004) and may modulate T-lymphocyte responses (Eugen-Olsen et al., 1997).

Paneth cells are secretory cells clustered in small groups at the base of the intestinal crypts. These cells contain several antimicrobial peptides (such as lysosyme, defensins, cryptidins, type II secretory phospholypase A, and angiogenin 4) in their large granules that are discharged into the intestine and limit bacterial growth (Bevins, 2004).

LP

The LP unit comprises the immune cells from the loose conjunctive tissue localized between the epithelial basal membrane and the smooth muscle cell layer muscularis mucosae.

Apart from fibroblasts, lymphatics, and blood vessels, the LP contains large numbers of immune cells, predominantly Mf, DC, and lymphocytes.

Mf capture food antigens transferred to the LP through enterocytes or M cells and can present them to the T cells among the LPL. These Mf are part of the mononuclear phagocyte system and exhibit a great degree of heterogeneity (their cytokine production was shown to follow M1-M2 polarization patterns analogous to Th1-Th2) (Mantovani et al., 2002). Thus, depending upon their effector molecules, Mf have been classified as M1 (producing large amounts of reactive nitrogen and oxygen free radicals but no polyamines when activated with IFN-γ and LPS) and M2 (that produce polyamines but no free radicals and are activated by IL-4–IL-13 or IL-10) (Mantovani et al., 2004). Different effector molecules may also reflect distinct roles played by these subsets in the modulation of specific immune responses; thus, M1 Mf secrete high levels of tumor necrosis factor alpha (TNF-α), IL-1, IL-12, and IL-23 and low levels of IL-10, while M2 Mf secrete predominantly IL-10 and little or no proinflammatory cytokines such as IL-12, IL-1, IL-6, and TNF.

In addition to these extremely polarized phenotypes, other intermediate Mf subsets have also been described (Mantovani et al., 2004): M2a (induced by IL-4 and IL-13), M2b (induced by immune complexes and LPS), and M2c (induced by IL-10). These subsets can be further identified according to their specific expression of chemokines, chemokine receptors, and other membrane proteins and may play different roles in the defense against bacteria and helminths or in the regulation of allergic and autoimmune reactions.

LP B cells play an important role in gut defense by locally producing the IgA antibodies that play an essential role in gut homeostasis and prevention of bacterial invasion. In the LP, there are around 10^{10} B-cell blasts and plasma cells per meter of gut; 75 to 90% secrete IgA, and the average daily production in humans is around 40 mg/kg of body weight (Brandtzaeg et al., 1999). While some IgA-secreting B cells differentiate in the PP and MLN, recent findings suggest that other LP B cells switch antibody class in the LP, due to direct contact with DC and stromal cells, without any T-cell-derived help (Fagarasan and Honjo, 2003). Such B cells then mature into plasma cells and secrete IgA specific for the bacterial antigens carried over by the activating DC. In the gut, IgA attenuates microorganisms' pathogenicity because it binds and neutralizes them but fails to activate complement and therefore does not promote inflammation.

LP T cells are mostly $CD4^+$ and mainly express the $\alpha\beta$ T-cell receptor in humans, while only 5 to 10% are $\gamma\delta$ T-cell receptor positive (MacDonald, 2003). LP T cells express the integrin $\alpha4\beta7$ that allows homing into the gut, due to its binding to endothelial mucosal addressin cell adhesion molecule 1 (MAdCAM1). $\alpha4\beta7$ expression, as well as the presence of other memory markers, suggests that LP T cells have been activated in the PP or MLN and have then migrated into the gut to encounter their specific antigen or die by apoptosis.

DC from the LP are predominantly iDC. As such, they actively capture antigens; upon maturation (a process that involves a decrease of the ability to take up and process new antigens and an increase of cell surface HLA II class and costimulatory molecules), iDC become mDC and start migrating through the LP toward the secondary lymphoid organs. It seems likely that, both during their residence in the LP and during their migration, DC interact with LPL, inducing immune responses. Interestingly, commensal and pathogenic bacteria induce distinct mDC functional phenotypes. Thus, both the probiotic (i.e., for life) bacterium *Lactobacillus rhamnosus* and the pathogen *Klebsiella pneumoniae* induce DC maturation in vitro, as shown by increased CD83 and CD86 expression. However,

L. rhamnosus-maturated DC produced less IL-12 and did not polarize T cells, while *K. pneumoniae*-maturated DC induced a strong level of Th1 polarization.

Furthermore, it has been shown that mDC may even activate B cells directly without requiring supplementary T-cell help and can induce IgA specific to the commensal bacteria often carried by the DC (Macpherson and Uhr, 2004). This might represent a positive feedback mechanism that would regulate responses to local microorganisms, since human iDC bind sIgA (but not serum IgA) through a carbohydrate-binding receptor and could thus take up IgA-bound antigens (Heystek et al., 2002).

Remarkably, if iDC migrate without differentiating into mDC they induce tolerance instead of immunity toward the antigens they present. Thus, it has been shown that expansion of iDC in vivo enhances oral tolerance induction (Viney et al., 1998), while the injection of iDC pulsed with an antigen leads to specific tolerance, which can further be exploited for therapeutic purposes (Xiao et al., 2003). Even stronger tolerance can be achieved if iDC are preexposed to anti-inflammatory cytokines such as IL-10 and transforming growth factor beta (TGF-β) (Link et al., 2003; Yarilin et al., 2002), which are normally produced in large amounts in the GALT upon exposure to tolerated food antigens (Gonnella et al., 1998).

Notably, a distinct DC population from the gut of rats was shown to carry apoptotic intestinal cells to the MLN (Huang et al., 2000). The existence of this population, which may have a human counterpart, suggests that constitutive antigen presentation by gut DC may play an important role in establishing tolerance to self antigens.

Other cells such as mast cells, eosinophils, neutrophils, and endothelial cells are also involved in the immune responses that occur in the intestinal LP.

Mast cells are involved in intestinal defense as they produce early-response cytokines (such as IL-4 and TNF-α) and chemokines upon interaction with bacteria (Marshall et al., 2003). Mast cells also bind opsonized bacteria through the IgG receptors and complement receptor 3. Bound bacteria are then phagocytosed and killed through oxidative and nonoxydative effector systems (Feger et al., 2002). Moreover, IFN-γ-activated mast cells become potent APCs for the IgE-bound antigens that they internalize using their Fc epsilon receptor (Tkaczyk et al., 1999). Thus, mast cells can amplify immune responses that comprise IgE production, as is the case with allergy and parasite infections. Mast cells may also be involved in the pathogenesis of chronic gut inflammation, since it was shown that they accumulate at sites of focal active gastritis, such as that found with Crohn's disease (CD) (Furusu et al., 2002).

Eosinophil infiltration of LP is minimal in baseline conditions but increases in the case of local Th2-dominant responses, due to local eotaxin production (Hogan et al., 2001; Rothenberg et al., 2001), unlike PP eosinophil infiltration, which is mainly caused by IL-5 (Mishra et al., 2000). With a mouse model of allergy to ovalbumin, it was found that eosinophil infiltration of LP peaked around 6 h after ovalbumin ingestion (Bae et al., 1999; Lee et al., 2004). Complex interactions among mast cells, histamine, and eotaxin were shown to modulate eosinophil infiltration of tissues in different forms of allergy (Menzies-Gow et al., 2004).

Neutrophils infiltrate the gut LP in active inflammatory conditions, for example, *Helicobacter pylori* gastritis (Eck et al., 2000). In this case, there is an interplay between invading *H. pylori* and neutrophils, characterized by the downregulation of neutrophil chemokine receptor CXCR1 and CXCR2 expression (Schmausser et al., 2004). Neutrophil killing of bacteria is crucial for preventing the invasion of deeper intestinal wall layers, but the release of free radicals may also lead to tissue damage. Therefore, the interaction between neutrophils and bacteria must be tightly regulated (Yoshikawa and Naito, 2000).

SECONDARY LYMPHOID STRUCTURES

The secondary lymphoid structures comprise the PP and lymphoid follicles, as well as the separate structures of the MLN. There are differences between them in respect to location and antigen uptake (M cells carry out these processes in the PP and lymphoid follicles but not for the MLN), but overall, both types of structures are designed to warrant rapid antigen presentation to naïve T cells, ensuring their subsequent differentiation to memory T cells and provision of help for the antigen-specific B cells.

The microfold (membranous) cells (M cells) are modified ECs found exclusively in the follicle-associated intestinal epithelium that covers the larger lymphoid follicles and the PP. Characteristically, M cells lack the brush border typical for EC and instead have only a few short microvilli. The basolateral surface of the M cells is deeply invaginated, resulting in pockets that are occupied by lymphocytes (mainly $CD4^+$ T cells and some B cells) or by prolongations of LP Mf and only rarely by neutrophils or plasma cells (Fawcett, 1994; Siebers and Finlay, 1996). The investigation of M-cell biology has been made easier in recent years by the finding that they could be produced in vitro by coculture of human enterocytes (the Caco-2 line) with PP lymphocytes (Kerneis et al., 1997). It has thus been shown that their phagocytic properties allow the translocation of bacteria from the gut lumen into the lymphoid structures. A recently described IgA receptor expressed on M cells (Mantis et al., 2002) also ensures the transport of IgA-bound bacteria into the lymphoid areas. Since IgA binding can neutralize pathogenic epitopes, these bacteria are presumably attenuated, so they may induce an immune response similar to the vaccination with attenuated pathogenic bacterial strains (natural vaccination).

The DC from the PP, the lymphocyte follicles, and the MLN, as well as the Mf and the follicular DC (FDC) from these structures, express HLA II molecules that are necessary to present antigens to the Th cells. In the PP, DC are concentrated in the dome zone that underlies the M cells, suggesting that antigen uptake occurs at that level (Telemo et al., 2003). Several DC subpopulations in the secondary lymphoid organs have been described, possibly achieving different functions. Although the precise differentiation pathway that DC undergo is still unclear, it was shown that in mice myeloid-like ($CD8\alpha^-$), lymphoid-like ($CD8\alpha^+$), and plasmacytoid-like ($B220^+$) DC all derive from a common precursor (Ardavin, 2003). In mouse PP, $CD11b^+$ $CD8\alpha^-$ DC induce Th2-skewed responses and secrete the chemokine CCL17 that selectively attracts Th2 cells, while $CD11b^-$ $CD8\alpha^+$ and double-negative $CD11b^-$ $CD8\alpha^-$ DC promote Th1 polarization (Niedergang et al., 2004). In human PP, one of the major DC populations is $CD8\alpha^+$ plasmacytoid, secreting IFN-α upon stimulation with CpG motif-containing molecules and inducing Treg (Bilsborough et al., 2003). Regarding DC migration to the MLN, labeling studies done with pigs suggest that MLN DC derive directly from the migrating LP DC and do not have to pass through the PP or the lymphoid follicles first (Bimczok et al., 2005).

T cells found in the PP and MLN can have either a naïve ($CD45RA^+$, L-selectin^{hi}) or a memory phenotype (characterized by the loss of L-selectin and the presence of CD45RO). T cells isolated from PP biopsies obtained from nonallergic children proliferated when stimulated with the cow's milk antigen β-lactoglobulin and had a Th1 phenotype, secreting IFN-γ but little or no IL-4, IL-5, IL-10, or TGF-β (Nagata et al., 2000). This result suggests that oral tolerance may be controlled by multiple antigen-specific subsets, some of which may not require classical suppressor cytokines. This finding was further supported by the fact that the rare food antigen-specific T cells that were found in the blood of nonallergic individuals had a Th1-skewed cytokine secretion phenotype (Turcanu et al., 2003).

The T-cell microenvironment from the PP and MLN is characterized by the production of large amounts of IL-10 and TGF-β in response to food antigens (Gonnella et al., 1998). The importance of these cytokines has been demonstrated by the abrogation of oral tolerance to ovalbumin when neutralizing anti-TGF-β was injected before oral ovalbumin feeding in rats (Lundin et al., 1999a). T cells appear to be the main source of TGF-β in the case of active suppression-mediated oral tolerance: TGF-β-secreting T cells appear in the MLN within 5 to 8 days after oral antigen administration (Lundin et al., 1999a). Other factors such as CC chemokine ligand 2 and its receptor are also involved in establishing oral tolerance, as suggested by the finding that the blockade of these signaling molecules abrogates oral tolerance (DePaolo et al., 2003).

Notably, homing receptors ensure that memory T cells that have first encountered their specific antigen in the PP or MLN are more likely to recirculate into these secondary lymphoid tissues and into gut LP. Thus, the MAdCAM1 that binds to the integrin $\alpha_4\beta_7$ is expressed by venular endothelial cells in the PP and MLN, as well as in the LP, possibly due to induction by LPS or IL-1β and TNF-α, as shown in vitro (Ogawa et al., 2005). The involvement of MAdCAM1 in lymphocyte homing in the gut is further confirmed by the finding that its in vivo blockade decreases lymphocyte infiltration and experimental colitis in mice (Farkas et al., 2005).

B cells and plasma cells are localized in the follicles of PP and MLN. Most PP contain at least five activated B-cell follicles whose germinal centers contain activated B cells that achieved different stages of maturation. Conversely, the primary follicles and the secondary follicle mantle zones contain predominantly naïve, $CD19^+$ $CD20^+$ IgM^+ IgD^+ recirculating B cells. IgA^+ B cells are well represented in PP, their progeny migrating to the LP and producing large amounts of IgA. A kinetic study of T- and B-cell migration in PP and MLS showed that both T and B cells enter the secondary lymphoid organs at the level of the high endothelial venules; B cells migrate through the T-cell area into the follicles, and then some of them return into the T-cell zone. On the contrary, another study showed that T cells remain in the outer zone and rarely migrate into the follicles (Blaschke et al., 1995). Interestingly, T cells that have been tolerized by oral administration of antigens remain able to enter B-cell follicles, following strong restimulation (with antigen in adjuvant), but are unable to provide B-cell help (Smith et al., 2002). This finding suggests that oral tolerance is not caused by defective follicular migration of T cells but rather by their incompetence in providing help for B cells.

In the follicle, B cells circulate through the network formed by the resident FDC that accumulate antigen-antibody complexes that can be presented to such B cells (Park and Choi, 2005). The overall role of FDC in the immune response is still unclear, but it has been recently shown that FDC from human tonsils secrete the B-cell-activating factor of the tumour necrosis factor family, which is an important apoptosis-preventing and differentiation-inducing molecule for B cells (Zhang et al., 2005) and also IL-15, which amplifies follicular proliferation (Park et al., 2004).

The initial B-cell activation event that triggers the formation of a secondary follicle probably occurs at the periphery of the follicles where CD4$^+$ T cells that were previously activated by DC from the extrafollicular area encounter naïve sIgD$^+$ IgM$^+$ CD38$^+$ antigen-specific B cells (Garside et al., 1998). Subsequently, B cells expand in the follicular dark zone (centroblasts), which then somatically hypermutate their Ig-variable region genes and differentiate into sIgD$^-$ IgM$^+$ CD38$^+$ centrocytes (Brandtzaeg et al., 1999). Centrocytes then take up antigen, process it, and present it to CD4$^+$ intrafollicularly or alternately die by apoptosis. Sustained CD40-CD40L interaction (on the B cells and Th cells) then promotes an Ig class switch. Interestingly, specific Ig switch patterns in allergic individuals suggest that it can be strictly sequential (IgM $\rightarrow$ G3 $\rightarrow$ G1 $\rightarrow$ A $\rightarrow$ G2 $\rightarrow$ G4 $\rightarrow$ IgE), partly sequential (IgM $\rightarrow$ G1 $\rightarrow$ IgE), or even nonsequential (IgM $\rightarrow$ IgE and IgM $\rightarrow$ G4) (Niederberger et al., 2002). After the Ig class switch, B cells further differentiate into memory B cells and plasma cells that leave the secondary lymphoid organs.

Recent findings suggest that B cells from the MLN are furthermore able to modulate T-cell function. Indeed, antigen uptake by membrane B-cell receptors (Ig's) is ~10,000-fold-more efficient than the nonspecific antigen capture existing in other APCs. Moreover, activated B cells secrete cytokines (IL-10) and express CD80 and CD86 (Lumsden et al., 2003) that engage costimulatory molecules (such as CD28) and inhibitory ones (such as CTLA-4), allowing B cells to decrease T-cell activation during cognate responses. Thus, in mice, the transfer of mesenteric B cells was shown to inhibit CD4$^+$ T-cell-driven colitis through the induction of a Treg cell subset (Wei et al., 2005).

Gut homing of IgA-secreting B cells and plasma cells that leave PP and MLN is ensured by mucosal homing receptors. Thus, antibody-secreting cells that have been activated by an orally administered vaccine (but not those induced by parenteral vaccination) express the gut-homing integrin $\alpha_4\beta_7$ that binds to MAdCAM1 (Kantele et al., 1997). Gut homing is also ensured by intestinal EC expression of the chemokine CCL25 that binds to the CCR9 receptor on the plasma cells. The role of this chemokine in plasma cell homing is revealed by the findings that CCR9$^{-/-}$ mice have low levels of IgA$^+$ plasma cell migration to the gut and cannot mount a normal IgA response to orally ingested antigens, although the PP and MLN architecture and differentiation of IgA plasma cells appear to be normal (Pabst et al., 2004).

LIVER

The liver has long been known to play an important role in the induction of tolerance, but recent findings suggest that it may also represent a nodal point for the induction of allergies to food antigens. Indeed, portal injection of ovalbumin in mice was shown to decrease subsequent delayed-type hypersensitivity reactions and specific T-cell proliferation (Chen et al., 2001). In the same model, however, portal administration of ovalbumin induced a shift from a Th1- to a Th2-driven immune response, reflected by increased levels of ovalbumin-specific IgG1 and decreased IgG2a antibodies. Moreover, it was found that the CD4$^+$ T-helper cells that support an IgE response to a dietary antigen develop in the liver and when transferred may induce strong IgE responses (Watanabe et al., 2003). With humans, it has also been reported that liver (but not kidney) transplantation transfers peanut allergy from the donor to the recipient (Legendre et al., 1997). Nevertheless, the liver has a crucial role in the establishment of oral tolerance to foods, since the portal inflow brings all the food antigens absorbed in the gut to the liver. Thus, oral tolerance to high-dose orally administered ovalbumin could be transferred by a liver graft (Li et al., 2004). While the mechanism of this effect is unclear, several liver nonparenchymal cells could be involved. In this respect, antibody-mediated deletion of the liver NK-T cells (that express the Fas receptor and may induce T-cell apoptosis) was shown to abrogate oral tolerance induction (Margenthaler et al., 2002). Also, the Kupffer cells that bind and clear IgA immune complexes in a highly effective way (up to 43% for a single passage in mice) through a specific receptor (Rifai and Mannik, 1984) can also present antigens and induce tolerance, since they secrete IL-10 when activated with endotoxin (Knolle et al., 1995) and create an immune suppressive microenvironment (Rai et al., 1997). Finally, liver sinusoidal endothelial cells (LSEC) that can present antigens to T cells may also regulate oral tolerance. Indeed, LSEC APC activity can lead to the deletion of the specific T cells (Limmer et al., 2000) or to their differentiation into Treg cells that produce IL-10 and IL-4 upon restimulation (Knolle et al., 1999b). Oral tolerance induction by the LSEC should be important, considering the fact that intact food antigens reach the circulation within 2 h after ingestion; this hypothesis is further supported by the findings that portosystemic venous shunting decreases oral tolerance and leads to increased production of antibodies to intestinal bacteria (Knolle and Limmer, 2001).

Interestingly, endotoxin (normally present at concentrations of 100 pg to 1 ng/ml in the portal blood) down-regulates LSEC activation of T cells by decreasing antigen processing and the constitutive expression of MHC-II, CD80, and CD86 (Knolle et al., 1999a), so LSEC tolerogenic activity can be modulated by bacterial pathogenic compounds.

All these findings indicate that the liver (being a final filter for all ingested antigens) might act as a final decision-making unit with respect to the induction and maintenance of tolerance to foods. Nevertheless, the multitude of processes and messengers (cellular and molecular signals, EC-derived exosomes, the small amounts of endotoxin normally found in the portal blood, etc.) involved in intestinal immunity versus tolerance decisions suggests that such responses are the outcome of integrated decisions taken at different levels by the decision-making units outlined.

OTHER SECONDARY LYMPHOID STRUCTURES

Other secondary lymphoid structures (the spleen) and other peripheral lymph nodes have also been shown to participate in immune responses to orally ingested antigens. Indeed, 1 week after weaning, β-lactoglobulin from cow's milk could be measured in the serum, with median concentrations of 7 μg/liter per gram of β-lactoglobulin ingested per kg (Kuitunen et al., 1994). Circulating food antigens can obviously be captured and then presented by resident DC from secondary lymphoid organs in a tolerogenic context, leading to the appearance of anergic or Treg cells and contributing to the establishment of specific tolerance.

In this respect, it has been shown that immune tolerance to nickel (induced by oral administration of nickel chloride) can be transferred from orally tolerized to naïve mice with as few as 100 purified splenic T cells, presumably containing Treg lymphocyes (Artik et al., 2001). Surprisingly, oral tolerance to nickel can also be transferred with 100 T-cell-depleted spleen cells, i.e., mainly APCs (Roelofs-Haarhuis et al., 2003). It seems unlikely that such low cell numbers could carry and transfer enough antigen to induce tolerance in a direct mode. Therefore, considering previous reports suggesting that Treg cells suppress T-cell responses indirectly by acting upon APCs (Taams et al., 2000; Taams et al., 1998), this finding implies that the spleen plays a role in establishing infectious tolerance (a Treg → regulatory APC [APCreg] → Treg circuit) for orally administered antigens.

MOTHER-CHILD INTEGRATED IMMUNOLOGICAL UNIT

The outcome (in terms of immunity or tolerance) of the gut immune responses to foods and microorganisms is determined by the interactions between these non-self antigens and the mother-child integrated immunological unit that occur from the fetal life to the end of the breast-feeding period. During the period, mothers transfer both antigens (food and microorganisms) and immune effectors (antibodies and cells) to their offspring, a Lamarckian (i.e., nongenetic) form of acquired phenotypical characteristic transmission between generations (Lemke et al., 2004), as opposed to the classical information transmission ensured by the transfer of genetic material.

One can distinguish several successive stages during the evolution of the mother-child integrated unit from fetal life to the end of the breast-feeding period: (i) in utero sensitization and maternal immunomodulation of subsequent child responses; (ii) colonization of the newborn at birth, predominantly with microorganisms that colonize the maternal mucosae; (iii) antibody and immune cell transfer by the colostrum and until perinatal gut closure; and (iv) further antibody and antigen transfer during lactation and natural vaccination with antibody-bound and attenuated bacteria until the end of breast-feeding.

In Utero Interactions

In humans, lymphocytes develop during the first trimester, so fetal immune responses can occur after around 20 weeks of gestation (Erbach et al., 1993; Settmacher et al., 1993; Settmacher et al., 1991).

Thus, during intrauterine life, the transplacental transfer of IgG and of environmental antigens that can be found in the blood ensures an early sensitization of the offspring with the antigenic constellation from the maternal milieu (Jones et al., 2002). Normally, this maternal microenvironment contains the same antigenic spectrum that the offspring would encounter after birth, so transplacental immune transfers (sometimes in the form of antigen-antibody complexes) will inform and modulate the postnatal immune responses. Indeed, the existence of antigen-specific memory ($CD45RO^+$) T lymphocytes and high specific proliferation in the cord blood suggests that an immune response to the respective antigens has occurred in utero (Devereux et al., 2001). In this respect, strong immune cord blood responses to food antigens such as cow's milk lactoglobulin have been previously described (Jones et al., 1996; Szepfalusi et al., 1997). Moreover, although maternal daily exposure to aeroallergens such as pollens and dust mite proteins has been estimated at approximately 10 ng/day (Platts-Mills and Woodfolk, 2000), low responses of the cord blood lymphocytes of the offspring to these antigens can still be detected (Szepfalusi et al., 2000). Although the amounts of inhaled allergen that might be transferred to the fetus could only be extremely small, maternal antibodies to these allergens could still play an important role because maternal IgG is able to suppress subsequent IgE responses in the offspring independently of allergen presence (Seeger et al., 1998). It has long been known that antigen-specific maternal IgG antibodies were shown to decrease subsequent IgE responses to the same antigens in the offspring (Jarrett and Hall, 1979). This finding was further investigated with several experimental models; thus, it was determined that in vivo transfer of dinitrophenyl-specific hybridomas (Hagen et al., 1992) or allergen immunization (Fusaro et al., 2002) in mice leads to a decrease of IgE production if the offspring were subsequently challenged with the same antigen. Other authors reached different conclusions, as they found that allergic sensitization and allergic exposure during pregnancy favored the development of atopy rather than tolerance in the offspring (Herz et al., 2001). In humans, the effects of maternal diet during late pregnancy and lactation upon subsequent children's development of atopy is less clear: a recent publication showed that cord blood levels of ovalbumin-specific IgG correlated with maternal levels and that the risk of atopy at 6 months of age was lower in infants with either relatively low or relatively high levels of IgG and higher in those with intermediate IgG levels (Vance et al., 2004).

Microbial Colonization at Birth

The process of microbial colonization of the newborn that occurs during the natural birth process ensures that the microoorganisms that the infant encounters are the same as those that the mother has already encountered and built up mucosal immune responses to, especially IgA. Indeed, previous studies of natural perinatal gut colonization of babies done by bacteriological matching of mother-baby pairs showed that in those infants whose gut flora matched those of their mother, the source of colonizing bacteria was the gut and not the birth canal (Kerr et al., 1976). In any case, the normal infant gut flora predominantly include *Bifidobacterium*-like and *Lactobacillus*-like bacteria; it has been observed though that caesarean section-delivered babies had fewer *Bacteroides fragilis*-like bacteria and an overall delay in gut colonization compared to those babies delivered vaginally (Gronlund et al., 1999). However, gut microflora diversify progressively with age, reaching 400 to 1,000 different species (Mackie et al., 1999; Macpherson and Harris, 2004; Zoetendal et al., 2004). This is why animal experiment data must be treated with a caveat, considering their extremely limited intestinal flora. Thus, most mice used in research are at present limited to the so-called Schaedler flora, established in the 1960s by Russell Schaedler and his colleagues (1965), starting from germ-free mice: a cocktail of *Escherichia coli* var. *mutabilis*, *Streptococcus faecalis*, *Bacteroides distasonis*, *Lactobacillus acidophilus*, *Lactobacillus salivarius*, group N *Streptococcus*, a *Clostridium* species, and a species of extremely oxygen-sensitive spiral-shaped fusiform bacteria (Macpherson and Harris, 2004). *Flexistipes* fusiform bacteria and three other extremely oxygen-sensitive fusiforms were added in 1987 (Dewhirst et al., 1999). Although it may be possible that in different experimental animal facilities, additional bacterial species may find their way to local colonies, this level of variability is far removed from that observed with normal humans.

Perinatal Mother-Child Immune Integrative Processes

Perinatal modulation of the infant's immune responses is most intense immediately after birth, due to the incomplete closure of the gut epithelium and the effects of colostrum. Although it is unclear how maternal lymphocytes in milk may survive ingestion, it appears that they are transferred to the infant gut where they may exert important immunomodulating roles (Hanson et al., 2000). Indeed, autoradiographic studies demonstrated that ingested lymphocytes could rapidly pass across the neonatal gastric mucosa in rats (Seelig and Head, 1987) and in newborn lambs (Tuboly et al., 1995), where they can remain immunologically active and transfer their immune functions. There are multiple mechanisms through which the maternally derived antibodies of the neonate can exert immunomodulatory effects. Thus, the idiotype–anti-idiotype responses were shown to be longer lasting in neonates than in adults and moreover could shape both the T-cell and the B-cell repertoire in the long-term response by inducing idiotype-specific Treg cells (Lemke et al., 2004). Such idiotype-specific Treg cells can completely silence the immune response to certain antigens that may be subsequently encountered by the child and thus represent a form of immnosuppression that is transferred from the mother to the offspring, depending upon the local environmental antigenic milieu. Moreover, antibodies to a capsular antigen of *E. coli* given orally to rats in the neonatal period were shown to enhance the specific immune response for two generations (Lundin et al., 1999b). Thus, the first 3 weeks of life are a window of opportunity for maternal immunomodulatory influences that subsequently shape the offspring immune responses for a long period of time.

Mother-Child Immune Integrative Processes during Breast-Feeding

Breast milk contains a large number of immunomodulatory factors that exert strong short-term but also long-term influences upon offspring; accordingly, breast-feeding is correlated with a subsequent lower incidence of allergy, IBD, autoimmune diabetes mellitus, and even some lymphomas (Garofalo and Goldman, 1999). Immunomodulatory factors found in human milk are prolactin (which auguments B- and T-cell development), IgA (anti-idiotypic IgA modulates subsequent immune responses), lactoferrin, nucleotides, prostaglandin E2, and cytokines (IL-1β, IL-6, IL-10, TNF-α, and IFN-γ) (Goldman et al., 1996). Human milk also contains chemokines (IL-8, GROα, monocyte chemoattractant protein 1, and RANTES), as well as colony-stimulating factors such as granulocyte colony-stimulating factor (CSF), macrophage CSF, and granulocyte-macrophage CSF (Garofalo and Goldman, 1998). Through its components, the general effect of human milk is anti-inflammatory, decreasing the immune responses to microoorganisms and promoting the growth and differentiation of the gut. Other substances that can be found in human milk and also have anti-inflammatory effects include IgE-binding factors (which inhibit IgE production and are antigenically related to CD23), platelet-activating factor–acethylhydrolase (inhibiting platelet-activating factor), protectin (which inhibits complement membrane attack complexes), TGFs and IL-10 (which suppress T-cell and phagocyte activation), antiproteases (α1-antitrypsin and α1-antichymotrypsin), and antioxidants (Garofalo and Goldman, 1999). Conversely, breast milk also contains antibacterial compounds that provide an efficient protection against pathogens colonization and invasion; breast-feeding was thus shown to prevent neonatal necrotizing enterocolitis (Lee and Polin, 2003). In conclusion, breast milk contains a large variety of antimicrobial, anti-inflammatory, and immunomodulating factors that protect the infant and also exert long-term effects upon immune responses to foods and gut microorganisms.

A full characterization of the normal and pathological immune responses to foods that occur in the gut should comprise normal responses to foods and commensal and pathogenic bacteria and the pathological gut responses: food allergies and autoimmune diseases.

NORMAL IMMUNE RESPONSES TO FOOD ANTIGENS

Foods are normally expected to elicit specific immune responses because food antigens are non-self, and their specific T cells are not deleted in the process of central tolerance. Immune responses to foods do indeed occur, but in healthy, nonallergic individuals, such responses are noninflammatory and represent oral tolerance: T-cell proliferation is decreased and the cytokines produced by the rare circulating specific Th cells are predominantly Th1 (IFN-γ; TNF-α) or Treg (IL-10), unlike those in allergic individuals whose responses are predominantly Th2 (Rustemeyer et al., 2004; Turcanu et al., 2003). Thus, in orally tolerant individuals, there is an active ongoing immune response to foods, as demonstrated by the existence of food-specific antibodies. The fact that this immune response is active and depends upon continuous antigen exposure is further supported by the finding that, during the course of an egg-free diet, the level of ovalbumin-specific IgG rapidly decreased in all but those who did not carefully exclude all egg-containing products from their meals (Vance et al., 2004). The development of IgG antibodies to foods has also been closely monitored in infants and children; it was shown that at 1 year of age the predominant ovalbumin- and casein-specific antibodies are IgG1, unlike with adults, where IgG4 predominates (Kemeny et al., 1991). This difference suggests that children develop a specific response, independent of maternal influences; moreover, since IgG production requires T-cell help, these data strongly indicate that food antigen-specific T cells are not completely deleted or anergized but conserve the ability to support B cells. When normal children were followed during their first 8 years of life, IgG1 and IgG3 subclasses to ovalbumin peaked at 18 months and then decreased; overall, higher levels of IgG correlated with IgE and with the incidence of allergic disease (Jenmalm and Bjorksten, 1999). Normal responses to commensal microorganisms permanently occur in the gut, achieving an equilibrium between the organism and the gut flora. The defining characteristics of commensals are their inability to escape being trapped in the mucus and invade EC, as well as their low endotoxicity, since many have pentacylated lipid A in LPS (Sansonetti, 2004). Nevertheless, commensals still express enough danger molecules, so a paradox of gut immunity is the lack of stronger responses directed against them. A possible explanation is that commensals have anti-inflammatory effects; indeed, nonvirulent salmonella was shown to inhibit NF-κB activation in EC (Neish et al., 2000). Another nonpathogenic bacterium often found in the gut, *Bacteroides thetaiotamicron*, also blocks this proinflammatory pathway, removing activated NF-κB complexes and inhibiting the induction of IL-8, TNF-α, and cyclooxygenase 2 induced by pathogenic *Salmonella enteridis* in EC lines (Kelly et al., 2004). The anti-inflammatory mechanism exerted by *B. thetaiotamicron* was based upon the induction of nuclear translocation of peroxysome proliferator activated receptor γ. Subsequently, peroxysome proliferator activated receptor γ enhances the nuclear export of NF-κB RelA protein, inhibiting inflammation (Kelly et al., 2004). The outcome of these interactions between commensal bacteria and the GALT is the emergence of specific Treg cells that inhibit inflammatory immune responses to commensals (Cong et al., 2002). By maintaining the activation state of such Treg (and consequently their secretion of suppressive cytokines such as IL-10 and TGF-β), commensals thus protect against both Th1- and Th2-mediated gut pathology (McGuirk and Mills, 2002).

Normal (defensive) responses to pathological microorganisms occur whenever such pathogens overcome the epithelial barrier and invade the subjacent tissues. In fact, the differentiation between commensal and pathogenic bacteria resides precisely in the pathogens' ability to overcome normal gut epithelial defenses and to invade tissues beyond the host microbial interface (Macpherson and Harris, 2004). Using this functional criterion to define bacterial pathogenicity explains why nonpathogens may become pathogenic upon barrier breakdown or immunodeficiency. Thus, when pathogenic bacteria (such as enteropathogenic *E. coli*, *Salmonella*, or *Shigella*) reach the gut, they attach themselves to EC and trigger a proinflammatory program in these cells by using several receptors such as Toll-like receptors, adhesion molecules, or the endocytosis of microbial products. Alternately, other pathogenic bacteria such as *Yersinia* spp. take advantage of the translocation mechanisms existing in M cells to cross the epithelial barrier, while other bacteria and viruses enter DC and are thus transported beyond the intestinal wall (Sansonetti, 2004). The effect of these pathogenic microorganisms upon the normal immune processes that occur in the gut is dramatic: intestinal parasitism with pinworm terminates self-tolerance and enhances the neonatal induction of autoimmunity in mice (Agersborg et al., 2001), while overall enteric infection inhibits oral tolerance and acts as an adjuvant for responses to food antigens (Shi et al., 2000). Moreover, bacterial products that have an adjuvant effect may abrogate oral tolerance, induce IgE antibody production and allergic sensitization to coadministered protein antigens (Snider et al., 1994), and amplify an ongoing autoimmune response (Xiao and Link, 1997).

Thus, pathology results whenever an adequate balance fails to be established between the environmental components (foods or microorganisms) and the systems that ensure the defense of the gastrointestinal tract. In this respect, the investigation of gastrointestinal diseases exposes the critical points of the immune regulation process where such failures are more likely to occur. Diseases can be seen therefore as natural experiments that reflect a pathological new balance, and their study may provide further insight into the normal mechanisms that underlie gut functions. We will further discuss the pathology caused by excessive Th1 and Th2 responses, i.e., autoimmune and allergic diseases.

AUTOIMMUNE DISEASES OF THE GUT

The major autoimmune diseases of the gut are IBD and celiac disease. IBD covers a spectrum of syndromes from CD to ulcerative colitis and is characterized by the presence of a complex, predominantly $CD4^+$ T-cell chronic inflammatory infiltrate in the gut wall. Because of their permanent activation state, the infiltrating T cells secrete large amounts of proinflammatory cytokines such as IL-15, which enhance tissue inflammation in CD (Monteleone et al., 2002). Conversely, ulcerative colitis pathogenesis is related to the production of large amounts of autoantibodies (anticolon tissue antigens and antineutrophil) that can also be detected in patient sera (Das et al., 1993). Experimental studies have confirmed that both Th1 and Th2 cells can be pathogenic and induce antigen-specific colitis in mice. While the histological characteristics differed depending upon the polarization type, the severity of the disease induced was similar (Iqbal et al., 2002). In any case, IBDs are ultimately determined by the balance between proinflammatory IFN-γ-dominated responses and anti-inflammatory TGF-β responses (Strober et al., 1997). This dichotomy is further emphasized by the fact that patients with IBD fail to develop oral tolerance to an ingested protein such as keyhole limpet hemocyanin, a function dependent upon TGF-β responses that occur in the gut mucosa (Kraus et al., 2004). Another factor that may contribute to the pathogenesis of IBDs is the breakdown of tolerance to the intestinal bacterial flora that also depends upon mucosal production of immunosuppressor cytokines, since tolerance can be restored with IL-10 or antibodies blocking IL-12 (Duchmann et al., 1996). The efficiency of anti-inflammatory and immunosuppressor drugs such as steroids, azathioprin, methotrexate, and anti-TNF-α (infliximab) for treating CD further demonstrates the role of pathological immune responses in this disease. Interestingly, it has been shown with animal models that intestinal helminth infections can prevent or cure CD (Elliott et al., 2005). Considering that helminths induce a strongly polarized Th2, this finding further supports the hypothesis regarding the pathogenic role of Th1 responses and might provide an insight into the causes of increases in IBD prevalence observed in westernized societies.

Celiac disease is also caused by a chronic lymphocyte infiltration in the LP, leading usually to abnormalities in the ECs and complete villus atrophy. Susceptibility to disease is caused by the HLA-DQ2 or DQ8 alleles, which can bind a gluten-derived peptide and present it to local T cells. The ensuing immune response leads to Th1 differentiation of the responding T cells and subsequent local secretion of Th1 cytokines, with characteristic tissue damage or antibodies against gliadin that might cross-react with cerebellar cells and trigger gluten ataxia (Kagnoff, 2005). Because it is caused by a single antigen, however, celiac disease can be reversed by the adoption of a gluten-free diet.

FOOD ALLERGIES

Allergy to foods results from a Th2-driven immune response, leading to the production of specific IgE that is then bound on the surface of mast cells, basophils, phagocytes, B cells, etc., by its specific receptors. Upon cross-linking by allergens, IgE triggers mast cell degranulation, release of histamine, allergic inflammation messengers, and anaphylaxis mediators. Therefore, the presence of IgE is defining for allergy, although non-IgE-mediated types of food intolerance have also been described (e.g., lactose intolerance due to lactase enzyme deficiency). The major clinical symptoms of food allergies are oral allergy syndromes, minor itching, urticaria, rashes, angioedema, bronchospasm, gastrointestinal or even generalized anaphylaxis that may lead to death. Interestingly, low levels of IgE can be found in healthy, nonallergic individuals (Clark and Ewan, 2003; Roberts and Lack, 2005). Indeed, a 95% diagnostic certainty for allergy is reached only if relatively high levels of food antigen-specific IgE are observed: 6 kU/liter for eggs, 32 kU/liter for milk, 15 kU/liter for peanuts, and 20 kU/liter for fish (Sampson and Ho, 1997). Therefore, other immunological factors (possibly inhibitory antibodies) interfere with the IgE triggering of immune reactions.

Food allergies usually develop in the first 2 years of life, reaching their peak prevalence around 1 year of age; most, however, are normally outgrown within a few years, with the notable exceptions of peanut and tree nut allergies (Hourihane et al., 1998). Several prevalence studies that have been performed to characterize the natural history of food allergies, as reviewed by Wood (2003), have demonstrated that almost any food can induce an allergy. Thus, in a Norwegian cohort of 3,623 children observed for 2 years, starting at birth, in whom allergy diagnosis was based on questionnaires, the cumulative food allergy prevalence was 35% at age 2 (Eggesbo et al., 1999). Milk allergy was the most frequent (cumulative prevalence, 11.6%); however, a subsequent diagnosis based on skin testing and oral challenges of 2,721 children from the original cohort found a point prevalence of only 1.1% at 2.5 years of age (Eggesbo et al., 2001a), while the point prevalence of egg allergy at the same age was 1.6% (Eggesbo et al., 2001b). Several studies showed that allergies to fruits (especially citrus fruits and strawberries) and vegetables (tomatoes) were also very frequent, causing, together with milk, almost two-thirds of all reported food reactions (Eggesbo et al., 1999). Peanut and tree nut allergies represent a major cause of concern because their prevalence appears to be rapidly rising, as shown by two sequential birth cohort studies performed with children born in 1989 compared with children born between 1994 and 1996 on the Isle of Wight, United Kingdom (Arshad et al., 2001; Tariq et al., 1996). It was found that peanut sensitization increased threefold (3.3% compared with 1.1%), while peanut allergy doubled (1% versus 0.5%) (Grundy et al., 2002). Moreover, as peanut allergy is outgrown in <20% of cases (Skolnick et al., 2001), peanuts are one of the most frequent causes of food-induced anaphylaxis in adults.

It is unclear at present why some food allergies are outgrown while others are not and, more generally, why some food proteins are potent allergens while other foods rarely cause allergies. The intrinsic allergenicity of some proteins might be related to their higher resistance to proteolysis (preserving allergenic epitopes), increased glycosylation, or specific enzymatic activity (Huby et al., 2000). It has been shown that the inhibition of fish protein digestion with antacids causes fish allergy in an experimental animal model (Untersmayr et al., 2003). Interestingly, the investigation of a cohort of 152 adult gastroenterological patients who were treated with antiulcer drugs that decrease gastric acid secretion showed an increase of preexisting food-specific IgE levels (in 10% of the subjects), as well as the induction of novel IgE responses (15%) (Untersmayr et al., 2005). Enzymatic or chemical activity may also determine higher allergenicity (Hilton et al., 1997). These findings lead to proposals for protein allergenicity prediction and risk assessment, especially in the case of transgenic plant development (Kimber et al., 1999), and for the design of hypoallergenic peanut proteins whose major IgE-binding epitopes are modified or deleted (Bannon et al., 1999). In the case of peanuts, although the allergy increase is mainly correlated with eczema and the use of skin preparations containing peanut oil (Lack et al., 2003), its allergenicity might also be increased due to the processing method. Thus, peanut consumption per capita is similar in China (where it consists mostly of boiled peanuts) and in the United States (where peanuts are mainly dry roasted at 180°C), but there is virtually no peanut allergy in China (Sampson, 2004). This observation led to studies that demonstrated that the Maillard reaction undergone by peanut proteins at high temperatures increased their allergenicity, so that after roasting they became more resistant to gastrointestinal enzyme digestion and bound 90-fold-more IgE from the sera of allergic individuals (Maleki and Hurlburt, 2004; Maleki et al., 2000). Notably, IgE from individuals whose peanut allergy was more severe bound to a higher diversity of peanut protein epitopes (Shreffler et al., 2004).

Nevertheless, the overall increase in food allergy prevalence can hardly be explained exclusively by some sudden change in dietary processing or habits, as it is part of the general increase in allergic diseases observed in recent decades (Bach, 2002). The hygiene hypothesis explains the allergy increase by a decrease in the incidence of common infections during early life, as the result of better hygiene and decreased household size (Strachan, 1989). It was observed that fewer allergies occurred in younger siblings, children who were placed in day care settings during their first year of life, children who lived in a farm environment, children whose mothers worked on a farm during pregnancy, or children who experienced many upper respiratory tract infections (Matricardi and Ronchetti, 2001). The immunological mechanism proposed is that infections induce strong Th1-polarized responses that lead to IL-12 and IFN-γ production that, as a consequence, inhibit Th2 proallergic responses. A special caveat should be considered in this situation for the viruses that may induce Th2 responses to themselves to evade defense mechanisms. For example, the respiratory syncytial virus elicits an IL-5-dependent eosinophilic inflammatory infiltrate and subsequent atopy (Schwarze and Gelfand, 2002; Schwarze et al., 1997), while the influenza virus induces an enhancement of allergic responses due to its action upon lung DC (Dahl et al., 2004). In an apparent paradox, protective effects against allergic diseases have been observed not only with individuals with Th1-inducing infections but also with those with chronic intestinal parasite infections that induce strong Th2 responses (Scrivener et al., 2001; van den Biggelaar et al., 2000). This finding is further supported by the observation that antihelminthic treatment increases atopic skin reactivity to a common allergen, such as house dust mites (van den Biggelaar et al., 2004).

A possible explanation for this paradox is that both Th1 inducers (viruses) and Th2 inducers (intestinal parasites) determine the development of pathogen-specific Treg cells (McGuirk and Mills, 2002) that inhibit allergy presumably through their production of IL-10, TGF-β, and other immunomodulating effects (Maizels and Yazdanbakhsh, 2003; Yazdanbakhsh et al., 2001). The conceptual advantage of this model is that it provides an explanation for the simultaneous increase in prevalence of Th1-mediated autoimmune diseases (such as diabetes mellitus and multiple sclerosis) and Th2-mediated allergic diseases that occurr in parallel with a decrease of Th1-inducing (viruses and tuberculosis) and Th2-inducing (intestinal parasite) infections in westernized countries (Bach, 2002). The immunological basis of this more encompassing version of the hygiene hypothesis, named the old friends hypothesis (Rook and Brunet, 2005; Rook et al., 2004), is that certain classes of nonpathogenic microorganisms (such as lactobacilli, helminths, and saprophytic mycobacteria) that have been present for a long time in the human (agricultural) habitat normally colonize the gut and induce an anti-inflammatory microenvironment dominated by Treg. By interacting with local APCs, gut Treg would turn them into APCreg, for example, by inducing their expression of active indoleamine 2,3-dioxygenase, thus further turning other T-helper cells into Treg. This would lead to the establishment of an inhibitory Treg-APCreg-Treg circuit that would prevent gut inflammation (Rook et al., 2004). As previously mentioned, similar self-perpetuating inhibitory circuits have been proposed to explain the infectious tolerance observed with experimental transplantation (Qin et al., 1993; Wood and Sakaguchi, 2003) and with oral tolerance to nickel (Roelofs-Haarhuis et al., 2003). Although Treg are supposed to mainly act upon the APC (Taams et al., 1998; Taams et al., 2005), direct inhibition of other T cells can also occur (Trzonkowski et al., 2004). Nevertheless, the injection of iDC induces antigen-specific tolerance (Dhodapkar et al., 2001) and the in vivo induction of Treg cells (Dhodapkar and Steinman, 2002). Conversely, Treg cells have been shown to tolerize DC (Chang et al., 2002) and, interestingly, endothelial cells (Cortesini et al., 2004) through the induction of inhibitory cell surface receptors Ig-like transcript 3 and Ig-like transcript 4. Therefore, it seems plausible that tolerogenic (immature) DC, Treg, and possibly other APCs are entangled into a regulatory circuit that achieves an inhibitory feedback loop. Recent progresses in the understanding of this self-sustaining suppressor circuit led to attempts to manipulate DC to induce Treg and thus achieve therapeutic tolerance in diseases such as experimental autoimmune thyroiditis (Verginis et al., 2005).

Regarding Treg, the cells belonging to this presumably heterogeneous population are currently defined mainly by their $CD4^+$ $CD25^+$ $Foxp3^+$ phenotype (Sakaguchi, 2005; Sakaguchi et al., 1995). Several (probably overlapping) Treg subsets have been described; some require cell-to-cell contact, possibly involving CTLA-4 (Liu et al., 2003). Other subsets may rely upon Notch signaling that was shown to transform immunogenic APCs into tolerogenic APCs (Hoyne et al., 2000), while others rely instead upon suppressor cytokines such as IL-10 and TGF-β (O'Garra et al., 2004). As proof of their activity, transferring Treg-depleted $CD4^+$ $CD25^-$ T cells to nude mice leads to autoimmunity (Singh et al., 2001), while the $CD4^+$ $CD25^+$ subset can cure colitis (Mottet et al., 2003).

The role of Treg cells in food allergies is still unclear: it has been reported that $CD4^+$ $CD25^+$ Treg function is not impaired in milk-allergic patients (Tiemessen et al., 2002); on the other hand, children who outgrew their milk allergy showed higher levels of Treg function (Karlsson et al., 2004). One possible explanation for this apparently contradictory result could be the existence of distinct natural and adaptive Treg subsets that differ with respect to their developmental pathways, inductive stimuli, antigen specificity, mechanism of activation, and effector molecules (Bluestone and Abbas, 2003). This unifying concept proposes that tolerance can be sustained by natural Treg cells that are always present and recognize self-antigens and/or an adaptive Treg cell subset differentiated during certain types of immune responses. Thus, the development of allergic and autoimmune diseases could result either from the absence of natural Treg or from the insufficient induction of adaptive Treg. However, a major theoretical difficulty in understanding Treg activity is still unsolved: considering that both pathogenic and nonpathogenic microorganisms contain danger molecules (such as LPS and CpG), where tolerated and immunogenic antigens coexist in that same microenvironment and are presented nonspecifically by the same APC, why do self-antigen and/or tolerated antigen-specific Treg not suppress responses directed to pathogens? Such suppression should theoretically prevent pathogen-specific defensive responses from proceeding. A potential explanation for this paradox is that in a steady, low-level activation state, Treg outnumber and overcome the responsive T cells because they are more active and express the CD25 receptor for IL-2 (Maloy and Powrie, 2001). Conversely, in a high-activation status when mDC stimulated by bacterial products enter the secondary lymph organs, the resulting intense activation leads to Treg proliferation and subsequently to a transient loss of their suppressor activity. This releases the responsive T cells from inhibition and consequently allows responses to pathogenic organisms to occur. Although by postulating the existence of activation-dependent loss of suppressor activity, this hypothesis could explain how immune responses to pathogens can still occur in the presence of Treg, it is still unclear why responses to ingested foods are not triggered during gastrointestinal infections, since in such a situation all potentially responsive T cells would be released from the control of Treg. An interesting variation of Treg activity was found in seasonal allergies such as grass pollen hay fever: $CD25^+$ $CD4^+$ T cells from atopic individuals had a lower allergen-specific effect during the grass pollen season than outside it (Ling et al., 2004). IL-10-secreting Treg can also inhibit allergy responses, although cell-to-cell contact also seems to be important for the suppressor mechanism (Robinson et al., 2004).

The practical applications of these new theoretical insights reside, though, in the use of probiotics for the prevention of allergy, the induction of oral tolerance for the treatment of autoimmune diseases and allergies, and the recent developments in mucosal vaccination.

PROBIOTICS

Probiotics are microbial preparations that have beneficial effects upon the health state of the host, are safe for human use, are stable in acid and bile, and can adhere to the intestinal mucosa; the most frequently used probiotics belong to the *Lactobacillus* and *Bifidobacterium* genera (Isolauri, 2001). Probiotic bacteria belong to the normal commensal flora (possibly old friends) and presumably exert a dual effect, preventing or decreasing intestinal colonization with pathogens through their existence (Arvola et al., 1999), on the other hand interacting with the GALT to prevent inflammatory responses and promote a state of tolerance to themselves and possibly to foods. Probiotic bacteria were used for the treatment of acute diarrhea and for gut inflammatory conditions (Saavedra, 2001); they were indeed shown to induce IL-10-producing Treg by acting upon DC (Smits et al., 2005). Likewise, the probiotic strain *L. rhamnosus* induces DC maturation toward a non-Th1/2-polarizing phenotype (Braat et al., 2004b) and decreases overall T-cell responses to itself (Braat et al., 2004a). A randomized placebo-controlled clinical trial using *L. rhamnosus* GG (ATCC 53103) showed that its perinatal administration to mothers and/or children halved the incidence of atopic eczema in children at high risk during the first 2 years of life (Kalliomaki et al., 2001) and that its protective effect persisted at least until 4 years of age (Kalliomaki et al., 2003).

THERAPEUTIC INDUCTION OF ORAL TOLERANCE

Oral administration of antigens has been shown to induce oral tolerance, a process defined as the loss of certain forms of reactivity to a subsequent classical immunization (i.e., subcutaneous or intraperitoneal, in the presence of adjuvants). It is worth noting that oral tolerance is experimentally assessed as a decrease in in vitro T-cell proliferation and lower-tissue T-cell infiltration (delayed-type hypersensitivity, usually measured as footpad or ear thickening). However, antigen-specific antibody responses are less affected, and since B-cell responses are T-cell dependent, these data suggest that some form of T-cell response must be conserved (otherwise, there would be no antibody production) but that the reacting T cells are skewed toward a different type of response.

At least two distinct oral tolerance forms seem to be possible, depending upon the antigen dose ingested (Faria and Weiner, 1999); low doses induce active suppression maintained by Treg cells that can be transferred to naïve recipients, while high doses of antigen induce specific anergy (Friedman and Weiner, 1994). Moreover, the presence of bacterial products in the gut also influences the establishment of oral tolerance. Thus, oral administration of allergen conjugated to cholera toxin B subunit (but not an admixture with the whole subunit) suppresses IgE responses and leads to oral tolerance (Marinaro et al., 1995; Rask et al., 2000). However, the induction of therapeutic tolerance during active autoimmune diseases by oral administration of antigens has been less successful than expected for the treatment of rheumatoid arthritis (Barnett et al., 1998) and multiple sclerosis (Bielekova et al., 2000; Kappos et al., 2000). The immunological substrate of these failures was explained by the fact that memory T cells may have become partially resistant to normal tolerance-inducing mechanisms and while delayed-type hypersensitivity, proliferation, and IL-5 or IFN-γ responses may be decreased, antibody production would still be maintained (Leishman et al., 1998; Leishman et al., 2000). Moreover, in such situations, oral antigen administration might even amplify existing immune responses (Blanas et al., 1996; Xiao and Link, 1997). Interestingly, oral vaccination against pathogens has received a boost due to the modern possibilities for genetically engineering vaccine production in plants (Mor et al., 1998); it has been recently shown that protection to tetanus could be achieved by a plant vaccine (Tregoning et al., 2005). While doubts still remain regarding the choice between tolerance and priming to induce protective immunity (McSorley and Garside, 1999), it has been shown with a mouse model that tolerization to a single leishmanial antigen prevents the disease (Julia et al., 1996). Recent reports have also outlined the potential manipulation of DC to induce antigen-specific tolerance (Verginis et al., 2005) or immunity (Nakamura et al., 2005).

All the same, sublingual specific immunotherapy (SIT) was shown in double-blind placebo-controlled studies to be efficient in treating allergic rhinoconjunctivitis (Khinchi et al., 2004; La Rosa et al., 1999). Sublingual SIT for grass pollen allergy was shown to increase the levels of specific IgG and IgG4 (Fanta et al., 1999); a similar effect was observed in the case of standard SIT with a subcutaneous allergen injection (Akdis and Blaser, 2000; Nouri-Aria et al., 2004). While the precise therapeutic mechanism is still unclear, it has been shown that IgG4 levels after 5 weeks of SIT are predictive of its long-term effect (Nanda et al., 2004), so it has been proposed that IgG4 acts as a blocking antibody that inhibits IgE-dependent immune responses (Wachholz and Durham, 2004). In conclusion, oral tolerance of allergic diseases (as represented by sublingual SIT) seems to exert therapeutic effects similar to those of classical, subcutaneous SIT.

On the subject of the unanswered questions we previously listed regarding the interactions between food antigens and the gut, we postulate that the gut immune system allows tolerance to food antigens that are mixed with commensal bacteria (containing LPS, CpG, and other proinflammatory adjuvants) by integrating the outcomes of multiple immune decision compartments such as those we proposed. Since several of these compartments (especially the intestinal epithelium and the liver) promote tolerance, this type of response will be dominant. The balance could theoretically be tilted toward immunity by danger-type signals, but these would presumably act at the same time and site as the first contact with the respective food, an unlikely occurrence that might explain why an individual only rarely becomes allergic to more than one food. It is, however, less clear how therapeutic oral tolerance should be induced in the presence of an active, ongoing immune response (such as with allergy or autoimmune diseases), given that ongoing responses in one immune decision unit would modulate subsequent responses in other such units.

In conclusion, the overall gut immune response to foods occurs in the wider context of the dynamic equilibrium between gut microorganisms and the GALT that has been progressively established under the changing influence of the early perinatal and postnatal environments. The requirement that the gut remain permeable for nutrients while being able to protect the internal environment against invasion might have been solved by establishing a multilayered defense barrier in which every unit is able to mount an independent response and at the same time inform and influence the other units so that an integrated reaction can be achieved. Since oral tolerance to new foods can be established at any age, it should therefore be possible to induce oral tolerance for therapeutic purposes in the treatment of autoimmune diseases and allergies if the appropriate immune mechanisms are appropriately targeted.

REFERENCES

Agersborg, S. S., K. M. Garza, and K. S. Tung. 2001. Intestinal parasitism terminates self tolerance and enhances neonatal induction of autoimmune disease and memory. *Eur. J. Immunol.* **31:**851–859.

Akdis, C. A., and K. Blaser. 2000. Mechanisms of allergen-specific immunotherapy. *Allergy* **55:**522–530.

Andre, C., F. Andre, L. Colin, and S. Cavagna. 1987. Measurement of intestinal permeability to mannitol and lactulose as a means of diagnosing food allergy and evaluating therapeutic effectiveness of disodium cromoglycate. *Ann. Allergy* **59:**127–130.

Ardavin, C. 2003. Origin, precursors and differentiation of mouse dendritic cells. *Nat. Rev. Immunol.* **3:** 582–590.

Arshad, S. H., S. M. Tariq, S. Matthews, and E. Hakim. 2001. Sensitization to common allergens and its association with allergic disorders at age 4 years: a whole population birth cohort study. *Pediatrics* **108:**E33.

Artik, S., K. Haarhuis, X. Wu, J. Begerow, and E. Gleichmann. 2001. Tolerance to nickel: oral nickel administration induces a high frequency of anergic T cells with persistent suppressor activity. *J. Immunol.* **167:** 6794–6803.

Arvola, T., K. Laiho, S. Torkkeli, H. Mykkanen, S. Salminen, L. Maunula, and E. Isolauri. 1999. Prophylactic Lactobacillus GG reduces antibiotic-associated diarrhea in children with respiratory infections: a randomized study. *Pediatrics* **104:**e64.

Bach, J. F. 2002. The effect of infections on susceptibility to autoimmune and allergic diseases. *N. Engl. J. Med.* **347:**911–920.

Bae, S. J., Y. Tanaka, J. Hakugawa, and I. Katayama. 1999.Interleukin-5 involvement in ovalbumin-induced eosinophil infiltration in mouse food-allergy model. *J. Dermatol. Sci.* **21:**1–7.

Bambou, J. C., A. Giraud, S. Menard, B. Begue, S. Rakotobe, M. Heyman, F. Taddei, N. Cerf-Bensussan, and V. Gaboriau-Routhiau. 2004. In vitro and ex vivo activation of the TLR5 signaling pathway in intestinal epithelial cells by a commensal Escherichia coli strain. *J. Biol. Chem.* **279:**42984–42992.

Bannon, G. A., D. Shin, S. Maleki, R. Kopper, and A. W. Burks. 1999.Tertiary structure and biophysical properties of a major peanut allergen, implications for the production of a hypoallergenic protein. *Int. Arch. Allergy Immunol.* **118:**315–316.

Barnett, M. L., J. M. Kremer, E. W. St. Clair, D. O. Clegg, D. Furst, M. Weisman, M. J. Fletcher, S. Chasan-Taber, E. Finger, A. Morales, C. H. Le, and D. E. Trentham. 1998. Treatment of rheumatoid arthritis with oral type II collagen. Results of a multicenter, double-blind, placebo-controlled trial. *Arthritis Rheumatol.* **41:**290–297.

Bevins, C. L. 2004. The Paneth cell and the innate immune response. *Curr. Opin. Gastroenterol.* **20:**572–580.

Bielekova, B., B. Goodwin, N. Richert, I. Cortese, T. Kondo, G. Afshar, B. Gran, J. Eaton, J. Antel, J. A. H. F. Frank, McFarland, and R. Martin. 2000. Encephalitogenic potential of the myelin basic protein peptide (amino acids 83–99) in multiple sclerosis: results of a phase II clinical trial with an altered peptide ligand. *Nat. Med.* **6:**1167–1175.

Bilsborough, J., T. C. George, A. Norment, and J. L. Viney. 2003. Mucosal CD8α^+ DC, with a plasmacytoid phenotype, induce differentiation and support function of T cells with regulatory properties. *Immunology* **108:** 481–492.

Bimczok, D., E. N. Sowa, H. Faber-Zuschratter, R. Pabst, and H. J. Rothkotter. 2005. Site-specific expression of CD11b and SIRPα (CD172a) on dendritic cells: implications for their migration patterns in the gut immune system. *Eur. J. Immunol.* **35:**1418–1427.

Blanas, E., F. R. Carbone, J. Allison, J. F. Miller, and W. R. Heath. 1996. Induction of autoimmune diabetes by oral administration of autoantigen. *Science* **274:**1707–1709.

Bland, P. W. 1998. Gut epithelium: food processor for the mucosal immune system? *Gut* **42:**455–456.

Bland, P. W., and L. G. Warren. 1986a. Antigen presentation by epithelial cells of the rat small intestine. I. Kinetics, antigen specificity and blocking by anti-Ia antisera. *Immunology* **58:**1–7.

Bland, P. W., and L. G. Warren. 1986b. Antigen presentation by epithelial cells of the rat small intestine. II. Selective induction of suppressor T cells. *Immunology* **58:**9–14.

Blaschke, V., B. Micheel, R. Pabst, and J. Westermann. 1995. Lymphocyte traffic through lymph nodes and Peyer's patches of the rat: B- and T-cell-specific migration patterns within the tissue, and their dependence on splenic tissue. *Cell Tissue Res.* **282:**377–386.

Bluestone, J. A., and A. K. Abbas. 2003. Natural versus adaptive regulatory T cells. *Nat. Rev. Immunol.* **3:** 253–257.

Boehme, S. A., F. M. Lio, L. Sikora, T. S. Pandit, K. Lavrador, S. P. Rao, and P. Sriramarao. 2004. Cutting edge: serotonin is a chemotactic factor for eosinophils and functions additively with eotaxin. *J. Immunol.* **173:**3599–3603.

Braat, H., J. van den Brande, E. van Tol, D. Hommes, M. Peppelenbosch, and S. van Deventer. 2004a. Lactobacillus rhamnosus induces peripheral hyporesponsiveness in stimulated CD4+ T cells via modulation of dendritic cell function. *Am. J. Clin. Nutr.* **80:**1618–1625.

Braat, H., E. C. de Jong, J. M. van den Brande, M. L. Kapsenberg, M. P. Peppelenbosch, E. A. van Tol, and S. J. van Deventer. 2004b. Dichotomy between Lactobacillus rhamnosus and Klebsiella pneumoniae on dendritic cell phenotype and function. *J. Mol. Med.* **82:**197–205.

Braley-Mullen, H., and G. C. Sharp. 1997. A thyroxine-containing thyroglobulin peptide induces both lymphocytic and granulomatous forms of experimental autoimmune thyroiditis. *J. Autoimmunol.* **10:**531–540.

Brandtzaeg, P. E. 2002. Current understanding of gastrointestinal immunoregulation and its relation to food allergy. *Ann. N. Y. Acad. Sci.* **964:**13–45.

Brandtzaeg, P., I. N. Farstad, and L. Helgeland. 1998. Phenotypes of T cells in the gut. *Chem. Immunol.* **71:**1–26.

Brandtzaeg, P., E. S. Baekkevold, I. N. Farstad, F. L. Jahnsen, F. E. Johansen, E. M. Nilsen, and T. Yamanaka. 1999.Regional specialization in the mucosal immune system: what happens in the microcompartments? *Immunol. Today* **20:**141–151.

Brunner, T., D. Arnold, C. Wasem, S. Herren, and C. Frutschi. 2001. Regulation of cell death and survival in intestinal intraepithelial lymphocytes. *Cell Death Differ.* **8:**706–714.

Caby, M. P., D. Lankar, C. Vincendeau-Scherrer, G. Raposo, and C. Bonnerot. 2005. Exosomal-like vesicles are present in human blood plasma. *Int. Immunol.* **17:**879–887.

Campbell, N., X. Y. Yio, So, L. P., Li, Y., and Mayer, L. 1999.The intestinal epithelial cell: processing and presentation of antigen to the mucosal immune system. *Immunol. Rev.* **172:**315–324.

Chang, C. C., R. Ciubotariu, J. S. Manavalan, J. Yuan, A. I. Colovai, F. Piazza, S. Lederman, M. Colonna, R. Cortesini, R. Dalla-Favera, and N. Suciu-Foca. 2002. Tolerization of dendritic cells by T(S) cells: the crucial role of inhibitory receptors ILT3 and ILT4. *Nat. Immunol.* **3:**237–243.

Chen, Y., C. R. Ong, G. J. McKenna, A. L. Mui, R. M. Smith, and S. W. Chung. 2001. Induction of immune hyporesponsiveness after portal vein immunization with ovalbumin. *Surgery* **129:**66–75.

Clark, A. T., and P. W. Ewan. 2003. Interpretation of tests for nut allergy in one thousand patients, in relation to allergy or tolerance. *Clin. Exp. Allergy* **33:**1041–1045.

Colgan, S. P., R. M. Hershberg, G. T. Furuta, and R. S. Blumberg. 1999. Ligation of intestinal epithelial CD1d induces bioactive IL-10: critical role of the cytoplasmic tail in autocrine signaling. *Proc. Natl. Acad. Sci. USA* **96:**13938–13943.

Colgan, S. P., R. S. Pitman, T. Nagaishi, A. Mizoguchi, E. Mizoguchi, L. F. Mayer, L. Shao, R. B. Sartor, J. R. Subjeck, and R. S. Blumberg. 2003. Intestinal heat shock protein 110 regulates expression of CD1d on intestinal epithelial cells. *J. Clin. Investig.* **112:**745–754.

Colman, R. W., R. B. Sartor, A. A. Adam, R. A. DeLa Cadena, and A. Stadnicki. 1998. The plasma kallikrein-kinin system in sepsis, inflammatory arthritis, and enterocolitis. *Clin. Rev. Allergy Immunol.* **16:**365–384.

Cong, Y., C. T. Weaver, A. Lazenby, and C. O. Elson. 2002. Bacterial-reactive T regulatory cells inhibit pathogenic immune responses to the enteric flora. *J. Immunol.* **169:**6112–6119.

Cortesini, N. S., A. I. Colovai, J. S. Manavalan, S. Galluzzo, A. J. Naiyer, J. Liu, G. Vlad, S. Kim-Schulze, L. Scotto, J. Fan, and R. Cortesini. 2004. Role of regulatory and suppressor T-cells in the induction of ILT3+ ILT4+ tolerogenic endothelial cells in organ allografts. *Transpl. Immunol.* **13:**73–82.

Dahl, M. E., K. Dabbagh, D. Liggitt, S. Kim, and D. B. Lewis. 2004. Viral-induced T helper type 1 responses enhance allergic disease by effects on lung dendritic cells. *Nat. Immunol.* **5:**337–343.

Das, K. M., A. Dasgupta, A. Mandal, and X. Geng. 1993. Autoimmunity to cytoskeletal protein tropomyosin. A clue to the pathogenetic mechanism for ulcerative colitis. *J. Immunol.* **150:**2487–2493.

DeMeo, M. T., E. A. Mutlu, A. Keshavarzian, and M. C. Tobin. 2002. Intestinal permeation and gastrointestinal disease. *J. Clin. Gastroenterol.* **34:**385–396.

DePaolo, R. W., B. J. Rollins, W. Kuziel, and W. J. Karpus. 2003. CC chemokine ligand 2 and its receptor regulate mucosal production of IL-12 and TGF-beta in high dose oral tolerance. *J. Immunol.* **171:**3560–3567.

Devereux, G., A. Seaton, and R. N. Barker. 2001. In utero priming of allergen-specific helper T cells. *Clin. Exp. Allergy* **31:**1686–1695.

Dewhirst, F. E., C. C. Chien, B. J. Paster, R. L. Ericson, R. P. Orcutt, D. B. Schauer, and J. G. Fox. 1999. Phylogeny of the defined murine microbiota: altered Schaedler flora. *Appl. Environ. Microbiol.* **65:** 3287–3292.

Dhodapkar, M. V., and R. M. Steinman. 2002. Antigen-bearing immature dendritic cells induce peptide-specific CD8+ regulatory T cells in vivo in humans. *Blood* **100:**174–177.

Dhodapkar, M. V., R. M. Steinman, J. Krasovsky, C. Munz, and N. Bhardwaj. 2001. Antigen-specific inhibition of effector T cell function in humans after injection of immature dendritic cells. *J. Exp. Med.* **193:** 233–238.

Du, G. C. Mont, R. C. Beach, and I. S. Menzies. 1984. Gastrointestinal permeability in food-allergic eczematous children. *Clin. Allergy* **14:**55–59.

Duchmann, R., E. Schmitt, P. Knolle, K. H. Meyer zum Buschenfelde, and M. Neurath. 1996. Tolerance towards resident intestinal flora in mice is abrogated in experimental colitis and restored by treatment with interleukin-10 or antibodies to interleukin-12. *Eur. J. Immunol.* **26:**934–938.

Dupont, C., E. Barau, P. Molkhou, F. Raynaud, J. P. Barbet, and L. Dehennin. 1989. Food-induced alterations of intestinal permeability in children with cow's milk-sensitive enteropathy and atopic dermatitis. *J. Pediatr. Gastroenterol. Nutr.* **8:**459–465.

Dwinell, M. B., P. A. Johanesen, and J. M. Smith. 2003. Immunobiology of epithelial chemokines in the intestinal mucosa. *Surgery* **133:**601–607.

Eastham, E. J., T. Lichauco, M. I. Grady, and W. A. Walker. 1978. Antigenicity of infant formulas: role of immature intestine on protein permeability. *J. Pediatr.* **93:**561–564.

Eck, M., B. Schmausser, K. Scheller, A. Toksoy, M. Kraus, T. Menzel, H. K. Muller-Hermelink, and R. Gillitzer. 2000. CXC chemokines Groα/IL-8 and IP-10/MIG in *Helicobacter pylori* gastritis. *Clin. Exp. Immunol.* **122:**192–199.

Eggesbo, M., R. Halvorsen, K. Tambs, and G. Botten. 1999. Prevalence of parentally perceived adverse reactions to food in young children. *Pediatr. Allergy Immunol.* **10:**122–132.

Eggesbo, M., G. Botten, R. Halvorsen, and P. Magnus. 2001a. The prevalence of CMA/CMPI in young children: the validity of parentally perceived reactions in a population-based study. *Allergy* **56:**393–402.

Eggesbo, M., G. Botten, R. Halvorsen, and P. Magnus. 2001b. The prevalence of allergy to egg: a population-based study in young children. *Allergy* **56:**403–411.

Elliott, D. E., R. W. Summers, and J. V. Weinstock. 2005. Helminths and the modulation of mucosal inflammation. *Curr. Opin. Gastroenterol.* **21:**51–58.

Erbach, G. T., J. P. Semple, R. Osathanondh, and J. T. Kurnick. 1993. Phenotypic characteristics of lymphoid populations of middle gestation human fetal liver, spleen and thymus. *J. Reprod. Immunol.* **25:**81–88.

Eugen-Olsen, J., P. Afzelius, L. Andresen, J. Iversen, G. Kronborg, P. Aabech, J. O. Nielsen, and B. Hofmann. 1997. Serotonin modulates immune function in T cells from HIV-seropositive subjects. *Clin. Immunol. Immunopathol.* **84:**115–121.

Fagarasan, S., and T. Honjo. 2003. Intestinal IgA synthesis: regulation of front-line body defences. *Nat. Rev. Immunol.* **3:**63–72.

Fanta, C., B. Bohle, W. Hirt, U. Siemann, F. Horak, D. Kraft, H. Ebner, and C. Ebner. 1999. Systemic immunological changes induced by administration of grass pollen allergens via the oral mucosa during sublingual immunotherapy. *Int. Arch. Allergy Immunol.* **120:**218–224.

Faria, A. M., and H. L. Weiner. 1999. Oral tolerance: mechanisms and therapeutic applications. *Adv. Immunol.* **73:**153–264.

Farkas, S., M. Hornung, C. Sattler, K. Edtinger, M. Steinbauer, M. Anthuber, H. J. Schlitt, H. Herfarth, and E. K. Geissler. 2006. Blocking MAdCAM-1 in vivo reduces leukocyte extravasation and reverses chronic inflammation in experimental colitis. *Int. J. Colorectal Dis.* **21:**71–78.

Fawcett, D. W. 1994. *A Textbook of Histology*, 12th ed. Chapman and Hall, New York, N.Y.

Feger, F., S. Varadaradjalou, Z. Gao, S. N. Abraham, and M. Arock. 2002. The role of mast cells in host defense and their subversion by bacterial pathogens. *Trends Immunol.* **23:**151–158.

Friedman, A., and H. L. Weiner. 1994. Induction of anergy or active suppression following oral tolerance is determined by antigen dosage. *Proc. Natl. Acad. Sci. USA* **91:**6688–6692.

Furusu, H., K. Murase, Y. Nishida, H. Isomoto, F. Takeshima, Y. Mizuta, B. R. Hewlett, R. H. Riddell, and S. Kohno. 2002. Accumulation of mast cells and macrophages in focal active gastritis of patients with Crohn's disease. *Hepatogastroenterology* **49:**639–643.

Fusaro, A. E., M. Maciel, J. R. Victor, C. R. Oliveira, A. J. Duarte, and M. N. Sato. 2002. Influence of maternal murine immunization with Dermatophagoides pteronyssinus extract on the type I hypersensitivity response in offspring. *Int. Arch. Allergy Immunol.* **127:**208–216.

Fusunyan, R. D., N. N. Nanthakumar, M. E. Baldeon, and W. A. Walker. 2001. Evidence for an innate immune response in the immature human intestine: toll-like receptors on fetal enterocytes. *Pediatr. Res.* **49:** 589–593.

Garofalo, R. P., and A. S. Goldman. 1998. Cytokines, chemokines, and colony-stimulating factors in human milk: the 1997 update. *Biol. Neonate* **74:**134–142.

Garofalo, R. P., and A. S. Goldman. 1999. Expression of functional immunomodulatory and anti-inflammatory factors in human milk. *Clin. Perinatol.* **26:**361–377.

Garside, P., E. Ingulli, R. R. Merica, J. G. Johnson, R. J. Noelle, and M. K. Jenkins. 1998. Visualization of specific B and T lymphocyte interactions in the lymph node. *Science* **281:**96–99.

Glaeser, H., S. Drescher, H. van der Kuip, C. Behrens, A. Geick, O. Burk, J. Dent, A. Somogyi, O. Von Richter, E. U. Griese, M. Eichelbaum, and M. F. Fromm. 2002. Shed human enterocytes as a tool for the study of expression and function of intestinal drug-metabolizing enzymes and transporters. *Clin. Pharmacol. Ther.* **71:**131–140.

Goldman, A. S., S. Chheda, R. Garofalo, and F. C. Schmalstieg. 1996. Cytokines in human milk: properties and potential effects upon the mammary gland and the neonate. *J. Mammary Gland Biol. Neoplasia* **1:**251–258.

Gonnella, P. A., Y. Chen, J. Inobe, Y. Komagata, M. Quartulli, and H. L. Weiner. 1998. In situ immune response in gut-associated lymphoid tissue (GALT) following oral antigen in TCR-transgenic mice. *J. Immunol.* **160:**4708–4718.

Groh, V., S. Bahram, S. Bauer, A. Herman, M. Beauchamp, and T. Spies. 1996. Cell stress-regulated human major histocompatibility complex class I gene expressed in gastrointestinal epithelium. *Proc. Natl. Acad. Sci. USA* **93:**12445–12450.

Gronlund, M. M., O. P. Lehtonen, E. Eerola, and P. Kero. 1999. Fecal microflora in healthy infants born by different methods of delivery: permanent changes in intestinal flora after cesarean delivery. *J. Pediatr. Gastroenterol. Nutr.* **28:**19–25.

Grundy, J., S. Matthews, B. Bateman, T. Dean, and S. H. Arshad. 2002. Rising prevalence of allergy to peanut in children: data from 2 sequential cohorts. *J. Allergy Clin. Immunol.* **110:**784–789.

Hackam, D. J., J. S. Upperman, A. Grishin, and H. R. Ford. 2005. Disordered enterocyte signaling and intestinal barrier dysfunction in the pathogenesis of necrotizing enterocolitis. *Semin. Pediatr. Surg.* **14:**49–57.

Hagen, M., B. Morrison, D. Robbinson, and G. H. Strejan. 1992. Effect of anti-DNP IgG1- and IgG2a-secreting hybridomas in vivo on the development of an anti-DNP IgE antibody response in mice. *Int. Arch. Allergy Immunol.* **97:**146–153.

Hanson, L., S. A. Silfverdal, L. Stromback, V. Erling, S. Zaman, P. Olcen, and E. Telemo. 2001. The immunological role of breast feeding. *Pediatr. Allergy Immunol.* **12**(Suppl. 14)**:**15–19.

Hanson, L. A., L. Ceafalau, I. Mattsby-Baltzer, M. Lagerberg, A. Hjalmarsson, R. Ashraf, S. Zaman, and F. Jalil. 2000. The mammary gland-infant intestine immunologic dyad. *Adv. Exp. Med. Biol.* **478:**65–76.

Hayday, A., E. Theodoridis, E. Ramsburg, and J. Shires. 2001. Intraepithelial lymphocytes: exploring the third way in immunology. *Nat. Immunol.* **2:**997–1003.

Herz, U., R. Joachim, B. Ahrens, A. Scheffold, A. Radbruch, and H. Renz. 2001. Allergic sensitization and allergen exposure during pregnancy favor the development of atopy in the neonate. *Int. Arch. Allergy Immunol.* **124:**193–196.

Heystek, H. C., C. Moulon, A. M. Woltman, P. Garonne, and C. van Kooten. 2002. Human immature dendritic cells efficiently bind and take up secretory IgA without the induction of maturation. *J. Immunol.* **168:** 102–107.

Hilton, J., R. J. Dearman, N. Sattar, D. A. Basketter, and I. Kimber. 1997. Characteristics of antibody responses induced in mice by protein allergens. *Food Chem. Toxicol.* **35:**1209–1218.

Hocker, M., and B. Wiedenmann. 1998. Molecular mechanisms of enteroendocrine differentiation. *Ann. N. Y. Acad. Sci.* **859:**160–174.

Hogan, S. P., A. Mishra, E. B. Brandt, M. P. Royalty, S. M. Pope, N. Zimmermann, P. S. Foster, and M. E. Rothenberg. 2001. A pathological function for eotaxin and eosinophils in eosinophilic gastrointestinal inflammation. *Nat. Immunol.* **2:**353–360.

Hourihane, J. O., S. A. Roberts, and J. O. Warner. 1998. Resolution of peanut allergy: case-control study. *BMJ* **316:**1271–1275.

Hoyne, G. F., M. J. Dallman, and J. R. Lamb. 2000. T-cell regulation of peripheral tolerance and immunity: the potential role for Notch signalling. *Immunology* **100:**281–288.

Huang, F. P., N. Platt, M. Wykes, J. R. Major, T. J. Powell, C. D. Jenkins, and G. G. MacPherson. 2000. A discrete subpopulation of dendritic cells transports apoptotic intestinal epithelial cells to T cell areas of mesenteric lymph nodes. *J. Exp. Med.* **191:**435–444.

Huby, R. D., R. J. Dearman, and I. Kimber. 2000. Why are some proteins allergens? *Toxicol. Sci.* **55:**235–246.

Husby, S., J. C. Jensenius, and S. E. Svehag. 1985. Passage of undegraded dietary antigen into the blood of healthy adults. Quantification, estimation of size distribution, and relation of uptake to levels of specific antibodies. *Scand. J. Immunol.* **22:**83–92.

Hwang, I., X. Shen, and J. Sprent. 2003. Direct stimulation of naive T cells by membrane vesicles from antigen-presenting cells: distinct roles for CD54 and B7 molecules. *Proc. Natl. Acad. Sci. USA* **100:**6670–6675.

Iqbal, N., J. R. Oliver, F. H. Wagner, A. S. Lazenby, C. O. Elson, and C. T. Weaver. 2002. T helper 1 and T helper 2 cells are pathogenic in an antigen-specific model of colitis. *J. Exp. Med.* **195:**71–84.

Isolauri, E. 2001. Probiotics in human disease. *Am. J. Clin. Nutr.* **73:**1142S–1146S.

Jackson, M. H. Lessof, R. W. Baker, J. Ferrett, and D. M. MacDonald. 1981. Intestinal permeability in patients with eczema and food allergy. *Lancet* **i:**1285–1286.

Jarrett, E., and E. Hall. 1979. Selective suppression of IgE antibody responsiveness by maternal influence. *Nature* **280:**145–147.

Jenmalm, M. C., and B. Bjorksten. 1999. Development of immunoglobulin G subclass antibodies to ovalbumin, birch and cat during the first eight years of life in atopic and non-atopic children. *Pediatr. Allergy Immunol.* **10:**112–121.

Jones, A. C., E. A. Miles, J. O. Warner, B. M. Colwell, T. N. Bryant, and J. A. Warner. 1996. Fetal peripheral blood mononuclear cell proliferative responses to mitogenic and allergenic stimuli during gestation. *Pediatr. Allergy Immunol.* **7:**109–116.

Jones, C. A., J. A. Holloway, and J. O. Warner. 2002. Fetal immune responsiveness and routes of allergic sensitization. *Pediatr. Allergy Immunol.* **13**(Suppl. 15)**:**19–22.

Julia, V., M. Rassoulzadegan, and N. Glaichenhaus. 1996. Resistance to Leishmania major induced by tolerance to a single antigen. *Science* **274**, 421–423.

Kagnoff, M. F. 2005. Overview and pathogenesis of celiac disease. *Gastroenterology* **128:**S10–S18.

Kalliomaki, M., S. Salminen, T. Poussa, H. Arvilommi, and E. Isolauri. 2003. Probiotics and prevention of atopic disease: 4-year follow-up of a randomised placebo-controlled trial. *Lancet* **361:**1869–1871.

Kalliomaki, M., S. Salminen, H. Arvilommi, P. Kero, P. Koskinen, and E. Isolauri. 2001. Probiotics in primary prevention of atopic disease: a randomised placebo-controlled trial. *Lancet* **357:**1076–1079.

Kantele, A., J. M. Kantele, E. Savilahti, M. Westerholm, H. Arvilommi, A. Lazarovits, E. C. Butcher, and P. H. Makela. 1997. Homing potentials of circulating lymphocytes in humans depend on the site of activation: oral, but not parenteral, typhoid vaccination induces circulating antibody-secreting cells that all bear homing receptors directing them to the gut. *J. Immunol.* **158:**574–579.

Kappos, L., G. Comi, H. Panitch, J. Oger, J. Antel, P. Conlon, L. Steinman, et al. 2000. Induction of a non-encephalitogenic type 2 T helper-cell autoimmune response in multiple sclerosis after administration of an altered peptide ligand in a placebo-controlled, randomized phase II trial. *Nat. Med.* **6:**1176–1182.

Karlsson, M. R., J. Rugtveit, and P. Brandtzaeg. 2004. Allergen-responsive CD4+CD25+ regulatory T cells in children who have outgrown cow's milk allergy. *J. Exp. Med.* **199:**1679–1688.

Kelly, D., J. I. Campbell, T. P. King, G. Grant, E. A. Jansson, A. G. Coutts, S. Pettersson, and S. Conway. 2004. Commensal anaerobic gut bacteria attenuate inflammation by regulating nuclear-cytoplasmic shuttling of PPAR-gamma and RelA. *Nat. Immunol.* **5:**104–112.

Kemeny, D. M., J. F. Price, V. Richardson, D. Richards, and M. H. Lessof. 1991. The IgE and IgG subclass antibody response to foods in babies during the first year of life and their relationship to feeding regimen and the development of food allergy. *J. Allergy Clin. Immunol.* **87:**920–929.

Kerneis, S., A. Bogdanova, J. P. Kraehenbuhl, and E. Pringault. 1997. Conversion by Peyer's patch lymphocytes of human enterocytes into M cells that transport bacteria. *Science* **277:**949–952.

Kerr, M. M., J. H. Hutchison, J. MacVicar, J. Givan, and T. A. McAllister. 1976. The natural history of bacterial colonization of the newborn in a maternity hospital (part II). *Scott Med. J.* **21:**111–117.

Khinchi, M. S., L. K. Poulsen, F. Carat, C. Andre, A. B. Hansen, and H. J. Malling. 2004. Clinical efficacy of sublingual and subcutaneous birch pollen allergen-specific immunotherapy: a randomized, placebo-controlled, double-blind, double-dummy study. *Allergy* **59:**45–53.

Kimber, I., N. I. Kerkvliet, S. L. Taylor, J. D. Astwood, K. Sarlo, and R. J. Dearman. 1999.Toxicology of protein allergenicity: prediction and characterization. *Toxicol. Sci.* **48:**157–162.

Knolle, P., J. Schlaak, A. Uhrig, P. Kempf, K. H. Meyer zum Buschenfelde, and G. Gerken. 1995. Human Kupffer cells secrete IL-10 in response to lipopolysaccharide (LPS) challenge. *J. Hepatol.* **22:**226–229.

Knolle, P. A., and A. Limmer. 2001. Neighborhood politics: the immunoregulatory function of organ-resident liver endothelial cells. *Trends Immunol.* **22:**432–437.

Knolle, P. A., T. Germann, U. Treichel, A. Uhrig, E. Schmitt, S. Hegenbarth, A. W. Lohse, and G. Gerken. 1999a. Endotoxin down-regulates T cell activation by antigen-presenting liver sinusoidal endothelial cells. *J. Immunol.* **162:**1401–1407.

Knolle, P. A., E. Schmitt, S. Jin, T. Germann, R. Duchmann, S. Hegenbarth, G. Gerken, and A. W. Lohse. 1999b. Induction of cytokine production in naive CD4(+) T cells by antigen-presenting murine liver sinusoidal endothelial cells but failure to induce differentiation toward Th1 cells. *Gastroenterology* **116:**1428–1440.

Kohler, H., B. A. McCormick, and W. A. Walker. 2003. Bacterial-enterocyte crosstalk: cellular mechanisms in health and disease. *J. Pediatr. Gastroenterol. Nutr.* **36:**175–185.

Kraus, T. A., L. Toy, L. Chan, J. Childs, and L. Mayer. 2004. Failure to induce oral tolerance to a soluble protein in patients with inflammatory bowel disease. *Gastroenterology* **126:**1771–1778.

Kuitunen, M., E. Savilahti, and A. Sarnesto. 1994. Human alpha-lactalbumin and bovine beta-lactoglobulin absorption in infants. *Allergy* **49:**354–360.

Lack, G., D. Fox, K. Northstone, and J. Golding. 2003. Factors associated with the development of peanut allergy in childhood. *N. Engl. J. Med.* **348:**977–985.

La Rosa, M., C. Ranno, C. Andre, F. Carat, M. A. Tosca, and G. W. Canonica. 1999. Double-blind placebo-controlled evaluation of sublingual-swallow immunotherapy with standardized Parietaria judaica extract in children with allergic rhinoconjunctivitis. *J. Allergy Clin. Immunol.* **104:**425–432.

Lee, J. B., T. Matsumoto, Y. O. Shin, H. M. Yang, Y. K. Min, O. Timothy, S. J. Bae, and F. S. Quan. 2004. The role of RANTES in a murine model of food allergy. *Immunol. Investig.* **33:**27–38.

Lee, J. S., and R. A. Polin. 2003. Treatment and prevention of necrotizing enterocolitis. *Semin. Neonatol.* **8:**449–459.

Legendre, C., S. Caillat-Zucman, D. Samuel, S. Morelon, H. Bismuth, J. F. Bach, and H. Kreis. 1997. Transfer of symptomatic peanut allergy to the recipient of a combined liver-and-kidney transplant. *N. Engl. J. Med.* **337:**822–824.

Leishman, A. J., P. Garside, and A. M. Mowat. 1998. Immunological consequences of intervention in established immune responses by feeding protein antigens. *Cell. Immunol.* **183:**137–148.

Leishman, A. J., P. Garside, and A. M. Mowat. 2000. Induction of oral tolerance in the primed immune system: influence of antigen persistence and adjuvant form. *Cell. Immunol.* **202:**71–78.

Lemke, H., A. Coutinho, and H. Lange. 2004. Lamarckian inheritance by somatically acquired maternal IgG phenotypes. *Trends Immunol.* **25:**180–186.

Li, W., S. T. Chou, C. Wang, C. S. Kuhr, and J. D. Perkins. 2004. Role of the liver in peripheral tolerance: induction through oral antigen feeding. *Am. J. Transplant.* **4:**1574–1582.

Limmer, A., J. Ohl, C. Kurts, H. G. Ljunggren, Y. Reiss, M. Groettrup, F. Momburg, B. Arnold, and P. A. Knolle. 2000. Efficient presentation of exogenous antigen by liver endothelial cells to CD8+ T cells results in antigen-specific T-cell tolerance. *Nat. Med.* **6:**1348–1354.

Ling, E. M., T. Smith, X. D. Nguyen, C. Pridgeon, M. Dallman, J. Arbery, V. A. Carr, and D. S. Robinson. 2004. Relation of CD4+CD25+ regulatory T-cell suppression of allergen-driven T-cell activation to atopic status and expression of allergic disease. *Lancet* **363:**608–615.

Link, H., Y. M. Huang, and B. Xiao. 2003. Suppression of EAMG in Lewis rats by IL-10-exposed dendritic cells. *Ann. N. Y. Acad. Sci.* **998:**537–538.

Liu, H., B. Hu, D. Xu, and F. Y. Liew. 2003. CD4+CD25+ regulatory T cells cure murine colitis: the role of IL-10, TGF-beta, and CTLA4. *J. Immunol.* **171:**5012–5017.

Lumsden, J. M., J. A. Williams, and R. J. Hodes. 2003. Differential requirements for expression of CD80/86 and CD40 on B cells for T-dependent antibody responses in vivo. *J. Immunol.* **170:**781–787.

Lundin, B. S., M. R. Karlsson, L. A. Svensson, L. A. Hanson, U. I. Dahlgren, and E. Telemo. 1999a. Active suppression in orally tolerized rats coincides with in situ transforming growth factor-beta (TGF-beta) expression in the draining lymph nodes. *Clin. Exp. Immunol.* **116:**181–187.

Lundin, B. S., A. Dahlman-Hoglund, I. Pettersson, U. I. Dahlgren, L. A. Hanson, and E. Telemo. 1999b. Antibodies given orally in the neonatal period can affect the immune response for two generations: evidence for active maternal influence on the newborn's immune system. *Scand. J. Immunol.* **50:**651–656.

MacDonald, T. T. 2003. The mucosal immune system. *Parasite Immunol.* **25:**235–246.

Mackie, R. I., A. Sghir, and H. R. Gaskins. 1999. Developmental microbial ecology of the neonatal gastrointestinal tract. *Am. J. Clin. Nutr.* **69:**1035S–1045S.

Macpherson, A. J., and N. L. Harris. 2004. Interactions between commensal intestinal bacteria and the immune system. *Nat. Rev. Immunol.* **4:**478–485.

Macpherson, A. J., and T. Uhr. 2004. Induction of protective IgA by intestinal dendritic cells carrying commensal bacteria. *Science* **303:**1662–1665.

Macpherson, A. J., M. M. Martinic, and N. Harris. 2002. The functions of mucosal T cells in containing the indigenous commensal flora of the intestine. *Cell. Mol. Life Sci.* **59:**2088–2096.

Maizels, R. M., and M. Yazdanbakhsh. 2003. Immune regulation by helminth parasites: cellular and molecular mechanisms. *Nat. Rev. Immunol.* **3:**733–744.

Maleki, S. J., and B. K. Hurlburt. 2004. Structural and functional alterations in major peanut allergens caused by thermal processing. *J. AOAC Int.* **87:**1475–1479.

Maleki, S. J., S. Y. Chung, E. T. Champagne, and J. P. Raufman. 2000. The effects of roasting on the allergenic properties of peanut proteins. *J. Allergy Clin. Immunol.* **106:**763–768.

Mallegol, J., G. van Niel, and M. Heyman. 2005. Phenotypic and functional characterization of intestinal epithelial exosomes. *Blood Cells Mol. Dis.* **35:**11–16.

Maloy, K. J., and F. Powrie. 2001. Regulatory T cells in the control of immune pathology. *Nat. Immunol.* **2:** 816–822.

Mantis, N. J., M. C. Cheung, K. R. Chintalacharuvu, J. Rey, B. Corthesy, and M. R. Neutra. 2002. Selective adherence of IgA to murine Peyer's patch M cells: evidence for a novel IgA receptor. *J. Immunol.* **169:**1844–1851.

Mantovani, A., S. Sozzani, M. Locati, P. Allavena, and A. Sica. 2002. Macrophage polarization: tumor-associated macrophages as a paradigm for polarized M2 mononuclear phagocytes. *Trends Immunol.* **23:**549–555.

Mantovani, A., A. Sica, S. Sozzani, P. Allavena, A. Vecchi, and M. Locati. 2004. The chemokine system in diverse forms of macrophage activation and polarization. *Trends Immunol.* **25:**677–686.

Margenthaler, J. A., K. Landeros, M. Kataoka, and M. W. Flye. 2002. CD1-dependent natural killer ($NK1.1^+$) T cells are required for oral and portal venous tolerance induction. *J. Surg. Res.* **104:**29–35.

Marinaro, M., H. F. Staats, T. Hiroi, R. J. Jackson, M. Coste, P. N. Boyaka, N. Okahashi, M. Yamamoto, H. Kiyono, H. Bluethmann, K. Fujihashi, and J. R. McGhee. 1995. Mucosal adjuvant effect of cholera toxin in mice results from induction of T helper 2 (Th2) cells and IL-4. *J. Immunol.* **155:**4621–4629.

Marshall, J. S., C. A. King, and J. D. McCurdy. 2003. Mast cell cytokine and chemokine responses to bacterial and viral infection. *Curr. Pharm. Des.* **9:**11–24.

Massey-Harroche, D. 2000. Epithelial cell polarity as reflected in enterocytes. *Microsc. Res. Tech.* **49:**353–362.

Matricardi, P. M., and R. Ronchetti. 2001. Are infections protecting from atopy? *Curr. Opin. Allergy Clin. Immunol.* **1:**413–419.

McGuirk, P., and K. H. Mills. 2002. Pathogen-specific regulatory T cells provoke a shift in the Th1/Th2 paradigm in immunity to infectious diseases. *Trends Immunol.* **23:**450–455.

McSorley, S. J., and P. Garside. 1999.Vaccination by inducing oral tolerance? *Immunol. Today* **20:**555-560.

Menzies-Gow, A., S. Ying, S. Phipps, and A. B. Kay. 2004. Interactions between eotaxin, histamine and mast cells in early microvascular events associated with eosinophil recruitment to the site of allergic skin reactions in humans. *Clin. Exp. Allergy* **34:**1276–1282.

Mishra, A., S. P. Hogan, E. B. Brandt, and M. E. Rothenberg. 2000. Peyer's patch eosinophils: identification, characterization, and regulation by mucosal allergen exposure, interleukin-5, and eotaxin. *Blood* **96:**1538–1544.

Monteleone, I., P. Vavassori, L. Biancone, G. Monteleone, and F. Pallone. 2002. Immunoregulation in the gut: success and failures in human disease. *Gut* **50**(Suppl. 3)**:**III60–III64.

Mor, T. S., M. A. Gomez-Lim, and K. E. Palmer. 1998. Perspective: edible vaccines—a concept coming of age. *Trends Microbiol.* **6:**449–453.

Mottet, C., H. H. Uhlig, and F. Powrie. 2003. Cutting edge: cure of colitis by CD4+CD25+ regulatory T cells. *J. Immunol.* **170:**3939–3943.

Nagata, S., C. McKenzie, S. L. Pender, M. Bajaj-Elliott, P. D. Fairclough, J. A. Walker-Smith, G. Monteleone, and T. T. MacDonald. 2000. Human Peyer's patch T cells are sensitized to dietary antigen and display a Th cell type 1 cytokine profile. *J. Immunol.* **165:**5315–5321.

Nakamura, M., M. Iwahashi, M. Nakamori, K. Ueda, T. Ojima, T. Naka, K. Ishida, and H. Yamaue. 2005. Dendritic cells transduced with tumor-associated antigen gene elicit potent therapeutic antitumor immunity: comparison with immunodominant peptide-pulsed DCs. *Oncology* **68:**163–170.

Nanda, A., M. O'Connor, M. Anand, S. C. Dreskin, L. Zhang, B. Hines, D. Lane, W. Wheat, J. M. Routes, R. Sawyer, L. J. Rosenwasser, and H. S. Nelson. 2004. Dose dependence and time course of the immunologic response to administration of standardized cat allergen extract. *J. Allergy Clin. Immunol.* **114:** 1339–1344.

Neish, A. S., A. T. Gewirtz, H. Zeng, A. N. Young, M. E. Hobert, V. Karmali, A. S. Rao, and J. L. Madara. 2000. Prokaryotic regulation of epithelial responses by inhibition of IκB-α ubiquitination. *Science* **289:**1560–1563.

Niederberger, V., B. Niggemann, D. Kraft, S. Spitzauer, and R. Valenta. 2002. Evolution of IgM, IgE and IgG(1-4) antibody responses in early childhood monitored with recombinant allergen components: implications for class switch mechanisms. *Eur. J. Immunol.* **32:**576–584.

Niedergang, F., A. Didierlaurent, J. P. Kraehenbuhl, and J. C. Sirard. 2004. Dendritic cells: the host Achille's heel for mucosal pathogens? *Trends Microbiol.* **12:**79–88.

Nouri-Aria, K. T., P. A. Wachholz, J. N. Francis, M. R. Jacobson, S. M. Walker, L. K. Wilcock, S. Q. Staple, R. C. Aalberse, S. J. Till, and S. R. Durham. 2004. Grass pollen immunotherapy induces mucosal and peripheral IL-10 responses and blocking IgG activity. *J. Immunol.* **172:**3252–3259.

O'Garra, A., P. L. Vieira, P. Vieira, and A. E. Goldfeld. 2004. IL-10-producing and naturally occurring CD4+ Tregs: limiting collateral damage. *J. Clin. Investig.* **114:**1372–1378.

Ogawa, H., D. G. Binion, J. Heidemann, M. Theriot, P. J. Fisher, N. A. Johnson, M. F. Otterson, and P. Rafiee. 2005. Mechanisms of MAdCAM-1 gene expression in human intestinal microvascular endothelial cells. *Am. J. Physiol. Cell Physiol.* **288:**C272–C281.

Pabst, O., L. Ohl, M. Wendland, M. A. Wurbel, E. Kremmer, B. Malissen, and R. Forster. 2004. Chemokine receptor CCR9 contributes to the localization of plasma cells to the small intestine. *J. Exp. Med.* **199:**411–416.

Park, C. S., and Y. S. Choi. 2005. How do follicular dendritic cells interact intimately with B cells in the germinal centre? *Immunology* **114:**2–10.

Park, C. S., S. O. Yoon, R. J. Armitage, and Y. S. Choi. 2004. Follicular dendritic cells produce IL-15 that enhances germinal center B cell proliferation in membrane-bound form. *J. Immunol.* **173:**6676–6683.

Perez-Machado, M. A., P. Ashwood, F. Torrente, C. Salvestrini, R. Sim, M. A. Thomson, J. A. Walker-Smith, and S. H. Murch. 2004. Spontaneous T(H)1 cytokine production by intraepithelial but not circulating T cells in infants with or without food allergies. *Allergy* **59:**346–353.

Platts-Mills, T. A., and J. A. Woodfolk. 2000. Cord blood proliferative responses to inhaled allergens: is there a phenomenon? *J. Allergy Clin. Immunol.* **106:**441–443.

Qin, S., S. P. Cobbold, H. Pope, J. Elliott, D. Kioussis, J. Davies, and H. Waldmann. 1993. "Infectious" transplantation tolerance. *Science* **259:**974–977.

Rai, R. M., S. Loffreda, C. L. Karp, S. Q. Yang, H. Z. Lin, and A. M. Diehl. 1997. Kupffer cell depletion abolishes induction of interleukin-10 and permits sustained overexpression of tumor necrosis factor alpha messenger RNA in the regenerating rat liver. *Hepatology* **25:**889–895.

Raposo, G., H. W. Nijman, W. Stoorvogel, R. Liejendekker, C. V. Harding, C. J. Melief, and H. J. Geuze. 1996. B lymphocytes secrete antigen-presenting vesicles. *J. Exp. Med.* **183:**1161–1172.

Rask, C., J. Holmgren, M. Fredriksson, M. Lindblad, I. Nordstrom, J. B. Sun, and C. Czerkinsky. 2000. Prolonged oral treatment with low doses of allergen conjugated to cholera toxin B subunit suppresses immunoglobulin E antibody responses in sensitized mice. *Clin. Exp. Allergy* **30:**1024–1032.

Rifai, A., and M. Mannik. 1984. Clearance of circulating IgA immune complexes is mediated by a specific receptor on Kupffer cells in mice. *J. Exp. Med.* **160:**125–137.

Roberton, D. M., R. Paganelli, R. Dinwiddie, and R. J. Levinsky. 1982. Milk antigen absorption in the preterm and term neonate. *Arch. Dis. Child* **57:**369–372.

Roberts, G., and G. Lack. 2005. Diagnosing peanut allergy with skin prick and specific IgE testing. *J. Allergy Clin. Immunol.* **115:**1291–1296.

Robinson, D. S., M. Larche, and S. R. Durham. 2004. Tregs and allergic disease. *J. Clin. Investig.* **114:** 1389–1397.

Roelofs-Haarhuis, K., Wu, X., M. Nowak, M. Fang, S. Artik, and E. Gleichmann. 2003. Infectious nickel tolerance: a reciprocal interplay of tolerogenic APCs and T suppressor cells that is driven by immunization. *J. Immunol.* **171:**2863–2872.

Rook, G. A., and L. R. Brunet. 2005. Microbes, immunoregulation, and the gut. *Gut* **54:**317–320.

Rook, G. A., V. Adams, J. Hunt, R. Palmer, R. Martinelli, and L. R. Brunet. 2004. Mycobacteria and other environmental organisms as immunomodulators for immunoregulatory disorders. *Springer Semin. Immunopathol.* **25:**237–255.

Rothenberg, M. E., A. Mishra, E. B. Brandt, and S. P. Hogan. 2001. Gastrointestinal eosinophils. *Immunol. Rev.* **179:**139–155.

Rudin, A., C. Macaubas, C. Wee, B. J. Holt, P. D. Slya, and P. G. Holt. 2001. "Bystander" amplification of PBMC cytokine responses to seasonal allergen in polysensitized atopic children. *Allergy* **56:**1042–1048.

Rustemeyer, T., B. M. von Blomberg, I. M. van Hoogstraten, D. P. Bruynzeel, and R. J. Scheper. 2004. Analysis of effector and regulatory immune reactivity to nickel. *Clin. Exp. Allergy* **34:**1458–1466.

Saavedra, J. M. 2001. Clinical applications of probiotic agents. *Am. J. Clin. Nutr.* **73:**1147S–1151S.

Sakaguchi, S. 2005. Naturally arising Foxp3-expressing CD25+CD4+ regulatory T cells in immunological tolerance to self and non-self. *Nat. Immunol.* **6:**345-352.

Sakaguchi, S., N. Sakaguchi, M. Asano, M. Itoh, and M. Toda. 1995. Immunologic self-tolerance maintained by activated T cells expressing IL-2 receptor alpha-chains (CD25). Breakdown of a single mechanism of self-tolerance causes various autoimmune diseases. *J. Immunol.* **155:**1151–1164.

Sampson, H. A. 2004. Update on food allergy. *J. Allergy Clin. Immunol.* **113:**805–819.

Sampson, H. A., and D. G. Ho. 1997. Relationship between food-specific IgE concentrations and the risk of positive food challenges in children and adolescents. *J. Allergy Clin. Immunol.* **100:**444–451.

Sansonetti, P. J. 2004. War and peace at mucosal surfaces. *Nat. Rev. Immunol.* **4:**953–964.

Schaedler, R. W., R. Dubs, and R. Costello. 1965. Association of germfree mice with bacteria isolated from normal mice. *J. Exp. Med.* **122:**77–82.

Schartz, N. E., N. Chaput, F. Andre, and L. Zitvogel. 2002. From the antigen-presenting cell to the antigen-presenting vesicle: the exosomes. *Curr. Opin. Mol. Ther.* **4:**372–381.

Schmausser, B., C. Josenhans, S. Endrich, S. Suerbaum, C. Sitaru, M. Andrulis, S. Brandlein, P. Rieckmann, H. K. Muller-Hermelink, and M. Eck. 2004. Downregulation of CXCR1 and CXCR2 expression on human neutrophils by *Helicobacter pylori*: a new pathomechanism in *H. pylori* infection? *Infect. Immun.* **72:**6773–6779.

Schwarze, J., and E. W. Gelfand. 2002. Respiratory viral infections as promoters of allergic sensitization and asthma in animal models. *Eur. Respir. J.* **19:**341–349.

Schwarze, J., E. Hamelmann, K. L. Bradley, K. Takeda, and E. W. Gelfand. 1997. Respiratory syncytial virus infection results in airway hyperresponsiveness and enhanced airway sensitization to allergen. *J. Clin. Investig.* **100:**226–233.

Scrivener, S., H. Yemaneberhan, M. Zebenigus, D. Tilahun, S. Girma, S. Ali, P. McElroy, A. Custovic, A. Woodcock, D. Pritchard, A. Venn, and J. Britton. 2001. Independent effects of intestinal parasite infection and domestic allergen exposure on risk of wheeze in Ethiopia: a nested case-control study. *Lancet* **358:**1493–1499.

Seeger, M., H. J. Thierse, H. Lange, L. Shaw, H. Hansen, and H. Lemke. 1998. Antigen-independent suppression of the IgE immune response to bee venom phospholipase A2 by maternally derived monoclonal IgG antibodies. *Eur. J. Immunol.* **28:**2124–2130.

Seelig, L. L., Jr., and J. R. Head. 1987. Uptake of lymphocytes fed to suckling rats. An autoradiographic study of the transit of labeled cells through the neonatal gastric mucosa. *J. Reprod. Immunol.* **10:**285–297.

Settmacher, U., H. D. Volk, R. von Baehr, H. Wolff, and S. Jahn. 1993. In vitro stimulation of human fetal lymphocytes by mitogens and interleukins. *Immunol. Lett.* **35:**147–152.

Settmacher, U., H. D. Volk, S. Jahn, K. Neuhaus, F. Kuhn, and R. von Baehr. 1991. Characterization of human lymphocytes separated from fetal liver and spleen at different stages of ontogeny. *Immunobiology* **182:**256–265.

Sharma, R., J. J. Tepas III, M. L. Hudak, P. S. Wludyka, D. L. Mollitt, R. D. Garrison, J. A. Bradshaw, and M. Sharma. 2005. Portal venous gas and surgical outcome of neonatal necrotizing enterocolitis. *J. Pediatr. Surg* **40:**371–376.

Shi, H. N., H. Y. Liu, and C. Nagler-Anderson. 2000. Enteric infection acts as an adjuvant for the response to a model food antigen. *J. Immunol.* **165:**6174–6182.

Shreffler, W. G., K. Beyer, T. H. Chu, A. W. Burks, and H. A. Sampson. 2004. Microarray immunoassay: association of clinical history, in vitro IgE function, and heterogeneity of allergenic peanut epitopes. *J. Allergy Clin. Immunol.* **113:**776–782.

Siebers, A., and B. B. Finlay. 1996. M cells and the pathogenesis of mucosal and systemic infections. *Trends Microbiol.* **4:**22–29.

Singh, B., S. Read, C. Asseman, V. Malmstrom, C. Mottet, L. A. Stephens, R. Stepankova, H. Tlaskalova, and F. Powrie. 2001. Control of intestinal inflammation by regulatory T cells. *Immunol. Rev.* **182:**190–200.

Skolnick, H. S., M. K. Conover-Walker, C. B. Koerner, H. A. Sampson, W. Burks, and R. A. Wood. 2001. The natural history of peanut allergy. *J. Allergy Clin. Immunol.* **107:**367–374.

Smith, K. M., F. McAskill, and P. Garside. 2002. Orally tolerized T cells are only able to enter B cell follicles following challenge with antigen in adjuvant, but they remain unable to provide B cell help. *J. Immunol.* **168:**4318–4325.

Smits, H. H., A. Engering, D. van der Kleij, E. C. de Jong, K. Schipper, T. M. van Capel, B. A. Zaat, M. Yazdanbakhsh, E. A. Wierenga, Y. van Kooyk, and M. L. Kapsenberg. 2005. Selective probiotic bacteria induce IL-10-producing regulatory T cells in vitro by modulating dendritic cell function through dendritic cell-specific intercellular adhesion molecule 3-grabbing nonintegrin. *J. Allergy Clin. Immunol.* **115:** 1260–1267.

Snider, D. P., J. S. Marshall, M. H. Perdue, and H. Liang. 1994. Production of IgE antibody and allergic sensitization of intestinal and peripheral tissues after oral immunization with protein Ag and cholera toxin. *J. Immunol.* **153:**647–657.

Stadnicki, A., R. B. Sartor, R. Janardham, A. Majluf-Cruz, C. A. Kettner, A. A. Adam, and R. W. Colman. 1998. Specific inhibition of plasma kallikrein modulates chronic granulomatous intestinal and systemic inflammation in genetically susceptible rats. *FASEB J.* **12:**325–333.

Stadnyk, A. W. 2002. Intestinal epithelial cells as a source of inflammatory cytokines and chemokines. *Can. J. Gastroenterol.* **16:**241–246.

Strachan, D. P. 1989. Hay fever, hygiene, and household size. *BMJ* **299:**1259–1260.

Strober, W., B. Kelsall, I. Fuss, T. Marth, B. Ludviksson, R. Ehrhardt, and M. Neurath. 1997. Reciprocal IFN-gamma and TGF-beta responses regulate the occurrence of mucosal inflammation. *Immunol. Today* **18:** 61–64.

Szepfalusi, Z., J. Pichler, S. Elsasser, van K. Duren, C. Ebner, G. Bernaschek, and R. Urbanek. 2000. Transplacental priming of the human immune system with environmental allergens can occur early in gestation. *J. Allergy Clin. Immunol.* **106:**530–536.

Szepfalusi, Z., I. Nentwich, M. Gerstmayr, E. Jost, L. Todoran, R. Gratzl, K. Herkner, and R. Urbanek. 1997. Prenatal allergen contact with milk proteins. *Clin. Exp. Allergy* **27:**28–35.

Taams, L. S., J. M. van Amelsfort, M. M. Tiemessen, K. M. Jacobs, E. C. de Jong, A. N. Akbar, J. W. Bijlsma, and F. P. Lafeber. 2005. Modulation of monocyte/macrophage function by human $CD4^+CD25^+$ regulatory T cells. *Hum. Immunol.* **66:**222–230.

Taams, L. S., E. P. Boot, W. van Eden, and M. H. Wauben. 2000. 'Anergic' T cells modulate the T-cell activating capacity of antigen-presenting cells. *J. Autoimmunol.* **14:**335–341.

Taams, L. S., A. J. van Rensen, M. C. Poelen, C. A. van Els, A. C. Besseling, J. P. Wagenaar, W. van Eden, and M. H. Wauben. 1998. Anergic T cells actively suppress T cell responses via the antigen-presenting cell. *Eur. J. Immunol.* **28:**2902–2912.

Tariq, S. M., M. Stevens, S. Matthews, S. Ridout, R. Twiselton, and D. W. Hide. 1996. Cohort study of peanut and tree nut sensitisation by age of 4 years. *BMJ* **313:**514–517.

Telemo, E., M. Korotkova, and L. A. Hanson. 2003. Antigen presentation and processing in the intestinal mucosa and lymphocyte homing. *Ann. Allergy Asthma Immunol.* **90:**28–33.

Telemo, E., B. Westrom, G. Dahl, and B. Karlsson. 1986. Transfer of orally or intravenously administered proteins to the milk of the lactating rat. *J. Pediatr. Gastroenterol. Nutr.* **5:**305–309.

Thery, C., L. Zitvogel, and S. Amigorena. 2002. Exosomes: composition, biogenesis and function. *Nat. Rev. Immunol.* **2:**569–579.

Tiemessen, M. M., E. Van Hoffen, A. C. Knulst, J. A. Van Der Zee, E. F. Knol, and L. S. Taams. 2002. CD4 CD25 regulatory T cells are not functionally impaired in adult patients with IgE-mediated cow's milk allergy. *J. Allergy Clin. Immunol.* **110:**934–936.

Tkaczyk, C., I. Villa, R. Peronet, B. David, and S. Mecheri. 1999. FcεRI-mediated antigen endocytosis turns interferon-gamma-treated mouse mast cells from inefficient into potent antigen-presenting cells. *Immunology* **97:**333–340.

Tregoning, J. S., S. Clare, F. Bowe, L. Edwards, N. Fairweather, O. Qazi, P. J. Nixon, P. Maliga, G. Dougan, and T. Hussell. 2005. Protection against tetanus toxin using a plant-based vaccine. *Eur. J. Immunol.* **35:** 1320–1326.

Trzonkowski, P., E. Szmit, J. Mysliwska, A. Dobyszuk, and A. Mysliwski. 2004. CD4+CD25+ T regulatory cells inhibit cytotoxic activity of T CD8+ and NK lymphocytes in the direct cell-to-cell interaction. *Clin. Immunol.* **112:**258–267.

Tuboly, S., S. Bernath, R. Glavits, A. Kovacs, and Z. Megyeri. 1995. Intestinal absorption of colostral lymphocytes in newborn lambs and their role in the development of immune status. *Acta Vet. Hung.* **43:** 105–115.

Turcanu, V., S. J. Maleki, and G. Lack. 2003. Characterization of lymphocyte responses to peanuts in normal children, peanut-allergic children, and allergic children who acquired tolerance to peanuts. *J. Clin. Investig.* **111:**1065–1072.

Untersmayr, E., N. Bakos, I. Scholl, M. Kundi, F. Roth-Walter, K. Szalai, A. B. Riemer, H. J. Ankersmit, O. Scheiner, G. Boltz-Nitulescu, and E. Jensen-Jarolim. 2005. Anti-ulcer drugs promote IgE formation toward dietary antigens in adult patients. *FASEB J.* **19:**656–658.

Untersmayr, E., I. Scholl, I. Swoboda, W. J. Beil, E. Forster-Waldl, F. Walter, A. Riemer, G. Kraml, T. Kinaciyan, S. Spitzauer, G. Boltz-Nitulescu, O. Scheiner, and E. Jensen-Jarolim. 2003. Antacid medication inhibits digestion of dietary proteins and causes food allergy: a fish allergy model in BALB/c mice. *J. Allergy Clin. Immunol.* **112:**616–623.

Vance, G. H., K. E. Grimshaw, R. Briggs, S. A. Lewis, M. A. Mullee, C. A. Thornton, and J. O. Warner. 2004. Serum ovalbumin-specific immunoglobulin G responses during pregnancy reflect maternal intake of dietary egg and relate to the development of allergy in early infancy. *Clin. Exp. Allergy* **34:**1855–1861.

van den Biggelaar, A. H., L. C. Rodrigues, R. van Ree, J. S. van der Zee, Y. C. Hoeksma-Kruize, J. H. Souverijn, M. A. Missinou, S. Borrmann, P. G. Kremsner, and M. Yazdanbakhsh. 2004. Long-term treatment of intestinal helminths increases mite skin-test reactivity in Gabonese schoolchildren. *J. Infect. Dis.* **189:**892–900.

van den Biggelaar, A. H., R. van Ree, L. C. Rodrigues, B. Lell, A. M. Deelder, P. G. Kremsner, and M. Yazdanbakhsh. 2000. Decreased atopy in children infected with Schistosoma haematobium: a role for parasite-induced interleukin-10. *Lancet* **356:**1723–1727.

van der Wouden, J. M., O. van Maier, I. S. van Ijzendoorn, and D. Hoekstra. 2003. Membrane dynamics and the regulation of epithelial cell polarity. *Int. Rev. Cytol.* **226:**127–164.

van Niel, G., G. Raposo, C. Candalh, M. Boussac, R. Hershberg, N. Cerf-Bensussan, and M. Heyman. 2001. Intestinal epithelial cells secrete exosome-like vesicles. *Gastroenterology* **121:**337–349.

Verginis, P., H. S. Li, and G. Carayanniotis. 2005. Tolerogenic semimature dendritic cells suppress experimental autoimmune thyroiditis by activation of thyroglobulin-specific CD4+CD25+ T cells. *J. Immunol.* **174:** 7433–7439.

Vezys, V., S. Olson, and L. Lefrancois. 2000. Expression of intestine-specific antigen reveals novel pathways of CD8 T cell tolerance induction. *Immunity* **12:**505–514.

Viney, J. L., A. M. Mowat, J. M. O'Malley, E. Williamson, and N. A. Fanger. 1998. Expanding dendritic cells in vivo enhances the induction of oral tolerance. *J. Immunol.* **160:**5815–5825.

Vukavic, T. 1984. Timing of the gut closure. *J. Pediatr. Gastroenterol. Nutr.* **3:**700–703.

Wachholz, P. A., and S. R. Durham. 2004. Mechanisms of immunotherapy: IgG revisited. *Curr. Opin. Allergy Clin. Immunol.* **4:**313–318.

Watanabe, M., Y. Ueno, T. Yajima, Y. Iwao, M. Tsuchiya, H. Ishikawa, S. Aiso, T. Hibi, and H. Ishii. 1995. Interleukin 7 is produced by human intestinal epithelial cells and regulates the proliferation of intestinal mucosal lymphocytes. *J. Clin. Investig.* **95:**2945–2953.

Watanabe, T., H. Katsukura, Y. Shirai, M. Yamori, T. Chiba, T. Kita, and Y. Wakatsuki. 2003. Helper CD4+ T cells for IgE response to a dietary antigen develop in the liver. *J. Allergy Clin. Immunol.* **111:** 1375–1385.

Wei, B., P. Velazquez, O. Turovskaya, K. Spricher, R. Aranda, M. Kronenberg, L. Birnbaumer, and J. Braun. 2005. Mesenteric B cells centrally inhibit CD4+ T cell colitis through interaction with regulatory T cell subsets. *Proc. Natl. Acad. Sci. USA* **102:**2010–2015.

Weiner, H. L. 2001. Induction and mechanism of action of transforming growth factor-beta-secreting Th3 regulatory cells. *Immunol. Rev.* **182:**207–214.

Wood, K. J., and S. Sakaguchi. 2003. Regulatory T cells in transplantation tolerance. *Nat. Rev. Immunol.* **3:** 199–210.

Wrenshall, L. E., J. D. Ansite, P. M. Eckman, M. J. Heilman, R. B. Stevens, and D. E. Sutherland. 2001. Modulation of immune responses after portal venous injection of antigen. *Transplantation* **71:**841–850.

Xiao, B. G., and H. Link. 1997. Mucosal tolerance: a two-edged sword to prevent and treat autoimmune diseases. *Clin. Immunol. Immunopathol.* **85:**119–128.

Xiao, B. G., R. S. Duan, H. Link, and Y. M. Huang. 2003. Induction of peripheral tolerance to experimental autoimmune myasthenia gravis by acetylcholine receptor-pulsed dendritic cells. *Cell. Immunol.* **223:**63–69.

Yarilin, D., R. Duan, Y. M. Huang, and B. G. Xiao. 2002. Dendritic cells exposed in vitro to TGF-β1 ameliorate experimental autoimmune myasthenia gravis. *Clin. Exp. Immunol.* **127:**214–219.

Yazdanbakhsh, M., A. van den Biggelaar, and R. M. Maizels. 2001. Th2 responses without atopy: immunoregulation in chronic helminth infections and reduced allergic disease. *Trends Immunol.* **22:**372–377.

Yoshikawa, T., and Y. Naito. 2000. The role of neutrophils and inflammation in gastric mucosal injury. *Free Radic. Res.* **33:**785–794.

Zhang, X., C. S. Park, S. O. Yoon, L. Li, Y. M. Hsu, C. Ambrose, and Y. S. Choi. 2005. BAFF supports human B cell differentiation in the lymphoid follicles through distinct receptors. *Int. Immunol.* **17:**779–788.

Zoetendal, E. G., C. T. Collier, S. Koike, R. I. Mackie, and H. R. Gaskins. 2004. Molecular ecological analysis of the gastrointestinal microbiota: a review. *J. Nutr.* **134:**465–472.

Food Allergy
Edited by S. J. Maleki et al.

Chapter 5

The Relationship of T-Cell Epitopes and Allergen Structure

Samuel J. Landry

This chapter will address the relationship of the CD4$^+$ helper T-cell epitope immunodominance to pathways of allergen-antigen processing, as they are directed by protein three-dimensional structures. Allergy is recognized as a failure of immune tolerance by CD4$^+$ T cells. Thus, it is critical that we understand the forces that shape the T-cell response. Antigen-allergen processing is the first step in a complex mechanism that leads to the development of helper T cells, and the antigens-allergens themselves are likely to play an important, if not paramount, role in molecular decision making. A number of allergens have been identified as enzymes or inhibitors (Breiteneder and Radauer, 2004; Maleki et al., 2003; Robinson et al., 1997; Untersmayr et al., 2003). Allergens also have been identified as ligand-binding proteins (Breiteneder and Mills, 2005). These properties have been suggested to have an influence on allergen processing and presentation, resulting in the failure to develop tolerance. However, many allergens probably do not have such activities, and thus more-subtle explanations must apply in these cases. This chapter will not attempt to address the reasons that some proteins are allergenic and others not. This chapter will illuminate how protein three-dimensional structure influences antigen processing, and it will highlight ways that allergen processing may interact with the environment during the development of allergy.

Antigen processing focuses immunity or tolerance on the dominant epitopes of an antigen. Circumstances may change the mechanisms of antigen processing and therefore alter the dominance pattern. For example, an epitope that is normally presented with low efficiency may be ignored during the development of immune tolerance. Subsequently, molecular or cellular circumstances may increase the presentation of that epitope, resulting in an allergic response. Thus, understanding the mechanisms that cause dominance may provide a basis for understanding how allergy arises and suggest ways to improve immunotherapy.

The organization of sections in this chapter is inspired by the steps in the development of immune tolerance and allergy (Fig. 1). The picture emerging suggests that there is

---

Samuel J. Landry • Department of Biochemistry SL43, Tulane University Health Sciences Center, 1430 Tulane Ave., New Orleans, LA 70112.

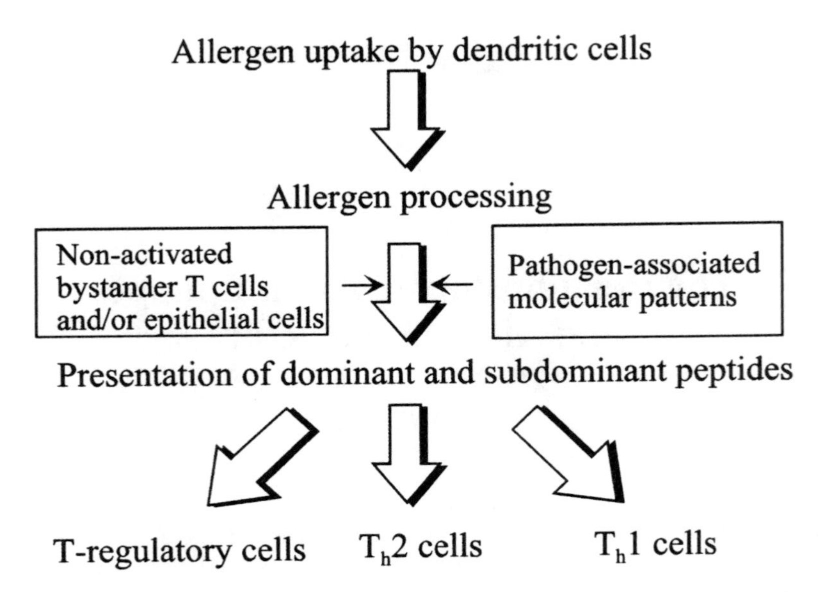

Figure 1. Schematic diagram indicating the steps involved in the development of regulatory, Th2, or Th1 cells, highlighting the idea that antigen processing and peptide presentation are essentially the same for presentation to all types of T cells. The developmental decision between cell types is made by DCs, which receive signals from preexisting T cells, epithelial cells, and pathogen-associated molecular patterns.

a complex interplay among allergen processing, T-cell activation, and the T-cell repertoire that shapes the T-cell response to the allergen. Several sections deal with general features of the process: a discussion of the cells that process and present food allergens; a discussion of the lysosomal enzymes that process allergens; a brief summary of evidence that proteolytic processing follows pathways that could depend on molecular and cellular contexts; evidence that antigen processing and peptide loading occur in the same compartment, where the two mechanisms may interact; the idea that tolerance and immunity represent different outcomes of the same mechanism of antigen processing; and a summary of the most direct evidence that the antigen-allergen structure has an influence on epitope dominance, including a discussion of the nature of structural data and how it is used.

The number of allergens for which there is high-resolution structural information and T-cell epitope mapping data is very small. As of January 2005, there are approximately 100 high-resolution three-dimensional structures of allergens available in the Research Collaboratory for Structural Bioinformatics Protein Data Bank, which represents 31 different proteins from 22 genera. Most of these are aeroallergens. T-cell epitopes have been mapped for even fewer allergens. For comparison, a search of the combined Swiss-Prot and Translated EMBL (Swiss-Prot/TrEMBL) sequence databases for allergens yielded 1,642 entries.

The next section in the chapter discusses the relationship of structure and epitope dominance for human immunodeficiency virus (HIV) gp120 and several food allergens: chicken lysozyme, chicken ovalbumin, bovine β-lactoglobulin (BLG), and the birch tree aeroallergen Bet v 1 (which is thought to induce allergy to the homologous protein in certain vegetables). HIV gp120 is used as a model antigen for this discussion because it is very well studied and has little, if any, similarity to self proteins. The lack of similarity to self proteins reduces the potential that tolerance constrains the T-cell repertoire; thus, the gp120 dominance profile more faithfully reflects epitope presentation than do the dominance profiles of proteins homologous to mammalian proteins, e.g., lysozyme.

The final section in this chapter discusses the implications of the relationship of structure to dominance on the development of allergy, evolution of the immune system, and allergy immunotherapy.

CELL TYPES INVOLVED IN PROCESSING AND PRESENTATION OF FOOD ALLERGENS

Several types of antigen-presenting cells (APCs) are potentially involved in allergy. Different APCs may have distinct roles because they have different types and/or levels of lysosomal enzymes. Different APCs also express different levels of the class II major histocompatibility antigen-presenting proteins (MHC-II) and costimulatory molecules that influence the development of T cells. Studies have shown that the efficiency of presentation for certain epitopes depends on the type of APCs (Gapin et al., 1998; Ma et al., 1999).

Orally administered antigens-allergens can be divided into two categories, soluble and particulate. Soluble antigens and even a soluble form of a particulate antigen are generally less immunogenic (Brewer et al., 2004; Raychaudhuri and Rock, 1998). The different responses may be attributable to the types of cells in the gut that take up the antigens. The three major routes for antigen uptake are epithelial cells of the lamina propria, dendritic cells (DCs), and Peyer's patch M cells. Whereas epithelial cells and DCs are associated with uptake of soluble antigens, M cells are associated with uptake of particulate antigens (Chehade and Mayer, 2005). Epithelial cells are themselves able to present antigen to T cells, but this presentation is not associated with priming of immune responses. M cells express MHC-II proteins and therefore could present antigen to T cells; however, their more important role is likely to be the transport of antigen to DCs in Peyer's patches of gut-associated lymphoid tissue. DCs also directly sample contents of the gut lumen by inserting themselves across the epithelium of the lamina propria (Rescigno et al., 2001). B cells are abundant in Peyer's patches, and they are able to present antigen to T cells. Nevertheless, they cannot prime naïve T cells (Fuchs and Matzinger, 1992), and they are not necessary for development of oral tolerance (Alpan et al., 2001).

Current theories about the nature of the immune response suggest that the decision between tolerance and Th1 cell-mediated or Th2 cell-mediated immunity is related to the presence or absence of danger signals, also known as pathogen-associated molecular patterns, which are recognized by pattern recognition receptors on DCs (Kapsenberg, 2003; Macdonald and Monteleone, 2005). In regard to orally administered antigens, DCs may be strongly biased toward Th2 cell-mediated immunity. Through the process of bystander suppression, DCs receive signals from preexisting T cells that cause them to prime the development of Th2 cells (Alpan et al., 2004). DCs also become "educated" toward Th2 orientation by exposure to gut epithelial cells (Rimoldi et al., 2005).

The development of allergy to some or all foods could involve interaction with the immune system outside of the gut. For example, allergy to a number of vegetables is strongly associated with allergy to the Bet v 1 aeroallergen of birch tree pollen. At the very least, allergy to Bet v 1 could involve interaction with immune cells of the nasal mucosa-associated lymphoid tissue, which is very similar in structure to gut-associated lymphoid tissue (Brandtzaeg et al., 1999). Bet v 1 might also interact with immune cells of the skin. In the skin, DCs known as Langerhans cells are well established as the APCs that make decisions on tolerance and immunity (Steinman et al., 2003).

Two postulates regarding T-cell development have profound consequences for understanding allergy. First, DCs are the only professional APCs, i.e., the only APCs that can prime naïve T cells (Alpan et al., 2004; Alpan et al., 2001; Fuchs and Matzinger, 1992). This postulate excludes the possibility that B cells prime T cells with distinct epitope specificities and cytokine profiles. Such T cells presumably could merge with the T-cell population that was primed by DCs and change the overall response to an antigen-allergen (i.e., cause immune deviation). Since DCs are the only professional APCs, all T-cell specificities originate from processing in these cells. Second, the cytokine phenotypes of individual T-cell clones are fixed when they are primed by DCs (Yip et al., 1999). This postulate excludes the notion that mature T cells can change their cytokine phenotype after priming by DCs, for example, by expressing more interleukin-4 (IL-4) in response to encounters with B cells acting as semiprofessional APCs. Such changes in T-cell properties also could change the overall immune response to an allergen. Since the phenotypes of individual T-cell clones are fixed, immune deviation only arises with the priming of T cells specific for epitopes that were ignored in previous exposures to the antigen-allergen. A critical analysis of the arguments for and against these postulates is beyond the scope of this chapter. The reader is directed to the primary references and recent reviews for treatment of these topics.

CONTENT OF THE ANTIGEN-PROCESSING COMPARTMENT

The enzymic content of the lysosome strongly influences the speed and potentially the pathways of antigen processing. More than a dozen proteases have been localized to the lysosome. However, only cathepsins B, D, F, K, L, and S and asparagine endopeptidase (AEP) have been implicated in antigen processing (Honey and Rudensky, 2003; Nakagawa and Rudensky, 1999; Watts, 2004). Proteases are generally classified by the type of active-site residue and homology to other proteases (Barrett et al., 1998). Cathepsins B, F, K, L, and S are members of a family of cysteine proteases that are homologous to papain. AEP is a cysteine protease that is homologous to caspases. Cathepsin D is an aspartic protease that is homologous to pepsin and HIV protease. Levels of expression and activity of the various enzymes are regulated in a manner that depends on the type of cell and activation state of the cell.

Analysis of protease specificity using peptide substrates and X-ray crystallography of enzyme-substrate complexes reveals that these classes of proteases bind four to six residues of the substrate in the active site (McGrath, 1999). Side chains of two or three substrate residues project into specificity pockets, and the discrimination of interactions in these sites accounts for substrate specificity as determined with peptide substrates. However, the cathepsins exhibit only weak sequence specificity. Cleavage efficiency of peptide substrates having different primary sequences typically ranges over a few fold, but cleavage efficiency of different sites in proteins can range over several orders of magnitude. The sensitivity of a given site to cleavage depends more on structure (exposure and flexibility) than on sequence. AEP cleaves almost exclusively after asparagine, but the sensitivity of sites in proteins to cleavage by this enzyme also depends greatly on local structure.

The individual proteases are differentially expressed in various immunological tissues, and their activities often are uniquely regulated by lysosomal acidification and the presence or absence of endogenous inhibitors (Honey and Rudensky, 2003; Lennon-Dumenil et al., 2002). Cathepsin L expression is found in all immune tissues, but its activity has not been observed with DCs or macrophages. Cathepsin S is present and active in most tissues. Cathepsins B and D are found in B cells, DCs, and macrophages. AEP is found in B cells and DCs. Cathepsin F is found in macrophages and epithelial cells, and cathepsin K is found in macrophages and osteoclasts.

Numerous studies have attributed special importance to particular proteases or classes of proteases for antigen processing. However, the specific roles of the proteases often have been confounded by the fact that inhibitors or knockouts block proteolytic cleavage of the invariant chain and therefore block maturation of the MHC-II (Fineschi and Miller, 1997; Villadangos et al., 1999). Studies combining knockouts of cathepsin L and an invariant chain in mice have established the importance of cathepsin L for both the maturation of MHC-II and peptide generation in thymic epithelial cells (Honey et al., 2002; Nakagawa et al., 1998). The cathepsin L knockout caused a 30% reduction in $CD4^+$ T cells, which is consistent with a role for epithelial cells in the positive selection of T cells. The knockout of cathepsin L also reduced the diversity of MHC-bound peptides and increased the display of MHC-bound invariant-chain fragments. The knockout of cathepsin S showed that this protease is important for maturation of MHC-II in B cells, DCs, and macrophages. Loss of cathepsin S affects the spectrum of peptides generated, and some of the effect could be due to its direct involvement in proteolytic processing of antigens (Hsieh et al., 2002; Pluger et al., 2002). Since knockout of neither cathepsin L nor cathepsin S prevented the development of a functional immune system, these enzymes are thought to have overlapping functions in antigen processing (Honey and Rudensky, 2003).

The role of AEP is interwoven with those of other lysosomal proteases. AEP is not required for maturation of invariant chain or MHC-II, but it is necessary for maturation of other lysosomal proteases (Maehr et al., 2005; Shirahama-Noda et al., 2003). Inhibition of AEP or elimination of AEP cleavage sites blocked the presentation of certain epitopes in tetanus toxoid (Antoniou et al., 2000; Manoury et al., 1998). Another study suggested that AEP is necessary for presentation of a tolerogenic epitope, rather than an encephalitogenic epitope, of myelin basic protein (Anderton et al., 2002; Manoury et al., 2002).

In some cases, disulfide bond cleavage may be necessary for antigen unfolding, proteolysis, and presentation. The gamma interferon (IFN-γ)-inducible lysosomal thiol reductase (GILT) is colocalized with MHC-II in a cellular compartment associated with peptide loading, and GILT knockout mice have a significantly reduced immune response to antigens that contain disulfide bonds (Maric et al., 2001). Presentation of certain epitopes in the egg allergen lysozyme are particularly dependent on the action of GILT (see below).

The activation of APCs by danger signals increases gene expression, increases enzyme activation, and modifies intracellular trafficking pathways (Chow and Mellman, 2005). APCs differ in their capacity for protein degradation and antigen presentation. Although immature DCs and B cells express higher levels of MHC-II than macrophages, they have lower levels of active proteases and slower rates of protein degradation (Delamarre et al., 2005). Since DCs are the most efficient APCs, the lower degradative rates favor the capture of antigenic peptides in MHC proteins, which subsequently present the peptides to T cells (Lennon-Dumenil et al., 2002). Thus, the enzymic content of the antigen-processing compartment is affected by activation. However, it is not yet clear to what extent these changes modify the pathways of antigen breakdown, the spectrum of epitopes presented to T cells, or the cytokine phenotype of the resulting T cells.

PATHWAYS OF ANTIGEN-ALLERGEN PROTEOLYSIS LEADING TO PRESENTATION

The extent to which proteolysis influences epitope presentation and immunodominance has been difficult to analyze, and it may depend greatly on the nature of the antigen. One early study concluded that the amino termini of myoglobin epitopes corresponded to cathepsin D cleavage sites and at least some of the carboxy termini corresponded to cathepsin B cleavage sites (Van Noort et al., 1991). Numerous studies have shown that the structural context of the epitope influences immunogenicity; in many cases, the differences have been attributed to effects on proteolytic processing (Chianese-Bullock et al., 1998; Eidem et al., 2000; Janssen et al., 1994; Li et al., 2002; Ma et al., 1999; Nayak, 1999; Phelps et al., 1998; Rouas et al., 1993; So et al., 1997; Thai et al., 2004; Vanegas et al., 1997).

Very few studies have investigated pathways of antigen breakdown. It is difficult to track the production and disappearance of proteolytic intermediates generated with purified antigen and protease in vitro and vastly more so for trace amounts of labeled antigen in cells. Early studies showed that antibodies (either prebound to the antigen or expressed by the B cell used for antigen processing) can change the response of T cells having specificity for certain epitopes in the antigen (Berzofsky, 1983; Davidson and Watts, 1989; Manca et al., 1988; Manca et al., 1985; Ozaki and Berzofsky, 1987; Simitsek et al., 1995). In some of these studies, the spectrum of proteolytic products was shown to be affected (Davidson and Watts, 1989; Simitsek et al., 1995). All of these studies used macrophages or B cells as APCs. It is not yet clear whether these conclusions are generalizable to DCs. Since DCs have generally lower protease activity (Delamarre et al., 2005), an antibody's negative impact on epitope processing could be exacerbated.

As noted above, one study illustrated the importance of AEP for the presentation of epitopes, and some of the affected epitopes were located well away from the asparagine-endopeptidase cleavage site (Antoniou et al., 2000). The authors concluded that an initial cleavage by this protease was necessary to globally unlock the structure of the antigen and allow MHC binding and other enzymatic cleavages to take place elsewhere in the molecule.

SIMULTANEOUS PROCESSING AND LOADING OF PEPTIDES

Proteolytic processing and peptide loading appear to occur in the same compartment and probably overlap to some extent. While it is clear that some epitopes are not available until a certain amount of proteolysis has occurred, other epitopes may be destroyed by proteolysis unless they are captured and protected by the MHC-II molecule (Donermeyer and Allen, 1989; Mouritsen et al., 1992). Analysis by mass spectrometry and amino acid sequencing of peptides generated from hen egg lysozyme (HEL) in B cells suggested that amino and carboxy peptidases digest the peptides from the termini after being loaded into the MHC-II protein (Gugasyan et al., 1998). The termini tended to be ragged, suggesting that proteolysis was completed to various extents on the population of MHC-peptide complexes. These results led Unanue and coworkers to conclude that MHC proteins assemble on the intact but unfolded polypeptide, and subsequently, endoproteolytic cleavages would separate the complexes prior to trimming of the peptides to the final size (Gugasyan et al. 1998). This picture is consistent with later work showing that the capture of two adjacent epitopes blocked their presentation until proteolysis allowed the MHC-peptide complexes to separate and be transported to the cell surface (Castellino et al., 1998). Thus, loading precedes proteolytic processing of some epitopes and follows proteolytic processing of other epitopes.

LIMITATIONS ON THE T-CELL REPERTOIRE BY NEGATIVE SELECTION

The notion that tolerance and immunity are mirror images of the same process has been around for a long time (Mamula et al., 1992; Moudgil and Sercarz, 1993). Antigen processing and presentation yield a determinant hierarchy that depends on antigen structure and the processing compartment, and the phenotype of the T-cell response depends on the circumstances of presentation (Lehmann et al., 1998). In a tolerogenic context, such as during presentation by gut epithelial cells, the T cells corresponding to the dominant epitope(s) are rendered anergic or deleted entirely. Anergized T cells may be indistinguishable from regulatory T cells (Takahashi et al., 1998; Thornton and Shevach, 1998). In an inflammatory context, such as during presentation by DCs that have been activated by invasive bacteria, the T cells corresponding to the dominant epitope(s) are stimulated to trigger strong immunity. The full implications of this parallelism in opposing responses cannot be fully appreciated until methods for systematically mapping tolerogenic responses have been worked out.

Responses of allergic individuals may represent the exceptional failure of tolerance to one or more dominant or subdominant epitopes. T cells that secrete tolerogenic cytokines (e.g., IL-10 or tumor growth factor β) are likely to dominate immune responses to proteins to which the individual is chronically exposed (Macdonald and Monteleone, 2005). Immune responses to such proteins are notoriously difficult to characterize, possibly because T-cell lines that secrete immune-activating cytokines are suppressed by coisolated T-cell lines that secrete tolerogenic cytokines. The coisolated T-cell lines may have overlapping determinants. Enzyme-linked immunospot assay (ELISPOT) studies indicate that individual T cells secrete only one cytokine; nevertheless, immunization appears to elicit a mixture of T cells that have various cytokine identities, even when stimulated by the same peptide (Karulin et al., 2000). The observed response to any given allergen segment may be a combination of tolerogenic and immunogenic T cells, and the ratio of the two types of cells may vary by epitope. This could be a setup for the development of allergy because a change in the type of exposure could shift the dominant response to an allergen segment with a different ratio of tolerogenic and immunogenic T cells.

RELATIONSHIP OF EPITOPE DOMINANCE TO ANTIGEN STRUCTURE

An initial study comparing T-cell epitope maps to antigen structural features found that dominant epitopes tended to occur on the carboxy-terminal flanks of locally unstable segments of the mycobacterial GroES and three model antigens (staphylococcal nuclease, HEL, and cytochrome *c*) (Landry, 1997). For the three model antigens, the best measurements of local instability were obtained by amide group hydrogen-deuterium exchange

(HX), coupled with nuclear magnetic resonance (NMR). (For brief explanations of biophysical methods used to study protein instability, see the bulleted list below.) Subsequently, the relationship of local instability and dominance in HIV gp120 was examined. For this work, local instability was equated with local disorder, as reported by X-ray crystallographic B factors. Two features of T-cell epitope immunodominance support the conclusion that local structural instability increases the immunogenicity of adjacent helper T-cell epitopes: (i) a consistent spatial relationship between epitopes and unstable segments and (ii) a correlation in the number of immunogenic segments and the number of unstable segments in a given antigen. Local instability may increase the immunogenicity of adjacent epitopes because the mechanisms of antigen-allergen processing and presentation act preferentially on unstable segments (Fig. 2). Unstable segments are preferred sites for proteolytic cleavage, and thus processing starts at these sites. Cleavage relieves conformational constraints near the polypeptide termini, and the increased conformational freedom could favor loading of the terminal segments into antigen-presenting proteins. For a discussion of how proteolytic sensitivity relates to structural instability and disorder, see the following list.

- Local instability and proteolytic sensitivity. Proteolytic antigen processing most likely begins with an endoproteolytic nick in the antigen-allergen polypeptide chain. The protease requires that the polypeptide conform to its active site, which immobilizes four to six residues in an extended conformation around the target peptide bond (McGrath, 1999). In the context of the antigen, the polypeptide segment including the target peptide bond often is not in an extended conformation. Moreover, it forms noncovalent bonds within its structure and with other parts of the polypeptide. These bonds must be broken for the polypeptide to bind to the protease. In the absence of an active unfolding machinery (such as the ATP-dependent activities of the cytosolic proteosome), these bonds will break only as often as the ambient thermal energy supports. Thus, proteases typically cleave proteins only at the structurally least stable sites, such as in linker segments between protein domains or in loops and turns between subdomains and secondary structural elements. It is fairly easy to predict approximate sites of protease sensitivity that occur between stable domains, such as in the linkers between Fab and Fc fragments of an immunoglobulin. In contrast, predicting the best of many potential protease cleavage sites within a domain is challenging.
- Predicting protease-sensitive sites. Hubbard et al. (1994) were able to predict the most sensitive cleavage site in a protein by calculating the thermodynamic cost of refolding all possible 10- to 12-residue segments of a protein into the conformation required by the protease active site. Subsequently, Hubbard (1998) showed that most probable cleavage sites were predicted reasonably well by a combination of local disorder (B factors) observed with the X-ray crystal structure, solvent accessibility, and other parameters for 10- to 12-residue protein segments. Crystallographic B factors provided the most valuable

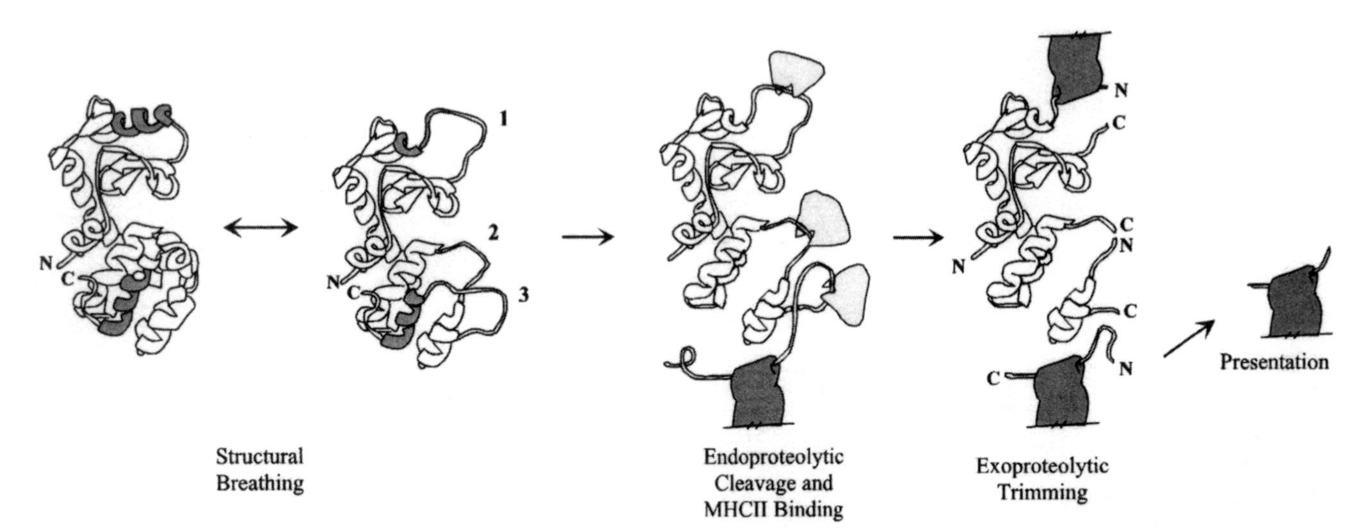

Figure 2. Model of antigen-allergen processing and presentation of dominant $CD4^+$ T-cell epitopes. Local structural breathing occurs in unstable loops between stable elements of secondary or tertiary structures. Unstable loops contain preferred sites for proteolytic cleavage because these segments of polypeptide most easily conform to protease-active sites. During or after cleavage, segments adjacent to cleavage sites become destabilized and bind to MHC-II antigen-presenting proteins. Further proteolytic trimming reduces the MHC-bound peptide to the typical size of 12 to 15 residues.

information. However, B factors are not a direct measure of structural instability.

- Protein structural stability-instability. For the present discussion, protein structure refers to the noncovalently bonded secondary, tertiary, and quaternary structures. Large forces of stabilization and destabilization poise the structure of a protein in the native conformation. On the one hand, hydrophobic interactions, hydrogen bonds, electrostatic forces, and Van der Waals contacts stabilize protein structure. On the other hand, the entropic cost of ordering the polypeptide backbone and side chains destabilizes protein structure. Stability-instability is a thermodynamic concept, a probability that something exists in one state relative to another. With regards to the folded structure of a protein, stability is the probability that the protein exists in the folded state. For most proteins, the probability of being folded is very high.
- Measuring structural stability. The classic method of measuring protein stability involves heating the protein or putting it in a solution of chaotrope in order to destabilize the protein enough to measure a significant quantity of the unfolded protein. Stability is then analyzed in terms of the fraction unfolded as a function of temperature or chaotrope concentration. In recent years, this analysis has been extended to the study of individual structural elements of a protein using a combination of amide-group hydrogen-deuterium exchange (HX) and NMR (Bai et al., 1994). Results of such studies have been very satisfying because they show that small, poorly organized, solvent-exposed elements of structure are the least stable, whereas larger, well-organized, buried elements are most stable. Local instability characterized by HX-NMR has been associated with proteolytic sensitivity (Wang et al., 1998). Unfortunately, this analysis is laborious and demands that the protein remain highly soluble in a variety of partially denaturing solutions. Thus, very few proteins have been studied so carefully.
- COREX/BEST prediction of local instability. The distribution of local structural stability in a protein can be predicted for any protein with a known three-dimensional structure using the COREX/BEST algorithm, which is accessible on the World Wide Web (http://www.best.utmb.edu/BEST/). The algorithm determines the relative free-energy cost of dissecting and unfolding all possible segments of a protein. The most important factor in the calculation is the change in solvent-accessible surface area. Residue stabilities from the COREX/BEST algorithm correlate reasonably well with results from HX-NMR (Hilser and Freire, 1996).
- Local structural disorder. All high-resolution protein structures archived in the Research Collaboratory for Structural Bioinformatics Protein Data Bank include a measure of uncertainty in the position of each atom. In X-ray crystal structures, the uncertainties are expressed as B factors, which weight the contribution of the atoms in

the backcalculation of electron density from atom positions in the structural model to maximize agreement with the experimental electron density (Ringe and Petsko, 1986). Larger B factors correspond to more conformational disorder. Generally, the peaks in backbone atom B factors occur in the loops and turns between secondary structure elements, where the polypeptide has a higher level of thermally activated conformational fluctuations. In NMR structures, the uncertainties are the root mean square deviations (RMSDs) in positions of the atoms for a group of models that satisfies the experimental interproton distance constraints. As in X-ray crystal structures, larger RMSDs correspond to more conformational disorder, and peaks occur in the loops and turns; the reasons for disorder may have as much to do with a lack of experimental distance restraints as with actual conformational disorder. Nevertheless, where structures have been determined by both crystallography and NMR, there is reasonably good agreement in the profiles of B factors and RMSDs.

The first aspect of epitope dominance that points to a relationship to antigen structure is that dominant epitopes tend to occur on the flanks of unstable segments. In an analysis of epitope-mapping data obtained with immunized mice, dominant epitopes were located an average of eight residues carboxy terminal from the center of unstable segments in HEL and in the outer domain of HIV gp120 (Dai et al., 2001; Landry, 2000). The profile of epitope dominance for the outer domain of gp120 was essentially the same in different mouse strains. In humans, the immunogenic segments of HIV gp120 were an average of 16 residues carboxy terminal from the center of unstable segments (data not shown). Again, the dominance pattern was independent of HLA (antigen-presenting protein) alleles present in these individuals.

Dominance that transcends the influence of peptide selectivity by antigen-presenting proteins has been observed in other studies (Caccamo et al., 2003; Diethelm-Okita et al., 2000; Dormann et al., 1998). For example, 86% of individuals immunized with tetanus toxoid respond to a so-called universal epitope that lies adjacent to an unstable loop in that antigen (Diethelm-Okita et al., 2000). These observations suggest that dominance is controlled by mechanisms of processing and presentation other than peptide selectivity by the antigen-presenting protein.

The second aspect of dominance that points to a relationship to antigen structure is that the number of immunogenic segments correlates with the number of flexible segments in the protein (Dai et al., 2001). This aspect of dominance becomes evident only when epitope maps are evaluated for a population of individuals. The number of immunogenic segments observed with a single individual may be much fewer than the number observed with a population, even of in-bred animals. Many types of stochastic events appear to influence the immunogenicity of a particular epitope in a given individual (Moudgil et al., 1996). Aside from the effect of preexisting tolerance on the repertoire of T cells (discussed above), the presentation of a given epitope is subject to a variety of mechanistic vagaries such as the interference by determinant capture, in which the epitope is blocked by loading of an overlapping epitope or changes in loading efficiency created by variances in the termini generated by proteolytic processing.

ANTIGEN-ALLERGEN T-CELL EPITOPE MAPS AND THREE-DIMENSIONAL STRUCTURES

T-cell epitopes have been mapped for a number of antigens and allergens whose three-dimensional structures have been determined by X-ray crystallography or NMR. T-cell epitope-mapping studies of allergens yield very few observations in comparison to mapping studies of antigens in immunized or infected individuals. Presumably, this is due to a mixed immune-tolerant response to the allergen. When T-cell specificities from immunized animals or infected humans are mapped, proliferative responses or cytokine secretion (immunospots) can be observed with freshly prepared lymphocytes. In contrast, T cells specific for the allergen are present in numbers too small to observe in freshly prepared samples. In the typical allergen epitope-mapping study, T-cell lines are established from the blood of a small group of allergic individuals. Subsequently, the lines are tested for proliferation in response to peptides spanning the length of the allergen. Generally, the number of lines that respond to a peptide (the number of epitope observations) is comparable to the number of individuals in the donor group. Thus, the density of epitope-mapping data is very low. When the data are used to score immunogenicity in the sequence of the allergen, it is likely that some epitopes have been missed, and some of the observed epitopes may be accorded more immunogenicity than is warranted.

HIV Envelope Glycoprotein

Immune responses to HIV envelope glycoprotein have received some of the most intense scrutiny by modern immunological techniques. HIV envelope glycoprotein was chosen as a model antigen for the present discussion because tolerance of self proteins should not limit the spectrum of T-cell specificities raised against it. A BLAST search of the entire glycoprotein against the combined Swiss-Prot/TrEMBL database yielded no human sequences longer than 40 amino acid residues with >34% identity. Thus, epitope immunogenicity in envelope glycoprotein is likely to track with the efficiency of peptide generation. This may not be the case for allergens that have human homologs.

In a number of respects, the immunology of this protein could be similar to that of allergens. Chronically infected patients receive long-term and massive exposure to the extracellular portion (gp120) of the protein. The $CD4^+$ T-cell response to epitopes of the antigen rapidly deteriorates after the initial exposure but long before $CD4^+$ T cells are depleted by viral toxicity. The reasons for the loss of this response are not well established but are likely to include the rapid accumulation of immune-evading mutations in the protein's epitopes and possibly the development of immune tolerance.

Epitope dominance in the helper T-cell epitope map of HIV gp120 was determined for 10 mice from each of two inbred strains (CBA and BALB/c) (Dai et al., 2001). By mapping responses in individual mice, it was possible to appreciate the variability in response to individual epitopes and also to directly measure epitope dominance. Animals were immunized by intranasal administration of recombinant gp120 with mutant (R192G) heat-labile toxin from enterotoxigenic *Escherichia coli* as the adjuvant. This adjuvant system allows the antigen to be administered in a soluble form that most likely preserves its native conformation. One week after the third weekly administration, aliquots of splenocytes were stimulated with a single member of a set of overlapping peptides spanning the full length of the immunogen. Since gp120 is fairly large (508 residues), the selection of peptide length (20-mer) and overlap (10 residues) represented a compromise between economy and the precision of epitope mapping. It was also important to ensure that there would not be too many peptides to obtain a complete map with the splenocytes from a single mouse. A response was graded positive if proliferation, measured by incorporation of tritiated thymidine, exceeded the unstimulated control by a factor of 4 (stimulation index of 4).

An epitope may be considered dominant if splenocytes from a majority of the mice (at least 6 of 10) proliferate in response to the corresponding peptide. Using that definition, 8 of 47 peptides would be classified as immunodominant in each of the two mouse strains. All 8 of these peptides were among the top 10 ranked by the average stimulation index for 10 mice. Generally, the peptides with the largest average stimulation index obtained responses in the largest number of mice. Thus, stimulation index probably is a good indicator of immunogenicity.

The relationship of antigen structure to T-cell epitope dominance emerges when dominance is compared for two mouse strains. Epitope clustering differs greatly in the N-terminal and C-terminal halves of gp120, which roughly correspond to the inner and outer structural domains of the protein. The designations inner and outer refer to the dispositions of the domains in the envelope glycoprotein trimer (Kwong et al., 1998). When the epitope maps for CBA and BALB/c strains were compared, it was evident that dominant segments in the inner domain were unique to the individual mouse strain, whereas dominant segments in the outer domain were common to both mouse strains (Fig. 3A and B). The inner domain undergoes major structural rearrangements when the protein binds to CD4 prior to viral entry into cells (Chen et al., 2005). The lack of stable structure in the inner domain allows access by the MHC proteins to a wide range of sequence, and so MHC-peptide-binding affinity largely determines epitope selection. In contrast, the outer domain is relatively rigid. Its stable structure limits access by the MHC to a narrow range of immunogenic segments. Proteolytic cleavage and MHC protein binding both require that antigen segments be exposed and in an extended conformation, and thus processing and presentation are both limited by stable antigen structure. As a result, epitopes cluster between elements of protein structure and near sites that are susceptible to proteolysis.

In HIV gp120, the intersection of structure and disorder determines the sites of strongest T-cell epitope immunogenicity. The most striking MHC-independent clustering of epitopes in gp120 occurs adjacent to and overlapping the V4 hypervariable loop in the outer domain. The V4 loop stretches from residues 396 to 405, and a majority of both CBA and BALB/c mice responded to a peptide in the region spanning residues 400 to 430 (Fig. 3C and D).

Studies of gp120 epitopes in humans are likely to have been hindered by generally poor T-cell responses, and so there have been very few systematic mapping efforts. However, one recent study utilized ELISPOT counting technique to quantify IFN-γ-secreting cells in freshly prepared peripheral blood mononuclear cells of seven HIV-positive individuals using the same set of overlapping peptides used in the mouse studies described above (Kleen et al., 2004). Since most of the IFN-γ-secreting cells were CD4$^+$, the immunodominance pattern should be comparable to the proliferative responses of mouse splenocytes. Indeed, the pattern of responses for a single individual, expressed in numbers of spot-forming units, resembles the immunodominance pattern obtained with immunized mice. Notably, there is a major peak in the segment spanning 400 to 430 (Fig. 3E). Moreover, six of the seven individuals responded to at least one peptide in this segment (data not shown).

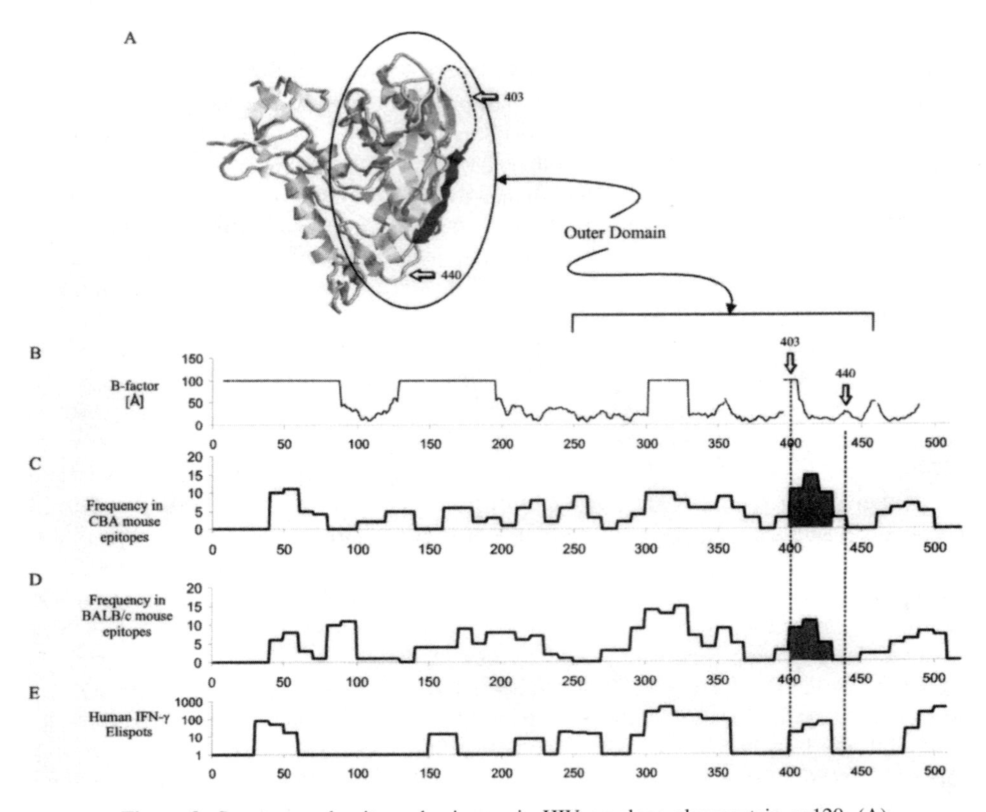

Figure 3. Structure and epitope dominance in HIV envelope glycoprotein gp120. (A) Ribbon diagram of the crystal structure of core gp120, showing the dominant epitope identified in CBA and BALB/c mice (highlighted in gray). (B) Profile of crystallographic B factors for nitrogen atoms in core gp120. (C) Epitope dominance determined by proliferation of splenocytes from gp120-immunized CBA mice, with the dominant epitope highlighted in gray. (D) Epitope dominance determined by proliferation of splenocytes from gp120-immunized BALB/c mice, with the dominant epitope highlighted in gray. (E) Epitope dominance determined by counting IFN-γ immunospots formed by freshly prepared T cells from a single HIV$^+$ donor. The single most immunogenic sequence of gp120 is the same in both CBA and BALB/c mice, and it corresponds to an epitope in the HIV$^+$ donor. Since this sequence is flanked by peaks in the B-factor profile, it is likely to be processed and presented most efficiently. Data sources were as follows: for crystal structure and B factors, Protein Data Bank (PDB) accession no. 1GC1; epitopes in mice were from Dai et al. (2001); and epitopes in humans were from Kleen et al. (2004). Data processing was as follows. Experimental B factors were not available for the hypervariable loops, either because these segments were removed from the protein prior to crystallization or because their disorder prevented electron density from being observed. For these segments, B factors were arbitrarily assigned a value of 100. Epitope frequencies and ELISPOT numbers were assigned to individual residues on the basis of the residue's presence in a peptide that stimulated proliferation or cytokine expression. A given residue was counted twice if it was present in two consecutive (overlapping) peptides that stimulated a response.

HIV gp120 was selected as a model antigen for the discussion of antigen structure and epitope dominance because gp120 has several features that may be useful for understanding dominance in allergens. (i) The lack of homologous proteins in mice presumably allows dominance to reflect peptide presentation. (ii) The availability of high-resolution structures allows the identification of locally unstable segments. (iii) The outer domain provides a good example of epitope dominance controlled by local instability. (iv) As for allergens, human exposure to gp120 is chronic and T-cell proliferation is weak, but the epitope map obtained by cytokine ELISPOT is similar to the epitope map obtained by T-cell proliferation in mice.

HEL and Mouse Lysozyme M

HEL is a significant egg allergen, and it is one of the best-studied proteins with regard to structure and experimental immunology. HEL-specific T-cell responses in allergic humans have been demonstrated (Holen and Elsayed, 1996), but so far none have been mapped. HEL-specific T-cell epitopes have been identified in multiple strains of immunized mice (Gammon et al., 1991; Moudgil, 1999; Moudgil et al., 1997). Proliferative responses have been mapped to the epitope determinant cores by peptide scanning in one-residue steps (Gammon et al., 1991).

Some of the research with HEL suggests that structure has less influence than primary sequence on T-cell epitope immunodominance. The dominant epitope in several I-A^k-containing strains of mice was mapped to a peptide spanning residues 52 to 61 (Fig. 4E). The most important feature of this epitope was said to be the presence of a single aspartic acid residue that binds to the P1 site of the I-A^k molecule (Nelson et al., 1996). A more recent study highlighted the role of H-2M in catalyzing the exchange of peptides in I-A^k, which results in the selection of peptides with the highest affinity for I-A^k (Lovitch et al., 2003). These studies emphasize peptide affinity as a major determinant of epitope immunodominance, and some investigators have concluded that the folded structure of HEL has little influence on epitope selection (Latek and Unanue, 1999; Moudgil et al., 1997).

One recent study using BALB/c mice found that denatured HEL selected the same T-cell specificities as native HEL but that the cytokine profile was skewed toward Th1 immunity (Ria et al., 2004). If HEL's four disulfide bonds are reduced and carboxymethylated, the protein loses almost all of its native three-dimensional structure (Gekko et al., 2003). As expected, the antibodies raised against HEL and reduced carboxymethylated HEL (RCM-HEL) were not cross-reactive. However, the T cells were cross-reactive and of the same clonotype by Vβ usage and recognized peptide-MHC complexes on the same APCs. The dominant I-E^d-restricted epitope in BALB/c mice is located in the region spanning residues 108 to 116 for native HEL, and it remains so for the RCM-HEL. In spite of these similarities, the T cells primed by RCM-HEL produced more IFN-γ and less IL-4 than those primed by HEL. The IgG1-IgG2a antibody isotype ratio also indicated a shift to Th1 for RCM-HEL. Although this study does not support the proposal that structure influences epitope dominance, it shows that structure can have a significant effect on the immunological outcome.

Nevertheless, other work suggests that structure stabilized by disulfide bonds may be profoundly important for T-cell epitope immunodominance in HEL. Early work in this area showed that the Cys^6-Cys^{127} disulfide bond stabilized HEL against proteolytic degradation

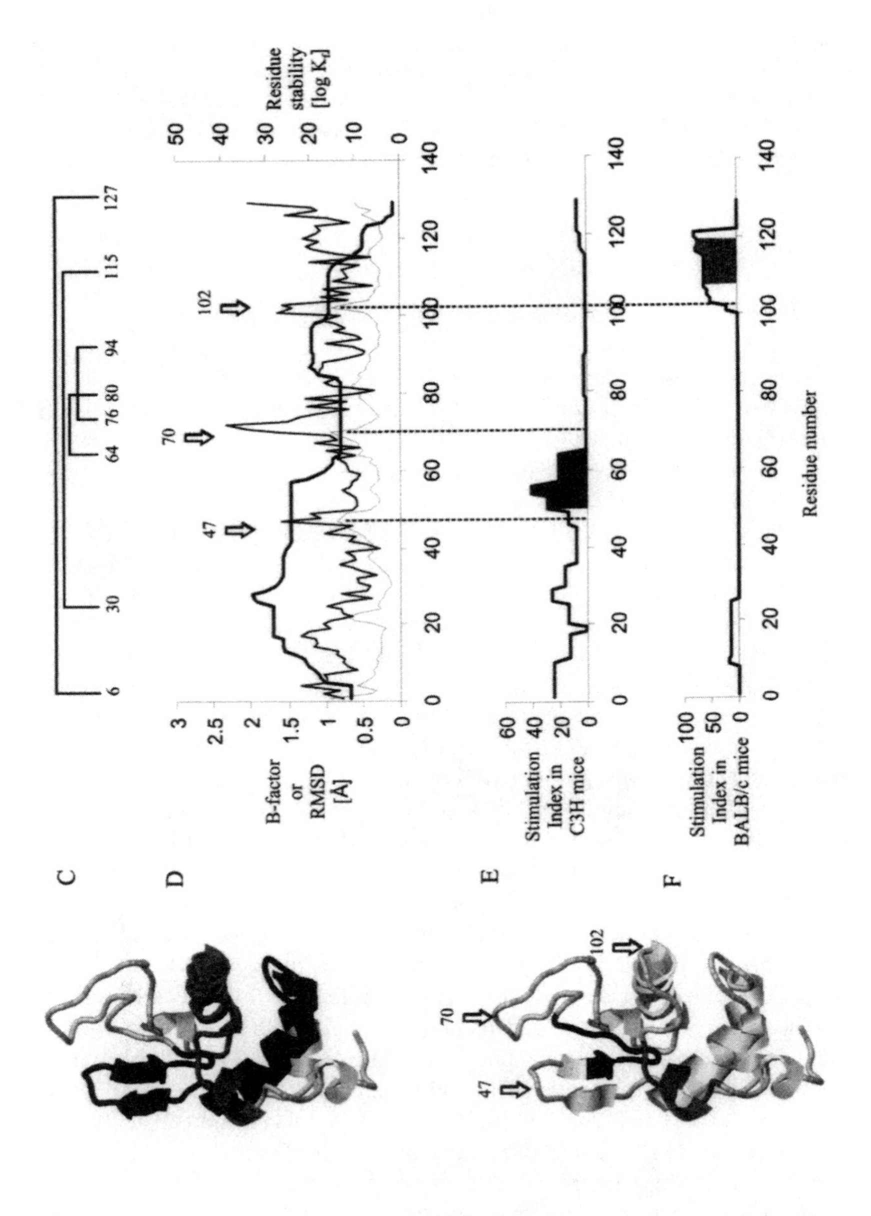

A
B
C
D
E
F
47
70
102
6
30
64
76
80
94
115
127
B-factor or RMSD [Å]
Residue stability [log K_d]
Stimulation Index in C3H mice
Stimulation Index in BALB/c mice
Residue number
0
20
40
60
80
100
120
140
0.5
1
1.5
2
2.5
3
10
30
50

Figure 4. Structure and epitope dominance in HEL. (A) Ribbon diagram of the NMR structure, showing large and small stable subdomains identified in the COREX/BEST residue stability profile (highlighted in black and gray, respectively). (B) Ribbon diagram showing the dominant epitopes in C3H and BALB/c mice (highlighted in black and gray, respectively). (C) Diagram of disulfide bond connectivity. (D) Three profiles of flexibility-stability: nitrogen atom RMSDs from NMR (thin line), nitrogen atom crystallographic B factors (medium line), and COREX/BEST residue stability profile (thick line). (E) Profile of stimulation index for proliferation of splenocytes from immunized C3H mice. (F) Profile of stimulation index for proliferation of splenocytes from immunized BALB/c mice. Arrows and dashed lines indicate peaks in the flexibility profile that may correspond to proteolytic cleavage sites. The dominant epitope in C3H mice lies between two peaks of flexibility and on the carboxy-terminal flank of a stable subdomain. Likewise, the dominant epitope in BALB/c mice lies between a peak of flexibility and the carboxy terminus and on the carboxy-terminal flank of a stable subdomain. Data sources are as follows: NMR structure and RMSDs, PDB accession no. 1E8L; B factors, PDB accession no. 2LYM; epitopes in C3H mice were from Sinha et al. (2004); and epitopes in BALB/c mice were from Ria et al. (2004). For data processing, RMSDs were calculated by the facility in MOLMOL (Koradi et al., 1996) for the mean of 50 NMR structures reported in the PDB file. Residue stabilities were calculated with COREX/BEST with the minimized average NMR structure (PDB accession no. 1GXV) and default values for all parameters, with an entropy weighting factor of 0.95.

and blocked presentation of the dominant epitope in BALB/c mice in vitro (So et al., 1997). More recently, the importance of disulfide reduction was demonstrated in mice in which the GILT had been knocked out (Maric et al., 2001). The response of GILT knockout mice to HEL was reduced to 1/10 that of wild-type mice. GILT appears to be essential for presentation of the I-A^b-restricted HEL epitope 46-61 but not epitopes 20-35 and 30-53. The authors suggested that the GILT-dependence of 46-61 could be related to the stability of the three-dimensional structure in proximity to the disulfide bond because structure blocks processing and presentation. Alternatively, GILT dependence could be related to the stability of the disulfide bond itself.

The distinctive profiles of proliferative responses for different mouse strains have often been attributed to sequence preferences of different MHC proteins, but experiments fail to show any relationship between peptide-binding affinity and the strength of immune response (Kim et al., 1997; Ma et al., 1999; Phelps et al., 1998). In one attempt at reducing dominance of a particular HEL epitope, peptide binding to the MHC protein was disrupted by mutations in the epitope, but the result was presentation of essentially the same peptide by a different MHC protein. When the dominant epitope observed with BALB/c mice (103-117) was mutated to destroy its binding to I-E^d, the resulting mutant HEL primed a dominance profile that was almost identical to that of HEL, except that the dominant epitope now utilized I-A^d instead of I-E^d (Gapin et al., 1997). Thus, it is possible that the wild-type and mutant 103-117 peptides are presented equally well, and I-A^d happens to do the presenting of the mutant sequence. (This study also provided an example wherein the epitope mapping of proliferative responses differed from epitope mapping of hybridomas. Most of the hybridomas prepared from mice immunized with the mutant HEL were specific for an epitope at 7-13 or 47-65, rather than the dominant epitope at 103-117. Thus, epitope maps obtained with cultured T cells could be a poor representation of epitope dominance in vivo.)

Distinctive epitope dominance profiles for different mouse strains could be due to differences in the relationship of immunogenicity to abundance of peptide presentation. Responses to epitopes that had been considered cryptic in BALB/c mice, including 11-25 and 46-61, were found to be immunogenic when the detection was by IFN-γ or IL-4 ELISPOT, rather than by cell proliferation (Peters et al., 2005). Thus, the dominance of the epitope at 108-116 in BALB/c mice could be due to a cytokine phenotype expressed by the T cells that favor proliferation. Other epitopes might be presented even more abundantly than 108-116, but the resulting T cells are less proliferative.

Dominance in the profile of tolerogenicity, rather than immunogenicity, may more faithfully represent the efficiency of peptide presentation. Transgenic mice expressing low levels of HEL (<2 ng/ml in serum) could be immunized with HEL (Gapin et al., 1997). The response could be recalled by intact HEL or the dominant 103-117 epitope. However, the response could not be recalled by mutant HEL in which the dominant I-E^d epitope had been disrupted. Thus, the dominant epitope at 103-117 was not responsible for tolerance of HEL. The authors concluded that epitopes in the segments 7-31 and 47-65 were responsible for the tolerance. If tolerance reflects presentation more faithfully than immunity, then epitopes at 7-31 and 47-65 are more efficiently presented than 103-117.

The epitope dominance profiles for tolerance of mouse (self) lysozymes potentially shape the epitope dominance profile for the immune response to HEL. The T-cell response to mouse lysozyme M has been examined for wild-type and lysozyme M knockout mice (Sinha et al., 2004). Mouse lysozyme M is 56% identical to HEL, and the backbone atoms of the

NMR-determined structures superimpose with a 1.27-Å average RMSD (Obita et al., 2003). As expected, a response to mouse lysozyme was not detected by lymphocyte proliferation after an attempted immunization with complete Freund's adjuvant. In contrast, significant proliferation was obtained with lymphocytes from immunized knockout mice. The lysozyme M response in knockout mice was much weaker than the HEL response (as gauged by stimulation index). The weaker response to lysozyme M could have been due to the nearly fivefold-lower amount of immunogen administered. Alternatively, the weaker response might be due to tolerance induced by other lysozymes that the mice continue to express.

Only one mouse lysozyme epitope (105-119) was immunogenic in the knockout mice (Fig. 5C). The dominance of this epitope is somewhat surprising because the C3H mice contain I-A^k and therefore might have been expected to respond to the dominant HEL peptide at 52-61. The 52-61 sequence is almost identical in mouse lysozyme M, differing only in the conservative substitution of Phe for Leu^{57}. Leu^{57} is solvent exposed in the crystal structure of the I-A^k–peptide complex (Fremont et al., 1998), and therefore its substitution is not expected to affect peptide binding to the MHC protein. The 105-119 sequence is different from the corresponding HEL epitope at 7 of 15 residues, but the changes are not expected to improve I-A^k-binding on the basis of the known preferences. Tolerance of the dominant epitope and of subdominant epitopes in flanking sequences (95-109 and 110-125) of lysozyme M was demonstrated by the successful immunization of knockout mice but not wild-type mice with the synthetic peptides. These data indicate that epitope dominance has shifted from 52-61 in HEL to 105-119 in mouse lysozyme M, but the shift cannot be explained by differences in the amino acid sequence. It is possible that tolerance of remaining lysozymes, including lysozyme P (>95% identity to lysozyme M), blocks the response to the more conserved epitope that was dominant in HEL.

The potential for tolerance of the mouse lysozyme M to influence epitope selection in the response to HEL was examined by comparing the epitope profiles in wild-type and the lysozyme M knockout mice. In spite of substantial overall identity between the endogenous lysozyme M and HEL and an almost identical sequence of the dominant HEL epitope at 52-61, the profile was completely unaffected. This result is difficult to reconcile with the observation that endogenous lysozyme M blocked the immune responses to the dominant lysozyme M epitope at 105-119 and the flanking subdominant epitopes. The best explanation may be that tolerance of one or more additional endogenous lysozymes, which remain present in the knockout mice, constrain epitope selection against HEL in a manner indistinguishable from tolerance of the full suite of lysozymes, including lysozyme M. Thus, the question as to what extent self tolerance influences the repertoire against HEL remains unanswered.

Regardless of how immune recognition is analyzed, dominant epitopes tend to be sandwiched between flexible loops that could be preferred sites for proteolytic cleavage of HEL (Fig. 4D to F). The I-A^k-restricted 52-61 epitope lies between peaks of local structural disorder at residues 47 and 70, as reported by B factors from the crystal structure or RMSDs from the NMR structure. Likewise, the I-E^d-restricted 108-116 epitope lies between a peak of structural disorder at residue 102 and the carboxy terminus.

Limited proteolysis of HEL confirms that preferential cleavage occurs near residue 47 (Fontana et al., 2004), which coincides with a peak of local structural disorder. An experiment with HEL with intact disulfide bonds was not successful because the protein is generally resistant to proteolytic cleavage. However, the oxidized form of the homologous bovine

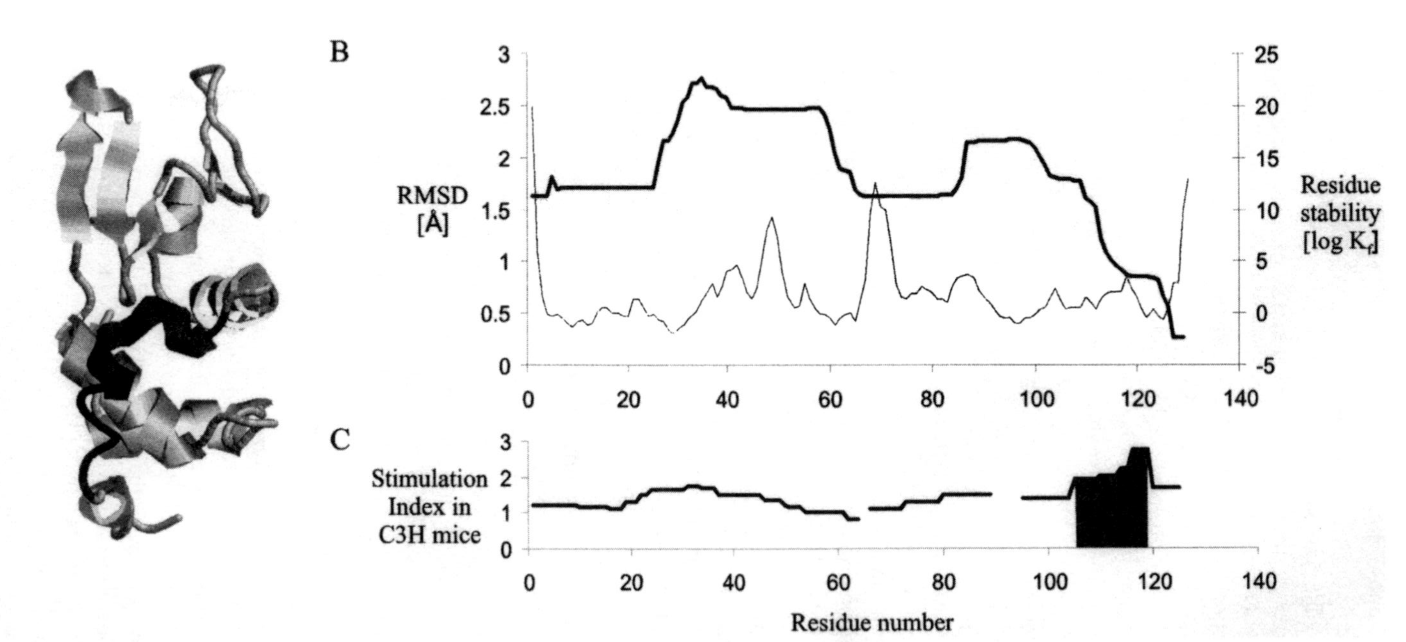

Figure 5. Structure and epitope dominance in mouse lysozyme M. (A) Ribbon diagram of the NMR structure, showing the dominant epitope in C3H mice (highlighted in gray). (B) Profiles of nitrogen atom RMSD from NMR (thin line), and COREX/BEST residue stability profile (thick line). (C) Profile of stimulation index for proliferation of splenocytes from immunized C3H mice. Lysozyme M does not have a strong response at residues 51 to 62, which was dominant for HEL in the C3H mice. The dominant epitope for lysozyme M in C3H mice corresponds to the dominant epitope for HEL in BALB/c mice. The sequence of residues 52 to 61 is nearly identical in HEL and lysozyme M; thus, the lack of response in lysozyme M probably is not due to peptide selectivity by the MHC-II protein. Profiles of flexibility-stability in HEL and lysozyme M are similar; thus, the pathways of proteolytic processing for the two proteins are likely to yield 52-61 in comparable abundance. The absence of response to 52-61 could be due to tolerance of this highly conserved segment in other mouse lysozymes, including lysozyme P. Data sources are as follows: for NMR structure and RMSDs, PDB accession no. 1IVM; epitopes in C3H mice were from Sinha et al. (2004). Data processing was as follows: RMSDs were calculated by the facility in MOLMOL (Koradi et al., 1996) for the mean of 20 structures reported in the PDB file. Residue stabilities were calculated using COREX/BEST with model 1 of the NMR structure (PDB accession no. 1IVM) and default values for all parameters, with an entropy weighting factor of 0.90.

lactalbumin was preferentially cleaved in the region spanning resides 34 to 57, which includes the peak in disorder at residue 47. Reduced HEL was efficiently digested by cathepsins D, L, S, and B (Pluger et al., 2002). Each enzyme produced a characteristic pattern of fragments, but several preferred cleavage sites were shared by two or more of the enzymes. Preferred cleavages were observed after residues 59 and 110 (cathepsin D); residues 30, 57, 76, and 85 (cathepsins B and L); and residues 15, 30, 45, and 57 (cathepsin S). In this same study, presentation of two I-A^b-restricted epitopes (30-44 and 46-49) to T-cell hybridomas was shown to be dependent on cathepsin S. The authors attributed the dependence to the uniquely cathepsin S-sensitive cleavage after residue 45.

The two dominant HEL epitopes 52-61 and 108-116 are located on the carboxy-terminal flanks of structurally stable subdomains. The COREX/BEST residue stability profiles (see the bulleted list above) for HEL and mouse lysozyme M exhibit two broad segments of high stability that are separated by the flexible loop centered at residue 70 (Fig. 4D and 5B, respectively). The larger and more stable of the two segments spans nearly the first 60 residues of HEL, including the most buried α helix and the two-stranded β sheet. HEL epitope 52-61 lies on the carboxy-terminal flank of this segment, including a small part of the second β strand and the adjacent loop up to its most flexible point (Fig. 5A and B). The second stable segment spans approximately 30 residues near the carboxy terminus, including the longest α helix, a short stretch of loop, and a single turn of the α helix. HEL epitope 108-116 lies on the carboxy-terminal side of this segment, spanning the loop and single turn of the α helix. These epitope locations are consistent with a model for processing and presentation, where the MHC protein binds near the ends of stable protein subdomains that correspond to major proteolytic fragments (Fig. 2).

Ovalbumin

Ovalbumin is another of the major egg allergens. Although it is a member of the serpin family of protease inhibitors, it is not known to inhibit any proteases. Serpins exhibit a distinctive conformational change when cleaved in the reactive center loop. A segment of the polypeptide that formed the reactive center loop inserts into the middle of a β sheet. Although native ovalbumin does not undergo this conformational change, a single amino acid substitution enables the insertion mechanism (Huntington et al., 1997). Thus, the native structure of ovalbumin represents the metastable form adopted by all serpins. A BLAST search with the complete sequence of chicken ovalbumin against the combined Swiss-Prot/TrEMBL database yielded human sequences with no more than 40% identity.

Many studies have utilized ovalbumin as a model antigen for understanding T-cell immunity, including some of the first evidence that proteolytic antigen processing is required for T-cell recognition. Initial epitope-mapping studies were undertaken with T-cell hybridomas generated from the splenocytes of immunized mice, and later studies used T-cell lines established by culturing peripheral blood mononuclear cells from allergic humans for a period of days or weeks in the presence of the antigen and recombinant IL-2.

The available T-cell epitope data for ovalbumin suggest that dominant epitopes occur between disordered loops that could be preferred sites of proteolytic cleavage. In an early study, fractions from the tryptic digestion of ovalbumin were tested for the ability to restimulate a mouse hybridoma (Shimonkevitz et al., 1984). A single epitope was identified in the peptide 323-339. That same peptide was later shown to be recognized by a T-cell line from an allergic human (Shimojo et al., 1994). This epitope lies on the amino-terminal flank of the reactive center loop, where there are several preferred sites of proteolytic cleavage (Fig. 6B, D, and E) (Takahashi et al., 1991). Subsequently, another study of mice identified eight I-A^s-restricted epitopes recognized by 1 or more of 19 hybridomas (Vidard et al., 1992). In this study, each hybridoma was tested against a full set of overlapping synthetic peptides. The hybridomas most frequently (6 of 19) recognized an epitope in the segment 233-249, which corresponds to a β strand in the middle of a large β sheet (Fig. 6B). Two studies with ovalbumin-specific T-cell lines generated from six allergic humans identified a total of six epitopes (Holen and Elsayed, 1996; Katsuki et al., 1996). Three epitopes were located between the amino terminus and a large flexible loop that contains a disulfide-sensitive subtilisin cleavage site (Fig. 6D and F) (Takahashi et al., 1991). Of the remaining three epitopes, two overlap a β strand that precedes the β strand bearing the dominant epitope in mice (Fig. 6A and B). The consecutive β strands are flanked by flexible loops centered at residues 204, 225, and 251 (Fig. 6D). Thus, dominant epitopes in mice and humans are sandwiched between known or potential proteolytic cleavage sites.

Immunogenic regions of ovalbumin tend to occur on the amino-terminal or carboxy-terminal flanks of structurally stable subdomains (Fig. 6C and D). Three epitopes associated with ovalbumin allergy occur on the amino-terminal or carboxy-terminal flank of a subdomain spanning residues 1 to 49 that has high stability, according to the COREX/BEST algorithm. The subdomain is composed of two long α helices that are separated by an intervening loop and a short β strand. Two other epitopes associated with allergy occur on the amino-terminal flank of a highly stable subdomain spanning residues 214 to 242 that is composed of three adjacent β strands in a six-stranded sheet. The dominant epitope in mice occurs on the carboxy-terminal flank of this subdomain. The relationship of these epitopes and stable subdomains in ovalbumin suggests that the pathway of antigen processing determines epitope immunodominance. In the end stage of proteolytic fragmentation, the MHC protein appears to select one or other of a subdomain's ragged ends.

BLG

BLG is a major allergen of cow's milk. BLG is a member of the lipocalin superfamily of proteins. Although humans are likely to have a number of homologous proteins, a BLAST search with the complete sequence of BLG against the combined Swiss-Prot/TrEMBL database yielded human sequences with no more than 46% identity. Likewise, there were no homologs in mice with >29% identity. Lipocalins are small single-domain proteins characterized by a meandering eight-stranded β sheet that forms a β barrel and a single carboxy-terminal α helix that is pressed to the outside of the barrel. Lipocalins transport and store lipophilic ligands inside the β barrel. The ligands for some lipocalins are known, such as that of serum retinol-binding protein; however, in the case of BLG, the physiologically relevant ligand has not been identified. Many biophysical studies have been performed on BLG in an effort to understand how its properties affect milk processing and allergenicity. BLG exists as a dimer (Brownlow et al., 1997). It is resistant to heat denaturation and proteolysis, especially at low pH (Iametti et al., 2002). Below

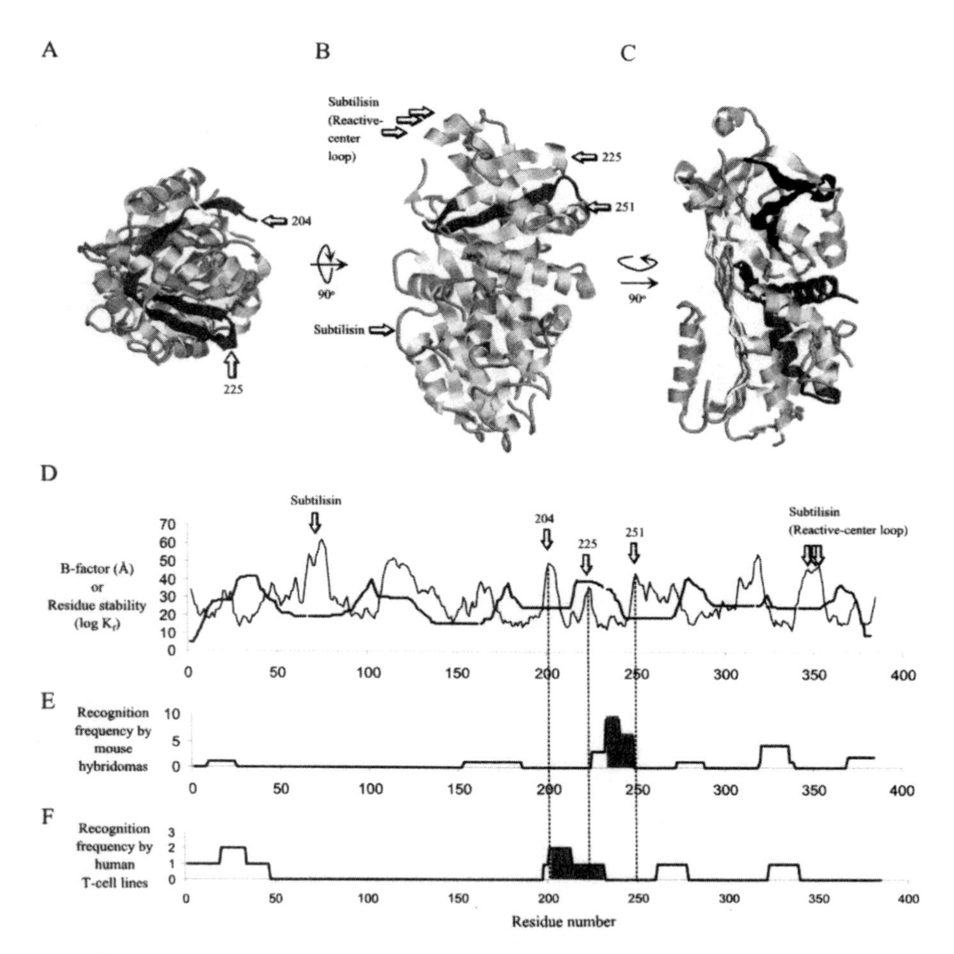

Figure 6. Structure and epitope dominance in chicken ovalbumin. (A) Ribbon diagram of the crystal structure, showing one of the dominant epitopes in allergic humans (highlighted in gray). (B) Ribbon diagram showing the dominant epitope in immunized mice (highlighted in gray). (C) Ribbon diagram showing two stable subdomains that may promote the dominance of epitopes in segments 1 to 49 and 204 to 251 (highlighted in gray and black, respectively). (D) Profiles of flexibility-stability, indicated by crystallographic B factor (thin line) and COREX/BEST residue stability (thick line). (E) Epitope dominance determined by frequency of recognition by hybridomas generated from an immunized mouse. (F) Epitope dominance determined by frequency of recognition by T-cell lines generated from allergic humans. Arrows and dashed lines indicate peaks in the flexibility profile that may correspond to proteolytic cleavage sites. The dominant epitope in mice lies between flexible sites and on the carboxy-terminal flank of a stable subdomain. One of two dominant epitopes associated with allergy lies on the amino-terminal flank of the same stable subdomain that contains the dominant mouse epitope. The other dominant epitope associated with allergy lies in a stable subdomain between the amino terminus and a protease-sensitive site. Data sources are as follows: crystal structure and B factors, PDB accession no. 1UHG; epitopes in mice were from Shimonkevitz et al. (1984) and Vidard et al. (1992); epitopes in humans were from Holen and Elsayed (1996) and Katsuki et al. (1996). Data processing was as follows: residue stabilities were calculated using COREX/BEST with the crystal structure (PDB accession no. 1UHG) and default values for all parameters, except with a window size of 27, temperature of 37°C, and an entropy weighting factor of 0.98.

pH 7.2, BLG undergoes the so-called Tanford transition, characterized by the protonation of Glu^{89} and opening of the barrel's lid, a loop that spans residues 85 to 90 (Qin et al., 1998).

BLG T-cell epitopes have been mapped with T-cell lines and clones from allergic humans (Inoue et al., 2001). Epitope determinant cores also have been mapped with proliferative responses of lymphocytes from multiple strains of immunized mice (Totsuka et al., 1997).

No MHC allele-independent dominant epitope stands out in the profile of BLG epitopes mapped in mice. Distinct patterns of determinant cores were mapped for each of three strains of mice (C57BL/6, C3H/HeN, and BALB/c) (Fig. 7E to G). This leads to two conclusions. (i) The various BLG epitopes are presented with nearly equal efficiency. (ii) MHC-peptide binding has a strong influence on epitope selection. Nevertheless, the determinant cores tend to be sandwiched between pairs of flexible loops (Fig. 7D), which suggests that structure specifies the location of epitopes.

With closer inspection of the mouse mapping data, structure emerges as the strongest influence on epitope dominance in BLG. The determinant cores observed with C57BL/6 and C3H/HeN mice are associated with the two stable subdomains (Fig. 7D). In contrast, the determinant cores observed with BALB/c mice are distributed throughout the sequence. The BALB/c profile was even more distinctive in that the epitopes in an unstable segment (residues 67 to 88) stimulated much higher levels of proliferation than epitopes from any other region of the protein (Totsuka et al., 1997). Thus, results with BLG resemble those with lysozyme. Dominant epitopes are located within or on the flanks of stable subdomains (Fig. 3E and 7E and F), except in BALB/c mice, where the dominant epitope is in a relatively unstable region (Fig. 4F and 7G).

Small differences in BLG stability can give rise to significant differences in T-cell responses. One study compared the ability of A and B variants of BLG to present epitopes in the sequence 25-40 (Ametani et al., 2003). The variants differ at only two positions in the amino acid sequences, Asp^{64} and Val^{118} in the A variant or Gly^{64} and Ala^{118} in the B variant. Epitopes were presented more efficiently by the B variant to T cells from each of three strains of mice that had been immunized with appropriate peptides. When the antigen was reduced and carboxymethylated, the difference in presentation was no longer observed. Thus, the difference in presentation by the intact variants depends on the three-dimensional structure. The authors note that the B variant has slightly higher levels of crystallographic B factors in the region of the epitopes, which could account for the more efficient presentation of epitopes in that region.

Epitopes associated with allergy occur most frequently near a flexible segment of BLG that is sensitive to proteolysis. Epitopes identified with T-cell lines from four allergic individuals were well distributed in the BLG sequence (Fig. 7H) (Inoue et al., 2001). Nevertheless, an amino-terminal segment spanning residues 14 to 21 appears to be immunodominant because three of the T-cell lines responded to peptides overlapping this sequence (Fig. 7B, C, and H). Although BLG is generally resistant to proteolysis, cleavage by chymotrypsin and trypsin becomes increasingly more efficient at elevated temperatures (Iametti et al., 2002). The preferred cleavage sites for the two proteases are after adjacent residues, Leu^{39} and Arg^{40}, respectively, which lie on the carboxy-terminal flank of a stable subdomain (Fig. 7C and D). Thus, the most allergenic segment of BLG lies in a stable subdomain that lies between the amino terminus and the most protease-sensitive site in the protein.

Birch Allergen Bet v 1

Bet v 1 is the major birch tree pollen allergen, and it is one of the best-studied aeroallergens. Bet v 1 is a member of the pathogenesis-related 10 proteins, which are ubiquitously expressed by plants in response to stress (Breiteneder and Ebner, 2000). Crystallographic and NMR structures of Bet v 1 have been determined. Bet v 1 resembles the lipocalins in two aspects. It is a small single-domain protein composed of a meandering β sheet and a small amount of α helix, and it is most likely a carrier molecule for hydrophobic ligands. In contrast to the lipocalins, the seven β strands of Bet v 1 do not complete a β barrel. The sheet partially enwraps a long α helix that forms a bundle with two short α helices. A large hydrophobic cavity between the sheet and long helix is thought to bind hydrophobic ligands, possibly two phytosteroid molecules (Markovic-Housley et al., 2003; Neudecker et al., 2001). A BLAST search with the complete sequence of Bet v 1 against human sequences in the combined Swiss-Prot/TrEMBL database yielded stretches of at least 50 residues with no more than 35% identity.

Sensitivity to Bet v 1 is associated with allergy to a number of fruits and vegetables, including celery, apples, hazelnuts, pears, cherries, and carrots (Fritsch et al., 1998; Neudecker et al., 2001; Osterballe et al., 2005). Sensitization to birch pollen is thought to initiate the sensitization to the related food allergens. Support for this conclusion has been obtained at the T-cell level. Many of the T cells induced by Bet v 1 cross-react with the homologous celery allergen Api g 1 in spite of only 41% overall amino acid sequence identity (Bohle et al., 2003).

The distribution of T-cell epitopes in Bet v 1 cannot be explained by peptide selectivity in the antigen-presenting proteins. One study examined the response to native and chemically modified Bet v 1 by 22 T-cell clones from nine allergic individuals (Dormann et al., 1998). Most of the Bet v 1 sequence was recognized by at least one of the clones, and several immunodominant segments reacted with four or more clones. The immunodominant segments did not associate with any particular HLA allele. Intramolecular cross-linking of Bet v 1 eliminated presentation or recognition of some of the epitopes. For at least one of the dominant segments (82-91), a defect in processing or presentation was the most likely explanation for the lack of response. There were few, if any, cross-linking sites in the epitope; modification of Bet v 1 with another reagent that does not cross-link but has similar specificity had almost no effect on presentation or recognition of the epitope. The identification of similar epitopes in other labs supports the lack of allele specificity in Bet v 1 epitope selection. One epitope observed with this study (10-19) corresponds to a highly promiscuous epitope (21-33) identified in allergic individuals by another study (Friedl-Hajek et al., 1999). Another segment (145-156) corresponds to the dominant epitope identified in BALB/c mice (139-152) (Bauer et al., 1997).

Dominant epitopes in Bet v 1 are situated between flexible sites and are associated with a highly stable subdomain. The two most immunogenic segments, 82-91 and 113-124, correspond roughly to the fourth and sixth β strands of the seven-stranded β sheet (Fig. 8A and D).

In the meandering sheet, each β strand is connected to the adjacent β strands by reverse turns, which are marked by high levels of crystallographic B factors and thus could be preferred protease cleavage sites (Fig. 8C). Each of the immunodominant segments is on one of the flanks of a large stable subdomain, marked by high residue stability in the COREX/BEST profile (Fig. 8C and D). The stable subdomain roughly corresponds to the

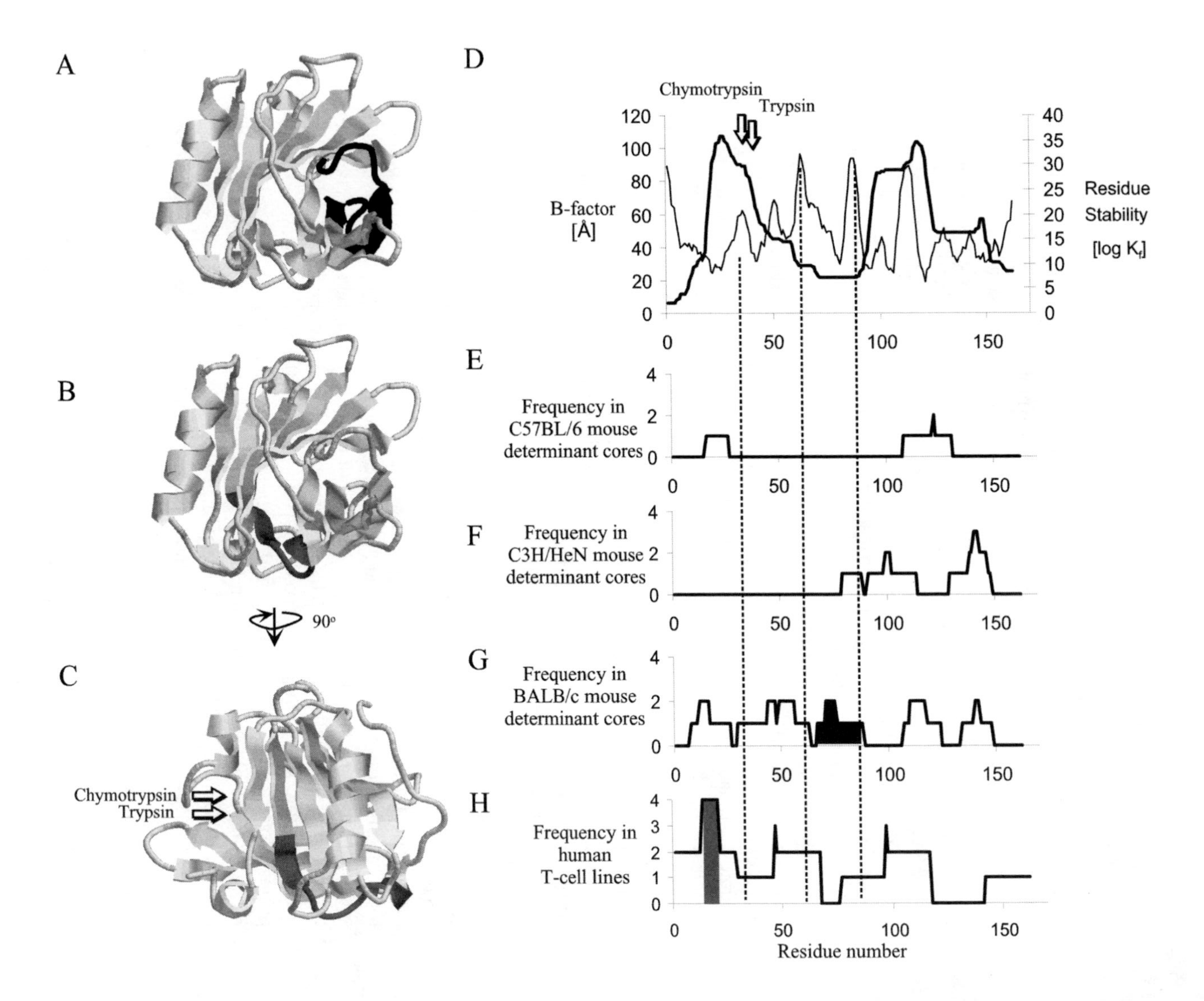
A
B
90°
C
Chymotrypsin
Trypsin
D
Chymotrypsin
Trypsin
B-factor [Å]
Residue Stability [log K_f]
E
Frequency in C57BL/6 mouse determinant cores
F
Frequency in C3H/HeN mouse determinant cores
G
Frequency in BALB/c mouse determinant cores
H
Frequency in human T-cell lines
Residue number

Figure 7. Structure and epitope dominance in BLG. (A) Ribbon diagram of the crystal structure, showing a structurally unstable segment that is immunodominant in BALB/c mice (highlighted in black). (B) Ribbon diagram of the crystal structure, showing the dominant epitope in allergic humans (highlighted in gray). (C) Ribbon diagram as in B rotated 90° about the vertical axis. (D) Profiles of flexibility-stability, indicated by nitrogen atom crystallographic B factor (thin line) and COREX/BEST residue stability (thick line). (E) Epitope dominance determined by frequency of occurrence in determinant cores observed with C57BL/6 mice. (F) Epitope dominance determined by frequency of occurrence in determinant cores observed with C3H/HeN mice. (G) Epitope dominance determined by frequency of occurrence in determinant cores observed with BALB/c mice. (H) Epitope dominance determined by frequency of recognition by T-cell lines generated from allergic humans. Arrows and dashed lines indicate peaks in the B-factor profile that may correspond to proteolytic cleavage sites. Mouse determinant cores tend to lie between peaks of flexibility and are associated with stable subdomains, except in BALB/c mice, where the epitope with strongest proliferative responses (highlighted in black) was associated with an unstable subdomain. The dominant epitope associated with human allergy (highlighted in gray) lies between the amino terminus and a protease-sensitive site. Data sources are as follows: crystal structure and B factors, PDB accession no. 1BSY; mouse determinant cores were from Totsuka et al. (1997); epitopes in humans were from Inoue et al. (2001). Data processing was as follows: residue stabilities were calculated using COREX/BEST with the crystal structure (PDB accession no. 1BSY) and default values for all parameters, except with an entropy-weighting factor of 0.95.

region of the β sheet that converges with the long α helix at the end of the hydrophobic cavity (Fig. 8B). A somewhat less immunogenic segment corresponds to the fifth β strand at the center of the stable subdomain. This pattern of immunodominance is consistent with a model for processing and presentation, in which proteolysis releases the stable subdomain from the larger molecule, and then relatively unstable segments at the ends of the subdomain are loaded into the MHC protein.

Subdominant epitopes in Bet v 1 are also associated with stable subdomains. Subdominant epitopes are located near the amino and carboxy termini of the molecule, 10-19 and 145-156, respectively (Fig. 8C). Each segment corresponds to a stable subdomain according to the COREX/BEST analysis. Each of the segments is separated from the rest of the molecule by a large flexible region marked by high levels of crystallographic B factors and/or low levels of residue stability in the COREX/BEST profile.

Epitope dominance in celery allergen Api g 1 appears to be governed by the same structural constraints as dominance in Bet v 1. The most immunogenic segment in Api g 1 is 115-126, which closely matches the dominant segment, 113-124, in Bet v 1 (Fig. 8E). A subdominant segment at 10-21 matches the subdominant segment 10-19 in Bet v 1. A lack of response to 82-91 in Api g 1 could have multiple explanations. It is possible that Api g 1 is less sensitive to proteolysis on one end or the other of this epitope. Since there is no crystal structure, the presence or absence of unstable sites cannot be compared to that of Bet v 1. Alternatively, the lack of response to 82-91 could be due to tolerance. Sequence identity of Api g 1 to Bet v 1 may be sufficiently low so that 82-91 of Api g 1 could be tolerated, while at the same time, the homologous sequence of Bet v 1 is allergenic.

Combining multiple profiles of flexibility (crystallographic B factor or NMR RMSD) with the analysis of structural stability (COREX/BEST) may compensate for the inadequacy of any individual dataset in the prediction of immunodominance. The B-factor profile from the crystal structure of Bet v 1 variant L (Fig. 8C) revealed strong peaks of flexibility on the carboxy-terminal sides of the two immunodominant segments (82-91 and 113-124), but it did not reveal such a strong peak on the carboxy-terminal side of 10-19. Nor could this particular profile of flexibility explain the weak immunogenicity in the broad region spanning residues 27 to 76. The COREX/BEST profile reveals a broad region of instability spanning residues 28 to 68, which closely matches the region of low immunogenicity (Fig. 8B and C). Instability in this region is corroborated by flexibility profiles from other crystallographic and NMR structures, which show two segments of very high flexibility within the unstable region (Fig. 8F).

These observations are consistent with a model for antigen-allergen processing and presentation in which the unstable protein segments are excluded from presentation, possibly because they are preferentially cleaved by proteases. Terminal segments of the resulting fragments are then loaded into MHC antigen-presenting proteins. Although somewhat resistant to proteolysis, the terminal segments may be sufficiently unstructured or unstable to conform to the peptide-binding groove and therefore are preferentially loaded into MHC antigen-presenting proteins.

IMPLICATIONS FOR THE DEVELOPMENT OF ALLERGY AND PROSPECTS OF IMMUNOTHERAPY

T-cell epitope immunodominance appears to be guided by the same forces, whether associated with immunity, tolerance, or allergy. For well-structured antigens, the mechanism

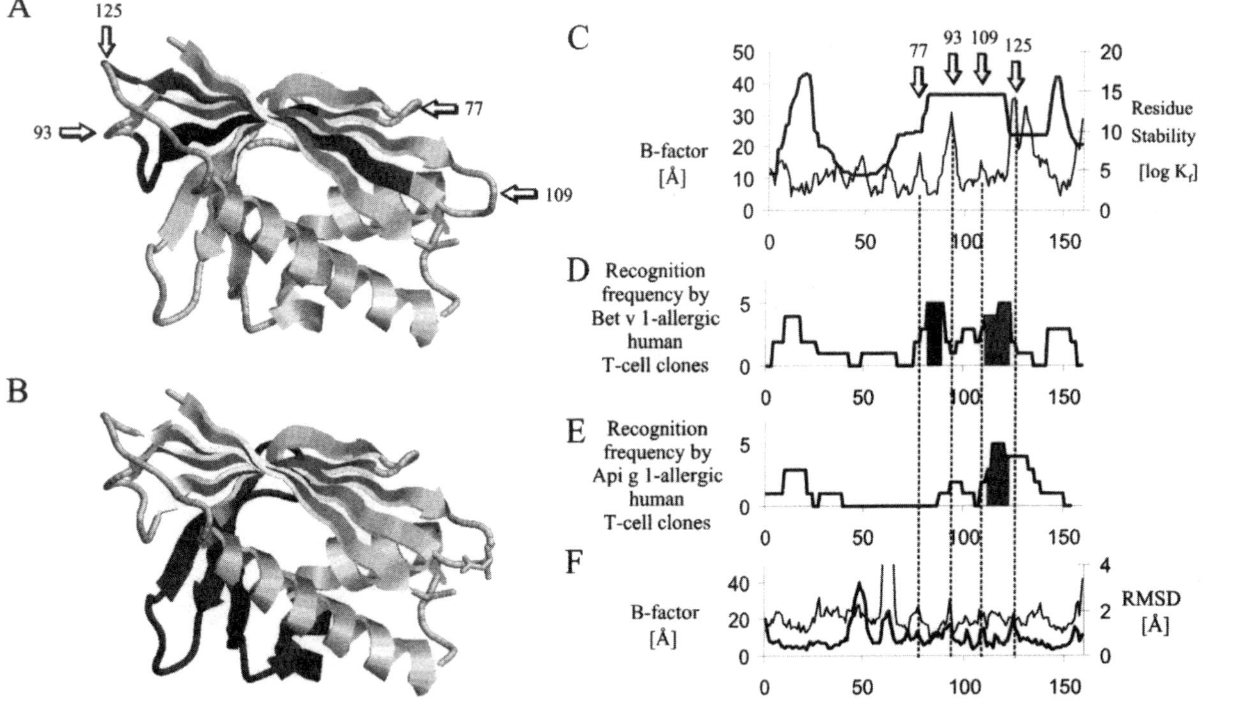

Figure 8. Structure and epitope dominance in birch pollen allergen Bet v 1. (A) Ribbon diagram of the crystal structure, showing two dominant epitopes in allergic humans (highlighted in black or gray). (B) Ribbon diagram showing a structurally unstable region that is poorly immunogenic. (C) Profiles of flexibility-stability for Bet v 1 variant L, indicated by crystallographic B factor (thin line) and COREX/BEST residue-stability (thick line). (D) Epitope dominance in Bet v 1 determined by frequency of recognition by T-cell lines generated from birch-allergic humans. (E) Epitope dominance in Api g 1 determined by frequency of recognition by T-cell lines generated from celery-allergic humans. (F) Profiles of flexibility for Bet v 1 variant A, indicated by nitrogen atom crystallographic B factor (thin line) and nitrogen atom RMSD from NMR (thick line). Arrows and dashed lines indicate peaks in flexibility profile that may correspond to proteolytic cleavage sites. Dominant epitopes associated with allergy lie between flexible sites and are excluded from the structurally unstable region. Data sources are as follows: crystal structure and B factors (A), PDB accession no. 1FM4; B factors (F), PDB accession no. 1BV1; RMSDs, PDB accession no. 1BTV; Bet v 1 epitopes were from Dormann et al. (1998); Api g 1 epitopes were from Bohle et al. (2003). Data processing is as follows: residue stabilities were calculated using COREX/BEST with the crystal structure (PDB accession no. 1BTV) and default values for all parameters, except with an entropy weighting factor of 0.90.

of antigen processing focuses the T-cell response on the intersection of stable and unstable antigen segments. Epitope dominance transcends the MHC-protein allelic background because antigen-presenting proteins can bind most sequences with adequate affinity and because antigen processing strongly biases which sequences are available for loading. When the antigen (or a large part of the antigen) is structurally unstable, e.g., the inner domain of HIV gp120, essentially the entire sequence is available for loading, and then epitope selection is largely controlled by MHC protein-peptide-binding affinity. However, MHC-allele-specific patterns of epitope dominance may not often be seen in food allergens because largely unstable proteins are destroyed in the digestive tract.

Since antigen processing determines which epitopes in food allergens will be primed, the structure of the allergen could be a major factor in the initiation of allergy. Pathways of allergen presentation may change depending on the circumstances of exposure, leading to the priming of new T cells with allergenic phenotypes. Preexisting Th2 cells may be able to skew the phenotype of new T cells toward Th2 through bystander mechanisms. If the epitopes to be primed were ignored in previous exposures with the same antigen, then the naïve T cells are likely to be primed toward Th2 unless there was a Th1-orienting influence present, such as one of the pathogen-associated molecular patterns.

Mechanisms of antigen processing focus the T-cell response on antigen-allergen sequences that are neither the most conserved nor the most divergent. Since structural stability correlates with sequence conservation and epitopes lie in the intersection of stable and unstable regions, the most immunogenic segments exhibit, on average, an intermediate level of sequence conservation. In evolutionary terms, this focusing of T-cell immunity might represent a compromise, namely, that the immune system targets features of pathogens that are not so well conserved that they are indistinguishable from self. An alternative possibility is that the most conserved, most stable segments (or segments with stability greater than that of a threshold level) are kinetically or thermodynamically inaccessible to the antigen-processing or antigen-presenting machinery.

The structure-dependent focusing of T-cell responses probably increases the success rate of allergy immunotherapy. If allergen structure did not influence epitope selection, then all sequences of the allergen would be available for binding to the MHC protein. With sufficient MHC diversity in a population, there would be little immunodominance; the effectiveness of immunotherapy in any particular individual would depend greatly on the MHC alleles. However, since structure focuses dominance onto certain sequences in an MHC-independent manner, sequences that have been ignored by the immune system remain available for recruitment as tolerogenic epitopes during immunotherapy. Epitopes most likely to have been ignored are found in the extremes of structural stability-instability. The least stable segments were cleaved in the digestive system or during processing, and the most stable segments resisted the conformational change necessary for binding to the MHC protein.

Knowledge of how structure focuses dominance could be used to improve immunotherapy. The presentation of previously ignored epitopes in an immunotherapeutic preparation could be increased by modifying the allergen structure or by changing the circumstances of administration. Numerous modifications of allergens have already been investigated, including proteolytic fragmentation or synthetic peptide derivatives (Alexander et al., 2002; Niederberger et al., 2004; Pecquet et al., 2000), heat treatment (Davis and Williams, 1998), and genetic modification (Bannon et al., 2001; Saarne et al., 2005). All of these modifications have the potential immunotherapeutic benefit of destroying antibody epitopes while preserving T-cell epitopes, but they may have very different T-cell dominance profiles.

Use of a peptide that contains a single epitope or very few epitopes for immunotherapy obviously is a choice for very narrow dominance. This strategy may be effective only when the allergy is characterized by strong, MHC allele-independent T-cell epitope dominance because the allergic individuals must be naïve to the therapeutic epitope(s). However, if dominance is too strong, then the epitope(s) would not be presented during a natural exposure, when the beneficial tolerogenic signals are desired.

Larger fragments, heat treatment, and genetic modifications can have highly variable consequences for antigen structure and epitope dominance. Most modifications would tend to cause some degree of global destabilization in the allergen and therefore decrease MHC-independent epitope dominance. This may be desirable because it would allow marginally accessible epitopes to be recruited for tolerance. Complete destabilization of structure may not be desirable because it would give the MHC protein equal access to all epitopes, and the resulting dominance profile would be controlled by MHC-peptide affinity.

REFERENCES

Alexander, C., A. B. Kay, and M. Larche. 2002. Peptide-based vaccines in the treatment of specific allergy. *Curr. Drug Targets Inflamm. Allergy* **1:**353–361.

Alpan, O., E. Bachelder, E. Isil, H. Arnheiter, and P. Matzinger. 2004. 'Educated' dendritic cells act as messengers from memory to naive T helper cells. *Nat. Immunol.* **5:**615–622.

Alpan, O., G. Rudomen, and P. Matzinger. 2001. The role of dendritic cells, B cells, and M cells in gut-oriented immune responses. *J. Immunol.* **166:**4843–4852.

Ametani, A., T. Sakurai, Y. Katakura, S. Kuhara, H. Hirakawa, T. Hosoi, S. Dosako, and S. Kaminogawa. 2003. Amino acid residue substitution at T-cell determinant-flanking sites in beta-lactoglobulin modulates antigen presentation to T cells through subtle conformational change. *Biosci. Biotechnol. Biochem.* **67:**1507–1514.

Anderton, S. M., N. J. Viner, P. Matharu, P. A. Lowrey, and D. C. Wraith. 2002. Influence of a dominant cryptic epitope on autoimmune T cell tolerance. *Nat. Immunol.* **3:**175–181.

Antoniou, A. N., S. L. Blackwood, D. Mazzeo, and C. Watts. 2000. Control of antigen presentation by a single protease cleavage site. *Immunity* **12:**391–398.

Bai, Y. W., J. S. Milne, L. Mayne, and S. W. Englander. 1994. Protein stability parameters measured by hydrogen exchange. *Proteins* **20:**4–14.

Bannon, G. A., G. Cockrell, C. Connaughton, C. M. West, R. Helm, J. S. Stanley, N. King, P. Rabjohn, H. A. Sampson, and A. W. Burks. 2001. Engineering, characterization and in vitro efficacy of the major peanut allergens for use in immunotherapy. *Int. Arch. Allergy Immunol.* **124:**70–72.

Barrett, A. J., N. D. Rawlings, and J. F. Woessner. 1998. *Handbook of Proteolytic Enzymes*. Academic Press, San Diego, Calif.

Bauer, L., B. Bohle, B. JahnSchmid, U. Wiedermann, A. Daser, H. Renz, D. Kraft, and C. Ebner. 1997. Modulation of the allergic immune response in BALB/c mice by subcutaneous injection of high doses of the dominant T cell epitope from the major birch pollen allergen Bet v 1. *Clin. Exp. Immunol.* **107:**536–541.

Berzofsky, J. A. 1983. T-B reciprocity. An Ia-restricted epitope-specific circuit regulating T cell-B cell interaction and antibody specificity. *Surv. Immunol. Res.* **2:**223.

Bohle, B., A. Radakovics, B. Jahn-Schmid, K. Hoffmann-Sommergruber, G. F. Fischer, and C. Ebner. 2003. Bet v 1, the major birch pollen allergen, initiates sensitization to Api g 1, the major allergen in celery: evidence at the T cell level. *Eur. J. Immunol.* **33:**3303–3310.

Brandtzaeg, P., I. N. Farstad, and G. Haraldsen. 1999. Regional specialization in the mucosal immune system: primed cells do not always home along the same track. *Immunol. Today* **20:**267–277.

Breiteneder, H., and C. Ebner. 2000. Molecular and biochemical classification of plant-derived food allergens. *J. Allergy Clin. Immunol.* **106:**27–36.

Breiteneder, H., and C. Radauer. 2004. A classification of plant food allergens. *J. Allergy Clin. Immunol.* **113:**821–830.

Breiteneder, H., and E. N. Mills. 2005. Molecular properties of food allergens. *J. Allergy Clin. Immunol.* **115:**14–23.

Brewer, J. M., K. G. Pollock, L. Tetley, and D. G. Russell. 2004. Vesicle size influences the trafficking, processing, and presentation of antigens in lipid vesicles. *J. Immunol.* **173:**6143–6150.

Brownlow, S., J. H. Morais Cabral, R. Cooper, D. R. Flower, S. J. Yewdall, I. Polikarpov, A. C. North, and L. Sawyer. 1997. Bovine beta-lactoglobulin at 1.8 A resolution—still an enigmatic lipocalin. *Structure* **5:** 481–495.

Caccamo, N., A. Barera, C. Di Sano, S. Meraviglia, J. Ivanyi, F. Hudecz, S. Bosze, F. Dieli, and A. Salerno. 2003. Cytokine profile, HLA restriction and TCR sequence analysis of human CD4+ T clones specific for an immunodominant epitope of Mycobacterium tuberculosis 16-kDa protein. *Clin. Exp. Immunol.* **133:** 260–266.

Castellino, F., F. Zappacosta, J. E. Coligan, and R. N. Germain. 1998. Large protein fragments as substrates for endocytic antigen capture by MHC class II molecules. *J. Immunol.* **161:**4048–4057.

Chehade, M., and L. Mayer. 2005. Oral tolerance and its relation to food hypersensitivities. *J. Allergy Clin. Immunol.* **115:**3–12.

Chen, B., E. M. Vogan, H. Gong, J. J. Skehel, D. C. Wiley, and S. C. Harrison. 2005. Structure of an unliganded simian immunodeficiency virus gp120 core. *Nature* **433:**834–841.

Chianese-Bullock, K. A., H. I. Russell, C. Moller, W. Gerhard, J. J. Monaco, and L. C. Eisenlohr. 1998. Antigen processing of two H2-IEd-restricted epitopes is differentially influenced by the structural changes in a viral glycoprotein. *J. Immunol.* **161:**1599–1607.

Chow, A. Y., and I. Mellman. 2005. Old lysosomes, new tricks: MHC-II dynamics in DCs. *Trends Immunol.* **26:**72–78.

Dai, G., N. K. Steede, and S. J. Landry. 2001. Allocation of helper T-cell epitope immunodominance according to three-dimensional structure in the human immunodeficiency virus type I envelope glycoprotein gp120. *J. Biol. Chem.* **276:**41913–41920.

Davidson, H. W., and C. Watts. 1989. Epitope-directed processing of specific antigen by B lymphocytes. *J. Cell Biol.* **109:**85–92.

Davis, P. J., and S. C. Williams. 1998. Protein modification by thermal processing. *Allergy* **53:**102–105.

Delamarre, L., M. Pack, H. Chang, I. Mellman, and E. S. Trombetta. 2005. Differential lysosomal proteolysis in antigen-presenting cells determines antigen fate. *Science* **307:**1630–1634.

Diethelm-Okita, B. M., D. K. Okita, L. Banaszak, and B. M. Conti-Fine. 2000. Universal epitopes for human CD4+ cells on tetanus and diphtheria toxins. *J. Infect. Dis.* **181:**1001–1009.

Donermeyer, D. L., and P. M. Allen. 1989. Binding to Ia protects an immunogenic peptide from proteolytic degradation. *J. Immunol.* **142:**1063–1068.

Dormann, D., C. Ebner, E. R. Jarman, E. Montermann, D. Kraft, and A. B. ReskeKunz. 1998. Responses of human birch pollen allergen-reactive T cells to chemically modified allergens (allergoids). *Clin. Exp. Allergy* **28:**1374–1383.

Eidem, J. K., I. B. Rasmussen, E. Lunde, T. F. Gregers, A. R. Rees, B. Bogen, and I. Sandlie. 2000. Recombinant antibodies as carrier proteins for sub-unit vaccines: influence of mode of fusion on protein production and T-cell activation. *J. Immunol. Methods* **245:**119–131.

Fineschi, B., and J. Miller. 1997. Endosomal proteases and antigen processing. *Trends Biochem. Sci.* **22:**377–382.

Fontana, A., P. P. de Laureto, B. Spolaore, E. Frare, P. Picotti, and M. Zambonin. 2004. Probing protein structure by limited proteolysis. *Acta Biochim. Pol.* **51:**299–321.

Fremont, D. H., D. Monnaie, C. A. Nelson, W. A. Hendrickson, and E. R. Unanue. 1998. Crystal structure of I-Ak in complex with a dominant epitope of lysozyme. *Immunity* **8:**305–317.

Friedl-Hajek, R., M. D. Spangfort, C. Schou, H. Breiteneder, H. Yssel, and R. J. Joost van Neerven. 1999. Identification of a highly promiscuous and an HLA allele-specific T-cell epitope in the birch major allergen Bet v 1: HLA restriction, epitope mapping and TCR sequence comparisons. *Clin. Exp. Allergy* **29:** 478–487.

Fritsch, R., B. Bohle, U. Vollmann, U. Wiedermann, B. Jahn-Schmid, M. Krebitz, H. Breiteneder, D. Kraft, and C. Ebner. 1998. Bet v 1, the major birch pollen allergen, and Mal d 1, the major apple allergen, cross-react at the level of allergen-specific T helper cells. *J. Allergy Clin. Immunol.* **102:**679–686.

Fuchs, E. J., and P. Matzinger. 1992. B cells turn off virgin but not memory T cells. *Science* **258:**1156–1159.

Gammon, G., H. M. Geysen, R. J. Apple, E. Pickett, M. Palmer, A. Ametani, and E. E. Sercarz. 1991. T cell determinant structures: cores and determinant envelopes in three mouse major histocompatibility complex haplotypes. *J. Exp. Med.* **173:**609–617.

Gapin, L., J. P. Cabaniols, R. Cibotti, D. M. Ojcius, P. Kourilsky, and J. M. Kanellopoulos. 1997. Determinant selection for T-cell tolerance in HEL-transgenic mice: dissociation between immunogenicity and tolerogenicity. *Cell. Immunol.* **177:**77–85.

Gapin, L., Y. B. deAlba, A. Casrouge, J. P. Cabaniols, P. Kourilsky, and J. Kanellopoulos. 1998. Antigen presentation by dendritic cells focuses T cell responses against immunodominant peptides: studies in the hen egg-white lysozyme (HEL) model. *J. Immunol.* **160:**1555–1564.

Gekko, K., A. Kimoto, and T. Kamiyama. 2003. Effects of disulfide bonds on compactness of protein molecules revealed by volume, compressibility, and expansibility changes during reduction. *Biochemistry* **42:**13746–13753.

Gugasyan, R., I. Vidavsky, C. A. Nelson, M. L. Gross, and E. R. Unanue. 1998. Isolation and quantitation of a minor determinant of hen egg white lysozyme bound to I-Ak by using peptide-specific immunoaffinity. *J. Immunol.* **161:**6074–6083.

Hilser, V. J., and E. Freire. 1996. Structure-based calculation of the equilibrium folding pathway of proteins. Correlation with hydrogen exchange protection factors. *J. Mol. Biol.* **262:**756–772.

Holen, E., and S. Elsayed. 1996. Specific T cell lines for ovalbumin, ovomucoid, lysozyme and two OA synthetic epitopes, generated from egg allergic patients' PBMC. *Clin. Exp. Allergy* **26:**1080–1088.

Honey, K., and A. Y. Rudensky. 2003. Lysosomal cysteine proteases regulate antigen presentation. *Nat. Rev. Immunol.* **3:**472–482.

Honey, K., T. Nakagawa, C. Peters, and A. Rudensky. 2002. Cathepsin L regulates $CD4^+$ T cell selection independently of its effect on invariant chain: a role in the generation of positively selecting peptide ligands. *J. Exp. Med.* **195:**1349–1358.

Hsieh, C. S., P. deRoos, K. Honey, C. Beers, and A. Y. Rudensky. 2002. A role for cathepsin L and cathepsin S in peptide generation for MHC class II presentation. *J. Immunol.* **168:**2618–2625.

Hubbard, S. J. 1998. The structural aspects of limited proteolysis of native proteins. *Biochim. Biophys. Acta* **1382:**191–206.

Hubbard, S. J., F. Eisenmenger, and J. M. Thornton. 1994. Modeling studies of the change in conformation required for cleavage of limited proteolytic sites. *Prot. Sci.* **3:**757–768.

Huntington, J. A., B. Fan, K. E. Karlsson, J. Deinum, D. A. Lawrence, and P. G. Gettins. 1997. Serpin conformational change in ovalbumin. Enhanced reactive center loop insertion through hinge region mutations. *Biochemistry* **36:**5432–5440.

Iametti, S., P. Rasmussen, H. Frokiaer, P. Ferranti, F. Addeo, and F. Bonomi. 2002. Proteolysis of bovine beta-lactoglobulin during thermal treatment in subdenaturing conditions highlights some structural features of the temperature-modified protein and yields fragments with low immunoreactivity. *Eur. J. Biochem.* **269:**1362–1372.

Inoue, R., S. Matsushita, H. Kaneko, S. Shinoda, H. Sakaguchi, Y. Nishimura, and N. Kondo. 2001. Identification of beta-lactoglobulin-derived peptides and class II HLA molecules recognized by T cells from patients with milk allergy. *Clin. Exp. Allergy* **31:**1126–1134.

Janssen, R., M. Wauben, R. van der Zee, M. de Gast, and J. Tommassen. 1994. Influence of amino acids of a carrier protein flanking an inserted T cell determinant on T cell stimulation. *Int. Immunol.* **6:**1187–1193.

Kapsenberg, M. L. 2003. Dendritic-cell control of pathogen-driven T-cell polarization. *Nat. Rev. Immunol.* **3:**984–993.

Karulin, A. Y., M. D. Hesse, M. Tary-Lehmann, and P. V. Lehmann. 2000. Single-cytokine-producing CD4 memory cells predominate in type 1 and type 2 immunity. *J. Immunol.* **164:**1862–1872.

Katsuki, T., N. Shimojo, K. Honma, H. Tsunoo, Y. Kohno, and H. Niimi. 1996. Establishment and characterization of ovalbumin-specific T cell lines from patients with egg allergy. *Int. Arch. Allergy Immunol.* **109:**344–351.

Kim, J., A. Sette, S. Rodda, S. Southwood, P. A. Sieling, V. Mehra, J. D. Ohmen, J. Oliveros, E. Appella, Y. Higashimoto, T. H. Rea, B. R. Bloom, and R. L. Modlin. 1997. Determinants of T cell reactivity to the *Mycobacterium leprae* GroES homologue. *J. Immunol.* **159:**335–343.

Kleen, T. O., R. Asaad, S. J. Landry, B. O. Boehm, and M. Tary-Lehmann. 2004. Tc1 effector diversity shows dissociated expression of granzyme B and interferon-gamma in HIV infection. *AIDS* **18:**383–392.

Koradi, R., M. Billeter, and K. Wuthrich. 1996. MOLMOL: a program for display and analysis of macromolecular structures. *J. Mol. Graph.* **14:**51–55, 29–32.

Kwong, P. D., R. Wyatt, J. Robinson, R. W. Sweet, J. Sodroski, and W. A. Hendrickson. 1998. Structure of an HIV gp120 envelope glycoprotein in complex with the CD4 receptor and a neutralizing human antibody. *Nature* **393:**648–659.

Landry, S. J. 2000. Helper T-cell epitope immunodominance associated with structurally stable segments of hen egg lysozyme and HIV gp120. *J. Theor. Biol.* **203:**189–201.

Landry, S. J. 1997. Local protein instability predictive of helper T-cell epitopes. *Immunol. Today* **18:** 527–532.

Latek, R. R., and E. R. Unanue. 1999. Mechanisms and consequences of peptide selection by the I-Ak class II molecule. *Immunol. Rev.* **172:**209–228.

Lehmann, P. V., O. S. Targoni, and T. G. Forsthuber. 1998. Shifting T-cell activation thresholds in autoimmunity and determinant spreading. *Immunol. Rev.* **164:**53–61.

Lennon-Dumenil, A. M., A. H. Bakker, R. Maehr, E. Fiebiger, H. S. Overkleeft, M. Rosemblatt, H. L. Ploegh, and C. Lagaudriere-Gesbert. 2002. Analysis of protease activity in live antigen-presenting cells shows regulation of the phagosomal proteolytic contents during dendritic cell activation. *J. Exp. Med.* **196:** 529–540.

Li, P., M. A. Haque, and J. S. Blum. 2002. Role of disulfide bonds in regulating antigen processing and epitope selection. *J. Immunol.* **169:**2444–2450.

Lovitch, S. B., S. J. Petzold, and E. R. Unanue. 2003. Cutting edge: H-2DM is responsible for the large differences in presentation among peptides selected by I-Ak during antigen processing. *J. Immunol.* **171:**2183–2186.

Ma, C., P. E. Whiteley, P. M. Cameron, D. C. Freed, A. Pressey, S. L. Chen, B. Garni-Wagner, C. Fang, D. M. Zaller, L. S. Wicker, and J. S. Blum. 1999. Role of APC in the selection of immunodominant T cell epitopes. *J. Immunol.* **163:**6413–6423.

Macdonald, T. T., and G. Monteleone. 2005. Immunity, inflammation, and allergy in the gut. *Science* **307:** 1920–1925.

Maehr, R., H. C. Hang, J. D. Mintern, Y. M. Kim, A. Cuvillier, M. Nishimura, K. Yamada, K. Shirahama-Noda, I. Hara-Nishimura, and H. L. Ploegh. 2005. Asparagine endopeptidase is not essential for class II MHC antigen presentation but is required for processing of cathepsin L in mice. *J. Immunol.* **174:**7066–7074.

Maleki, S. J., O. Viquez, T. Jacks, H. Dodo, E. T. Champagne, S. Y. Chung, and S. J. Landry. 2003. The major peanut allergen, Ara h 2, functions as a trypsin inhibitor, and roasting enhances this function. *J. Allergy Clin. Immunol.* **112:**190–195.

Mamula, M. J., R.-H. Lin, C. A. Janeway, Jr., and J. A. Hardin. 1992. Breaking T cell tolerance with foreign and self co-immunogens. *J. Immunol.* **149:**789–795.

Manca, F., D. Fenoglio, A. Kunkl, C. Cambiaggi, M. Sasso, and F. Celada. 1988. Differential activation of T cell clones stimulated by macrophages exposed to antigen complexed with monoclonal antibodies. *J. Immunol.* **9:**2893–2898.

Manca, F., A. Kunkl, D. Fenoglio, A. Fowler, E. Sercarz, and F. Celada. 1985. Constraints in T-B cooperation related to epitope topology on E. coli beta-galactosidase. I. The fine specificity of T cells dictates the fine specificity of antibodies directed to conformation-dependent determinants. *Eur. J. Immunol.* **15:** 345–350.

Manoury, B., E. W. Hewitt, N. Morrice, P. M. Dando, A. J. Barrett, and C. Watts. 1998. An asparaginyl endopeptidase processes a microbial antigen for class II MHC presentation. *Nature* **396:**695–699.

Manoury, B., D. Mazzeo, L. Fugger, N. Viner, M. Ponsford, H. Streeter, G. Mazza, D. C. Wraith, and C. Watts. 2002. Destructive processing by asparagine endopeptidase limits presentation of a dominant T cell epitope in MBP. *Nat. Immunol.* **3:**169–174.

Maric, M., B. Arunachalam, U. T. Phan, C. Dong, W. S. Garrett, K. S. Cannon, C. Alfonso, L. Karlsson, R. A. Flavell, and P. Cresswell. 2001. Defective antigen processing in GILT-free mice. *Science* **294:** 1361–1365.

Markovic-Housley, Z., M. Degano, D. Lamba, E. von Roepenack-Lahaye, S. Clemens, M. Susani, F. Ferreira, O. Scheiner, and H. Breiteneder. 2003. Crystal structure of a hypoallergenic isoform of the major birch pollen allergen Bet v 1 and its likely biological function as a plant steroid carrier. *J. Mol. Biol.* **325:**123–133.

McGrath, M. E. 1999. The lysosomal cysteine proteases. *Annu. Rev. Biophys. Biomol. Struct.* **28:**181–204.

Moudgil, K. D. 1999. Determinant hierarchy: shaping of the self-directed T cell repertoire, and induction of autoimmunity. *Immunol. Lett.* **68:**251–256.

Moudgil, K. D., and E. E. Sercarz. 1993. Dominant determinants in hen eggwhite lysozyme correspond to the cryptic determinants within its self-homologue, mouse lysozyme: implications in shaping of the T cell repertoire and autoimmunity. *J. Exp. Med.* **178:**2131–2138.

Moudgil, K. D., H. K. Deng, N. K. Nanda, I. S. Grewal, A. Ametani, and E. E. Sercarz. 1996. Antigen processing and T cell repertoires as crucial aleatory features in induction of autoimmunity. *J. Autoimmun.* **9:**227–234.

Moudgil, K. D., D. Sekiguchi, S. Y. Kim, and E. E. Sercarz. 1997. Immunodominance is independent of structural constraints: each region within hen eggwhite lysozyme is potentially available upon processing of native antigen. *J. Immunol.* **159:**2574–2579.

Mouritsen, S., M. Meldal, O. Werdelin, A. S. Hansen, and S. Buus. 1992. MHC molecules protect T cell epitopes against proteolytic destruction. *J. Immunol.* **149:**1987–1993.

Nakagawa, T. Y., and A. Y. Rudensky. 1999. The role of lysosomal proteinases in MHC class II-mediated antigen processing and presentation. *Immunol. Rev.* **172:**121–129.

Nakagawa, T., W. Roth, P. Wong, A. Nelson, A. Farr, J. Deussing, J. A. Villadangos, H. Ploegh, C. Peters, and A. Y. Rudensky. 1998. Cathepsin L: critical role in Ii degradation and CD4 T cell selection in the thymus. *Science* **280:**450–453.

Nayak, B. P. 1999. Differential sensitivities of primary and secondary T cell responses to antigen structure. *FEBS Lett.* **443:**159–162.

Nelson, C. A., N. J. Viner, S. P. Young, S. J. Petzold, and E. R. Unanue. 1996. A negatively charged anchor residue promotes high affinity binding to the MHC class II molecule I-Ak. *J. Immunol.* **157:**755–762.

Neudecker, P., K. Schweimer, J. Nerkamp, S. Scheurer, S. Vieths, H. Sticht, and P. Rosch. 2001. Allergic cross-reactivity made visible: solution structure of the major cherry allergen Pru av 1. *J. Biol. Chem.* **276:** 22756–22763.

Niederberger, V., F. Horak, S. Vrtala, S. Spitzauer, M. T. Krauth, P. Valent, J. Reisinger, M. Pelzmann, B. Hayek, M. Kronqvist, G. Gafvelin, H. Gronlund, A. Purohit, R. Suck, H. Fiebig, O. Cromwell, G. Pauli, M. van Hage-Hamsten, and R. Valenta. 2004. Vaccination with genetically engineered allergens prevents progression of allergic disease. *Proc. Natl. Acad. Sci. USA* **101**(Suppl. 2)**:**14677–14682.

Obita, T., T. Ueda, and T. Imoto. 2003. Solution structure and activity of mouse lysozyme M. *Cell. Mol. Life Sci.* **60:**176–184.

Osterballe, M., T. K. Hansen, C. G. Mortz, and C. Bindslev-Jensen. 2005. The clinical relevance of sensitization to pollen-related fruits and vegetables in unselected pollen-sensitized adults. *Allergy* **60:**218–225.

Ozaki, S., and J. A. Berzofsky. 1987. Antibody conjugates mimic specific B cell presentation of antigen: relationship between T and B cell specificity. *J. Immunol.* **138:**4133–4142.

Pecquet, S., L. Bovetto, F. Maynard, and R. Fritsche. 2000. Peptides obtained by tryptic hydrolysis of bovine beta-lactoglobulin induce specific oral tolerance in mice. *J. Allergy Clin. Immunol.* **105:**514–521.

Peters, N. C., D. H. Hamilton, and P. A. Bretscher. 2005. Analysis of cytokine-producing Th cells from hen egg lysozyme-immunized mice reveals large numbers specific for "cryptic" peptides and different repertoires among different Th populations. *Eur J. Immunol.* **35:**56–65.

Phelps, R. G., V. L. Jones, M. Coughlan, A. N. Turner, and A. J. Rees. 1998. Presentation of the Goodpasture autoantigen to CD4 T cells is influenced more by processing constraints than by HLA class II peptide binding preferences. *J. Biol. Chem.* **273:**11440–11447.

Pluger, E. B., M. Boes, C. Alfonso, C. J. Schroter, H. Kalbacher, H. L. Ploegh, and C. Driessen. 2002. Specific role for cathepsin S in the generation of antigenic peptides in vivo. *Eur. J. Immunol.* **32:**467–476.

Qin, B. Y., M. C. Bewley, L. K. Creamer, H. M. Baker, E. N. Baker, and G. B. Jameson. 1998. Structural basis of the Tanford transition of bovine beta-lactoglobulin. *Biochemistry* **37:**14014–14023.

Raychaudhuri, S., and K. L. Rock. 1998. Fully mobilizing host defense: building better vaccines. *Nat. Biotechnol.* **16:**1025–1031.

Rescigno, M., M. Urbano, B. Valzasina, M. Francolini, G. Rotta, R. Bonasio, F. Granucci, J. P. Kraehenbuhl, and P. Ricciardi-Castagnoli. 2001. Dendritic cells express tight junction proteins and penetrate gut epithelial monolayers to sample bacteria. *Nat. Immunol.* **2:**361–367.

Ria, F., A. Gallard, C. R. Gabaglia, J. C. Guery, E. E. Sercarz, and L. Adorini. 2004. Selection of similar naive T cell repertoires but induction of distinct T cell responses by native and modified antigen. *J. Immunol.* **172:**3447–3453.

Rimoldi, M., M. Chieppa, V. Salucci, F. Avogadri, A. Sonzogni, G. M. Sampietro, A. Nespoli, G. Viale, P. Allavena, and M. Rescigno. 2005. Intestinal immune homeostasis is regulated by the crosstalk between epithelial cells and dendritic cells. *Nat. Immunol.* **6:**507–514.

Ringe, D., and G. A. Petsko. 1986. Study of protein dynamics by X-ray diffraction. *Methods Enzymol.* **131:**389–433.

Robinson, C., N. A. Kalsheker, N. Srinivasan, C. M. King, D. R. Garrod, P. J. Thompson, and G. A. Stewart. 1997. On the potential significance of the enzymatic activity of mite allergens to immunogenicity. Clues to structure and function revealed by molecular characterization. *Clin. Exp. Allergy* **27:**10–21.

Rouas, N., S. Christophe, F. Housseau, D. Bellet, J. G. Guillet, and J. M. Bidart. 1993. Influence of protein-quaternary structure on antigen processing. *J. Immunol.* **150:**782–792.

Saarne, T., L. Kaiser, H. Gronlund, O. Rasool, G. Gafvelin, and M. van Hage-Hamsten. 2005. Rational design of hypoallergens applied to the major cat allergen Fel d 1. *Clin. Exp. Allergy* **35:**657–663.

Shimojo, N., T. Katsuki, J. E. Coligan, Y. Nishimura, T. Sasazuki, H. Tsunoo, T. Sakamaki, Y. Kohno, and H. Niimi. 1994. Identification of the disease-related T cell epitope of ovalbumin and epitope-targeted T cell inactivation in egg allergy. *Int. Arch. Allergy Immunol.* **105:**155–161.

Shimonkevitz, R., S. Colon, J. W. Kappler, P. Marrack, and H. M. Grey. 1984. Antigen recognition by H-2-restricted T cells II. A tryptic ovalbumin peptide that substitutes for processed antigen. *J. Immunol.* **133:** 2067–2074.

Shirahama-Noda, K., A. Yamamoto, K. Sugihara, N. Hashimoto, M. Asano, M. Nishimura, and I. Hara-Nishimura. 2003. Biosynthetic processing of cathepsins and lysosomal degradation are abolished in asparaginyl endopeptidase-deficient mice. *J. Biol. Chem.* **278:**33194–33199.

Simitsek, P. D., D. G. Campbell, A. Lanzavecchia, N. Fairweather, and C. Watts. 1995. Modulation of antigen processing by bound antibodies can boost or suppress class II major histocompatibility complex presentation of different T cell determinants. *J. Exp. Med.* **181:**1957–1963.

Sinha, P., H. H. Chi, H. R. Kim, B. E. Clausen, B. Pederson, E. E. Sercarz, I. Forster, and K. D. Moudgil. 2004. Mouse lysozyme-M knockout mice reveal how the self-determinant hierarchy shapes the T cell repertoire against this circulating self antigen in wild-type mice. *J. Immunol.* **173:**1763–1771.

So, T., H. Ito, T. Koga, S. Watanabe, T. Ueda, and T. Imoto. 1997. Depression of T-cell epitope generation by stabilizing hen lysozyme. *J. Biol. Chem.* **272:**32136–32140.

Steinman, R. M., D. Hawiger, and M. C. Nussenzweig. 2003. Tolerogenic dendritic cells. *Annu. Rev. Immunol.* **21:**685–711.

Takahashi, N., T. Koseki, E. Doi, and M. Hirose. 1991. Role of an intrachain disulfide bond in the conformation and stability of ovalbumin. *J. Biochem.* (Tokyo) **109:**846–851.

Takahashi, T., Y. Kuniyasu, M. Toda, N. Sakaguchi, M. Itoh, M. Iwata, J. Shimizu, and S. Sakaguchi. 1998. Immunologic self-tolerance maintained by CD25+CD4+ naturally anergic and suppressive T cells: induction of autoimmune disease by breaking their anergic/suppressive state. *Int. Immunol.* **10:**1969–1980.

Thai, R., G. Moine, M. Desmadril, D. Servent, J. L. Tarride, A. Menez, and M. Leonetti. 2004. Antigen stability controls antigen presentation. *J. Biol. Chem.* **279:**50257–50266.

Thornton, A. M., and E. M. Shevach. 1998. CD4+CD25+ immunoregulatory T cells suppress polyclonal T cell activation in vitro by inhibiting interleukin 2 production. *J. Exp. Med.* **188:**287–296.

Totsuka, M., A. Ametani, and S. Kaminogawa. 1997. Fine mapping of T-cell determinants of bovine beta-lactoglobulin. *Cytotechnology* **25:**101–113.

Untersmayr, E., I. Scholl, I. Swoboda, W. J. Beil, E. Forster-Waldl, F. Walter, A. Riemer, G. Kraml, T. Kinaciyan, S. Spitzauer, G. Boltz-Nitulescu, O. Scheiner, and E. Jensen-Jarolim. 2003. Antacid medication inhibits digestion of dietary proteins and causes food allergy: a fish allergy model in BALB/c mice. *J. Allergy Clin. Immunol.* **112:**616–623.

Vanegas, R. A., N. E. Street, and T. M. Joys. 1997. In a vaccine model, selected substitution of a highly stimulatory T cell epitope of hen's egg lysozyme into a Salmonella flagellin does not result in a homologous, specific, cellular immune response and may alter the way in which the total antigen is processed. *Vaccine* **15:**321–324.

Van Noort, J. M., J. Boon, A. C. M. Van der Drift, J. P. A. Wagenaar, A. M. H. Boots, and C. J. P. Boog. 1991. Antigen processing by endosomal proteases determines which sites of sperm-whale myoglobin are eventually recognized by T cells. *Eur. J. Immunol.* **21:**1989–1996.

Vidard, L., K. L. Rock, and B. Benacerraf. 1992. Diversity in MHC class II ovalbumin T cell epitopes generated by distinct proteases. *J. Immunol.* **149:**498–504.

Villadangos, J. A., R. A. Bryant, J. Deussing, C. Driessen, A. M. Lennon-Dumenil, R. J. Riese, W. Roth, P. Saftig, G. P. Shi, H. A. Chapman, C. Peters, and H. L. Ploegh. 1999. Proteases involved in MHC class II antigen presentation. *Immunol. Rev.* **172:**109–120.

Wang, L., R. X. Chen, and N. R. Kallenbach. 1998. Proteolysis as a probe of thermal unfolding of cytochrome c. *Proteins* **30:**435–441.

Watts, C. 2004. The exogenous pathway for antigen presentation on major histocompatibility complex class II and CD1 molecules. *Nat. Immunol.* **5:**685–692.

Yip, H. C., A. Y. Karulin, M. TaryLehmann, M. D. Hesse, H. Radeke, P. S. Heeger, R. P. Trezza, F. P. Heinzel, T. Forsthuber, and P. V. Lehmann. 1999. Adjuvant-guided type-1 and type-2 immunity: infectious/noninfectious dichotomy defines the class of response. *J. Immunol.* **162:**3942–3949.

Section III

IMMUNOTHERAPY AND THE ROLE OF ANIMAL MODELS

Food Allergy
Edited by S. J. Maleki et al.

Chapter 6

Vaccines and Immunotherapies for Future Treatment of Food Allergy

Wesley Burks, Ariana Buchanan, and Laurent Pons

Food allergy is a major cause of life-threatening hypersensitivity reactions (Sampson, 1999). Food-induced anaphylaxis is the most common reason for someone to present to the emergency department for an anaphylactic reaction (Sampson, 2003b). Currently, the avoidance of the allergenic food is the only method of preventing further reactions for allergic patients. Even with good educational information, 50% of allergic patients have accidental ingestions and allergic reactions over a 24-month period (Sicherer et al., 1998). With better characterization of allergens and an understanding of the immunologic mechanism involved in this reaction, investigators have developed several therapeutic modalities potentially applicable to the treatment and eventual prevention of food allergy (Burks et al., 2004). Among the therapeutic options currently under investigation, there are peptide immunotherapy, traditional Chinese medicine, mutated protein immunotherapy, DNA immunization, immunization with immunostimulatory sequences (ISSs), and anti-immunoglobulin E (anti-IgE) therapy. These novel forms of treatment for allergic disease hold promise for the safe and effective treatment of food-allergic individuals and the prevention of food allergy in the future.

Desirable therapeutic strategies for the treatment and prevention of food allergies must be safe, relatively inexpensive, and easily administered. Our recent advances in the understanding of the immunological mechanisms underlying allergic disease and the better characterization of food allergens have greatly expanded the potential therapeutic options for future use.

TRADITIONAL THERAPIES

Immunotherapy for IgE-mediated disease has been used since its first description nearly a century ago (Noon and Cantab, 2005). Although injection immunotherapy has traditionally been employed in the treatment of inhalant allergies, such as allergic rhinitis, it has also been used with success in the treatment of food allergy. Injection immunotherapy was first used in the treatment of food allergy when a young child was successfully desensitized to fish (Freeman, 1930). In patients with allergic rhinitis experiencing oral allergy symptoms

Wesley Burks, Ariana Buchanan, and Laurent Pons • Pediatric Allergy and Immunology, Duke University Medical Center, Durham, NC 27710.

with the ingestion of cross-reacting allergens in fresh fruits, nuts, and vegetables, traditional injection immunotherapy has also been successful in many cases in ameliorating the oral allergy symptoms with minimal adverse reactions (Asero, 1998; Asero, 2000).

Traditional injection immunotherapy for food allergy is currently not recommended because of the allergic side effects of the therapy (Burks, 2003). Studies have been conducted of a double-blind, placebo-controlled trial of rush immunotherapy for the treatment of anaphylactic hypersensitivity to peanuts (Oppenheimer et al., 1992). Patients in the treatment group were able to tolerate increased amounts of peanuts in post-treatment food challenges. Unfortunately, an unacceptably high rate of adverse systemic reactions was associated with the rush immunotherapy and maintenance protocols (Oppenheimeret al., 1992; Nelson et al., 1997). The actual peanut immunotherapy produced so many side effects of treatment that it would be practically impossible to perform on a routine basis.

While food desensitization in the oral allergy syndrome has been generally successful and well tolerated with cross-reacting pollen immunotherapy (Asero, 2003), the practice of injection immunotherapy for food allergy has been largely abandoned, due to the associated risk of serious systemic reactions. Since traditional immunotherapy has been largely impractical for the treatment of most food allergies, several novel therapies are now being explored (Table 1).

Table 1. Potential immunotherapeutic strategies for the treatment of food allergy

Therapy	Type of allergy	Route(s)	Immunologic mechanism	Risk(s)
Traditional injection immunotherapy	Oral allergy syndrome	Subcutaneous	Increased IgG-blocking antibodies; decreases specific-IgE	Safe when performed properly
Peptide immunotherapy	IgE-mediated food allergy	Subcutaneous	Immune deviation from Th2 to Th1	Appears safe
Traditional Chinese medicine	IgE-mediated food allergy (asthma)	Subcutaneous	Immune deviation from Th2 to Th1	Appears safe
Fusion proteins	IgE-mediated (all types)	Subcutaneous	Blocks IgE-mediated signaling	Unknown now
Mutated protein immunotherapy	IgE-mediated food allergy	Subcutaneous, intranasal and oral	Immune deviation from Th2 to Th1 (?)	Appears safe
DNA immunization	IgE-mediated food allergy	Subcutaneous and oral	(Subcutaneous) immune deviation from Th2 to Th1 (oral) increases levels of allergen specific secretory IgA in gut and systemic IgG	Unknown long-term
Immunostimulatory sequences	IgE-mediated food allergy	Subcutaneous	Immune deviation from Th2 to Th1	Appears safe
Anti-IgE therapy	IgE-mediated food allergy	Subcutaneous	Depletes IgE; blocks IgE from binding to high-affinity IgE-receptor (FcεRI); down-regulates IgE receptor production	Appears safe

NOVEL THERAPIES

Techniques under current investigation for the treatment of food allergy include peptide immunotherapy, traditional Chinese medicine, mutated protein immunotherapy, allergen DNA immunization, vaccination with immunostimulatory DNA sequences, and anti-IgE therapy.

Peptide Immunotherapy

Peptide immunotherapy utilizes peptide fragments containing T-cell-reactive epitopes rather than complete protein molecules. Theoretically, these peptide fragments are unable to cross-link two IgE molecules required to activate mast cells causing clinical allergic symptoms. These peptides do appear to render T cells unresponsive to subsequent allergen exposure (Briner et al., 1993). In one study utilizing peptide immunotherapy, the investigators injected cat-allergic patients with T-cell-reactive peptide fragments of a dominant cat allergen (Norman et al., 1996). Patients tolerated the injections with few, mild adverse reactions. The investigators demonstrated that cat-allergic patients treated with the peptide fragments had statistically significant improvement in nasal, lung, and total symptom scores compared to placebo-treated subjects after exposure to cats (Norman et al., 1996). A similarly related approach that has involved larger numbers of peptides from cats has also been shown to be effective (Oldfield et al., 2002). Cat-allergic patients treated with intradermal cat peptides developed a significant immune response, as well as clinical improvement in their symptoms.

Another study demonstrated that pepsin-digested peanut allergen containing T-cell epitopes but no IgE-binding epitopes induced gamma interferon (IFN-γ), a Th1 cytokine, in a concentration-dependent manner (Hong et al., 1999). While peptide immunotherapy for food allergy has not yet reached clinical trials, studies utilizing the peptides of the peanut allergens are interesting and suggest a possible role for peptide immunotherapy in the future therapy of food allergy.

Traditional Chinese Medicine

Herbal remedies have been used in Asia for centuries for the treatment of allergic diseases (Li et al., 2001; Li and Sampson, 2004). They have a favorable safety profile and the advantage of low cost. Recent studies have utilized traditional Chinese medicine in a mouse model of peanut allergy (Li and Sampson, 2004; Srivastav et al., 2005). IgE levels were significantly reduced by food allergy herbal formula 2 (FAHF-2) treatment and remained significantly lower for as long as 5 weeks posttherapy. Splenocytes from FAHF-2-treated mice showed significantly reduced interleukin 4 (IL-4), IL-5, and IL-13 levels and enhanced IFN-γ production. Work is now underway to delineate the active ingredients in FAHF-2 and understand the mechanism of action for this compound.

Fusion Proteins

In another concept to prevent IgE-mediated symptoms, investigators have shown that cross-linking of FcγRIIb to FcεRI on human mast cells and basophils by a genetically engineered Fcγ-Fcε protein (GE2) leads to the inhibition of mediator release upon FcεRI challenge (Saxon et al., 2004). GE2 protein was shown to inhibit cord blood-derived mast cell and peripheral blood basophil mediator release in vitro in a dose-dependent fashion, including inhibition of human IgE reactivity to cat allergens. The mechanism of inhibition in mast cells included alterations in IgE-mediated Ca^{2+} mobilization, tyrosine kinase phosphorylation, and the formation of downstream complexes. Proallergic effects of Langerhan's cell-like dendritic cells and B-cell IgE switching were also inhibited by GE2. In vivo, GE2 was shown to block passive cutaneous anaphylaxis driven by human IgE in mice expressing human FcεRI and to inhibit skin test reactivity to dust mite antigens in rhesus monkeys in a dose-dependent manner (Saxon et al., 2004). The balance between positive and negative signaling controls of mast cell and basophil reactivity is critical in the expression of human allergic diseases. This approach using a human Fcγ-Fcε fusion protein to coaggregate FcεRI with FcγRII holds promise as a new therapeutic platform for the immunomodulation of allergic diseases and potentially other mast cell- and basophil-dependent disease states.

Mutated Protein Immunotherapy

With greater understanding of the allergic mechanism and better characterization of allergens, we and our collaborators (Li et al., 2003a; Burks et al., 2004) are developing techniques that appear to be both safe and effective in the treatment of food allergy. This type of immunotherapy can be achieved through modification of the primary amino acid sequences of IgE-binding epitopes of the major allergens present in foods. Specific IgEs typically bind to the linear amino acid sequences of food allergens, whereas, in inhaled allergens, conformational sites are mainly recognized (Li et al., 2003a). By altering the primary amino acid sequence of the IgE epitopes, researchers have been able to substantially alter IgE antibody binding with the major peanut and shrimp allergens, which resulted in either reduced IgE binding or complete abolishment of binding.

The major peanut allergens, Ara h 1, Ara h 2, and Ara h 3, have had mutations made to their cDNA; the mutated less-allergenic proteins were then expressed in *Escherichia coli* (Stanley et al., 1996; Stanley et al., 1997). Mutating the IgE-binding sites did not significantly alter the T-cell epitopes; however, patients with peanut hypersensitivity had dramatically reduced IgE binding to the mutated proteins compared to binding for the wild-type peanut proteins. In animal studies with peanut-allergic mice, the mutated proteins significantly reduced the amount of peanut-specific IgE antibody produced by peanut-allergic animals, as well as reducing their clinical symptoms following a peanut challenge (Li et al., 2000; Li, 2003b). The treated animals had a reduction in Th2 cytokines over the course of treatment, with data suggesting an increase in T regulatory cells producing IL-10.

Designing molecules that have reduced or abolished IgE binding, while preserving their ability to stimulate T cells, has been shown to be effective treatment in the case of inhaled allergens and is expected to result in the availability of safe and effective therapeutics for the treatment of food allergy. Alteration of such molecules in foods theoretically could substantially reduce the incidence of food-induced anaphylactic reactions in sensitized patients. However, this approach currently is very unlikely because of the large numbers of allergens in some of the most prevalent allergenic foods such as egg, milk, peanuts, and tree nuts.

DNA Immunization

Another novel approach to the treatment of food allergy currently under investigation is DNA immunization (Spiegelberg et al., 1997; Slater and Colberg-Poley, 1997). DNA immunization employs the subcutaneous injection of a plasmid DNA (pDNA) vector encoding a specific allergenic protein. The pDNA sequence is taken up by antigen-presenting cells (APCs). Once inside the cell, the DNA-encoding allergen is transcribed and translated. The allergen is then presumably presented on the surface of the APC, in the context of the major histocompatibility complex (MHC), to T cells. This endogenously produced allergenic protein or protein fragment induces a Th1 phenotypic response with upregulation of IFN-γ, an increase in IgG2, and suppression of allergen-specific IgE production (Roy et al., 1999; Chu et al., 1997). Oral delivery of DNA immunizations has also been described. It has been utilized as an immunoprophylactic strategy to modulate peanut antigen-induced anaphylaxis (Roy et al., 1999). In this model, the oral delivery of DNA complexed to chitosan, a biocompatible polysaccharide, also favored a Th1 response and suppressed the Th2 allergic immune response.

Immunization with ISSs

Another DNA immunization technique under investigation is immunization with pDNA conjugated to ISSs. These ISSs contain unmethylated cytosine and guanine dinucleotide repeat motifs. These CpG motifs stimulate APCs and natural killer cells to secrete IFN-γ and IL-12, cytokines that promote immune deviation toward the Th1 phenotype and away from the allergic Th2 phenotype (Chu et al., 1997). These ISSs stimulate immune deviation to the Th1 phenotype when administered in several ways. They can be administered with DNA encoding the allergen (DNA immunization) (Hsu et al., 1996), given alone (Bohle et al., 1999), or conjugated with allergen. Although the majority of these DNA immunization techniques have been studied with a mouse model, a recent report describes enhanced immune deviation to the Th1 phenotype and reduced allergenicity after injection immunotherapy with the major ragweed allergen, Amb a 1, conjugated to ISSs in mice and rabbits, as well as in primates (Tighe et al., 2000).

Anti-IgE Immunotherapy

In general, the immune response to allergen immunotherapy is allergen specific, making treatment of multiple allergies complex and expensive. Often, patients are allergic to more than one allergen. Ideally, one would prefer to utilize a single therapy to combat the entire allergic mechanism, one therapeutic option that would be effective for multiple inhalant allergies and food allergies. One such therapy, already in clinical trials, involves the use of anti-IgE antibodies (Leung et al., 2003). IgE freely circulates or is attached to mast cells and basophils. When two mast cell-bound IgE molecules are cross-linked by antigen, the cell is activated, and vasoactive mediators are released. Anti-IgE, a humanized mouse monoclonal IgG antibody directed against the human IgE molecule, binds to freely circulating IgE and not to mast cell- or basophil-bound IgE. Once circulating IgE is bound by anti-IgE, it is unable to bind to its specific high-affinity receptor (FcεRI) on mast cells and basophils. These antigen-antibody complexes are then cleared from the circulation. Several clinical trials involving patients with allergic rhinitis (Nayak et al., 2003) and allergic asthma (Milgrom et al., 1999) have been completed. Statistically significant reductions in medication use, allergic symptoms, and airway hyperreactivity have been described following anti-IgE treatment. Reductions in total serum IgE levels to <1% of pretreatment levels were noted, in addition to down-regulation of the high-affinity IgE-receptor (FcεRI) (Saini et al., 1999).

A recent study of 84 patients with peanut hypersensitivity treated with anti-IgE has been completed (Leung et al., 2003). A significant number of patients, receiving the highest dose of anti-IgE monthly for 4 months, had a significant decrease in their clinical symptoms following peanut challenge posttreatment. Patients received 150, 300, or 450 mg of anti-IgE monthly for 4 months. After treatment, the subjects had a peanut challenge. Those subjects receiving the highest dose had an increased tolerance of eight peanuts compared to pre- and postfood challenge results. One problem with the study, however, is that 25% of the subjects even in the high-dose group had no change in their tolerance to peanuts after treatment.

Although anti-IgE therapy is expensive and requires frequent administration to maintain the IgE-deficient state, it could theoretically be used quite effectively in conjunction with other therapeutic modalities. For example, pretreatment with anti-IgE can potentially prevent the systemic reactions experienced with peanut immunotherapy (Oppenheimer et al., 1992) while achieving the beneficial effects of desensitization. This promising technique is currently under investigation in patients with food allergy.

CONCLUSIONS

Food allergy affects approximately 6% of children and 3.5% of adults (Sampson, 2003a). For those patients and the families of patients who have anaphylactic reactions, food allergy can be devastating. Reactions can range from mild urticarial reactions to severe anaphylactic shock and death from allergenic foods consumed unknowingly. The only preventative measure currently available for food allergy is strict avoidance of the incriminating food, which is often very difficult. It is likely that immunotherapy will be available in the near future as a safe and potentially effective therapy for the treatment of food allergy. New therapies currently under investigation should help the physician greatly improve care for food-induced allergic reactions while reducing the risk of anaphylaxis in these patients.

REFERENCES

Asero, R. 1998. Effects of birch pollen-specific immunotherapy on apple allergy in birch pollen-hypersensitive patients. *Clin. Exp. Allergy* **28:**1368–1373.

Asero, R. 2000. Fennel, cucumber, and melon allergy successfully treated with pollen-specific injection immunotherapy. *Ann. Allergy Asthma Immunol.* **84:**460–462.

Asero, R. 2003. How long does the effect of birch pollen injection SIT on apple allergy last? *Allergy* **58:** 435–438.

Bohle, B., B. Jahn-Schmid, D. Maurer, D. Kraft, and C. Ebner. 1999. Oligodeoxynucleotides containing CpG motifs induce IL-12, IL-18 and IFN-gamma production in cells from allergic individuals and inhibit IgE synthesis in vitro. *Eur. J. Immunol.* **29:**2344–2353.

Briner, T. J., M. C. Kuo, K. M. Keating, B. L. Rogers, and J. L. Greenstein. 1993. Peripheral T-cell tolerance induced in naive and primed mice by subcutaneous injection of peptides from the major cat allergen Fel d I. *Proc. Natl. Acad. Sci. USA* **90:**7608–7612.

Burks, A. W. 2003. Classic specific immunotherapy and new perspectives in specific immunotherapy for food allergy. *Allergy* 67:121–124.

Burks, W., S. B. Lehrer, and G. A. Bannon. 2004. New approaches for treatment of peanut allergy: chances for a cure. *Clin. Rev. Allergy Immunol.* **27:**191–196.

Chu, R. S., O. S. Targoni, A. M. Krieg, P. V. Lehmann, and C. V. Harding. 1997. CpG oligodeoxynucleotides act as adjuvants that switch on T helper 1 (Th1) immunity. *J. Exp. Med.* **186:**1623–1631.

Freeman, J. 1930. Rush inoculation, with a special reference to hay fever treatment. Lancet i:744–747.

Hong, S. J., J. G. Michael, A. Fehringer, and D. Y. Leung. 1999. Pepsin-digested peanut contains T-cell epitopes but no IgE epitopes. *Allergy Clin. Immunol.* **104:**473–478.

Hsu, C. H., K. Y. Chua, M. H. Tao, Y. L. Lai, H. D. Wu, S. K. Huang, and K. H. Hsieh. 1996. Immunoprophylaxis of allergen-induced immunoglobulin E synthesis and airway hyperresponsiveness in vivo by genetic immunization. *Nat. Med.* **2:**540–544.

Leung, D. Y., H. A. Sampson, J. W. Yunginger, A. W. Burks, Jr., L. C. Schneider, C. H. Wortel, F. M. Davis, J. D. Hyun, and W. R. Shanahan, Jr. 2003. Effect of anti-IgE therapy in patients with peanut allergy. *N. Engl. J. Med.* **348:**986–993.

Li, X. M., and H. A. Sampson. 2004. Novel approaches to immunotherapy for food allergy. *Clin. Allergy Immunol.* **18:**663–679.

Li, X. M., D. Serebrisky, S. Y. Lee, C. K. Huang, L. Bardina, B. H. Schofield, J. S. Stanley, A. W. Burks, G. A. Bannon, and H. A. Sampson. 2000. A murine model of peanut anaphylaxis: T- and B-cell responses to a major peanut allergen mimic human responses. *J. Allergy Clin. Immunol.* **106:**150–158.

Li, X. M., K. Srivastava, A. Grishin, C. K. Huang, B. Schofield, W. Burks, and H. A. Sampson. 2003a. Persistent protective effect of heat-killed *Escherichia coli* producing "engineered," recombinant peanut proteins in a murine model of peanut allergy. *J. Allergy Clin. Immunol.* **112:**159–167.

Li, X. M., K. Srivastava, J. W. Huleatt, K. Bottomly, A. W. Burks, and H. A. Sampson. 2003b. Engineered recombinant peanut protein and heat-killed *Listeria monocytogenes* coadministration protects against peanut-induced anaphylaxis in a murine model. *J. Immunol.* **170:**3289–3295.

Li, X. M., T. F. Zhang, C. K. Huang, K. Srivastava, A. A. Teper, L. Zhang, B. H. Schofield, and H. A. Sampson. 2001. Food allergy herbal formula-1 (FAHF-1) blocks peanut-induced anaphylaxis in a murine model. *J. Allergy Clin. Immunol.* **108:**639–646.

Milgrom, H., R. B. Fick, Jr., Su, J. Q., J. D. Reimann, R. K. Bush, M. L. Watrous, W. J. Metzger, et al. 1999. Treatment of allergic asthma with monoclonal anti-IgE antibody. *N. Engl. J. Med.* **341:**1966–1973.

Nayak, A., T. Casale, S. D. Miller, J. Condemi, M. McAlary, A. Fowler-Taylor, C. G. Della, and N. Gupta. 2003. Tolerability of retreatment with omalizumab, a recombinant humanized monoclonal anti-IgE antibody, during a second ragweed pollen season in patients with seasonal allergic rhinitis. *Allergy Asthma Proc.* **24:**323–329.

Nelson, H. S., J. Lahr, R. Rule, A. Bock, and D. Leung. 1997. Treatment of anaphylactic sensitivity to peanuts by immunotherapy with injections of aqueous peanut extract. *J. Allergy Clin. Immunol.* **99:**744–751.

Noon, L., and B. Cantab. 2005. Prophylactic inoculation against hay fever. *Lancet* **i:**1572.

Norman, P. S., J. L. Ohman, Jr., A. A. Long, P. S. Creticos, M. A. Gefter, Z. Shaked, R. A. Wood, P. A. Eggleston, K. B. Hafner, P. Rao, L. M. Lichtenstein, N. H. Jones, and C. F. Nicodemus. 1996. Treatment of cat allergy with T-cell reactive peptides. *Am. J. Respir. Crit. Care Med.* **154:**1623–1628.

Oldfield, W. L., M. Larche, and A. B. Kay. 2002. Effect of T-cell peptides derived from Fel d 1 on allergic reactions and cytokine production in patients sensitive to cats: a randomised controlled trial. *Lancet* **360:**47–53.

Oppenheimer, J. J., H. S. Nelson, S. A. Bock, F. Christensen, and D. Y. Leung. 1992. Treatment of peanut allergy with rush immunotherapy. *J. Allergy Clin. Immunol.* **90:**256–262.

Roy, K., H. Q. Mao, S. K. Huang, and K. W. Leong. 1999. Oral gene delivery with chitosan-DNA nanoparticles generates immunologic protection in a murine model of peanut allergy. *Nat. Med.* **5:**387–391.

Saini, S. S., D. W. MacGlashan, Jr., S. A. Sterbinsky, A. Togias, D. C. Adelman, L. M. Lichtenstein, and B. S. Bochner. 1999. Down-regulation of human basophil IgE and FC epsilon RI alpha surface densities and mediator release by anti-IgE-infusions is reversible in vitro and in vivo. *J. Immunol.* **162:**5624–5630.

Sampson, H. A. 1999. Food allergy. Part 1: immunopathogenesis and clinical disorders. *J. Allergy Clin. Immunol.* **103:**717–728.

Sampson, H. A. 2003a. 9. Food allergy. *J. Allergy Clin. Immunol.* **111**(Suppl. 2)**:**S540–S547.

Sampson, H. A. 2003b. Anaphylaxis and emergency treatment. *Pediatrics* **111:**1601–1608.

Saxon, A., D. Zhu, K. Zhang, L. C. Allen, and C. L. Kepley. 2004. Genetically engineered negative signaling molecules in the immunomodulation of allergic diseases. *Curr. Opin. Allergy Clin. Immunol.* **4:**563–568.

Sicherer, S. H., A. W. Burks, and H. A. Sampson. 1998. Clinical features of acute allergic reactions to peanut and tree nuts in children. *Pediatrics* **102:**e6.

Slater, J. E., and A. Colberg-Poley. 1997. A DNA vaccine for allergen immunotherapy using the latex allergen Hev b 5. *Arb. Paul Ehrlich Inst. Bundesamt Sera Impfstoffe Frankf. A M* **1997:**230–235.

Spiegelberg, H. L., E. M. Orozco, M. Roman, and E. Raz. 1997. DNA immunization: a novel approach to allergen-specific immunotherapy. *Allergy* **52:**964–970.

Srivastava, K. D., J. D. Kattan, Z. M. Zou, Li, J. H., L. Zhang, S. Wallenstein, J. Goldfarb, H. A. Sampson, and X. M. Li. 2005. The Chinese herbal medicine formula FAHF-2 completely blocks anaphylactic reactions in a murine model of peanut allergy. *J. Allergy Clin. Immunol.* **115:**171–178.

Stanley, J. S., R. M. Helm, G. Cockrell, A. W. Burks, and G. A. Bannon. 1996. Peanut hypersensitivity. IgE binding characteristics of a recombinant Ara h I protein. *Adv. Exp. Med. Biol.* **409:**213–216.

Stanley, J. S., N. King, A. W. Burks, S. K. Huang, H. Sampson, G. Cockrell, R. M. Helm, C. M. West, and G. A. Bannon. 1997. Identification and mutational analysis of the immunodominant IgE binding epitopes of the major peanut allergen Ara h 2. *Arch. Biochem. Biophys.* **342:**244–253.

Tighe, H., K. Takabayashi, D. Schwartz, G. Van Nest, S. Tuck, J. J. Eiden, A. Kagey-Sobotka, P. S. Creticos, L. M. Lichtenstein, H. L. Spiegelberg, and E. Raz. 2000. Conjugation of immunostimulatory DNA to the short ragweed allergen amb a 1 enhances its immunogenicity and reduces its allergenicity. *J. Allergy Clin. Immunol.* **106:**124–134.

Food Allergy
Edited by S. J. Maleki et al.

Chapter 7

Animal Models for Food Allergy

Ricki M. Helm

Animal models have been used to provide insight into the complex immunological and pathophysiological mechanisms of human type I allergic diseases. Research efforts that include mechanistic studies in search of new therapies and screening models for hazard identification of potential allergens in animals have received considerable attention in the past decade (Knippels et al., 2004; Fritsche, 2003; Helm et al., 2003; Ladics et al., 2003; Kimber et al., 2003). Mechanistic studies include designs on allergen structure, organ-immune system processing, digestion stability, and genetics. Therapeutic studies include allergen modifications, route of exposure, tolerance development with bacterial agents and/or herbal medicines, and cytokine skewing. Allergenic potential of novel proteins has received substantial interest in recent years, aimed at predicting allergenicity for genetically modified foods based upon known allergens, rare allergens, and nonallergens in different animal models. The present chapter will take information from these sources and others to provide the reader with the author's perspective on animal models and food allergy that could extrapolate to human type I allergic disease.

FOOD ALLERGENS

Reviews relevant to the question addressing "What makes a protein an allergen?" can be found related to structural aspects (Aalberse, 2000), cross-reactivity (Jenkins et al., 2005), functional aspects (Bredehort and David, 2001), and biochemical characteristics (Bannon, 2004), which have been summarized by Breiteneder and Mills (2005). Factors that contribute to allergenicity of a protein or otherwise development of food allergy include (i) the genetics of the individual, (ii) structural and/or functional characteristics of the allergen, and (iii) biochemical and physical properties that do not necessarily distinguish a food component as an allergen. Plant and animal food allergens belong to only a few of the several thousand known protein families; however, food allergens consist of only a few of the proteins in any one food source. As more information is obtained to define the characteristics

Ricki M. Helm • Department of Microbiology/Immunology, University of Arkansas for Medical Sciences, Arkansas Children's Hospital Research Institute, Arkansas Children's Nutrition Center, 1120 Marshall St., Little Rock, AR 72202.

of known food allergens, investigators are in a position to further define what makes a protein an allergen.

ANIMAL MODELS

It is neither ethical nor efficacious to sensitize or challenge allergic individuals to investigate the role of dietary proteins that result in immediate hypersensitivity. Therefore, a number of animal models have been investigated that provide significant information to our understanding of what makes a protein an allergen. However, interpretation of experimental animal models and the resultant research data require an understanding of the differences between immune systems of humans and laboratory animal models. An ideal food allergy animal model should include the following features. (i) It should resemble the major features of human type I hypersensitivity disease, either regional (oral allergy and gastrointestinal symptoms) or systemic (skin and bronchial manifestations). (ii) There should be a predisposed genetic background, with high or low immunoglobulin E (IgE)-IgG responders. (iii) There should be innate or adoptive features of response. (iv) An oral route of allergen sensitization should be available. (v) There should be a rank order of well-known high, medium, and weak food allergens and non-allergen proteins.

Laboratory Animal Models

Historically, reports of studies using guinea pigs (Devy et al., 1976; Poulsen and Hau, 1987; Piacentini et al., 1994), rats (Fritsché and Bonzon, 1990; Atkinson and Miller, 1994; Knippels et al.; 1998; Knippels et al., 2003; Madsen and Pilegaard, 2003; Ogawa et al., 2004; Knippels et. al, 1999), and mice (Li et. al., 1999; Roy et. al 1999; Nguyen et. al, 2001; Srivistava et al., 2005; Li et al., 2000; Li et al., 2001; Matysiak-Budnik et al., 2003; Lee et al., 2001; Dearman et al., 2003; Dearman et al., 2001; Hsieh et al., 2003; Li et al., 2003a; Li et al., 2003b; Chatel et al., 2003; Morafo et al., 2003) have represented the bulk of available literature sources that contribute significant information to the study of food allergy. There are some generalizations that should be taken into account when small animals are used. The advantages of the small-animal model include the following: (i) extensive understanding of the murine genome; (ii) engineered models, where the knockout can be compared to the wild type; (iii) extensive knowledge regarding technology for developing a disease state; (iv) an extensive repertoire of immunological reagents; and (v) the low cost of the model and the availability of large numbers of subjects.

The guinea pig can be easily sensitized by the oral route without the use of adjuvants; however, the reaginic antibody response is of the IgG1a subtype, limiting its use as a suitable model. The brown Norway rat requires intraperitoneal sensitization or prolonged daily low-dose antigen and generates both IgG and IgE to different antigens in the food source. In mice, there is a significant and strong tendency to develop oral tolerance that must be overcome by antigen, dose, and use of adjuvants. Moreover, murine intraepithelial lymphocytes predominantly express gamma-delta T-cell receptors, whereas human gamma-delta T cells make up only a small proportion of intraepithelial lymphocytes (Viney et al., 1990; Guy-Grand et al., 1991).

Domestic Animal Models

Domestic animals include companion, food-producing, and wild animals, which are normally directly linked to veterinary medicine (Charley, 1996). The pathogenesis and immune mechanisms involved in domestic animal diseases (autoimmunity and hypersensitivity reactions), immunologically based diagnostic tools (monoclonal antibodies and enzyme-linked immunosorbent assay), vaccination strategies and products (vectors, immunoadjuvants, and antigen delivery systems), genetic selection of resistant animals based on immune parameters, and immunotherapy or gene therapy against animal diseases provide both basic and applied research objectives to immunologists (Charley, 1996; Charley and Wilkie, 1994). Domestic animals supply many immunological research opportunities with fundamental and specific immunological features of spontaneously occurring diseases that may be more readily extrapolated to human diseases with increased in vivo relevance. Moreover, anatomical and physiological characteristics of domestic animals are much more closely related to human features than those of rodents. The immunological features or advantages of domestic animals are as follows: (i) ileal Peyer's patch accessibility to B-cell differentiation; (ii) immunoglobulin gene diversification by somatic hypermutation; (iii) role of unusually high numbers of gamma or delta T cells and their role in defense mechanisms; (iv) coexistence of T cells expressing CD4 and CD8 molecules; (v) impermeability of placentas in ungulates, with ontogeny of fetal immunity, development of immunocompetence, and passive transfer of maternal immunity; and (vi) large size, which offers easy access for cannulation studies, with dietary effects, lymphocyte traffic, and homing.

Recently, dogs (Buchanan and Frick, 2002) were added to the repertoire of animal models used to investigate food allergy. Kennis (2002) provides insight into the veterinary perspective of atopic dogs to investigate adverse reactions to food. Using the knowledge that swine develop transient soybean hypersensitivity when weaned onto soybean meal, Helm et al. (2002) developed a neonatal swine peanut allergy model.

The major advantage of domestic animals lies in the fact they provide spontaneously occurring animal diseases as models for human disease. Additional advantages include the following. (i) Dogs and swine are predisposed to developing high or low IgE concentrations. (ii) They can be used for both skin and gastroscopic injection of selected antigens, observations for the onset of vomiting or diarrhea, and clinical improvement when fed elimination diets. (iii) They offer a "physiological immune system that is more closely related to humans and can be manipulated and analyzed at regional and systemic immune levels in a living animal" (Charley, 1996).

MECHANISTIC MODELS

Investigations into immunologic mechanisms responsible for tolerance or sensitization to food proteins using animal models and human clinical trials are reaching new heights. In this chapter, IgE-mediated gastrointestinal food hypersensitivity disorders (gastrointestinal anaphylaxis; oral allergy syndrome) will be highlighted. The normal immune response in animals to dietary proteins is oral tolerance; however, abrogation of active immune suppression can result in adverse reactions, such as IgE-mediated food allergy. Continued research into the immunological mechanisms that contribute to the sensitization by food proteins resulting in IgE-mediated clinical disorders, prediction of neoallergens, and immunotherapy models of gastrointestinal food allergy has received major efforts in recent years (Sampson, 2005).

The gastrointestinal tract has the distinction of being the mucosa-associated lymphoid tissue (MALT) responsible for distinguishing foreign (harmful) antigens from nutrients and beneficial effects of commensal bacteria. Genetic factors, environment, gut microflora (commensal bacteria that compete for residence and nutrition with pathogens), and digestive processes of the host can lead to proinflammatory and/or anti-inflammatory products. These physiological processes, combined with immunologic processing events by resident antigen-presenting cells of the regional gastrointestine-associated lymphoid tissue (GALT), provide an intricate balance that results in tolerance or food-induced disorders. A shift in the Th1/Th2 T-cell balance under T regulatory control (Th3, T-reg, Tr1, or $CD4^{+}$-$CD25^{+}$ cells) normally brings about immune tolerance. However, an imbalance in regulatory cells (and therefore a shift in favor of a Th2 response) is considered to be the most significant mechanism for developing IgE to individual food proteins.

Li et al. (2000) showed peanut-specific IgE to be induced in C3H/HeJ mice following intragastric administration of peanut extract in combination with cholera toxin. The IgE antibodies to Ara h 2 identified the same IgE epitopes reflected with human serum IgE from peanut-sensitive individuals, indicating that this model could be useful in further characterizing peanut allergens in humans. van Witk et al. (2004) and van Witk et al. (2005), using a modified oral sensitization in the C3H/HeJ model, indicated that both Th1 and Th2 profiles were involved in the induction of peanut allergy. In both studies, cholera toxin was used with slightly differing results; Ara h 1 and 2 responses were prominent in the study by Li et al. (2000), whereas antibody responses were directed to Ara h 3 > Ara h 1 > Ara h 2 > Ara h 6 in the van Witk studies.

Immunomodulatory responses to food proteins can also be influenced by adjuvant. Feeding a whey protein mixture induced divergent responses, following immunization with complete Freund's adjuvant (tolerance) or alum (priming) of BALB/c mice (Afuwape et al., 2004). Conflicting results regarding the BALB/c and C3H/HeJ strains in their susceptibility to oral or intraperitoneal administration of food allergens suggest that studies of the immunopathogenic mechanisms applied in various mouse models will still need further refinement (Adel-Patient et al., 2005). Another consideration is that many proteins used in allergen sensitization are contaminated with endotoxins, which are normally found in food and drinking water and are handled without a problem (Petsch and Anspach, 2000); however, intravenous administration has a profound biological effect. Thus, route of allergen administration, adjuvant, and mouse strain must be taken into consideration when results with food allergy models are compared.

Digestion of peanuts following gastric lumen (Rich et al., 2003a) and duodenal (Rich et al., 2003b) environmental models and transfer to murine intestinal loops resulted in large numbers of protein bodies and soluble proteins. Digested peanut material was then introduced into intestinal loops of mice, and protein bodies were shown to be exclusively transported across intestinal epithelium by specialized antigen-sampling M cells (Chambers et al., 2004). The transfer of potentially highly immunogenic forms of peanut proteins to inductive sites (Peyer's patches) suggests that in the presensitized gut, an induction of immune responses rather than oral tolerance may play a significant role in the genesis of allergic reactions. How the GALT differentiates soluble proteins as either tolerogenic or immunogenic will have important implications for the nature of immune responses that develop. Sensitization to peanut and other food allergens will require future investigations to determine the relative role of each of the quantity of allergens in foods, antigen processing, costimulatory molecules (native or added as adjuvants), and antibody and cytokine profiles.

Classically, type I hypersensitivities including food allergy are regarded as Th2-induced responses counterbalanced by a Th1 response. However, the complexity of the immune response to sensitization of food allergens and IgE-mediated hypersensitivities is increasingly being recognized as a much more complex disease. Mixed cytokines (interleukin 4 [IL-4] and gamma interferon [IFN-γ]), as well as antibody (IgA, IgG1, IgG2a, and IgE) responses, were identified in mice against peanut allergens (Ara h 1, -2, and -3), suggesting that the Th2-mediated response may be oversimplistic (van Witk et al., 2004). This study further indicated the selection of cholera toxin as a Th2 inducer was also controversial, as a clear mixed Th1-Th2-induced response to peanut extract was identified during the course of sensitization. In addition, the model was shown to involve the cytotoxic lymphocyte-associated protein 4 signal pathways that regulated the intensity of the hypersensitivity response to peanut sensitization (van Witk et al., 2004). These studies and others demonstrate that the relative concentrations of specific food allergens, the use of an adjuvant (alum or cholera toxin) to induce a Th2-mediated response, and a mixed cytokine (IL-4 and INF-γ) and antibody (IgA, IgG1, IgG2a, and IgE) response, combined with additional accessory costimulatory signals (e.g., CTL-4), are involved in food allergen sensitization.

The immune system can be divided into regions depending upon the MALT. These divisions include the GALT, the bronchus-associated lymphoid tissue, the skin-associated lymphoid tissue, and the nose-associated lymphoid tissue, all of which can interact via migrating lymphocytes following sensitization in the respective tissue site. Proteins from different food sources can sensitize individuals by a variety of these MALT interactions that can ultimately lead to allergic disorders. Interactions with proteins and skin-associated lymphoid tissue, nose-associated lymphoid tissue, and GALT add to the complexity of determining what makes a protein an allergen (Breiteneder et al., 2005). Distinguishing conformational versus linear IgE- and IgG-binding epitopes (Vila et al., 2001); cross-reacting pollen-, fruit-, and vegetable-specific antibodies and/or allergens (Aalberse, 2000; van Ree et al., 2000); and processing events (Maleki et al., 2000; Beyer et al., 2001) contributes to unifying characteristics of major food allergens (Jenkins et al., 2005). Protein molecular mimicry between pollen and food allergens as primary sensitizers is an area that continues to be a major source of controversy (Valenta and Kraft, 1996; Pauli et al., 1992). Jenson-Jarolim et al. (1999) suggest that a possible modification of the conformation of Bet v 1 through its binding to BIP1, a monoclonal antibody to Bet v 1, increased the accessibility of epitopes for IgE binding. Thus, mimics of allergenic epitopes present in the diet could trigger or enhance symptoms of birch pollinosis. Conjecture could certainly be made that the reversal represented by inhalant epitopes sensitizes the atopic individual, leading to pollinosis and vegetable and fruit allergy syndromes. Bacterial and worm antigens merit a certain degree of interest, as they are a major focus of the mucosal barrier of the immune system.

Exposure to bacteria or bacterial products that act as adjuvants may be a factor contributing to food allergy sensitizations. Nanogram-sized quantities of pertussis toxin, when administered with a food protein, resulted in long-term sensitization to antigen and altered intestinal neuroimmune function (Kosecka et al., 1999). *Helicobacter* infection was shown to inhibit the development of oral tolerance to ovalbumin (OVA) by increasing the absorption of proteins across the digestive barrier (Matysiak-Budnik et al., 2003). These data suggest that chronic exposure to bacterial pathogens may prolong the normally transient immune responsiveness to inert food antigens. Discussions of the toxic effects regarding lectins on the GALT remain controversial with respect to intestinal epithelial integrity. The effects of dietary wheat germ agglutinin were investigated in the brown Norway rat (Watzl et al., 2001). In OVA-immunized rats, wheat germ agglutinin treatment resulted in increased mast cell protease concentration and reduced allergen-specific IgE. This suggests that factors or agents have the potential to break the mucosal barrier, leading to allergen sensitization, and therefore should receive more critical evaluation in the study of antigen-allergen presentation to the immune system.

Alternate pathways of sensitization to peanuts have received recent attention because peanut allergy is increasing in the United Kingdom. The absence of an association with maternal consumption of peanuts during pregnancy and an inability to detect specific IgE to peanuts led to the suggestion that sensitization occurred by transmission of peanut proteins in topical preparations of lotions containing peanut oil (Lack et al., 2003). Factors believed to have contributed to sensitization included absorption of soy and low doses of peanut antigens through inflamed skin. Low-dose oral exposure to peanut antigens in breast creams containing peanut oil was not believed to be involved. Animal models often prove useful in assessing such issues. Strid et al. (2004a) utilized the link between food allergies and atopic dermatitis to investigate the role of barrier properties in the stratum corneum layer of three strains of mice. Aqueous solutions of peanut protein or OVA were applied to epidermal surfaces, following abrasion by removal of adhesive tape that resulted in primary antigen specific and systemic immune responses which were strongly Th2 biased. In contrast, subcutaneous immunizations led to Th1-type immune responses. The results suggest that a response to barrier disruption leaving the subdermal skin layer intact results in cytokine production in the skin that promotes Th2 responses, as opposed to deeper immunization of dermal layer. Implied in these results is the differential stimulation-activation of Langerhan cells in an abraded antigen response (versus a dermal injection antigen response). Using a combination of intragastric feeding, intravenous immunization with adjuvant, and hind footpad injection of antigen, Strid et al. (2004b) demonstrated oral tolerance to be antigen specific. Oral tolerance was shown to be antigen specific between peanut and OVA and dose dependent, with tolerance induced when both Th1 and Th2 were suppressed. Future designs will need to be performed to define the regulatory mechanisms following the oral, mucosal, and systemic immune responses leading to tolerance or sensitization.

The hygiene hypothesis has led to several studies that minimize immunopathology and reduce host susceptibility to allergy. As such, experimental models are probing immune modulation as one mechanism by which the regulatory environment of the host can be manipulated toward a protective and/or tolerant situation toward food allergens (Wilson and Maizels, 2004; Yazdankbakhsh and Matricardi, 2004). Kalliomaki and Isolauri (2002) suggest putative mechanisms for the action of gut commensals in host-microbe interactions: lipopolysaccharide portions of gram-negative bacteria and specific CpG motifs of bacterial DNA activate immunomodulatory genes via Toll-like receptors, which in turn control the physiological balance in the gut. Thus, the use of probiotics (live microorganisms administered in adequate amounts which confer a beneficial health effect on the host) in association with foods may promote local antigen-specific events that shift the balance between tolerance and sensitization (Majamma and Isolauri, 1977). Supplementation of infant formulas with viable but not heat-inactivated probiotic bacteria is beneficial in the management of atopic dermatitis and cow's milk allergy (Fritsche, 2003; Pohjavuari et al., 2004). The in vitro effects of 100 strains of lactic acid bacteria on murine OVA Th2-polarized cells showed that the Th1-Th2 balance was dependent upon the lactic acid bacillus strain rather than the species (Fujiwara et al., 2004). In the above investigation, oral administration of KW3110 lactic acid bacillus directed the balance to a Th1 maturation of antigen-presenting cells and the inhibition of serum IgE production. Ishida et al. (2003) obtained similar results by feeding different lactobacilli and bifidobacterium-fermented milk to mice with OVA-specific elevations of IgE. The active preventative role of probiotics in the allergy remains to be elucidated.

Despite the controversy of helminths and their role in the production of polyclonal IgE and atopy, the stimulation of host immunoregulatory networks leading to the synthesis of anti-inflammatory cytokines (IL-10; transforming growth factor beta) could provide new approaches to atopy prevention (Ferreira et al., 2003). Different levels of immunological homeostasis patterns between the expulsions of worm infestations compared to the pathologic consequences of allergic reactions could lead to new concepts for the role of IgE, parasites, and allergies. With their mouse model of peanut allergy, Bashir et al. (2002) speculate that parasite-induced immunoregulation is mediated in part by cytokines IL-10 and IL-13, immunoregulatory mechanisms blocking the production of allergen-specific IgE, which occurs at the level of allergen presentation. Speculations concerning the polyclonal IgE induced by helminths that saturate the Fcε receptor response (Lynch et al., 1993) have also received attention. On the flip side, a secretory protein of *Nocardia brasiliensis* (NES) has been shown to act as an intrinsic adjuvant when coadministered with hen egg lysozyme in the generation of hen egg lysozyme lymphocyte proliferation, IL-4 release, and IgG1 antibody responses, suggesting that NES is responsible for third-party antigen presentations via an IgE polyclonal response (Holland et al., 2000). NES apparently cleared the gut of parasites; however, a Th2 response that could activate antigen-presenting cells was in evidence. Further studies with parasite and/or parasite proteins considered harmful and which induce a protective mode of IgE parasite clearance are warranted. In light of the fact that a dysregulation of the IgE network in gastrointestinal diseases (food allergy and parasite clearance) and a clear association with the immunoregulatory events of cytokines (Th1- versus Th2-induced responses), intrinsic adjuvanticity merits further attention (Pritchard et al., 1997). The true significance of polyclonal IgE production has not been critically evaluated in light of a parasite burden clearance, third-party antigen-allergen sensitization with respect to allergen-specific IgE production, or polyclonal IgE protection by Fcε receptor blockade.

In a murine model of fish allergy, antacids were shown to have a critical impact on the pathophysiology of this food allergy (Untersmayer et al., 2003). A mechanism speculated for the effects observed included the ability of antacids to hinder peptic digestion, thus preserving conformational epitopes that could be recognized by gamma-/delta T cells in the intestinal mucosa, leading to secretion of IL-4 and the induction of IgE. In this scenario, there is evidence that digestion of novel proteins, as well as known food allergens, may play a significant role in the induction of food allergy. Evidence related to conformational versus linear epitope recognition of food allergens and potential novel proteins should receive additional research with regards to sensitization. This topic is explored in more detail in Chapter 5 of this book.

In a murine model described as reproducing clinical and pathological changes in mild chronic food allergy, regulatory mechanisms (mucus hypersecretion) occurred in the intestinal mucosa, preventing overt pathology (Saldanha et al., 2004). The interesting phenomenon reported by this group was the observation for a taste aversion to egg white water in the subcutaneous-chronic orally treated group, which was also observed with previous studies (Cara et al., 1994, 1997). Although the mechanisms are not known, the evidence supports a brain-gut communication that could be crucial for the manifestation of functional gastrointestinal disorders such as those reported by Basso et al. (2004) and possibly responsible for the induction of symptoms in peanut-sensitive individuals responding to the smell of peanuts. Clearly, this is an intriguing aspect of neuronal behavior associated with food allergy that needs additional investigation.

Immunological opportunities with farm animals include studies of the pathogenesis and immune mechanisms that are of fundamental interest (Hein and Greibel, 2003). For instance, the placenta of ungulates (swine) is impermeable to large molecules, simplifying studies of the development of immunocompetence and passive transfer of immunity to offspring in the absence of external antigenic or antibody influence. The size of domestic animals in comparison to that of small laboratory animals makes surgical access and in vivo cannulation models in the newborn or adult more accessible for long-term studies of lymphocyte maturation, trafficking, and homing. As characterized by Hein (1995), the major advantage of domestic animals for immunological research may well rest upon their increased relevance in vivo; the physiology of the immune system can be manipulated and analyzed, at a regional level, in the living animal.

IMMUNOTHERAPEUTIC MODELS

Contrary to inhalant allergy, there is no satisfactory treatment of gastrointestinal food allergy. However, novel treatments with animal models have been a major focus and were reviewed by Li and Sampson (2002). Cloning of food allergens has made experimental DNA-based therapy possible, as shown by the intramuscular introduction of plasmid DNA encoding Ara h 2 prior to intraperitoneal sensitization in mice (Li et al., 1999). Protective effects were variable with AKR/J mice receiving moderate protection and C3H/HEN mice undergoing analphylaxis following peanut challenges. Oral administration of chitosan-associated pAra h 2 provided substantial protection in AKR mice (Roy et al., 1999). Data using synthetic immunostimulatory oligodeoxynucleotides suggest this avenue may have a prophylactic effect against food allergy similar to that observed in the treatment of allergic asthma (Nguyen et al., 2001). Methods of desensitization, dose effects, and strain response variability under the influence of cloned allergens or immunostimulatory sequences such as CpG motifs suggest much work is still required for optimizing gene therapy and/or immunostimulatory sequence delivery as a protective therapy.

Mutated protein and peptide immunotherapy appears to be more promising (Li et al., 2003a; Li et al., 2003b). Peanut allergens Ara h 1, Ara h 2, and Ara h 3 were modified to reduce IgE binding and coadministered subcutaneously with heat-killed *Listeria* (HKL) or *Esherichia coli*, which reduced peanut anaphylaxis in a C3H/HeJ murine model of peanut allergy. Reduction in symptom scores, specific IgE levels, and histamine release;

decreased levels of Th2-associated cytokines IL-4, IL-5, and IL-13; and increased levels of Th1-associated cytokine IFN-γ were identified at various time points following therapy treatment. The results suggest that combined adjuvant activity of heat-killed bacteria and modified allergens may be useful in the desensitization protocol in this animal model. Future dose effects, route of administration, strain comparisons, and different food allergen will need to be investigated, as will safety issues, prior to clinical trials in humans (Chatel et al., 2003).

Helm and Burks (2004) summarize the use of three murine models and the neonatal swine model that are used to address different immunotherapeutic strategies and mechanistic studies for food allergic sensitization. Briefly, Ara h 1 with mutated linear IgE-binding epitopes for the preparation of hypoallergenic transgenic peanut plants and immunotherapeutic agents to alleviate allergic reactions and the legume homologs as desensitizing reagents are being used in various murine models. A comparison of the different legume sensitization-clinical symptom responses in the neonatal swine model is also highlighted.

An herbal formula containing nine herbs was shown to provide extended protection from allergic symptoms in a murine model of peanut anaphylaxis (Li et al., 2001; Srivastava et al., 2005). These protective effects included a decrease in serum peanut-specific IgE levels and significantly reduced peanut-induced splenocyte production of IL-4, IL-5, and IL-13, with modest increases in IFN-ã production. The active ingredients, as well as the extent of active components in food allergy herbal formula 2-induced anti-IgE levels to other foods, are yet to be defined. A Th1-skewing effect was noted with a BALB/c murine OVA model subjected to an immunomodulatory protein isolated from *Flammulina velutipes* mushrooms (Hsieh et. al, 2003). Although these experiments demonstrate novel and interesting dramatic down-regulations of allergic responses in allergen-sensitized mice, which could lead to both new classes of drug therapy and further insights into the immunopathogenesis of allergic reactions, much work remains to be performed prior to clinical trials.

The conversion of Th2 to Th1 was investigated by subjecting allergic dogs to the protective immune response afforded by HKL coadministered subcutaneously with peanuts, wheat, or cow's milk (Frick et al., 2005). Vaccination with peanut extract with HKL into highly food allergic dogs resulted in improvements in skin test reactivity and loss of Ara h 1-specific IgE in peanut-allergic animals. Results similar to those with milk-allergic dogs in the above study were seen with a mouse model of asthma (Yeung et al., 1998) and peanut-sensitized mice (Li et al., 2001), where HKL administered with recombinant Ara h 1 and 3 prevented peanut-induced anaphylaxis. Kim et al. (2005) utilized a C3H/HeJ murine mouse model to determine the relative effects of probiotic treatment (hygiene hypothesis) on OVA-induced food allergy. Administration of *E. coli*, *Bifidobacterium bifidum*, and *Lactobacillus casei* decreased OVA-induced allergy, suggesting that oral administration bacteria may prevent oral sensitization to food allergens. Thus, coadministration of HKL, other probiotics, and components of herbal medicines in conjunction with other food allergens may be an effective curative immunotherapy based upon a Th1-Th2 cytokine balance.

RISK ASSESSMENT/ALLERGENICITY PREDICTION MODELS

It is now evident that no single criterion will be sufficient to predict protein allergenicity of novel foods or genetically modified organisms expressing transgenic proteins. The Food and Agriculture Organization of the United Nations-World Health Organization Codex Alimentarius Commission Ad Hoc Intergovernmental Task Force on Foods Derived from Biotechnology (FAO/WHO, 2001) revisited the decision tree approach and recommended that a risk assessment process adopt an integrated step-by-case approach to determining potential allergenicity. Additional insights into the implication of genetically modified foods and predictive testing of novel proteins as potential sources of allergens, which have not involved prior exposure to consumers, are topics of risk assessment (Ladics et al., 2003; Taylor and Hefle, 2002; Oehlschalger et al., 2001).

A number of animal models have been used to assess the potential allergenicity of novel proteins by establishing sensitization-allergic responses to both known food allergens and non-food allergens (Knippels et al., 2004; Fritche 2003; Helm et al., 2003; Ladics et al., 2003; Kimber et al., 2003). The group led by Knippels and Penninks (Knippels et al., 1998; Knippels and Penninks, 2003) successfully used a brown Norway rat model by sensitizing animals by the oral route, albeit by a prolonged low-dose sensitization (≥43 days). Clinical symptoms (elevated allergen-specific IgE levels, gut permeability, respiratory function, and blood pressure) were evident in allergen-sensitized groups compared to those non-allergen-sensitized animals, following oral challenge. In a brown Norway rat model developed to assess OVA-specific IgE, a comparison of age, sex, dosing volume, and allergen preparation led to moderate numbers of IgE responders in contrast to the results of other investigators using this model (Pilegaard and Madsen, 2004). In addition to a lower number of IgE responders, a simultaneous lower level of OVA-specific IgG was reflected, suggesting that IgG responses may not be predictive of allergy. The group led by Dearman and Kimber (Dearman et al., 2003; Dearman and Kimber, 2001) has examined immune characteristics of proteins based upon IgG antibody responses compared to IgE antibody responses. Systemic exposure of BALB/c mice to a range of allergens and non-food allergen proteins administered intraperitoneally revealed that the proteins under investigation were immunogenic (IgG responses) and that the measurement of IgE antibody responses could be used to identify allergens accurately, distinguishing them from nonallergenic proteins (Dearman et al., 2003). Recent evidence from this group, using mice exposed to peanut lectin by intradermal administration, resulted in a robust induction of IgE antibody response and was shown to be associated with a Th2-type cytokine expression profile at both the mRNA and secreted protein levels (Betts et al., 2004). Importantly, culture of naïve lymph node cells with this lectin failed to stimulate proliferation or cytokines in BALB/c mice. Results suggest that it may be possible to distinguish proteins that have an inherent potential to induce allergic sensitization in their murine model; however, additional work with all predictive models based on allergen-specific IgE and IgG levels needs careful evaluation and design before any consensus can be made regarding their use as predictors of novel protein allergenicity.

Birmingham et al. (2002) measured specific IgG1 antibody serum levels to evaluate the immunogenicity of various food sources. Results led the authors to conclude the following. (i) Food types vary widely with regard to their relative immunogenicity levels in mice. (ii) Strong immunogenicity is not a feature shared by commonly allergenic food types. (iii) Weak immunogenicity is not a feature shared by rarely allergenic or nonallergenic food types. (iv) Therefore, immunogenicity per se does not distinguish commonly allergenic foods from rarely or nonallergenic food in the C57BL animal model. The relative immunogenicity levels revealed the following: almonds = filberts > spinach (Rubisco) > peanuts ≥ sweet potatoes > cherries > lettuce > walnuts > chicken eggs > carrots ≥ white potatoes > wheat = coffee = soybeans. On the basis of skin test reactivity responses to allergen sensitization in high-level IgE responders, a clinical spectrum of peanuts > tree nuts > wheat > soy > barley was similar to that observed with humans (Teuber et al., 2002).

In summary, there is no approved valid animal model for the prediction of potential allergenicity of novel proteins. There is still a need to evaluate animal differences, sensitivity, dose, route of sensitization, and the relative profile of known allergens and nonallergens for any proposed animal model. Confirmation of risk factors, modes of sensitization, and antigen concentration are but a few of the issues in a complex immune system that leads to distinguishing proteins involved in food allergy. In each case, controversial reports exist with a multitude of animal models being used to define the mechanisms of IgE-mediated diseases and various appropriate therapeutic options. Not least, however, is the attempt to predict a novel protein as an allergenic source being introduced into our food source as a neoallergen in as-yet-unvalidated animal models. In conclusion, the most difficult task from one or more of these promising animal studies should be to extrapolate successfully to human disease. Experiments must be designed that bridge the characterization of immune responses, including an understanding of the respective mechanisms favoring tolerance versus sensitization, application of potential therapeutic agents without overt side effects, and determination of the basic characteristics and/or features of allergens for use in the prediction of potential neoallergens.

REFERENCES

Aalberse, R. C. 2000. Structural biology of allergens. *J. Allergy Clin. Immunol.* **106:**228–238.

Adel-Patient K., H. Bernard, S. Ah-Leung, C. Creminon, and J.-M. Wal. 2005. Peanut- and cow's milk-specific IgE, Th2 cells and local anaphylactic reaction are induced in BALB/c mice orally sensitized with cholera toxin. *Allergy* **60:**658–667.

Afuwape, A. O., M. W. Turner, and S. Strobel. 2004. Oral administration of bovine whey proteins in mice elicits opposing immunoregulatory responses and is adjuvant dependent. *Clin. Exp. Immunol.* **136:**40–48.

Atkinson, H. A., and K. Miller. 1994. Assessment of the brown Norway rat as a model for the investigation of food allergy. *Toxicology* **91:**281–288.

Bannon, G. A. 2004. What makes a food protein an allergen? *Curr. Allergy Asthma Rep.* **4:**43–46.

Bashir, M. E. H, P. Andersen, I. J. Fuss, H. N. Shi, and C. Nagler-Anderson. 2002. An effect of helminth infection protects against an allergic response to dietary antigen. *J. Immunol.* **169:**3284–3292.

Basso, A. S., F. A. Costa-Pinto, L. R. Britto, L. C. de Sa-Rocha, and J. Palermo-Neto. 2004. Neural pathways involved in food allergy signaling in the mouse brain: role of capsaicin-sensitive afferents. *Brain Res.* **1009:**181–188.

Betts, C. J., B. F. Flanagan, H. T. Caddick, R. J. Dearman, and I. Kimber. 2004. Intradermal exposure of BALB/c strain mice to peanut protein elicits a type-2 cytokine response. *Food Toxicol.* **42:**1589–1599.

Beyer, K., E. Morrow, X. M. Li, L. Bardina, G. A. Bannon, and A. W. Burks. 2001. Effects of cooking methods on peanut allergenicity. *J. Allergy Clin. Immunol.* **107:**1077–1081.

Birmingham, N., S. Thanesvorakul, and V. Gangur. 2002. Relative immunogenicity of commonly allergenic foods versus rarely allergenic and nonallergenic foods in mice. *J. Food Prot.* **65:**1988–1991.

Bredehort, R., and K. David. 2001. What establishes a protein as an allergen? *J. Chromatogr. B. Biomed. Sci. Appl.* **756:**33–40.

Breiteneder, H., and M. N. Clare Mills. 2005. Molecular properties of food allergens. *J. Allergy Clin. Immunol.* **115:**14–23.

Buchanan, B. B., and O. L. Frick. 2002. The dog as a model for food allergy. *Ann. N. Y. Acad. Sci.* **964:** 173–183.

Cara, D. C., A. A. Conde, and N. M. Vaz. 1994. Immunological induction of flavor aversion in mice. *Br. J. Med. Biol. Res.* **27:**1331–1341.

Cara, D. C., A. A. Conde, and N. M. Vaz. 1997. Immunological induction of flavour aversion in mice. II. Passive/adoptive transfer and pharmacological inhibition. *Scand. J. Immunol.* **45:**16–20.

Chambers, S. J., M. S. J. Wickman, M. Regoli, E. Bertelli, P. A. Gunning, and C. Nicoletti. 2004. Rapid in vivo transport of proteins from digested allergen across pre-sensitized gut. *Biochem. Biophys. Res. Commun.* **325:**1258–1263.

Charley, B. 1996. The immunology of domestic animals: its present and future. *Vet. Immunol. Immunopathol.* **54:**3–6.

Charley, B., and B. N. Wilkie. 1994. Why study the immunology of domestic animals. *Immunologist* **2:**103–105.

Chatel, J. M., L. Song, B. Bhogal, and F. M. Orson. 2003. Various factors (allergen nature, mouse strain, CpG/recombinant protein expressed) influence the immune response elicited by genetic immunization. *Allergy* **58:**641–647.

Dearman, R. J., and I. Kimber. 2001. Determination of protein allergenicity: studies in mice. *Toxicol. Lett.* **120:** 181–186.

Dearman, R. J., S. Stone, H. T. Caddick, D. A. Basketter, and I. Kimber. 2003. Evaluation of protein allergenic potential in mice: dose-response analyses. *Clin. Exp. Allergy* **33:**1586–1594.

Devey, M. E., K. J. Anderson, R. R. Coombs, M. S. Henschell, and M. E. Coates. 1976. The modified anaphylaxis hypothesis for cot death. Anaphylactic sensitization in guinea-pigs fed cow's milk. *Clin. Exp. Immunol.* **26:**542–548.

FAO/WHO. 2001. Evaluation of allergenicity of genetically modified foods: report of a joint FAO/WHO expert consultation on allergenicity of foods derived from biotechnology. FAO, Rome, Italy.

Ferreira, M. B., S. L. da Silva, and A. G. Carlos. 2002. Atopy and helminths. *Allerg. Immunol.* **34:**10–12.

Frick, O. L., S. S. Teuber, B. B. Buchanan, S. Morigasaki, and D. Umetsu. 2005. Allergen immunotherapy with heat-killed *Listeria monocytogenes* alleviates peanut and food induced anaphylaxis in dogs. *Allergy* **60:**243–250.

Fritsché, R. 2003. Animal models in food allergy: assessment of allergenicity and preventive activity of infant formulas. *Toxicol. Lett.* **140–141:**303–309.

Fritsché, R., and M. Bonzon. 1990. Determination of cow milk formula allergenicity in the rat model by in vitro mast cell triggering and in vivo IgE induction. *Int. Arch. Allergy Appl. Immunol.* **93:**289–293.

Fujiwara, D., S. Inoue, H. Wakabayashi, and T. Fujii. 2004. The anti-allergic effects of lactic acid bacterial are strain dependent and mediated by effects on both Th1/Th2 cytokine expression and balance. *Int. Arch. Allergy Immunol.* **135:**205–215.

Guy-Grand, D., N. Cerf-Bensussan, B. Malassis-Seris, C. Briottet, and P. Vassalli. 1991. Two gut epithelial CD8+ lymphocyte populations with different T cell receptors: a role for the gut epithelium in T cell differentiation. *J. Exp. Med.* **173:**471–481.

Hein, W. R. 1995. Sheep as experimental animals for immunological research. *Immunologist* **3:**12–18.

Hein, W. R., and P. J. Greibel. 2003. A road less traveled: large animal models in immunological research. *Nat. Rev. Immunol.* **3:**79–84.

Helm, R. M., and A. W. Burks. 2004. Sensitization and allergic response and intervention therapy in animal models. *J. AOAC Int.* **87:**1441–1447.

Helm, R. M., R. W. Ermel, and O. L. Frick. 2003. Non murine animal models of food allergy. *Environ. Health Perspect.* **111:**239–244.

Helm, R. M., G. T. Furuta, J. S. Stanley, J. Ye, G. Cockrell, C. Connaughton, P. Simpson, G. A. Bannon, and A. W. Burks. 2002. A neonatal swine model for peanut allergy. *J. Allergy Clin. Immunol.* **109:**136–142.

Holland, M. J., Y. M. Harcus, P. L. Riches, and R. M. Maizel. 2000. Proteins secreted by the parasitic nematode Nippostrongylus brasiliensis act as adjuvants for Th2 responses. *Eur. J. Immunol.* **30:**1977–1987.

Hsieh, K. Y., C. I. Hsu, J. Y. Lin, C. C. Tsai, and R. H. Lin. 2003. Oral administration of an edible-mushroom-derived protein inhibits the development of food-allergic reactions in mice. *Clin. Exp. Allergy* **33:**1595–1602.

Ishida, Y., I. Bandou, H. Kanzato, and N. Yamamoto. 2003. Decrease in ovalbumin IgE of mice serum after oral uptake of lactic acid bacteria. *Biosci. Biotechnol. Biochem.* **67:**951–957.

Jenkins, J. A., S. Griffiths-Jones, P. R. Shewry, H. Pretender, and E. N. Clare Mills. 2005. Structural relatedness of plant food allergens with specific reference to cross-reactive allergens: an in silico analysis. *J. Allergy Clin. Immunol.* **115:**163–170.

Jensen-Jarolim, E., U. Wiedermann, E. Ganglberger, A. Zurcher, B. M. Stadler,G. Boltz-Nitulescu, O. Scheiner O, and H. Breiteneder. 1999. Allergen mimotopes in food enhance type I allergic reactions in mice. *FASEB J.* **13:**1586–1592.

Kalliomaki, M., and E. Isolauri. 2002. Pandemic of atopic diseases—a lack of microbial exposure in early infancy. *Curr. Drug Targets Infect. Disord.* **2:**193–199.

Kennis, R. A. 2002. Use of atopic dogs to investigate adverse reactions to food. *J. Am. Vet. Med. Assoc.* **221:** 638–640.

Kim, H., K. Kwack, D-Y. Kim, and G. E. Ji. 2005. Oral probiotic bacterial administration suppressed allergic responses in an oval-induced allergy mouse model. *FEMS Immunol. Med. Microbiol.* **45:**259–267.

Kimber, I., R. J. Dearman, A. H. Penninks, L. M. J. Knippels, R. B. Buchanan, B. Hammerberg, H. A. Jackson, and R. M. Helm. 2003. Assessment of protein allergenicity on the basis of immune reactivity: animal models. *Environ. Health Perspect.* **111:**1125–1130.

Knippels, L. M. J., and A. H. Penninks. 2003. Assessment of the allergic potential of food protein extracts and proteins on oral application using the brown Norway rat model. *Environ. Health Perspect.* **111:**233–238.

Knippels, L. M. J., W. Femke, and A. H. Penninks. 2004. Food allergy: what do we learn from animal models? *Curr. Opin. Allergy Clin. Immunol.* **4:**205–209.

Knippels, L. M., J., A. H. Penninks, J. J. Smit, and G. F. Houben. 1999. Immune-mediated effects upon oral challenge of ovalbumin-sensitized brown Norway rats: further characterization of a rat food allergy model. *Toxicol. Appl. Pharmacol.* **156:**161–169.

Kosecka, U., M. C. Berin, and M. H. Perdue. 1999. Pertussis adjuvant prolongs intestinal hypersensitivity. *Int. Arch. Allergy Immunol.* **119:**205–211.

Knippels, L. M. J., A. H. Penninks, S. Spanhaak, and G. F. Houben. 1998. Oral sensitization to food proteins: a brown Norway rat model. *Clin. Exp. Allergy* **28:**368–375.

Lack, G., D. Fox, K. Northstone, and J. Golding. 2003. Factors associated with the development of peanut allergy in childhood. *N. Engl. J. Med.* **348:**977–985.

Ladics, G. S., M. P. Holsapple, J. D. Astwood, I. Kimber, L. M. J. Knippels, R. M. Helm, and W. Dong. 2003. Workshop overview: approaches to the assessment of the allergenic potential of food from genetically modified crops. *Toxicol. Sci.* **73:**8–16.

Lee, S. Y., C. K. Huang, T. F. Zhang, B. H .Schofield, A. W. Burks, G. A. Bannon, H. A. Sampson, and X. M. Li. 2001. Oral administration of IL-12 suppresses anaphylactic reactions in a murine model of peanut hypersensitivity. *Clin. Immunol.* **101:**220–228.

Li, X. M., and H. A. Sampson. 2002. Novel approaches for the treatment of food allergy. *Curr. Opin. Allergy Clin. Immunol.* **2:**273–278.

Li, X. M., C. K. Huang, B. H. Schofield, A. W. Burks, G. A. Bannon, K. H. Kim, S. K. Huang, and H. A. Sampson. 1999. Strain-dependent induction of allergic sensitization caused by DNA immunization in mice. *J. Immunol.* **162:**3045–3052.

Li, X. M., D. Serebrisky, S. Y. Lee, C. K. Huang. L. Bardina, B. H. Schofield, J. S. Stanley, A. W. Burks, G. A. Bannon, and H. A. Sampson. 2000. A murine model of peanut anaphylaxis: T- and B-cell responses to major peanut allergens mimic human responses. *J. Allergy Clin. Immunol.* **106:**150–158.

Li, X. M., T. F. Zhang, C. K. Huang, K. Srivastava, A. A. Teper, L. Zhang, B. H. Schofield, and H. A. Sampson. 2001. Food allergy herbal formula-1 (FAHF-1) blocks peanut-induced anaphylaxis in a murine model. *J. Allergy Clin. Immunol.* **108:**639–646.

Li, X. M., K. Srivastava, A. Grishin, C. K . Huang. B. Schofield, A. W. Burks, and H. A. Sampson. 2003a. Persistent protective effect of heat-killed *Escherichia coli* producing 'engineered,' recombinant peanut proteins in a murine model of peanut allergy. *J. Allergy Clin. Immunol.* **112:**159–167.

Li, X. M., K. Srivastava, J. W. Huleatt, K. Bottomly, A. W. Burks, and H. A. Sampson. 2003b. Engineered recombinant peanut protein and heat-killed Listeria monocytogenes coadministration protects against peanut-induced anaphylaxis in a murine model. *J. Immunol.* **170:**3289–3295.

Lynch, N. R., I. Hagel, M. Perez, R. Prisco, R. Lopez, and N. Alvarez. 1993. Effect of anti-helminthic treatment on the allergic reactivity of children in tropical slum. *J. Allergy Clin. Immunol.* **92:**404–411.

Madsen, C., and K. Pilegaard. 2003. No priming of the immune response in newborn brown Norway rats dosed with ovalbumin in the mouth. *Int. Arch. Allergy Immunol.* **130:**66–72.

Majamma, H., and E. Isolauri. 1997. Probiotics: a novel approach in the management of food allergy. *J. Allergy Clin. Immunol.* **99:**179–185.

Maleki, S. J., S. Y. Chung, E. T. Champagne, and J. P. Raufman. 2000. The effects of roasting on the allergenic properties of peanut proteins. *J. Allergy Clin. Immunol.* **106:**763–768.

Matysiak-Budnik, T., G. van Niel, F. Megraud, K. Mayo, C. Bevilacqua, V. Gaboriau-Routhiau, M. C. Moreau, and M. Heyman. 2003. Gastric Helicobacter infection inhibits development of oral tolerance to food antigens in mice. *Infect. Immunol.* **71:**5219–5224.

Morafo, V., K. Srivastava, C. K. Huang, G. Kleiner, S. Y. Lee, H. A. Sampson, and X. M. Li. 2003. Genetic susceptibility to food allergy is linked to differential Th2-Th1 responses in C3H/HeJ and BALB/c mice. *J. Allergy Clin. Immunol.* **111:**1122–1128.

Nguyen, M. D., N. Cinman, J. Yen, and A. A. Horner. 2001. DNA-based vaccination for the treatment of food allergy. *Allergy* **56:**127–130.

Oehlschlager, S., P. Reece, A. Brown, E. Hughson, H. Hird, J. Chisholm, J. Atkinson, C. Meredith, R. Pumphrey, P. Wilson, and J. Sunderland. 2001. Food allergy—towards predictive testing for novel foods. *Food Addit. Contam.* **18:**1099–1107.

Ogawa, T., S. Miura, Y. Tsuzuki, T. Ogino, K. Teramoto, T. Inamura, C. Watanabe, R. Hokari, H. Nagata, and H. Ishii. 2004. Chronic allergy to dietary ovalbumin induces lymphocyte migration to rat small intestinal mucosa that is inhibited by MAdCAM-1. *Am. J. Physiol. Gastrointest. Liver Physiol.* **6:**702–710.

Pauli, G., F. de Blay, J. C. Bessot, and A. Dietemann. 1992. The association between respiratory allergies and food hypersensitivities. *ACI News* **4:**43–47.

Petsch, D., and F. B. Ansparch. 2000. Endotoxin removal from protein solutions. *J. Biotechnol.* **76:**97–119.

Piacentini, G., L., A. Bertolini, E. Spezia, T. Piscione, and A. L. Boner. 1994. Ability of a new infant formula prepared from partially hydrolyzed bovine whey to induce anaphylactic sensitization: evaluation in a guinea pig model. *Allergy* **49:**361–364.

Pilegaard, K., and C. Madsen. 2004. An oral brown Norway rat model for food allergy: comparison of age, sex, dosing volume, and allergen preparation. *Toxicology* **196:**247–257.

Pohjavuari, E., M. Viljanen, R. Korpela, M. Kuitunen, M. Tittanen, O. Vaarala, and E. Savilahti. 2004. Lactobacillus GG effect in increasing IFN-gamma production in infants with cow's milk allergy. *J. Allergy Clin. Immunol.* **114:**131–136.

Poulsen, O. M., and J. Hau. 1987. Murine passive cutaneous anaphylaxis test (PCA) for 'all or none' determination of allergenicity of bovine whey proteins and peptides. *Clin. Allergy* **17:**75–83.

Pritchard, D. I., C. Hewitt, and R. Moqbel. 1997. The relationship between immunological responsiveness controlled by T-helper 2 lymphocytes and infections with parasitic helminths. *Parasitology* **115:**S33–S44.

Rich, G. T., R. M. Bailey, M. L. Parker, M. S. J. Wickman, and A. J. Fillery-Travis. 2003a. Solubilization of carotenoids from carrot juice and spinach in lipid phases. I. Modeling the gastric lumen. *Lipids* **38:**933–945.

Rich, G. T., R. M. Faulks, M. S. J. Wickman, and A. J. Fillery-Travis. 2003b. Solubilization of carotenoids form carrot juice and spinach in lipid phases. II. Modeling the duodenal environment. *Lipids* **38:**947–956.

Roy, K., H. Q. Mao, S. K. Huang, and K. W. Leong. 1999. Oral gene delivery with chitosan-DNA nanoparticles generates immunologic protection in a murine model of peanut allergy. *Nat. Med.* **5:**387–391.

Saldanha, J. C. S., D. L. Garfiulo, S. S. Silva, F. H. Carmo-Pinto, M. C. Andrade, J. I. Alvarez-Leite, M. M. Teixeira, and D. C. Cara. 2004. A model of chronic IgE-mediated food allergy in ovalbumin-sensitized mice. *Braz. J. Med. Biol. Res.* **37:**809–816.

Sampson, H. A. 2005. Food allergy: when mucosal immunity goes wrong. *J. Allergy Clin. Immunol.* **115:**139–141.

Srivistava, K. D., J. D. Kattan, Z. M . Zou, J. H. Li, L. Zhang, and S. Wallenstein. 2005. The Chinese herbal medicine formula FAHF-2 completely blocks anaphylactic reactions in a murine model of peanut allergy. *J. Allergy Clin. Immunol.* **115:**171–178.

Strid, J., J. Hourihane, I. Kimber, R. Callard, and S. Strobel. 2004a. Disruption of the stratum corneum allows potent epicutaneous immunization with protein antigens resulting in a dominant systemic Th2 response. *Eur. J. Immunol.* **34:**2100–2109.

Strid, J., M. Thomson, J. Hourihane, I. Kimber, and S. Strobel. 2004b. A novel model of sensitization and oral tolerance to peanut protein. *Immunology* **113:**293–303.

Taylor, S. E., and S. L. Hefle. 2002. Genetically engineered foods: implications for food allergy. *Curr. Opin. Allergy Clin. Immunol.* **2:**249–252.

Teuber, S. S., G. Del Val, S. Morigasaki, H. R. Jung, P. H. Eisele, O. L. Frick, and B. B. Buchanan. 2002. The atopic dog as a model of peanut and tree nut allergy. *J. Allergy Clin. Immunol.* **110:**921–927.

Untersmayer, E., I. Scholl, I. Swoboda, W. J. Beil, E. Forster-Wadl, F. Walter, A. Riemer, G. Kraml,T. Kinaciyan, S, Spitzauer, G. Boltz-Nitulescu, O. Scheiner, and E. Jensen-Jarolim. 2003. Antacid medication inhibits digestion of dietary proteins and causes food allergy: a fish allergy model in BALB/c mice. *J. Allergy Clin. Immunol.* **112:**616–623.

Valenta, R., and D. Kraft. 1996. Type 1 allergic reactions to plant-derived food: a consequence of primary sensitization to pollen allergens. *J. Allergy Clin. Immunol.* **97:**893–895.

van Ree, R., M. Cabanes-Macheteau, J. Akkerdaas, J-P. Milazzo, C. Loutelier-Bourhis, C. Rayon, M. Villalba, S. J . Koppelman, R. C. Aalberse, R. Rodriquez, L. Faye, and P. Lerouge. 2000. β(1,2)-xylose and α(1,3)-fucose residues have a strong contribution in IgE binding to plant glycoproteins. *J. Biol. Chem.* **275:**11451–11458.

van Witk, F., D. Hartgring, S. J. Koppelman, R. Pierters, and L. M. J. Knippels. 2004. Mixed antibody and T cell responses to peanut and the peanut allergens Ara h 1, Ara h 2, Ara h 3, and Ara h 6 in an oral sensitization. *Clin. Exp. Allergy* **34:**1422–1428.

van Witk, F., S. Hocks, S. Nierkens, S. J. Koppelman, P. van Kooten, L. Boon, L. M. J. Knippels, and R. Pieters. 2005. CTLA-4 signaling regulates the intensity of hypersensitivity responses to food antigens, but is not decisive in the induction of sensitization. *J. Immunol.* **174:**174–179.

Vila, L., K. Beyer, K. M . Jarvinen, P. Chatchatee, L. Bardina, and H. A. Sampson. 2001. Role of conformational and linear epitopes in the achievement of tolerance in cow's milk allergy. *Clin. Exp. Allergy* **31:** 1599–1606.

Viney, J., T. T. McDonald, and J. Spencer. 1990. Gamma/delta T cells in the gut epithelium. *Gut* **31:**841–844.

Watzl, B., C. Neudecker, G. M. Hansch, G. Rechkemmer, and B. L. Pool-Zobell. 2001. Dietary wheat germ agglutinin modulates ovalbumin-induced immune responses in brown Norway rats. *Br. J. Nutr.* **85:**483–490.

Wilson, M. S., and R. M. Maizels. 2004. Regulation of allergy and autoimmunity in helminth infection. *Clin. Rev. Allergy Immunol.* **26:**35–50.

Yazdankbakhsh, M., and P. M. Matricardi. 2004. Parasites and the hygiene hypothesis: regulating the immune response. *Clin. Rev. Allergy Immunol.* **26:**15–24.

Yeung, V. P., R. S. Gieni, D. Umetsu, and R. H. DeKruyff. 1998. Heat killed Listeria monocytogenes as an adjuvant converts established Th2-dominant immune responses into Th1-dominated responses. *J. Immunol.* **161:**4146–4152.

Zacharia, B., and P. Sherman. 2003. Atopy, helminths, and cancer. *Med. Hypotheses* **60:**1–5.

Section IV

IDENTIFYING AND PREDICTING POTENTIAL ALLERGENS: DIRECT TESTING AND BIOINFORMATICS

Food Allergy
Edited by S. J. Maleki et al.

Chapter 8

Approaches to the Detection of Food Allergens, from a Food Science Perspective

Carmen D. Westphal

Food allergies have become an increasingly important food safety issue in recent years. The reasons for this concern are the greater number of reported cases of allergic reactions, including fatalities due to anaphylactic reactions and the increased publicity (Sampson, 2004). The current approaches for the prevention of future reactions are focused on the evaluation of the inadvertent presence of allergenic material in food and the identification of the allergen on labels by the consumer governed by the implementation of new regulations in several countries (Table 1).

At the present time, there are no curative or prophylactic treatments for food allergies. Therefore, the only option to prevent an allergic reaction is the total avoidance of the offending food. For this reason, allergic individuals have to rely fully on the accuracy of the ingredients listed on the labels. It has been reported that many allergic reactions occur in places other that the home, such as schools (Munoz-Furlong, 2004), restaurants, or catered locations (Sicherer et al., 2001), where there is no control over the ingredient content. However, adverse events still may occur at home for the following reasons.

- The language utilized in the label is so complex or ambiguous that the patient cannot recognize the allergen (Joshi et al., 2002).
- The use of precautionary labeling such as "may contain" leads the affected person to risk eating the product (Hefle and Taylor, 2004).
- The allergens are inadvertently present in trace amounts in the food, due to mislabeling and/or contamination of raw material, cross-contact due to line sharing between allergen- and non-allergen-containing food or rework procedures, increased allergenicity due to protein modification by processing, and/or use of collective ingredient labeling such as "natural flavor" or "spices."

Carmen D. Westphal • Center for Food Safety and Applied Nutrition, U.S. Food and Drug Administration, Laurel, MD 20708.

Table 1. Mandatory or recommended labeling of allergenic ingredients

Country (URL)	Labeling	
	Mandatory	Recommended
United States (http://www.cfsan.fda.gov/≈acrobat/alrgact.pdf)	Milk, egg, fish (e.g., bass, flounder, cod), crustacean shellfish (e.g., crab, lobster, shrimp), tree nuts (e.g., almonds, pecans, or walnuts), wheat, peanuts, and soybeans, derived products	
European Union (http://europa.eu.int/eur-lex/pri/en/oj/dat/2003/l_308/l_30820031125en00150018.pdf)	Cereals containing gluten (i.e., wheat, rye, barley, oats, spelt, kamut, or their hybridized strains), crustaceans, eggs, fish, peanuts, soybeans, milk, nuts (i.e., almonds, hazelnuts, walnuts, cashews, pecan nuts, Brazil nuts, pistachio nuts, macadamia nuts, and Queensland nuts), celery, mustard, sesame seeds, sulfur dioxide, and sulfites (>10 mg/kg or 10 mg/liter) and derived products	
Canada (http://www.hc-sc.gc.ca/fn-an/alt_formats/hpfb-dgpsa/pdf/securit/allergen_paper-evaluation_allergene_e.pdf)		Peanuts, tree nuts, sesame seeds, soy, cow's milk, eggs, fish, crustaceans, shellfish, wheat, added sulfites (>10 ppm), and derived products
Australia and New Zealand (http://www.foodstandards.gov.au/_srcfiles/ACFA91C.pdf)	Cereals containing gluten (wheat, rye, barley, oats, spelt, and their hybridized strains), cereals, egg, fish, milk, peanut, soybeans, added sulfites (≥10 mg/kg), tree nuts, sesame seeds, and derived products	
Japan (http://www.mhlw.go.jp/english/topics/qa/allergies/al2.html)	Wheat, buckwheat, eggs, peanuts	Abalone, squid, salmon roe, shrimp-prawns, oranges, crabs, kiwi fruit, beef, tree nuts, salmon, mackerel, soybeans, chicken, pork, matsutake mushrooms, peaches, yams, apples, gelatin

Regulatory officials from different countries are working on local legislation (Table 1) to implement and update regulations to include labeling of the most important allergens for their respective countries. The criteria used to determine which allergenic foods will be included in the regulations are based on published information regarding severity and incidence of allergic reactions to a particular allergen in a particular geographical area (Yeung et al., 2000a). Some countries, such as the United States, only include the so-called big eight allergenic foods, which are responsible for about 90% of the allergic reactions in the United States (Food and Agriculture Organization of the United Nations, 1995). Other countries, such as those in the European Union (EU), have expanded the list to include other allergenic foods, such as sesame seeds, celery, and mustard. It is probable that in the coming years current regulations will be modified to include new allergenic foods and exemptions, as new scientific data become available.

To comply with new regulations and before they become fully implemented, the food industry must review its procedure protocols through good manufacturing practices and hazard analysis and critical control points plans or some similar approach (Crevel, 2002) to identify and prevent allergen contamination. To evaluate the magnitude of different undeclared allergens, the U.S. Food and Drug Administration, in partnership with both the Minnesota Department of Agriculture and the Wisconsin Department of Agriculture, Trade, and Consumer Protection, conducted a number of inspections of bakeries and ice cream and candy manufacturing installations in the states of Minnesota and Wisconsin in 1999 (U.S. Food and Drug Administration, 2001). About 25% of undeclared peanuts correlated with deficiencies in the facilities. The report included a list of activities and areas of the facility where the inadvertent introduction of an allergen may occur and made recommendations for correcting the deficiencies, which are listed below.

- Raw materials must be properly labeled, and the composition must be verified by the certificate of analysis or guarantee. Also, materials should be stored in areas or containers separate from allergenic ingredients.
- Ideally, utensils and equipment should be located and dedicated to allergen-containing foods only. However, for practical and economical reasons, this is impossible for many small food processors to achieve, and only large companies have the resources to do so.
- Processing activities should be properly scheduled in shared lines so that nonallergenic products are processed before allergen-containing products. Products and materials used in rework activities must be properly documented, labeled, and tracked. Reusable materials, such as tray liners, should be avoided, and single-use materials employed instead.
- Sanitation procedures must be adapted for every product type and equipment design. Additional attention must be paid to hard-to-clean areas where it is more likely that remaining food residues will be found. Special attention is needed to types of equipment where aqueous solutions cannot be used.
- Processed foods must be in the correct packaging materials. In situations where bulk products contain multiple packages, each wrapped item should also be labeled appropriately.

- According to the corresponding local regulations, labeling must be accurate and must contain all of the allergenic ingredients; allergenic ingredients must be described in a language that is easy for the consumer to understand.
- Employees play a key role in preventing cross-contact. They must understand the basics of food allergies, critical points of cross-contact, and how to prevent such contact. Also, they must recognize and report deficiencies or problems during manufacturing operations to minimize production losses.

Although the food industry develops plans to avoid cross-contact, the only manner of verifying the absence of traces of allergenic components in manufactured products and processing lines is by the use of detection methods.

Validated detection methods are needed not only by the food industry to comply with local and foreign regulations but also by government agencies for enforcement purposes and complaint assistance. Since regulations establish the labeling of several allergenic foods and their derived products, validated detection methods must be available (Lauwaars and Anklam, 2004).

Analytical techniques must fulfill a number of requirements to be valid for the detection of allergenic foods.

- The techniques must be specific for the target compound(s) in the food.
- The effect of the matrix or any interfering compound present in processed food on the detection of the allergen should be minimized.
- The techniques must be sensitive enough to detect trace amounts of the allergen or allergenic food.
- Techniques must avoid false positives and negatives.

Depending on the type of target molecule, protein, or DNA, the current detection methods for food allergens can be classified as immunoassay or PCR. Immunoassay is a very versatile technique because it uses antibodies that have the ability to detect a very specific protein(s) (allergen[s]) within a complex mixture of compounds. On the other hand, PCR targets the gene (DNA fragment) encoding the protein or allergen of interest. From the point of view of food safety, DNA is considered to be safe (nonallergenic); its use as a marker for the presence of allergenic material in food might not necessarily correlate with exposure to the allergenic protein (Poms et al., 2004b).

Detection methods that can be used for rapid screening within the facility are usually portable devices, are easy to use, do not require additional equipment, and provide qualitative results (i.e., lateral flow devices, rapid enzyme-linked immunosorbent assay [ELISA]). Moreover, more-complex formats are available that require certain expertise from the operator or special equipment for completion of the analysis process (i.e., PCR or, ELISA). Results can be qualitative, quantitative, or semiquantitative.

This chapter will provide an overview of the current trends for the detection of food allergens, based on published information and the factors and steps involved in the analysis process. Moreover, special attention will be focused on commercially available kits. A brief section will be dedicated to future approaches, such as the applicability of proteomics and genomics in the field, automation of analysis procedures, and confirmatory methods.

IMMUNOASSAY

Immunoassay has been the method of choice for the analysis of food allergens, primarily because immunoassays target specific proteins and because they are sensitive and versatile. Immunoassay is an antibody-based analytical technique. Its versatility has enabled the detection of any kind of molecule, even when a molecule is part of a complex mixture of compounds, because of the specificity of the antibodies. Antibodies, or immunoglobulins (Igs), are specialized proteins that are produced by the immune system after being exposed to foreign compounds (antigens). Every antibody binds noncovalently to only a few amino acids of the antigen (epitope) and with a determined strength (affinity).

Depending on the assay design, there are two types of antibodies that can be used, (i) polyclonal antibodies and (ii) monoclonal antibodies. Polyclonal antibodies are those produced in vivo by immunization of an animal, usually a goat, sheep, or rabbit, over time with a determined concentration of the antigen of interest. The serum, collected from the animal, will contain a heterogeneous mixture of antibodies with different specificities and affinities for the same antigen. The fact that polyclonal antibodies can recognize different proteins or epitopes within the same protein has made this type of antibody preferred for the detection of food allergens. Monoclonal antibodies are those produced in vitro by hybridoma cells. These cells result from the fusion of antibody-producing.B cells and immortalized myeloma cells. A single hybridoma cell will produce antibodies with the same specificity and affinity. Although monoclonal antibodies have not been commonly used for the detection of food allergens, the careful selection of antibodies has been used to develop assays for the detection of peanut residues in different food products (Hefle et al., 1994; Pomés et al., 2003; Pomés et al., 2004).

A good immunoassay will depend greatly on the quality of the antibodies. Affinity and specificity can be optimized through the manipulation of the immunization schedule and careful characterization and selection of the best performing antibodies. Specificity can also be improved by purifying antibodies from the rest of the animal sera components with an IgG-binding protein A or protein G column. Undesirable cross-reacting or nonspecific antibodies that also bind other proteins present in the food can be removed from allergen-specific polyclonal antibodies with an immunoaffinity column (Drs et al., 2004).

The terminology commonly used in the field of food allergy and allergen detection by immunoassays creates some confusion among people not familiar with the topic. An antigen is any molecule that has the ability to stimulate the production of antibodies (mostly, IgG). An allergen is a molecule that is able to elicit the production of IgE . Food allergens are proteins, but not all of the proteins are allergens. It is not fully understood what makes a protein an allergen, but many of the allergenic proteins seem to share some common features, which are not necessarily unique to allergenic proteins. They are usually a major component of the protein fraction of the food. The molecular mass of most allergens ranges from 10,000 to 70,000 Da. However, there are allergens outside this range, such as the peanut allergen Ara h 1, a trimer in its native form, that has a molecular mass of >250,000 Da (Maleki et al., 2000b). Most allergenic proteins have acidic isoelectric points and contain posttranslational modifications like N-termination modification or the addition of glycosylated moieties to the structure (Taylor and Lehrer, 1996; Bayard and Lottspeich, 2001; Bredehorst and David, 2001; Burks et al., 2001). Most of the allergens also have in common the ability to be resistant to heat and chemical denaturation (Hefle, 1999; Besler et al., 2001).

Nonallergenic proteins are antigenic because they are able to elicit the production of antibodies but not of the IgE isotype. From the point of view of detection of food allergens by immunoassay, allergenic proteins are comparable to nonallergenic proteins in that they are both antigens that are detected by IgG antibodies. There is a situation where IgE can be used for the detection of food allergens. When an allergic reaction occurs due to the consumption of a product containing hidden allergens, serum from the patient is used to identify whether or not the eliciting allergen is present in the sample. The use of IgE for the detection of hazelnut and almond residues in chocolate samples (Scheibe et al., 2001) and hazelnut residues in diverse food products (Koppelman et al., 1999) has been previously reported. However, the availability of sera from allergic individuals is the main limitation for the use of IgE for the detection of traces of allergens in food samples. Therefore, detection by IgE is basically restricted to the clinical research area and for diagnostic purposes.

Immunoassay Formats

There are several immunochemical techniques used for the detection of food allergens. Although they are all based on the antibody-antigen-binding principle, there are some formats more commonly used for diagnostic purposes for the detection of allergen-specific IgE. These techniques, such as the radioallergosorbent test, enzyme allergosorbent test, and rocket immunoelectrophoresis (RIE), are discussed in detail in Chapter 2. For the detection of allergens in food products or processing equipments, the types of immunoassays most commonly used are ELISAs, dipsticks, and lateral flow devices.

The term ELISA refers to the visualization of antibody-antigen binding by the formation of a colored compound catalyzed by an enzyme, which usually is either horseradish peroxidase or alkaline phosphatase. These assays require a supporting material to which capture antibodies or target antigens are bound. The most common supporting type is the polystyrene 96-well microtiter plate, but there are other types of supports, such as nitrocellulose or polyvinylidene fluoride membranes, used for dipsticks. The use of the acronym ELISA is commonly associated with the use of 96-well microtiter plates only.

ELISA

The following are the three types of enzyme-linked immunosorbent assay (ELISA) designs most commonly used for the detection of allergens in food (Fig. 1).

Sandwich. The allergen is trapped between the first antibody (capture antibody) bound to the plate and the second antibody (tracer or detecting antibody), which has an enzyme attached that catalyzes the substrate to give a colored compound. Color development is proportional to the concentration of allergen in the sample.

Enhanced. The enhanced ELISA design is similar to sandwich ELISA, but it involves an amplification system. For example, an enzyme-conjugated second antibody can be used to bind the detecting antibody. Another common amplification system uses a detecting antibody bound to a molecule of biotin instead of being coupled to an enzyme directly. Biotin has the ability to bind four molecules of streptavidin conjugated to the enzyme. Color development is proportional to the concentration of allergen in the sample.

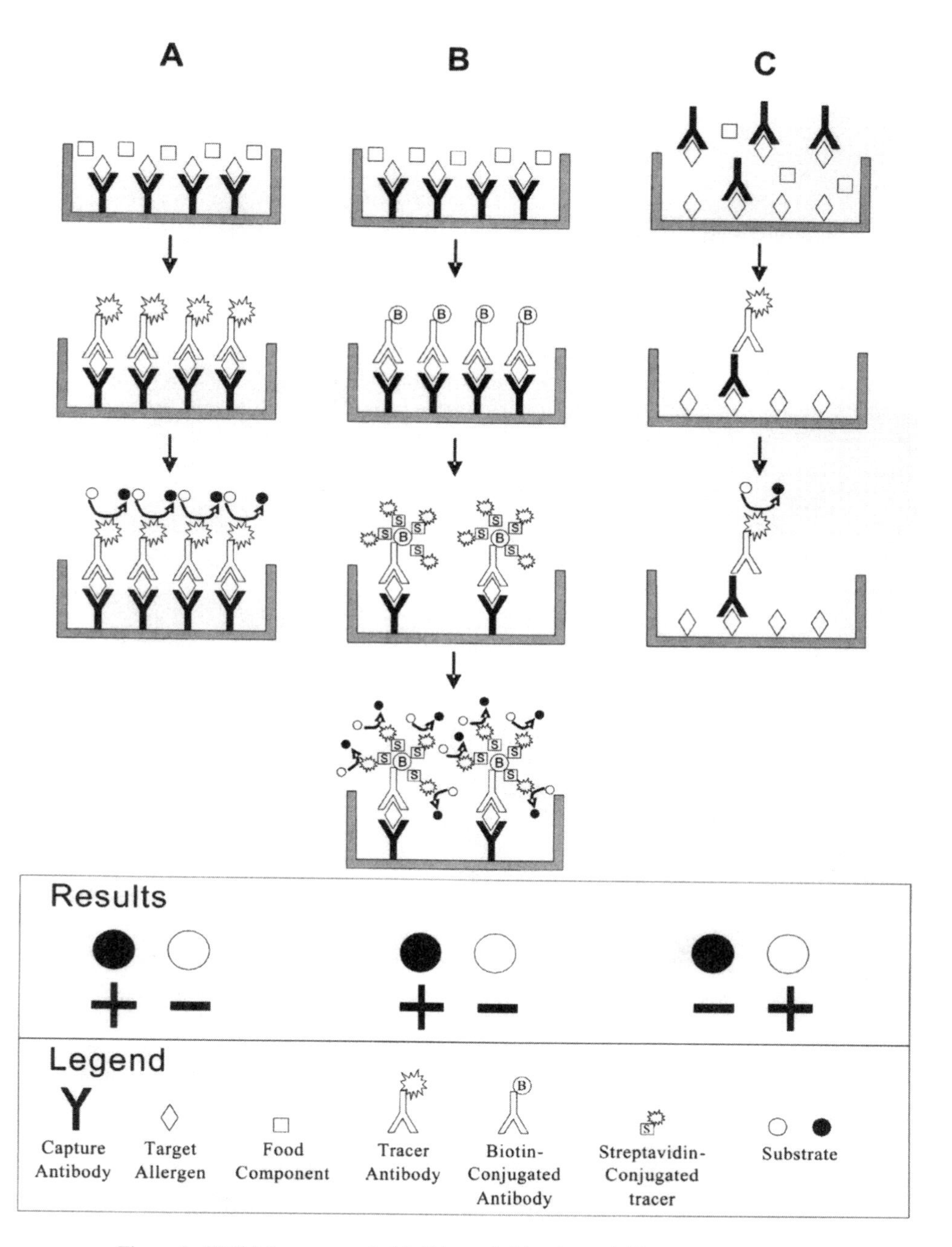

Figure 1. ELISA formats: sandwich (A), sandwich enhanced (B), and competitive (C).

Competitive (Inhibition). In this case, the allergen or food proteins are bound to the plate. Depending on the concentration of the allergen in the sample, the first antibody will bind either the allergen in the sample or the allergen bound to the plate. The higher the concentration of the allergen in the sample, the less binding of the antibody to the allergen in the plate will occur. A second antibody attached to an enzyme will bind the first antibody that bound the allergen on the plate. Color development is inversely proportional to the concentration of allergen in the sample.

The term direct ELISA or indirect ELISA refers to whether or not the detecting antibody is identified with a second antibody conjugated to the enzyme (indirect assay). In quantitative ELISA, color development is measured with a spectrophotometer for 96-well plates (ELISA reader). In a semiquantitative ELISA, the color developed for a sample is compared by eye to the different color shades of the calibration curve. One of the major advantages of ELISA is that several samples can be processed simultaneously.

Several ELISA tests have been developed in house by different laboratories for the detection of the most common food allergens (Table 2). Although some of them detect the same allergenic food, they use different approaches because of the versatility of immunoassays. Most of them use polyclonal antibodies raised in rabbits, goats, or sheep. Approaches include raising antibodies against whole-food extracts, as well as targeted allergens or other specific food proteins.

Polyclonal antibodies raised in chickens have also been successfully used for the development of ELISAs for the detection of peanuts (de Meulenaer et al., 2005), Brazil nuts (Blais et al., 2002), and hazelnuts (Drs et al., 2004). Monoclonal antibodies are less frequently used, but they have been applied in assays for the detection of peanut allergens (Hefle et al., 1994; Pomés et al., 2003; Pomés et al., 2004), native and denatured α-lactoblobulin (Negroni et al., 1998), soy bean allergens Gly m Bd 30 K (Tsuji et al., 1995) and Gly Bd 28K (Bando et al., 1998), gluten proteins (Skerrit and Hill, 1991; Sorell et al., 1998), and the shrimp allergen Pen a 1 (tropomyosin) (Jeoung et al., 1997).

Dipstick and Lateral Flow Tests

Detection of food allergens by dipstick is performed by immersing the device in a sample or reagents, following the same sequence of steps as ELISA. The main difference between ELISA and the dipstick test is the type of supporting material. The dipstick is developed on a support made of nitrocellulose, polyvinylidene fluoride, or polyester cloth. The dipstick is a qualitative (i.e., yes/no) assay. A positive sample is observed when the enzyme catalyzes a chromogenic substrate that precipitates on the dipstick. Dipsticks do not require special equipment to be performed, making them portable devices that can be used anywhere during manufacturing or at the point of sale.

The fact that the dipstick is a rapid method does not mean that it is less sensitive than conventional ELISAs. Dipstick tests have been developed for peanuts (Mills et al., 1997) and for peanuts and hazelnuts with polyclonal antibodies (Stephan et al., 2002), with reported detection limits ranging from 2 to 10 ppm of food. A dipstick for egg white proteins that uses polyclonal antibodies with a detection limit of 20 μg/kg of food has also been reported (Baumgartner et al., 2002). A cloth-based immunoassay was developed for the detection of peanut proteins with chicken IgY (equivalent to IgG) antibodies (Blais and Phillippe, 2000). This assay can detect 0.03 μg of peanut protein/ml. Detection of multiple allergens simultaneously by cloth immunoassay has been reported. This approach allows the detection of peanut, hazelnut, and Brazil nut proteins in the same sample, with positive results as low as 0.01 μg/ml for peanut and 0.03 μg/ml for hazelnut and Brazil nut (Blais et al., 2003).

The lateral flow test (Fig. 2) is another type of immunoassay. It has been successfully used for years for pregnancy testing. Like the sandwich ELISA, the target allergen is trapped between two antibodies, but the sequence in which the binding occurs takes place in a reverse fashion. The sample is applied to one edge of the device where the allergen will first bind the detecting antibody. The complex will travel along the surface (membrane) of the device until it reaches a band with capture antibodies. The development of this type of assay is more complex than a typical ELISA because it includes the additional variable of controlling the sample flow along the membrane. The visualization of the binding does not require the participation of any enzyme because the detecting antibody is attached to a colored compound, usually colored latex particles or nanogold. As a consequence, it does not require additional equipment to be developed, which makes it very suitable for use within food production installations. Another important advantage of lateral flow is that it does not require specially trained personnel for its use. Lateral flow is used for qualitative analysis only and is very useful for screening purposes. Because of its advantages, lateral flow has become a device in high demand; in recent years, manufacturers of detection methods for food allergens have begun to market it (Table 3).

Biosensors

Real-time monitoring of traces of allergen contaminants during food processing would be the ideal screening system in the food industry, in that the economic loss resulting from the holding and storage of suspicious batches until results are released from the laboratory could be minimized. Biosensors, among other type of sensors, have been widely used by the food industry for different applications, such as detection of microorganisms and toxins (Baeumner, 2003). There are a few reports of biosensors applied to the detection of food allergens.

They have in common the use of allergen-specific antibodies coated on the surface of a chip loaded in the device. The surface plasmon resonance measures changes in the refractive index resulting from antibody-antigen binding (Mullett et al., 2000). The surface plasmon resonance biosensor has been developed for the detection of peanut residues and has a detection limit of 0.7 μg/ml (Mohammed et al., 2001). Ovomucoid (egg), β-lactoglobulin (milk), and corylin (hazelnut) have detection limits of 10 ng/ml (Jonsson and Hellenäs, 2001); parvalbumin can be detected at concentrations as low as 0.11 mg/kg (Lu et al., 2004). Another type of biosensor, the evanescent wave fluoroimmunosensor, uses a pair of antibodies, capture and detecting. The detecting antibody is labeled with a fluorescent probe. The assay has been developed for ovalbumin (Shriver-Lake et al., 2004; Williams et al., 2004).

Limits of Detection

Current regulations from different countries do not cover limits of tolerance for any food allergen; therefore, it is understood that zero tolerance is in place. However, this is not really practical or feasible for the food industry. In theory, detection methods should be able to detect the lowest eliciting concentration of the offending food allergen. Currently, threshold concentrations for the different food allergens have not been established,

Table 2. Published ELISA protocols for the detection of allergenic foods[a]

Allergenic food	ELISA type	Specificity	Antibody or antibodies	LOD	Reference
Peanut	Enhanced sandwich	Peanut proteins	MAb, PAb	200 μg/g	Hefle et al., 1994
	Direct sandwich	Ara h 1	PAb	0.1 μg of peanut/g	Koppelman et al., 1996
	Competitive	Extract (crude, roasted)	PAb	400 ng/g	Yeung and Collins, 1996
	Indirect competitive	Native peanut protein	PAb	2 μg/g	Holzhauser and Vieths, 1999a
	Sandwich	Ara h 1	MAb	30 ng/ml	Pomés et al., 2003;
	Sandwich	Peanut proteins	PAb	0.07 ppm of peanut protein	Pomés et al., 2004
	Competitive inhibition	Ara h 1	PAb	12 ng/ml	Stephan and Vieths, 2004
		Ara h 2	PAb	0.4 ng/ml	Schmitt et al., 2004
	Sandwich	Peanut	PAb	1.5 μg/g	Immer et al., 2004
	Indirect competitive	Peanut proteins	PAb	1 μg/g	de Meulenaer et al., 2005
Almond	Competitive inhibition	Purified major allergen	PAb	300 ng/ml	Acosta et al., 1999
	Sandwich	Roasted almond meal, ground almonds	PAb	1 μg/g	Hlywka et al., 2000
	Competitive	Amandin	PAb	87 ng/ml; 5 ppm in food	Roux et al., 2001
Hazelnut	Sandwich	Crude extract	PAb	5 ng/ml	Koppelman et al., 1999
	Sandwich	Native, heated corylin	PAb	1 μg/g	Holzhauser and Vieths, 1999b
	Sandwich	Globulin fraction	PAb	0.03 μg/ml	Blais and Phillippe, 2001
	Competitive	Roasted hazelnut extract	PAb	0.45 ng/ml	Benrejeb et al., 2003
	Indirect competitive	Roasted hazelnut extract	PAb	10 ng/ml	Drs et al., 2004
	Sandwich	Pepsin-resistant protein	PAb	0.7 ng/ml	Akkerdaas et al., 2004
Brazil nut	Sandwich	Extract	PAb	0.015–0.030 μg/ml	Blais et al., 2002
	Indirect competitive	2S Brazil nut protein	PAb	1 ppm	Clemente et al., 2004
Cashew	Sandwich	Anacardein	PAb	20 ng/ml	Wei et al., 2003

Milk	Two-site	Native BLG	MAb	30 pg/ml	Negroni et al., 1998
		Denatured BLG		200 pg/ml	
	Enhanced sandwich	Native and denatured BLG	PAb	0.3 ng/ml	Rosendal and Barkholt, 2000
	Sandwich	Casein	PAb	0.5 ppm	Hefle and Lambrecht, 2004
Egg	Indirect	EW proteins	PAb	0.03% (wt/vol) raw-pasteurized, 0.125% sterilized products	Leduc et al., 1999
	Competitive	Cooked whole egg	PAb	0.2 ppm	Yeung et al., 2000b
	Sandwich	OVA, dehydrated EW	PAb	1 ppm dried whole egg	Hefle et al., 2001
Soybean	Sandwich	Gly m Bd 30K	MAb	10 ng/well	Tsuji et al., 1995
	Competitive	Denatured soy proteins	PAb	2 µg/g	Yeung and Collins, 1997
	Sandwich	Gly m Bd 28K	MAb	0.2 ng/well	Bando et al., 1998
	Competitive	KSTI	PAb	3 ng/ml	Rosendal and Barkholt, 2000
Gluten-containing cereals	Sandwich	Gliadin	MAb	0.158 µg/ml	Skerrit and Hill, 1991
	Sandwich	Gliadins	MAb	1.5 ng/ml	Sorell et al., 1998
		Hordeins		0.05 ng/ml	
		Secalins		0.15 ng/ml	
		Avenins		12 ng/ml	
	Sandwich	Tri a Bd 17K	MAb	4 ng/well	Yamashita et al., 2001
	Sandwich	Gliadins	MAb	1.56 µg/g	Valdes et al., 2003
Shrimp	Enhanced sandwich	Pen a 1 (tropomysin)	MAb	1 ng/ml	Jeoung et al., 1997

[a]MAb, monoclonal antibody; PAb, polyclonal antibody; EW, egg white; BLG, β-lactoglobulin; KSTI, Kunitz soy trypsin inhibitor; LOD, limit of detection.

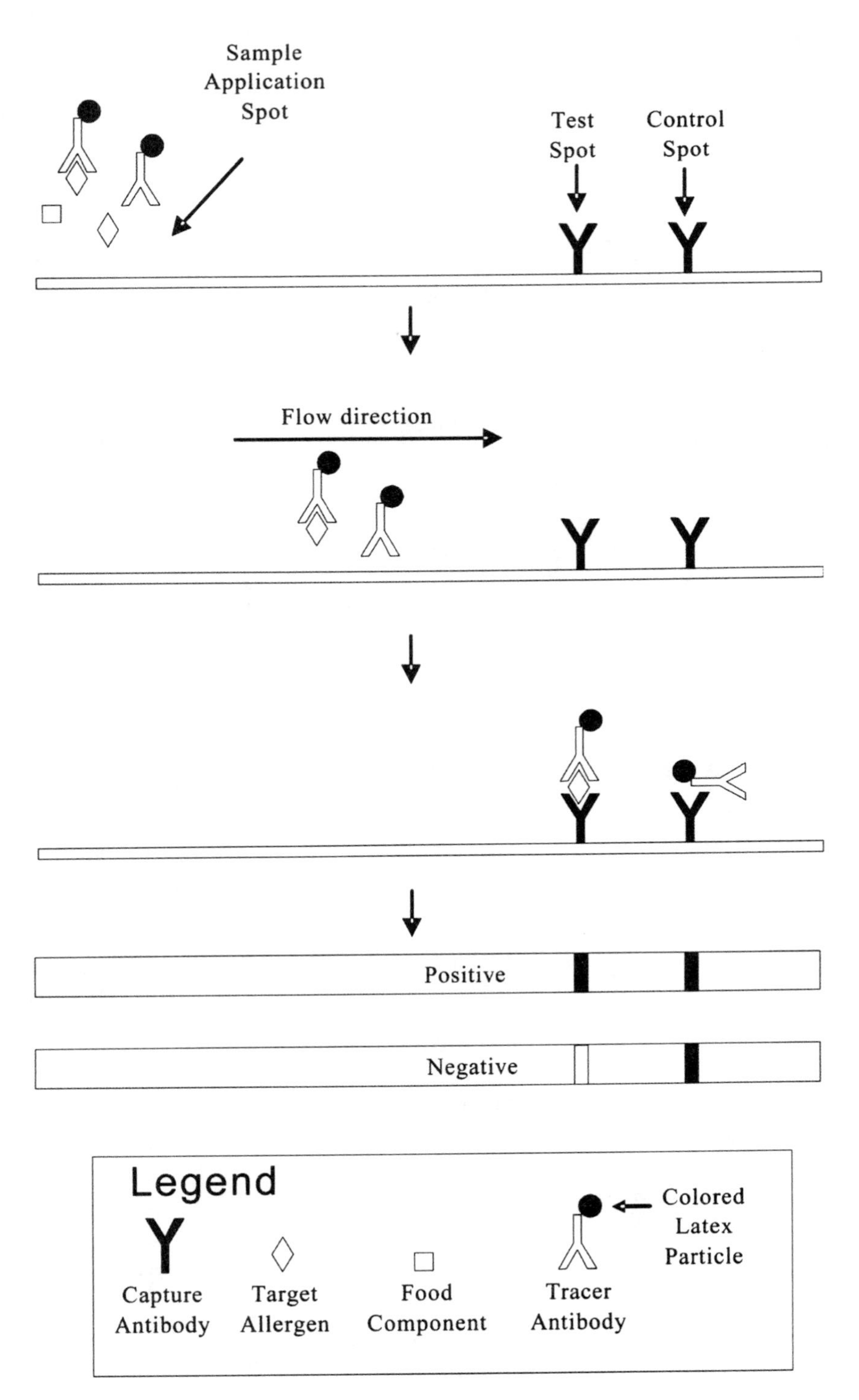

Figure 2. Scheme of a lateral flow test.

although there are studies that provide an estimate of the minimal threshold dose. Double-blind placebo control food challenge (DBPCFC) is considered the most appropriate test to help establish threshold doses for food allergens. Because DBPCFC has usually been used for diagnostic purposes, high doses of the allergens have usually been used. In addition, there been no uniform protocol to follow. In recent years, experts in food allergy from different countries have been working together to standardize the DBPCFC protocol to be used in the establishment of threshold doses (Taylor et al., 2002; Taylor et al., 2004).

To establish thresholds for food allergens, the challenge of the most-sensitive individuals seems to be very relevant because it has been reported that they suffer the most-severe reactions (Wensing et al., 2002; Perry et al., 2004). However, they might not be included in the studies because of the risk associated with the challenges (Perry et al., 2004; Hourihane and Knulst, 2005).

Only the most recent studies of milk, egg, peanut, and sesame have included a range of doses to determine risk assessment parameters such as the lowest-observed adverse effect level. The lowest-observed adverse effect level for food allergy is considered to be a few hundreds of micrograms of protein (Taylor et al., 2002) or low numbers of milligrams of the offending food (Morisset et al., 2003). Another risk assessment parameter, the no observed adverse effect level, could be useful to determine thresholds as well, but no data are available at this time (Taylor et al., 2002; Morisset et al., 2003; Moneret-Vautrin and Kanny, 2004). Until threshold values are established, the lower limit of quantification (LOQ) of the analytical technique, between 0.3 and 2.5 ppm of allergenic food, can be useful as an analytical threshold. This level is probably below the dose needed to elicit an allergic reaction in the majority of the population, but it is unknown whether it is low enough to protect the most sensitive individuals.

The lack of established threshold doses is opening a controversial debate regarding the possible presence of traces of allergenic proteins in refined oils, soy lecithin, fish gelatin, and some alcoholic beverages. These proteins are thought to be removed during processing, but it is possible that traces of allergenic proteins remain. It has been shown that oil refining procedures reduce protein content in oils by >100 times from crude peanut and sunflower oils that contained about 100 to 300 μg protein/ml (Crevel et al., 2000). Although traces of proteins were found in some soy oils and soy lecithin samples, they did not bind serum IgE from patients allergic to soybean (Awazuhara et al., 1998). In the processing of some alcoholic beverages, allergenic proteins may be added for technological purposes. Caseins and egg proteins are usually used in the production of wine in a procedure called fining, to remove unwanted compounds such as tannins and other phenolic compounds. These proteins are removed later in the production process. To date, there are no reports of allergic reactions due to egg or milk residues in wine. However, additional studies are needed to demonstrate that these products lack the potential to elicit allergic reactions in sensitive individuals.

Other Considerations

Apart from the development of analytical techniques for the detection methods, the final user also plays an important role in the detection of food allergens. The analyst must be familiar with all of the steps involved in the assay and the importance they have to provide high-quality results, from sample preparation to interpretation of results.

Table 3. Commercial kits for the detection of food allergens[a]

Kit	Manufacturer and food tested					
	Tepnel Biosystems	r-Biopharm	Neogen Corporation	Pro-Lab Diagnostics	ELISA Systems	Morinaga Institute of Biological Science
ELISA	Peanut Casein β-Lactoglobulin Egg Soy Wheat/gluten Sesame seed	Peanut β-Lactoglobulin Egg Hazelnut Almond Wheat, gluten	Peanut Milk Egg Almond Gliadin	Peanut Milk Egg	Peanut Milk Casein Egg Hazelnut Almond Crustacean Soy Sesame	Peanut Milk Egg Wheat
Rapid Methods[a]	Peanut Gluten	Peanut β-Lactoglobulin Egg Almond Gluten	Peanut Milk Egg Almond Gliadin			
PCR	Peanut Milk Soya	Peanut Hazelnut Almond Soy				
PCR-ELISA		Peanut Hazelnut Almond Soy				

[a]Includes rapid ELISA and lateral flow devices.

Sample Preparation

The sample must be processed so that the maximum amount of the allergen is extracted from the matrix with the minimum rate of interference from other compounds. The sensitivity of the assay is irrelevant if the sample is not properly prepared. It is very important to avoid cross-contamination during sample preparation to reduce the potential for false positives. Devices used for sample preparation, like grinders, must be thoroughly cleaned between samples. It is advisable to use disposable pipette tips and single-use extraction containers, chemicals, or detergents to remove protein residues. It is also advisable to analyze extraction buffers to verify the absence of allergen contamination.

The presence of food allergens as contaminants can be analyzed directly in raw ingredients and final products or indirectly from processing equipment after cleaning procedures have been applied to verify sanitation efficiency.

There are two types of samples that can be used to ensure cleaning effectiveness: (i) rinsing solutions obtained after cleaning detergents or chemicals have been applied to the equipment and (ii) swab samples of equipment surfaces. Swabbing is applied to equipment where cleaning is incompatible with aqueous solutions, such as shared lines used to manufacture chocolate. In this type of production line, chocolate is pushed through the line to clean any residue of allergen.

To extract the target allergenic residue from the sample, an appropriate extraction solution must be used. To improve extraction efficiency, solid samples must be either ground to a fine powder or melted, as in the case with chocolate or ice cream. Liquid samples do not require any further preparation, although extremely diluted samples could be concentrated. The extraction buffer must be compatible with the immunoassay. This means that the buffer used must favor antibody-antigen binding. Because most of the antibodies are found naturally in the body, physiological saline solutions of neutral pH are usually used. Depending on solubility, proteins can be classified as globulins (soluble in saline solutions), albumins (soluble in aqueous solutions), or prolamines (soluble in an alcohol-aqueous mixture).

The extraction and immunoassay analysis of allergens from food are more complex than the extraction and analysis of other components and contaminants in food for the following reasons.

Food allergens are proteins. Most of the allergenic foods contain more than one allergen (Table 4) of different physical-chemical characteristics. Thus, extraction procedures and the detection method can be developed for a mixture of allergenic and nonallergenic proteins, for several allergenic proteins, or for specific allergens. The sample extraction buffer must be optimized for every allergenic food or every allergen, depending on what is being analyzed. Both qualitative and quantitative differences found in protein extractability have been previously reported when the same allergenic foods were extracted with different extraction solutions (Kopper et al., 2004; Poms et al., 2004a; Westphal et al., 2004). The pH of the buffer is one of the factors that seem to have an important impact on the ability of extraction solutions to solubilize proteins from the matrix. To illustrate this fact, Fig. 3 shows peanut protein extracted with four different extraction solutions from light and dark roasted peanut flours.

Food processing affects the solubility and antigenicity of food proteins. Proteins have a complex structure which can be affected by food processing in many different ways (Davis and Williams, 1998; Besler et al., 2001; Maleki, 2004; Poms and Anklam, 2004). Depending on the degree and type of manufacturing process, proteins can be broken down or denatured by disruption of their tertiary and secondary structure, resulting in the modification of antigenic determinants (epitopes). Yeung and colleagues (Yeung and Collins, 1996, 1997; Yeung et al., 2000b) have evaluated the suitability of antibodies raised against native and heated or roasted peanut, soybean, and egg extracts for the development of ELISAs to detect traces of these allergenic materials in processed foods. Moreover, some proteins can be rendered insoluble and therefore nonextractable, due to the exposure of hydrophobic amino acid residues that are typically located internally in the native protein or due to protein aggregation (Davis et al., 2001; Koppelman et al., 2002). Progressive roasting decreases the solubility of peanut proteins (Kopper et al., 2004), including some of the major peanut allergens such as Ara h 1 (Maleki et al., 2000a). Also, proteins can react covalently with other components of food. The best characterized is the reaction of peanut allergens with sugars via the Maillard reaction (Davis et al., 2001). It is very important to understand the analysis of food allergens from this perspective. In fact, the target protein can be very different from the native form the analyst initially intends to analyze.

The detection of multiple proteins (allergenic and nonallergenic) to determine the presence of allergenic material in foods seems to be a good approach if their antigenicity remains

Table 4. Nomenclature of allergenic proteins[a]

Allergen (species)	Allergenic protein(s)
Peanut (*Arachis hypogaea*)	Ara h 1, Ara h 2, Ara h 3, Ara h 4, Ara h 5, Ara h 6, Ara h 7, Ara h 8
Hen's egg	
Egg white	Ovomucoid (Gal d 1), ovalbumin (Gal d 2), ovotransferrin (Gal d 3), lysozyme (Gal d 4)
Egg yolk	Chicken albumin or α-livetin (Gad d 5), apovitellenin, vitellogenin
Milk	
Casein	α-, β-, κ-, Bos d 8
Whey fraction	β-Lactoglobulin (Bos d 5), α-lactalbumin (Bos d 4), bovine serum albumin (Bos d 6), bovine immunoglobulins (Bos d 7)
Soybean (*Glycine max*)	Gly m Bd 30K, glycinin, conglycinin, Gly m 1, Gly m 2, Gly m3, Gly m 4, Kunitz-trypsin inhibitor, trypsin inhibitor, pathogenesis related protein, Bd 28 kDa
Tree nuts	
Almond (*Prunus dulcis*)	Amandin, profiling, lipid transfer protein
Brazil nut (*Bertholletia excelsa*)	Ber e 1, Ber e 2
Cashew (*Anacardium occidentale*)	Ana o 1, Ana o 2, Ana o 3
Hazelnut (*Corylus avellana*)	Cor a 1, Cor a 2, Cor a 8, Cor a 9, Cor a 10, Cor a 11, oleosin
Pecan (*Carya illinoinensis*)	Seed storage protein
Walnut (*Juglans regia*)	Jur r 1, Jur r 2, Jur r 3
Gluten-containing cereals	Gliadins (α-, γ-, ω_{12}-, ω_5-), trypsin α-amylase inhibitor, profiling, glutenins (HMW, LMW)
Buckwheat (*Fagopyrum esculentum*)	Fag e 1, Fag e 8kD
Fish	Common food allergen is parvalbumin
Cod (Several species)	Gad c 1, Gad m 1
Salmon (*Salmo salar*)	Sal s 1
Mackerel (several species)	Sco a 1, Sco j 1, Sco s 1
Shellfish	Common food allergen is tropomyosin
Shrimp (several species)	Pen a 1, Pen i 1, Pen m 1, Pen m 2, Pen o 1, Met e 1
Crab (*Charybdis feriatus*)	Cha f 1
Lobster (several species)	Pan s 1, Hom a 1
Sesame seeds (*Sesame indicum*)	Ses i 1, Ses i 2, Ses i 3
Celery (*Apium graveolens*)	Api g 1, Api g 3, Api g 4, Api g 5
Mustard	
White mustard (*Sinapis alba*)	Sin a 1
Oriental mustard (*Brassica juncea*)	Bra j 1

[a]Sources of information: http://www.allergenonline.com/, http://www.ifrn.bbsrc.ac.uk/protall/, and http://fermi.utmb.edu/SDAP/sdapf_src.html. HMW, high molecular weight; LMW, low molecular weight.

unchanged after food processing. However, it has been documented that proteins are affected by different food processes and to different degrees (Maleki, 2004; Poms and Anklam, 2004). This means that a particular detection method could perform differently and therefore would provide a different result when the same sample was processed by different technological treatments. To overcome this problem, various approaches have been suggested, such

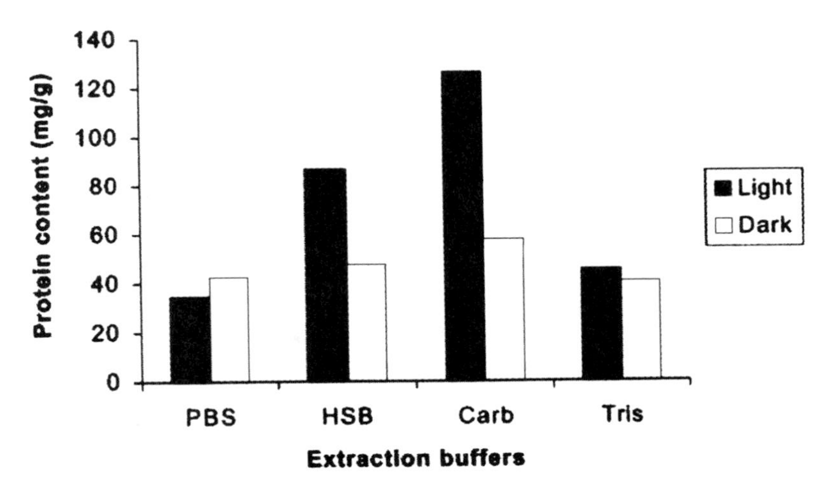

Figure 3. Peanut protein extracted from light and dark roasted peanut flour with phosphate buffer (pH 7.4) (PBS), high-salt buffer (pH 7.4) (HSB), carbonate buffer (pH 9.5) (Carb), and Tris buffer (pH 8.2) (Tris). (Reproduced with permission from the *Journal of AOAC International* [Westphal et al. 2004].)

as targeting specific proteins that are resistant to food processing (Westphal et al., 2004). This approach has already been applied for the detection of pepsin-resistant hazelnut protein, which allows the detection of trace amounts of hazelnut in processed foods and raw ingredients (Akkerdaas et al., 2004). Another approach is to use antibodies raised against extracts from processed foods (Yeung and Collins, 1996 and 1997; Yeung et al., 2000b).

Chapter 13 focuses specifically on the effects of food processing on the antigenicity and allergenicity of food allergens. The lack of solubility of food allergens has an important impact from the perspective of food safety. The nonextractable allergens that remain in food, undetected, may have an adverse effect on sensitive individuals. Using stronger extraction protocols may not be feasible because the antigenicity of the allergen and the structure of the antibodies used in the immunoassay may be modified. It is important that all of the extraction procedure reagents are compatible with the components of the immunoassay.

Some known food components interfere with food allergen extraction. For example, tannins and some other types of phenols present in chocolate bind to peanut proteins, rendering them nonextractable (Keck-Gassenmeier et al., 1999).

To minimize this problem, some in-house and commercial procedures use or suggest the addition of an excess of protein such as fish gelatin or dry skimmed milk to the extraction buffer. In addition, it is also recommended by some kit manufacturers that extraction procedures be carried out at 60 to 65°C. However, the extraction of some allergenic proteins from the food matrix is more efficient at room temperature, as has been demonstrated with the extraction of the major peanut allergen Ara h 1 (Pomés et al., 2004).

Interpretation of Results

For quantitative assays carried out with ELISA plates, color development of samples must be compared with a curve built with known concentrations of the analyte (Fig. 4).

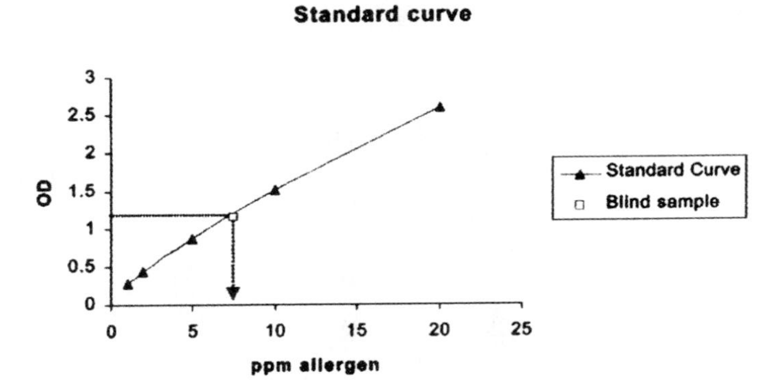

Figure 4. Typical standard curve in sandwich ELISA and direct ELISA, used to determine the concentration of an allergen in a blind sample.

The plates are read in an ELISA reader, and the resulting measurements are expressed as units of optical density (OD) at the given wavelength recommended for the substrate used. The calibration curve or standard curve is obtained by plotting the ODs corresponding to each calibrator (y axis) against their concentration (x axis).

Curve fitting will show a relationship between the resulting ODs and their corresponding concentration. A mathematical algorithm resulting from the application of regression analysis to the standard curve will allow the operator to extrapolate the values for each unknown. There are some commercial software products available that perform this type of analysis. Some kit manufacturers also provide software to analyze the data obtained with their kits.

Assay Validation

Analytical techniques used for the detection of food allergens have to be fully validated by a study involving different laboratories. The following are the most important parameters that need to be evaluated, which are commonly confused concepts.

Specificity

Specificity describes the ability of an antibody to exclusively bind the compound of interest. This parameter is related to cross-reactivity, which refers to the binding of antibody to compounds other than the target analyte.

Accuracy

Accuracy defines how close the average measured value is to the true concentration of the analyte in the sample. Recovery studies are used to evaluate this parameter. For this purpose, a known amount of the compound of interest is added to a food matrix. Recovery values can be affected by the presence of endogenous components of the food matrix that interfere with the assay. This is called matrix effect. Matrix components cause interference by blocking the antibody-analyte binding or by direct interaction with the analyte. Alteration of the protein conformation or modification of the protein during processing will also affect assay accuracy.

Precision

Precision describes the reproducibility of results by a single assay. Precision is really a measure of the error (coefficient of variation) when the same sample is analyzed several times within the same assay lot (intraassay variation), when the same sample is analyzed in several runs and on different days (interassay variability), or when the sample is analyzed with different reagent lots (interlot variability). Not only is the source of error inherent to the assay but additional error can result from mishandling of the assay by the end user. Poor pipetting techniques, poor control of incubation times, and reagent temperatures are among those factors or activities that adversely impact precision. Complete training on immunoassay technique is advisable to make the analyzer aware of the negative effects a deficient manipulation can have on the final results.

Sensitivity

Sensitivity is equivalent to the lower limit of detection, which indicates the lowest concentration of the analyte different from zero that the assay can achieve. It should not be confused with the lower LOQ, which is the lowest concentration of the analyte that can be measured with an acceptable level of precision (coefficient of variation of <15 to 20%) and accuracy. For practical purposes, the LOQ of an immunoassay is the lowest calibrator, different from zero, used to prepare the standard curve.

To perform validation assays, acceptance criteria must be set for each parameter by the study director, based on his or her scientific judgment and purpose of the assay. Validation studies also require the availability of stable standards and controls.

Commercial Kits

Several companies have developed kits for the detection of food allergens (Table 3). New or updated regulations regarding the labeling of food allergens have increased the need to validate existing assays and to develop new assays for additional food allergens. Commercial kits have been validated in house; those for the detection of peanuts have been the only ones studied more extensively. An effort has been made by agencies from different countries, kit manufacturers, and some private organizations to validate these kits. A peanut suspension made from National Institute of Standards and Technology peanut butter was evaluated as a reference material (Trucksess et al., 2004). This suspension was further used in a performance evaluation of commercial kits for peanuts sponsored by the AOAC Research Institute (Park et al., 2005). Veratox (Neogen Corp), Biokit (Tepnel Life Sciences), and Ridascreen (r-Biopharm) were the kits and kit manufacturers that participated in the study. The purpose of this study was to evaluate sensitivity, selectivity, robustness, cross-reactivity, and interference. All three kits have been certified as performance-tested methods by the AOAC Research Institute. The Institute for Reference Material and Measurements from the EU also carried out an interlaboratory validation study of five peanut kits to quantify peanut residues in two different matrices, biscuits and dark chocolate (Poms et al., 2005).

The ability of four commercial kits to detect different peanut concentration in four different matrices has been evaluated. Analytical variability was observed among the kits and increased with the increase in peanut concentration (Whitaker et al., 2005). These differences may result from the fact that different kits use antibodies against different antigen preparations. It is also known that most of the manufacturers use peanut extracts as cali-

Table 5. Characteristics of commercial kits for the detection of egg or egg proteins[a]

Characteristic	Tepnel	r-Biopharm	Neogen	ELISA Systems
Ab specificity	OM	EWP	EWP	OM OVA
Antibody type	Polyclonal	Polyclonal	Polyclonal	Polyclonal
Result	EWP (ppm)	EWP (ppm)	Egg (ppm)	EWP (ppm)
Extraction buffer	HSB + protein	Unknown	PBS + additive	Unknown
LOQ	0.5 ppm EWP	2 ppm EWP	2.5 ppm egg	1 ppm EWP

[a]HSB, high-salt buffer; PBS, phosphate buffer; LOQ, lower limit of quantification; OM, ovomucoid; OVA, ovoalbumin; EWP, egg white proteins.

brators. These extracts are not well characterized and contain allergenic and nonallergenic proteins. Peanut ELISA kits from Tepnel Biosystems and ELISA Systems are the only kits that use a purified fraction of peanut that includes the major allergens Ara h 1 and Ara h 2, respectively. A study has shown that four kits were able to detect both Ara h 1 and Ara h 2 peanut allergens (Nogueira et al., 2004).

Moreover, kits use different extraction solutions, some of them of unknown composition. All of these factors make it difficult to compare and evaluate results given by the different kits. Another controversial issue about these kits is the manner in which results are expressed. Results can be given in micrograms of protein per gram or milliliter of sample or as micrograms of whole allergenic food per gram or milliliter of sample. Most of the peanut kits use a constant factor to convert the concentration of peanut proteins in the sample into whole peanut material, and this factor is based on the ratio of water-soluble protein per total peanut. However, it has been shown that this relationship depends greatly on the effects of processing on protein solubility and stability (Westphal et al., 2004). Therefore, because the conversion factor does not seem to be a reliable constant, it is not suitable for use with all types of food and all types of processing. To illustrate how difficult kit comparison can become, Table 5 compares some of the characteristics of commercial kits for the detection of egg.

DNA-BASED ANALYTICAL TECHNIQUES FOR THE DETECTION OF FOOD ALLERGENS

An alternative to the detection of allergenic proteins in food by immunoassays is the detection of the DNA from an allergenic food. The technique most commonly used for this purpose is PCR. PCR allows the copying of a section of a gene of interest millions of times. This makes PCR a very sensitive technique. Obviously, it requires the identification of the gene of interest. PCR is also a very specific technique because it amplifies a segment of DNA among the complex mixture of DNA present in food. To perform PCR, a DNA template is isolated from the food source. The concept of amplification of the target DNA by PCR is very simple, requiring three basic steps repeated a determined number of times (cycles) under controlled times and temperatures (Fig. 5), as follows.

1. Denaturing step. The two strands of the DNA template are separated into single strands by the application of high temperatures.

2. Annealing step. The DNA template is then incubated with complementary short DNA sequences or primers, which will hybridize with each DNA strand. The primers will limit the size of DNA to be amplified and also will be the starting point for the elongation step.
3. Elongation step. The DNA amplification is enzymatically carried out by the DNA polymerase that will recognize the nucleotide in the DNA template and will insert its complement.

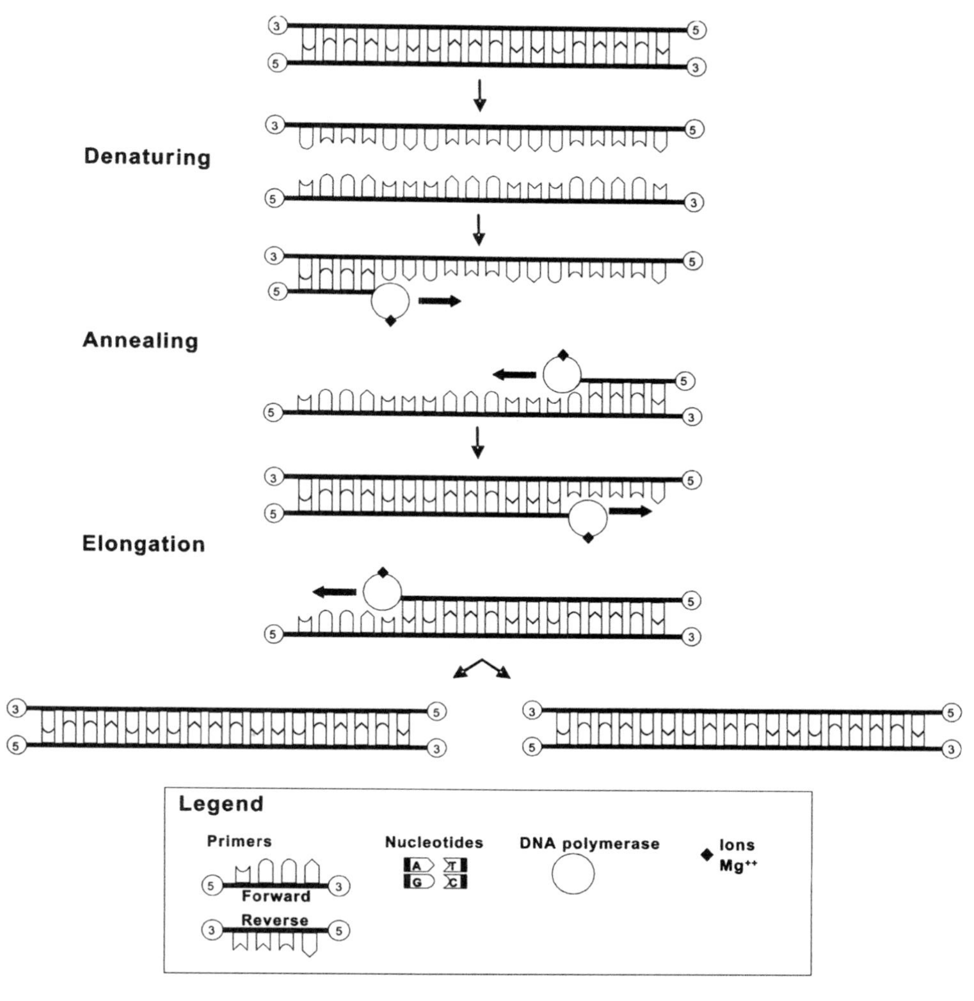

Figure 5. Steps of a conventional PCR cycle.

Resulting PCR amplicons or PCR products are used as a DNA template for the subsequent cycles (usually between 30 and 40). PCR products are separated in agarose gel by electrophoresis and visualized by ethidium bromide staining. The number of base pairs will determine the size of the PCR amplicon and the location of the amplicon band in the gel. This is compared to a standard DNA ladder of known base pairs.

The optimization of the PCR is critical for the success of the technique. The selection of a good set of primers will define the specificity of the technique, while the time, temperature, and the concentration of the different PCR reagents will determine its sensitivity. The success of PCR also depends greatly on external factors such as the sample size; the quantity, quality, and purity of the isolated DNA; and the type of matrix from which the DNA is extracted. There are a wide variety of physical-chemical processes, such as heat treatments, enzymatic activities, or use of acid pH, commonly used by the food industry that can affect DNA integrity (Holzhauser et al., 2000). Food processing procedures can modify the DNA structure or fragment it to levels where PCR will not perform efficiently or at all. As with ELISA, some food components, such as polyphenols and heteropolysaccarides present in cocoa and cereals, can act as PCR inhibitors (De Boer et al., 1995).

Traditional PCR is a qualitative technique, since it only detects the presence of the DNA of interest. However, new technologies have allowed the development of the quantitative version of PCR, called real-time PCR (RT-PCR), which uses labeled DNA probes with the aforementioned reagents. These probes provide a measurable fluorescence signal every time DNA amplification takes place. This allows the monitoring of the reaction after every cycle has been completed. The signal increases proportionally with the concentration of DNA in the sample, and it is compared to a standard curve built with a known concentration of DNA target. The results are expressed as a percentage of DNA target in the sample.

PCR or RT-PCR can target genes encoding specific allergenic proteins (Table 6), such as *Cor a* 1.0401 from hazelnut (Holzhauser et al., 2000), *Ara h 2* from peanut (Stephan and Vieths, 2004), and *Api g 1* from celery (Stephan et al., 2004). However, the detection of allergenic ingredients is not limited only to the amplification of genes encoding allergens. Other sections of DNA can also be amplified and used as markers for the presence of an allergenic ingredient (Table 6). For instance, James and Schmidt (2004) developed a PCR to amplify the intron region of a chloroplast tRNA gene (*trnL*). Using species-specific forward primers and a common reverse primer, they determined the presence of peanuts, soybeans, and wheat. With a modification of this assay, it could be possible to detect multiple allergenic ingredients simultaneously in a multiplex PCR. Multiplex PCR assays have already been developed for the detection of multiple transgenic crops (James et al., 2003). A similar approach targeting 5S rRNA gene intergenic spacer (NTS) is used to detect mackerel ingredients in seafood (Aranishi and Okimoto, 2004). Another type of PCR, called nested PCR, has also been applied to the detection of soya in processed meats (Meyer et al., 1996). This PCR consists of two consecutive PCRs, where the second PCR uses a second pair of primers to amplify a sequence within the PCR product resulting from the first PCR.

Although immunoassays still lead the market as the analytical techniques of choice for the detection of food allergens, new interest has been shown in PCR, leading to new kits being developed and marketed (Table 3). Tepnel Biosystems offers PCR kits for soya (soya lectin), cow's milk, and peanut. r-Biopharm has developed a quantitative RT-PCR for soya, hazelnut, peanut, and almond. All of them have limits of detection of <10 ppm.

PCR-ELISA combines the traditional PCR as a first step, followed by the detection of PCR products by ELISA (Holzhauser et al., 2002). In the second step, the detection of

Table 6. Published PCR and RT-PCR protocols for the detection of allergenic foods[a]

Food	PCR type	Gene	Primer or probe (5′–3′) and sequence	bp	Reference
Peanut	RT-PCR	*Ara h 2*	FP AR-58 F: GCAGCACTGGGA ACTCCAAGGAGACA	86	Stephan and Vieths, 2004
			RP AR-143 R: GCATGAGATGTTGCTCGCAG		
			Probe AR-103: CGAGAGGGCGAACCTGAGGCC		
Peanut	RT-PCR	*Ara h 2*	FP GCTCGAGAGGCCGAACCT	ND	Hird et al., 2003
			RP TCCTCGTCACGTTGGATCTTC		
			Probe AGGCCCTGCGAGCAACATCTCATG		
Celery	RT-PCR	*Api g 1*	FP Api-346F: ACCACTGCCATCTTTCACA	145	Stephan et al., 2004
			RP Api-490R: TGACATTACAACAAGACATTCC		
			Probe Api-391T; CTTCTGGAACCACAGCATCACCT		
Wheat	RT-PCR	ω-Gliadin gene	FP TAG2315F: CAGAAAGCGAGTGGAAAGATGAAAG	181	Sandberg et al., 2003
			RP TAG2473R: GCAAGGAGGACAAAGATGAGGAA		
Rye		ω-Secalin gene	FP SEC2599F: TTTTTCAGAAAGCGAGTTCAATGATG	181	
			RP SEC2756R: CGAGGACAAAGATGAGGAAGGTCT		
Barley		Hordein gene	FP HVH1618F: ATTAATTCCCAAACTGAACGACTA	164	
			RP HVH1763R: CATGGCGAACAATGTGAAC		
Oats		Avenin gene	FP ASA 1097F: CGCTCAGTGGCTTCTAAGA	104	
			RP ASA1177R: TTTTATTTTATTTGTCACCGCTAC		
Soybean	PCR	*trnL* intron region	FP soybean SN-2F: AATAATAGAATCCTTCCGTC	343	James and Schmidt, 204
Wheat			FP wheat WT-1F: GAGGGGTTTTATACCTTATAC	397	
Peanut			FP peanut PT-4F: AGGAATCCTTCTGATACAAATG	403	
			Common RP PL-2D: GGGGATAGAGGGACTTGAAC		
Soybean	Nested PCR	*Le1*	Pair 1: FP GM01: TGCCGAAGCAACCAAACATGATCCT	414	Meyer et al., 1996
			RP GM02: TGATGGATCTGATAGAATTGACGTT		
			Pair 2: FP GM03: GCCCTCTACTCCACCCCCATCC	118	
			RP GM04: GCCCATCTGCAAGCCTTTTTGTG		
Hazelnut	PCR	*Cor a 1.0401*	FP CAF02: GGAGATCGACCACGCAAACTTCAA	182	Holzhauser et al., 2000
			RP CAR03: CCTCCTCATTGATTGAAGCGTTG		
Hazelnut	PCR	*Nad1*	FP NADH7: AAACCAGGGCAACAGATGT	294	Herman et al., 2003
			RP NADH2: AGCGAATCGCGAACACGA		
Mackerel	PCR	5S rDNA NTS	FP Saba-18F: GGGCGCTGTTGCTCCATC	359	Aranishi and Okimoto, 2004
			RP Saba-20R: ATGCTGTGACACCACTGACA	311	

[a]ND, not defined; FP, forward primer; RP, reverse primer; bp, base pairs.

amplicons occurs by their binding to a complementary sequence of DNA, which is labeled with a compound detectable by enzyme-conjugated antibodies. The use of ELISA to detect PCR products ensures that amplified DNA corresponds to the target gene encoding the allergen of interest, minimizing false positives. It also makes the traditional PCR a quantitative technique. This technique has been developed and marketed by r-Biopharm under the name of SureFood ELISA-PCR, and it is available for peanut, almond, soy, and hazelnut. Among the advantages of the assay is the possibility to analyze several samples simultaneously, and it has been shown to eliminate the cross-reactivity seen with ELISA methods for hazelnut (Holzhauser et al., 2002).

Although RT-PCR and PCR-ELISA have been shown to have a good correlation with ELISA results (Holzhauser et al., 2002; Stephan and Vieths, 2004; Stephan et al., 2004), there is an associated risk when DNA-based methods are used to evaluate the presence of allergenic components in food. This risk is described in the two following situations.

First, the absence of DNA does not guarantee the absence of the allergenic protein, leading to false negatives. Processing can degrade DNA, or the presence of PCR inhibitors can lead to false-negative results, while the allergenic protein or allergenic peptides remain intact (Holzhauser et al., 2000). Moreover, a food product may be fortified with isolated allergenic protein fractions (Stephan and Vieths, 2004). In this situation, the DNA encoding the added protein would not be present, also leading to a false negative result.

Second, the presence of DNA does not necessarily indicate the presence of the allergenic protein; it only shows that an allergenic ingredient was used in the food formulation. Food allergens can be degraded by food processing, while the DNA remains relatively intact. This can lead to unnecessary labeling of the product.

CONFIRMATORY METHODS—NEW TECHNOLOGIES

Current challenges in the detection of food allergens are focusing on the development of confirmatory methods. The confirmation process involves the use of an analytical technique other than an immunoassay, which corroborates the presence of the allergen or protein previously detected by immunoassay. Currently, there are some promising methodologies under investigation. Liquid chromatography-tandem mass spectrometry has been used for the confirmation of the presence of the major protein Ara h 1 in ice cream (Shefcheck and Musser, 2004) and for the identification of milk proteins (Natale et al., 2004).

Advancements in the fields of genomics and proteomics and their associated technologies successfully used for clinical and new drug development are being applied to the food allergy field (Beyer et al., 2002a; Beyer et al., 2002b; Pineiro et al., 2003; Yu et al., 2003). Epitope mapping, one- and two-dimensional electrophoresis, and immunoblotting, among others, have allowed the characterization and identification of known and new food allergens, as well as their isoforms (Hird et al., 2000; Sen et al., 2002; Gruber et al., 2004; Natale et al., 2004). Among genomic techniques, DNA sequencing of allergens, the construction of cDNA libraries, and the production of recombinant allergens in prokaryotes are the techniques most commonly used. Recombinant allergenic proteins are useful in that they can provide a better understanding about the role of the allergen structure in the development of the allergic disease and their susceptibility or resistance to digestion and food processing. Recombinant techniques can provide a consistent source of characterized proteins that are useful for the development of prophylactic and curative therapies and new methodologies for the detection of food allergens (Lorenz et al., 2001).

Laboratories that perform a large number of analyses each day will benefit from the use of high-throughput testing and automated equipment. High-throughput tests, widely applied for drug discovery, are being investigated for their use in the food allergy area. Although initial studies are being developed for the clinical aspects of food allergy, such as detection of allergen-specific IgE (Hiller et al., 2002; Fall et al., 2003; Harwanegg et al., 2003; Jahn-Schmid et al., 2003; Bacarese-Hamilton et al., 2004; Shreffler et al., 2004), the application of high-throughput testing to the detection of allergens in food will be probably addressed sometime in the future.

Automated equipment is being developed to perform the same steps an operator would do and to process a larger number of samples simultaneously. Equipment such as sample dispensers, mechanical washers, and multiplate readers not only speed up the analysis steps but also help minimize the variability that is potentially introduced by the operator.

The large amount of information obtained for each allergen analyzed by genomic and proteomic analytical techniques is difficult to handle with conventional analysis tools. Bioinformatics tools provide the algorithms necessary for an efficient and rapid processing of data and organization of data in databases (Gendel, 2004).

CONCLUSIONS

Immunoassays have been demonstrated to be a suitable analytical technique for the detection of hidden food allergens. The different formats allow the final user the possibility of using qualitative portable tests that analyze the allergen in the field, while quantitative formats simultaneously analyze multiple samples down to the low parts per million. PCR is gaining importance as an alternate detection method, although it does not directly detect the allergenic protein. Specific genes can be targeted and be used as markers for the presence of allergenic residues in food. As technology advances, new analytical techniques are being applied to the field of detection of food allergens (e.g., biosensor or microarray) or the characterization of new ones (e.g., two-dimensional electrophoresis or mass spectrometry).

EU and U.S. government regulations will be fully implemented by late 2005 and early 2006, respectively. This means that it will be imperative for the food industry and government to have detection methods for all of the allergenic ingredients that need to be labeled by law to ensure the health of sensitive individuals and their food choices. New labeling demands underscore the need for validated analytical techniques and reference materials. Also, kit manufacturing companies should harmonize the way the results are reported for all of the allergenic ingredients, in parts per million of allergenic protein (e.g., ovomucoid) or parts per million of allergenic food (e.g., egg). In addition to this work agenda, a threshold for every food allergen still needs to be determined by standardized protocols. In the meantime, zero tolerance, i.e., the complete absence of food allergens, is in place. However, the complete absence of allergen is technically impossible to determine, since commercial kits have cutoff values or LOQ in the low parts per million. Any quantity of allergen residue below that value will remain undetected.

REFERENCES

Acosta, M. R., K. H. Roux, S. S. Teuber, and S. K. Sathe. 1999. Production and characterization of rabbit polyclonal antibodies to almond (*Prunus dulcis* L.) major storage protein. *J. Agric. Food Chem.* **47:**4053-4059.

Akkerdaas, J. H., M. Wensing, A. Knulst, O. Stephan, S. L. Hefle, R. C. Aalberse, and R. van Ree. 2004. A novel approach for the detection of potentially hazardous pepsin stable hazelnut proteins as contaminants in chocolate-based food. *J. Agric. Food Chem.* **52:**7726–7731.

Aranishi, F., and T. Okimoto. 2004. PCR-based detection of allergenic mackerel ingredients in seafood. *J. Genet.* **83:**193–195.

Awazuhara, H., H. Kawai, M. Baba, T. Matsui, and A. Komiyama. 1998. Antigenicity of the proteins in soy lecithin and soy oil in soybean allergy. *Clin. Exp. Allergy* **28:**1559–1564.

Bacarese-Hamilton, T., A. Ardizzoni, J. Gray, and A. Crisanti. 2004. Protein arrays for serodiagnosis of disease. *Methods Mol. Biol.* **264:**271–283.

Baeumner, A. J. 2003. Biosensors for environmental pollutants and food contaminants. *Anal. Bioanal. Chem.* **377:**434–445.

Bando, N., H. Tsuji, M. Hiemori, K. Yoshizumi, R. Yamanishi, M. Kimoto, and T. Ogawa. 1998. Quantitative analysis of Gly m Bd 28K in soybean products by a sandwich enzyme-linked immunosorbent assay. *J. Nutr. Sci. Vitaminol.* **44:**655–664.

Baumgartner, S., I. Steiner, S. Kloiber, D. Hirmann, R. Krska, and J. M. Yeung. 2002. Towards the development of a dipstick immunoassay for the detection of trace amounts of egg proteins in food. *Eur. Food Res. Technol.* **214:**168–170.

Bayard, C., and F. Lottspeich. 2001. Bioanalytical characterization of proteins. *J. Chromatogr. B Biomed. Sci. Appl.* **756:**113–122.

Benrejeb, B. S., M. Abbott, D. Davies, J. Querry, C. Cleroux, C. Streng, P. Delahaut, and J. M. Yeung. 2003. Immunochemical-based method for the detection of hazelnut proteins in proteins in processed foods. *J. AOAC Int.* **86:**557–563.

Besler, M., H. Steinhart, and A. Paschke. 2001. Stability of food allergens and allergenicity of processed foods. *J. Chromatogr. B Biomed. Sci. Appl.* **756:**207–228.

Beyer, K., L. Bardina, G. Grishina, and H. A. Sampson. 2002a. Identification of sesame seed allergens by 2-dimensional proteomics and Edman sequencing: seed storage proteins as common food allergens. *J. Allergy Clin. Immunol.* **110:**154–159.

Beyer, K., G. Grishina, L. Bardina, A. Grishin, and H. A. Sampson. 2002b. Identification of an 11S globulin as a major hazelnut food allergen in hazelnut-induced systemic reactions. *J. Allergy Clin. Immunol.* **110:**517–523.

Blais, B. W., and L. Phillippe. 2001. Detection of hazelnut proteins in foods by enzyme immunoassay using egg yolk antibodies. *J. Food Prot.* **64:**895-898.

Blais, B. W., and L. M. Phillippe. 2000. A cloth-based enzyme immunoassay for detection of peanut proteins in foods. *Food Agric. Immunol.* **12:**243–248.

Blais, B. W., M. Gaudreault, and L. Phillippe. 2003. Multiplex enzyme immunoassay system for the simultaneous detection of multiple allergens in foods. *Food Control* **14:**43–47.

Blais, B. W., M. Omar, and L. Phillippe. 2002. Detection of Brazil nut proteins in foods by enzyme immunoassay. *Food Agric. Immunol.* **14:**163–168.

Bredehorst, R., and K. David. 2001. What establishes a protein as an allergen. *J. Chromatogr. B Biomed. Sci. Appl.* **756:**33–40.

Burks, W., R. Helm, S. Stanley, and G. A. Bannon. 2001. Food allergens. *Curr. Opin. Allergy Clin. Immunol.* **1:**243–248.

Clemente, A., S. J. Chambers, F. Lodi, C. Nicoletti, and G. M. Brett. 2004. Use of the indirect competitive ELISA for the detection of Brazil nut in food products. *Food Control* **15:**65.

Crevel, R. 2002. Industrial dimensions of food allergy. *Biochem. Soc. Trans.* **30:**941–944.

Crevel, R. W., M. A. Kerkhoff, and M. M. Koning. 2000. Allergenicity of refined vegetable oils. *Food Chem. Toxicol.* **38:**385–393.

Davis, P. J., C. M. Smales, and D. C. James. 2001. How can thermal processing modify the antigenicity of proteins? *Allergy* **56**(Suppl. 67)**:**56–60.

Davis, P. J., and S. C. Williams. 1998. Protein modification by thermal processing. *Allergy* **53**(Suppl. 46)**:** 102–105.

De Boer, S. H., L. J. Ward, X. Li, and S. Chittaranjan. 1995. Attenuation of PCR inhibition in the presence of plant compounds by addition of BLOTTO. *Nucleic Acids Res.* **23:**2567–2568.

de Meulenaer, B., M. de La Court, D. Acke, T. de Meyere, and A. van de Keere. 2005. Development of an enzyme-linked immunosorbent assay for peanut proteins using chicken immunoglobulins. *Food Agric. Immunol.* **16:**129–148.

Drs, E., S. Baumgartner, M. Bremer, A. Kemmers-Voncken, N. Smits, W. Hassnoot, J. Banks, P. Reece, C. Danks, V. Tomkies, U. Immer, K. Schmitt, and R. Krska. 2004. Detection of hidden hazelnut protein in food by IgY-based indirect competitive enzyme-immunoassay. *Anal. Chim. Acta* **520:**223–228.

Fall, B. I., B. Eberlein-Konig, H. Behrendt, R. Niessner, J. Ring, and M. G. Weller. 2003. Microarrays for the screening of allergen-specific IgE in human serum. *Anal. Chem.* **75:**556–562.

Food and Agriculture Organization of the United Nations. 1995. *Report of the FAO Technical Consultation on Food Allergies,* p. 13–14. Food and Agriculture Organization of the United Nations, Rome, Italy.

Gendel, S. M. 2004. Bioinformatics and food allergens. *J. AOAC Int.* **87:**1417–1422.

Gruber, P., M. Suhr, A. Frey, W. M. Becker, and T. Hofmann. 2004. Development of an epitope-specific analytical tool for the major peanut allergen Ara h 2 using a high-density multiple-antigenic peptide strategy. *Mol. Nutr. Food Res.* **48:**449–458.

Harwanegg, C., S. Laffer, R. Hiller, M. W. Mueller, D. Kraft, S. Spitzauer, and R. Valenta. 2003. Microarrayed recombinant allergens for diagnosis of allergy. *Clin. Exp. Allergy* **33:**7–13.

Hefle, S. L. 1999. Impact of processing on food allergens. *Adv. Exp. Med. Biol.* **459:**107–119.

Hefle, S. L., and D. M. Lambrecht. 2004. Validated sandwich enzyme-linked immunosorbent assay for casein and its application to retail and milk-allergic complaint foods. *J. Food Prot.* **67:**1933–1938.

Hefle, S. L., and S. L. Taylor. 2004. Food allergy and the food industry. *Curr. Allergy Asthma Rep.* **4:**55–59.

Hefle, S. L., R. K. Bush, J. W. Yunginger, and F. S. Chu. 1994. A sandwich enzyme-linked immunosorbent assay (ELISA) for the quantitation of selected peanut proteins in food. *J. Food Prot.* **57:**419–423.

Hefle, S. L., E. Jeanniton, and S. L. Taylor. 2001. Development of a sandwich enzyme-linked immunosorbent assay for the detection of egg residues in processed foods. *J. Food Prot.* **64:**1812–1816.

Herman, L., J. D. Block, and R. Viane. 2003. Detection of hazelnut DNA traces in chocolate by PCR. *Int. J. Food Sci. Technol.* **38:**633–640.

Hiller, R., S. Laffer, C. Harwanegg, M. Huber, W. M. Schmidt, A. Twardosz, B. Barletta, W. M. Becker, K. Blaser, H. Breiteneder, M. Chapman, R. Crameri, M. Duchene, F. Ferreira, H. Fiebig, K. Hoffmann-Sommergruber, T. P. King, T. Kleber-Janke, V. P. Kurup, S. B. Lehrer, J. Lidholm, U. Muller, C. Pini, G. Reese, O. Scheiner, A. Scheynius, H. D. Shen, S. Spitzauer, R. Suck, I. Swoboda, W. Thomas, R. Tinghino, M. Van Hage-Hamsten, T. Virtanen, D. Kraft, M. W. Muller, and R. Valenta. 2002. Microarrayed allergen molecules: diagnostic gatekeepers for allergy treatment. *FASEB J.* **16:**414–416.

Hird, H., J. Lloyd, R. Goodier, J. Brown, and P. Reece. 2003. Detection of peanut using real-time polymerase chain reaction. *Eur. Food Res. Technol.* **217:**265–268.

Hird, H., R. Pumphrey, P. Wilson, J. Sunderland, and P. Reece. 2000. Identification of peanut and hazelnut allergens by native two-dimensional gel electrophoresis. *Electrophoresis* **21:**2678–2683.

Hlywka, J. J., S. L. Hefle, and S. L. Taylor. 2000. A sandwich enzyme-linked immunosorbent assay for the detection of almonds in foods. *J. Food Prot.* **63:**252–257.

Holzhauser, T., and S. Vieths. 1999a. Indirect competitive ELISA for determination of traces of peanut (*Arachis hypogaea* L.) protein in complex food matrices. *J. Agric. Food Chem.* **47**(Suppl. 2)**:**603–611.

Holzhauser, T., and S. Vieths. 1999b. Quantitative sandwich ELISA for determination of traces of hazelnut (*Corylus avellana*) protein in complex food matrixes. *J. Agric. Food Chem.* **47:**4209–4218.

Holzhauser, T., O. Stephan, and S. Vieths. 2002. Detection of potentially allergenic hazelnut (Corylus avellana) residues in food: a comparative study with DNA PCR-ELISA and protein sandwich-ELISA. *J. Agric. Food Chem.* **50:**5808–5815.

Holzhauser, T., A. Wangorsh, and S. Vieths. 2000. Polymerase chain reaction (PCR) for detection of potentially allergenic hazelnut residues in complex food matrixes. *Eur. Food Res. Technol.* **211:**360–365.

Hourihane, J. O., and A. C. Knulst. 2005. Thresholds of allergenic proteins in foods. *Toxicol. Appl. Pharmacol.* **207**(Suppl. 2)**:**152–156.

Immer, U., B. Reck, S. Lindeke, and S. J. Koppelman. 2004. RIDASCREEN FAST PEANUT, a rapid and safe tool to determine peanut contamination in food. *I. J. Food Sci. Technol.* **39:**869–871.

Jahn-Schmid, B., C. Harwanegg, R. Hiller, B. Bohle, C. Ebner, O. Scheiner, and M. W. Mueller. 2003. Allergen microarray: comparison of microarray using recombinant allergens with conventional diagnostic methods to detect allergen-specific serum immunoglobulin E. *Clin. Exp. Allergy* **33:**1443–1449.

James, D., and A. M. Schmidt. 2004. Use of an intron region of a chloroplast tRNA gene (*trn*L) as a target for PCR identification of specific food crops including sources of potential allergens. *Food Res. Int.* **37:**395–402.

James, D., A. M. Schmidt, E. Wall, M. Green, and S. Masri. 2003. Reliable detection and identification of genetically modified maize, soybean, and canola by multiplex PCR analysis. *J. Agric. Food Chem.* **51:**5829–5834.

Jeoung, B. J., G. Reese, P. Hauck, J. B. Oliver, C. B. Daul, and S. B. Lehrer. 1997. Quantification of the major brown shrimp allergen Pen a 1 (tropomyosin) by a monoclonal antibody-based sandwich ELISA. *J. Allergy Clin. Immunol.* **100:**229–234.

Jonsson, H., and K. E. Hellenäs. 2001. Optimizing assay conditions in the detection of food allergens with Biacore's SPR technology. *Biacore J.* **2:**16–18.

Joshi, P., S. Mofidi, and S. H. Sicherer. 2002. Interpretation of commercial food ingredient labels by parents of food-allergic children. *J. Allergy Clin. Immunol.* **109:**1019–1021.

Keck-Gassenmeier, B., S. Bénet, C. Rosa, and C. Hischenhuber. 1999. Determination of peanut traces in food by a commercially-available ELISA test. *Food Agric. Immunol.* **11:**243–250.

Koppelman, S. J., H. Bleeker-Marcelis, G. Duijn, and M. Hessing. 1996. Detecting peanut allergens. The development of an immunochemical assay for peanut proteins. *World Ingredients* **12:**35–38.

Koppelman, S. J., A. C. Knulst, W. J. Koers, A. H. Penninks, H. Peppelman, R. A. Vlooswijk, I. Pigmans, G. van Duijn, and M. Hessing. 1999. Comparison of different immunochemical methods for the detection and quantification of hazelnut proteins in food products. *J. Immunol. Methods* **229:**107–120.

Koppelman, S. J., G. A. van Koningsveld, A. C. Knulst, H. Gruppen, I. G. Pigmans, and H. H. de Jongh. 2002. Effect of heat-induced aggregation on the IgE binding of patatin (*Sol t 1*) is dominated by other potato proteins. *J. Agric. Food Chem.* **50:**1562–1568.

Kopper, R. A., N. J. Odum, M. Sen, R. M. Helm, J. S. Stanley, and A. W. Burks. 2004. Peanut protein allergens: the effect of roasting on solubility and allergenicity. *Int. Arch. Allergy Immunol.* **136:**16–22.

Lauwaars, M., and E. Anklam. 2004. Method validation and reference materials. *Accredit. Qual. Assur.* **9:**253–258.

Leduc, V., C. Demeulemester, B. Polack, C. Guizard, L. Le Guern, and G. Peltre. 1999. Immunochemical detection of egg-white antigens and allergens in meat products. *Allergy* **54:**464–472.

Lorenz, A. R., S. Scheurer, D. Haustein, and S. Vieths. 2001. Recombinant food allergens. *J. Chromatogr. B Biomed. Sci. Appl.* **756:**255–279.

Lu, Y., T. Oshima, and H. Ushio. 2004. Rapid detection of fish major allergen parvalbumin by surface plasmon resonance biosensor. *J.Food Chem. Toxicol.* **69:**C652–C658.

Maleki, S. J. 2004. Food processing: effects on allergenicity. *Curr. Opin. Allergy Clin. Immunol.* **4:**241–245.

Maleki, S. J., S. Y. Chung, E. T. Champagne, and J. P. Raufman. 2000a. The effects of roasting on the allergenic properties of peanut proteins. *J. Allergy Clin. Immunol.* **106:**76–768.

Maleki, S. J., R. A. Kopper, D. S. Shin, C. W. Park, C. M. Compadre, H. Sampson, A. W. Burks, and G. A. Bannon. 2000b. Structure of the major peanut allergen Ara h 1 may protect IgE-binding epitopes from degradation. *J. Immunol.* **164:**5844–5849.

Meyer, R., F. Chardonnens, P. Hubner, and J. Luthy. 1996. Polymerase chain reaction (PCR) in the quality and safety assurance of food: detection of soya in processed meat products. *Z. Lebensm. Unters. Forsch.* **203:**339–344.

Mills, E. N., A. Potts, G. W. Plumb, N. Lambert, and M. R. A. Morgan. 1997. Development of a rapid dipstick immunoassay for the detection of peanut contamination of food. *Food Agric. Immunol.* **9:**37–50.

Mohammed, I., W. M. Mullett, E. P. C. Lai, and J. M. Yeung. 2001. Is biosensor a viable method for food allergen detection? *Anal. Chim. Acta* **444:**97–102.

Moneret-Vautrin, D. A., and G. Kanny. 2004. Update on threshold doses of food allergens: implications for patients and the food industry. *Curr. Opin. Allergy Clin. Immunol.* **4:**215–219.

Morisset, M., D. A. Moneret-Vautrin, G. Kanny, L. Guenard, E. Beaudouin, J. Flabbee, and R. Hatahet. 2003. Thresholds of clinical reactivity to milk, egg, peanut and sesame in immunoglobulin E-dependent allergies: evaluation by double-blind or single-blind placebo-controlled oral challenges. *Clin. Exp. Allergy* **33:**1046–1051.

Mullett, W. M., E. P. Lai, and J. M. Yeung. 2000. Surface plasmon resonance-based immunoassays. *Methods* **22:**77–91.

Munoz-Furlong, A. 2004. Food allergy in schools: concerns for allergists, pediatricians, parents, and school staff. *Ann. Allergy Asthma Immunol.* **93**(Suppl. 3)**:**S47–S50.

Natale, M., C. Bisson, G. Monti, A. Peltran, L. P. Garoffo, S. Valentini, C. Fabris, E. Bertino, A. Coscia, and A. Conti. 2004. Cow's milk allergens identification by two-dimensional immunoblotting and mass spectrometry. *Mol. Nutr. Food Res.* **48:**363–369.

Negroni, L., H. Bernard, G. Clement, J. M. Chatel, P. Brune, Y. Frobert, J. M. Wal, and J. Grassi. 1998. Two-site enzyme immunometric assays for determination of native and denatured β-lactoglobulin. *J. Immunol. Methods* **220:**25–37.

Nogueira, M. C. L., R. McDonald, C. D. Westphal, S. J. Maleki, and J. M. Yeung. 2004. Can commercial peanut assay kits detect peanut allergens? *J. AOAC Int.* **87:**1480–1484.

Park, D. L., S. Coates, V. A. Brewer, E. A. E. Garber, M. Abouzied, K. Johnson, B. Ritter, and D. McKenzie. 2005. Performance tested method multiple laboratory validation study of ELISA-based assays for the detection of peanuts in food. *J. AOAC Int.* **88:**156–160.

Perry, T. T., E. C. Matsui, M. K. Conover-Walker, and R. A. Wood. 2004. Risk of oral food challenges. *J. Allergy Clin. Immunol.* **114:**1164–1168.

Pineiro, C., J. Barros-Velazquez, J. Vazquez, A. Figueras, and J. M. Gallardo. 2003. Proteomics as a tool for the investigation of seafood and other marine products. *J. Proteome Res.* **2:**127–135.

Pomés, A., R. M. Helm, G. A. Bannon, A. W. Burks, A. Tsay, and M. D. Chapman. 2003. Monitoring peanut allergen in food products by measuring Ara h 1. *J. Allergy Clin. Immunol.* **111:**640–645.

Pomés, A., R. Vinton, and M. D. Chapman. 2004. Peanut allergen (Ara h 1) detection in foods containing chocolate. *J. Food Prot.* **67:**793–798.

Poms, R. E., and E. Anklam. 2004. Effects of chemical, physical and technological processes on the nature of food allergens. *J. AOAC Int.* **87:**1466–1474.

Poms, R. E., C. Capelletti, and E. Anklam. 2004a. Effect of roasting history and buffer composition on peanut protein extraction efficiency. *Mol. Nutr. Food Res.* **48:**459–464.

Poms, R. E., C. L. Klein, and E. Anklam. 2004b. Methods for allergen analysis in food: a review. *Food Addit. Contam.* **21:**1–31.

Poms, R. E., M. E. Agazzi, A. Bau, M. Brohee, C. Capelletti, J. V. Norgaard, and E. Anklam. 2005. Interlaboratory validation study of five commercial ELISA test kits for the determination of peanut proteins in biscuits and dark chocolate. *Food Addit. Contam.* **22:**104–112.

Rosendal, A., and V. Barkholt. 2000. Detection of potentially allergenic material in 12 hydrolyzed milk formulas. *J. Dairy Sci.* **83:**2200–2210.

Roux, K. H., S. S. Teuber, J. M. Robotham, and S. K. Sathe. 2001. Detection and stability of the major almond allergen in foods. *J. Agric. Food Chem.* **49:**2131–2136.

Sampson, H. A. 2004. Update on food allergy. *J. Allergy Clin. Immunol.* **113:**805–819.

Sandberg, M., L. Lundberg, M. Ferm, and I. M. Yman. 2003. Real time PCR for the detection and discrimination of cereal contamination in gluten free foods. *Eur. Food Res. Technol.* **217:**344–349.

Scheibe, B., W. Weiss, F. Ruëff, B. Przybilla, and A. Görg. 2001. Detection of trace amounts of hidden allergens: hazelnut and almond proteins in chocolate. *J. Chromatogr. B Biomed. Sci. Appl.* **756:**229–237.

Schmitt, D. A., H. Cheng, S. J. Maleki, and A. W. Burks. 2004. Competitive inhibition ELISA for quantification of Ara h 1 and Ara h 2, the major allergens of peanuts. *J. AOAC Int.* **87:**1492–1502.

Sen, M., R. Kopper, L. Pons, E. C. Abraham, A. W. Burks, and G. A. Bannon. 2002. Protein structure plays a critical role in peanut allergen stability and may determine immunodominant IgE-binding epitopes. *J. Immunol.* **169:**882–887.

Shefcheck, K. J., and S. M. Musser. 2004. Confirmation of the allergenic peanut protein, Ara h 1, in a model food matrix using liquid chromatography/tandem mass spectrometry (LC/MS/MS). *J. Agric. Food Chem.* **52:**2785–2790.

Shreffler, W. G., K. Beyer, T. H. Chu, A. W. Burks, and H. A. Sampson. 2004. Microarray immunoassay: association of clinical history, in vitro IgE function, and heterogeneity of allergenic peanut epitopes. *J. Allergy Clin. Immunol.* **113:**776–782.

Shriver-Lake, L. C., C. R. Taitt, and F. S. Ligler. 2004. Application of an array biosensor for detection of food allergens. *J. AOAC Int.* **87:**1498–1502.

Sicherer, S. H., J. DeSimone, and T. J. Furlong. 2001. Peanut and tree nut allergic reactions in restaurant and food establishments. *J. Allergy Clin. Immunol.* **107:**S231.

Skerrit, J. H., and A. S. Hill. 1991. Enzyme immunoassay for determination of gluten in foods: collaborative study. *J. AOAC Int.* **74:**257–264.

Sorell, L., J. A. López, I. Valdés, P. Alfonso, E. Camafeita, B. Acevedo, F. Chirdo, J. Gavilondo, and E. Méndez. 1998. An innovative sandwich ELISA system based on an antibody cocktail for gluten analysis. *FEBS Lett.* **439:**46–50.

Stephan, O., N. Möller, S. Lehmann, T. Holzhauser, and S. Vieths. 2002. Development and validation of two dipstick type immunoassays for determination of trace amounts of peanut and hazelnut in processed foods. *Eur. Food Res. Technol.* **215:**431–436.

Stephan, O., and S. Vieths. 2004. Development of a real-time PCR and a sandwich ELISA for detection of potentially allergenic trace amounts of peanut (*Arachis hypogaea*) in processed foods. *J. Agric. Food Chem.* **54:**3754–3760.

Stephan, O., N. Weisz, S. Vieths, T. Weiser, B. Rabe, and W. Vatterott. 2004. Protein quantification, sandwich ELISA, and real-time PCR used to monitor industrial cleaning procedures for contamination with peanut and celery allergens. *J. AOAC Int.* **87:**1448–1457.

Taylor, S. L., and S. B. Lehrer. 1996. Principles and characteristics of food allergens. *Crit. Rev. Food Sci. Nutr.* **36**(Suppl.)**:**S91–S118.

Taylor, S. L., S. L. Hefle, C. Bindslev-Jensen, F. M. Atkins, C. Andre, C. Bruijnzeel-Koomen, A. W. Burks, R. K. Bush, M. Ebisawa, P. A. Eigenmann, A. Host, J. O. Hourihane, E. Isolauri, D. J. Hill, A. Knulst, G. Lack, H. A. Sampson, D. A. Moneret-Vautrin, F. Rance, P. A. Vadas, J. W. Yunginger, R. S. Zeiger, J. W. Salminen, C. Madsen, and P. Abbott. 2004. A consensus protocol for the determination of the threshold doses for allergenic foods: how much is too much? *Clin. Exp. Allergy* **34:**689–695.

Taylor, S. L., S. L. Hefle, C. Bindslev-Jensen, S. A. Bock, A. W. Burks, Jr., L. Christie, D. J. Hill, A. Host, J. O. Hourihane, G. Lack, D. D. Metcalfe, D. A. Moneret-Vautrin, P. A. Vadas, F. Rance, D. J. Skrypec, T. A. Trautman, I. M. Yman, and R. S. Zeiger. 2002. Factors affecting the determination of threshold doses for allergenic foods: how much is too much? *J. Allergy Clin. Immunol.* **109:**24–30.

Trucksess, M. W., V. A. Brewer, K. M. Williams, C. D. Westphal, and J. T. Heeres. 2004. Preparation of peanut butter suspension for determination of peanuts using enzyme-linked immunoassay kits. *J. AOAC Int.* **87:**424–428.

Tsuji, H., N. Okada, R. Yamanishi, N. Bando, M. Kimoto, and T. Ogawa. 1995. Measurement of Gly m Bd 30K, a major soybean allergen, in soybean products by a sandwich enzyme linked immunosorbent assay. *Biosci. Biotechnol. Biochem.* **59:**150–151.

U.S. Food and Drug Administration. 2001. Food allergen partnership. [Online.] http://vm.cfsan.fda.gov/~dms/alrgpart.html.

Valdes, I., E. Garcia, M. Llorente, and E. Mendez. 2003. Innovative approach to low-level gluten determination in foods using a novel sandwich enzyme-linked immunosorbent assay protocol. *Eur. J. Gastroenterol. Hepatol.* **15:**465–474.

Wei, Y., S. K. Sathe, S. S. Teuber, and K. H. Roux. 2003. A sensitive sandwich ELISA for the detection of trace amounts of cashew (*Anacardium occidentale* L.) nut in foods. *J. Agric. Food Chem.* **51:**3215–3221.

Wensing, M., A. H. Penninks, S. L. Hefle, S. J. Koppelman, C. A. Bruijnzeel-Koomen, and A. C. Knulst. 2002. The distribution of individual threshold doses eliciting allergic reactions in a population with peanut allergy. *J. Allergy Clin. Immunol.* **110:**915–920.

Westphal, C. D., M. R. Pereira, R. B. Raybourne, and K. M. Williams. 2004. Evaluation of extraction buffers using the current approach of detecting multiple allergenic and nonallergenic proteins in food. *J. AOAC Int.* **87:**1458–1465.

Whitaker, T. B., K. M. Williams, M. W. Trucksess, and A. B. Slate. 2005. Immunochemical analytical methods for the determination of peanut proteins in foods. *J. AOAC Int.* **88:**161–174.

Williams, K. M., L. C. Shriver-Lake, and C. D. Westphal. 2004. Determination of egg proteins in snack food and noodles. *J. AOAC Int.* **87:**1485–1491.

Yamashita, H., M. Kimoto, M. Hiemori, M. Okita, K. Suzuki, and H. Tsuji. 2001. Sandwich enzyme-linked immunosorbent assay system for micro-detection of the wheat allergen, Tri a Bd 17 K. *Biosci. Biotechnol. Biochem.* **65:**2730–2734.

Yeung, J. M., and P. G. Collins. 1996. Enzyme immunoassay for determination of peanut proteins in food products. *J. AOAC Int.* **79:**1411–1416.

Yeung, J. M., and P. G. Collins. 1997. Determination of soy proteins in food products by enzyme immunoassay. *Food Technol. Biotechnol.* **35:**209–214.

Yeung, J. M., R. S. Applebaum, and R. Hildwine. 2000a. Criteria to determine food allergen priority. *J. Food Prot.* **63:**982-986.

Yeung, J. M., W. H. Newsome, and M. Abbott. 2000b. Determination of egg proteins in food products by enzyme immunoassay. *J. AOAC Int.* **83:**139–143.

Yu, C. J., Y. F. Lin, B. L. Chiang, and L. P. Chow. 2003. Proteomics and immunological analysis of a novel shrimp allergen, Pen m 2. *J. Immunol.* **170:**445–453.

Food Allergy
Edited by S. J. Maleki et al.

Chapter 9

Predicting the Allergenicity of Novel Proteins in Genetically Modified Organisms

Richard E. Goodman and John Wise

The allergenicity assessment of genetically modified (GM) crops or GM organisms (GMOs) is one of the important steps in evaluating whether food and feed products from new varieties of plants and animals developed using biotechnology should be safe to eat or would pose a real health risk to the consumer (Konig et al., 2004; U.S. Food and Drug Administration, 1992; National Academy of Sciences, 2004; European Food Safety Authority, 2004). The possibility that the introduced gene might encode an allergen was seen as a possible risk of GMOs, although of low probability, as few of the hundreds or thousands of proteins in food products or common environmental organisms are known to be allergenic. However, if the introduced gene encoded a major allergen that was transferred into a different food crop, the risk would be substantial for individuals with existing allergies to that protein. This chapter is devoted primarily to evaluating whether the introduced protein is an allergen or is sufficiently similar to suspect potential cross-reactivity, with strong emphasis on the use of computer sequence comparisons between the introduced protein and known allergens.

An example of the creation of a GMO that would present a risk associated with the unintended introduction of an allergen occurred in the early 1990s when Pioneer Hi-Bred International introduced a gene encoding a methionine-rich protein from the 2S albumin fraction of the Brazil nut tree (*Bertholletia excelsa*) into soybeans to improve the nutritional content of animal feed. At the time, the protein had not been identified as an allergen. However, since the Brazil nut was recognized as a potent allergen for a few consumers, the potential product was tested for immunoglobulin E (IgE) binding using sera from individuals with diagnosed allergies to Brazil nuts. In that study, eight of nine individuals were found to have IgE that bound to the protein in laboratory tests; some were also skin test positive to the GM soybeans but not to non-GM soybeans (Nordlee et al., 1996). As a result, development of the potential product was stopped. The Brazil nut protein is now known as the allergen Ber e 1. If the GM soybeans had entered commerce,

Richard E. Goodman and John Wise • Food Allergy Research and Resource Program, Department of Food Science and Technology, University of Nebraska, Lincoln, NE 68583–0955.

those with severe allergies to Brazil nuts may have experienced severe reactions or even death following the consumption of foods containing the modified soybeans.

Following that experience, a panel of experts developed a strategy to systematically evaluate the potential allergenicity of proteins encoded by genes introduced into GMOs, based on the current understanding of allergenic proteins (Metcalfe et al., 1996). Since currently there is insufficient knowledge to determine a priori whether any protein might become a food allergen, the assessment focuses on reducing the probability of the greatest risks and whether the protein is a known allergen or is sufficiently similar to a known allergen that allergic cross-reactivity is a reasonable concern. A critical step in the assessment is the use of computer sequence comparisons or bioinformatics to evaluate the similarity of the introduced protein to those of known allergens. If a significant similarity is identified in the sequence comparison, the protein should be rigorously tested with sera from specifically allergic individuals, or the product should not be allowed on the market. If there is no indication that the protein is an allergen or is likely to be cross-reactive, the characteristics common to many major food allergens are evaluated to judge potential allergenicity. Those include evaluation of the digestibility in pepsin at acidic pH, abundance, and stability in heating and processing (Codex Alimentarius Commission, 2003; Goodman et al., 2005).

FOOD ALLERGY

Substantial knowledge about the risks of food allergy, characteristics of food and other allergens, and testing methods is essential for defining an effective allergen assessment strategy for GMOs. Many people experience very mild allergic reactions to inhaled or ingested proteins. It is also important to recognize that a significant risk of serious or life-threatening allergic reactions is faced by only a small percentage of highly allergic individuals in the population when they consume the specific allergens to which they are sensitized, such as peanuts (Sachs et al., 1981), milk (Tabar et al., 1996), eggs (Bernhisel-Broadbent et al., 1994), wheat (Kushimoto et al., 1985), and soybeans (Burks and Sampson, 1993). While the prevalence of food allergy is not precisely known, estimates from clinical surveys indicate that 1 to 3% of adults and 6 to 8% of young children experience some food allergic reactions with an overall prevalence of up to 4% of the population of the United States having IgE-mediated food allergies (Sampson, 2004). However, as many as 20 to 30% of individuals responding to questionnaires believe they occasionally experience allergic reactions, because symptoms of food intolerance or food poisoning can be similar to allergic reactions (Roehr et al., 2004). At this time, the only effective strategy for those with food allergies is to avoid the specific allergenic food. While most allergic reactions to foods are mild, extrapolation of limited data led to an estimate that, among the total population in the United States of 280 million, there may be as many as 30,000 individual cases of IgE-mediated food allergic reactions requiring emergency room treatment per year and as many as 150 to 200 deaths from all sources of food allergy (Sampson, 2003). Most allergic individuals react to a single allergen or a small number of allergens. However, a few individuals experience reactions to many foods. Eight allergenic foods or groups of foods (peanuts, soybeans, tree nuts, milk, eggs, fish, crustaceans, and wheat) are thought to cause nearly 90% of food allergic reactions in the United States (Metcalfe et al., 1996). Similar numbers have been estimated from other cultures and countries, although with some differences in the list of dominant allergens (Kanny et al., 2001; Roehr et al., 2004). Because of differences in diets across cultures, there are some additional foods that may be responsible for many reactions in specific geographies; e.g., in some parts of the European Union, sesame seeds, celery root, and mustard are now viewed as significant allergens (Beyer et al., 2002; Vieths et al., 2002; Morisset et al., 2003). In Asia, buckwheat, which is commonly consumed, is now recognized as a major food allergen that can cause severe reactions (Park et al., 2000).

Most cases of serious reactions occur when the individual consumes an allergen that he or she knows should be avoided but which he or she did not know was in the food that caused the reaction. Therefore, it is extremely important to avoid the transfer of major allergens into a different food source that allergic individuals would assume are safe. Even minor allergens affecting a small number of people could cause serious reactions in a few highly allergic individuals. As we consider risks, it should also be recognized that the allergenicity of a food crop may be significantly reduced by creating a GM crop with reduced or blocked synthesis of the major allergens (Shewry et al., 2001; Herman et al., 2003; Gilissen et al., 2005).

ALLERGENIC PROTEINS

The number of allergenic proteins identified grew rapidly in the past 15 years. Relatively few allergenic proteins had been purified and characterized by the early 1990s. The identification of specific allergenic proteins in complex allergenic material is not simple. Appropriately diagnosed serum donors (Chapters 1, 2, and 3) are required to provide IgE that is used to test binding to isolated proteins or for screening of cDNA expression libraries to identify recombinant allergens. Diagnosis of food allergy in patients typically requires careful clinical histories to identify potential sources, evaluate exposure, and characterize symptoms. Suspected allergens are eliminated from the diet to clear symptoms. Then, the suspected food may be added back in the diet, and skin prick tests (SPTs) with an extract of the relevant source material or laboratory testing for allergen-specific IgE can be performed to confirm the diagnosis (Burks and Sampson, 1993; Sampson, 2004; Bindslev-Jensen et al., 2004). For airway or contact allergens, verification of allergenic sources is often more difficult, as sources are much harder to control. For instance, cat dander, house dust mites, and pollen are rather ubiquitous. Therefore, elimination of the allergen for periods sufficient to clear symptoms is hard to achieve, and controlled reintroduction is impossible. SPTs and specific laboratory IgE tests are thus given much more weight. The results require appropriate interpretation, and the predictive value of the tests depends greatly on the specificity of test methods, materials, and the experience of the practitioner (Hamilton et al., 2002; Ricci et al., 2003; Schafer et al., 2003; Williams et al., 2003).

The sera from selected allergic and control donors are then used to evaluate protein-specific IgE binding with samples of protein purified from the source or produced through recombinant means and characterized to ensure purity and identity. Similar methods have been used to identify major allergenic proteins from a number of foods and contact and airway allergen sources. For example, with peanuts, the most important allergens to be identified are Ara h 1, Ara h 2, and Ara h 3 (Hales et al., 2004; Koppelman et al., 2004; Shreffler et al., 2004). An additional eight peanut allergens have been identified in the literature and are listed in the Allergome database (http://www.allergome.org/), primarily based on limited in vitro IgE-binding studies, but few have been purified or characterized. The clinical significance of the additional eight proteins is unknown.

The amino acid sequences of hundreds of allergenic proteins have now been reported in the literature or are listed in general or allergen specific databases on the internet (Chapters 10 and 11). The current version of the allergen database maintained at the University of Nebraska (http://www.allergenonline.com) includes 1,191 protein sequence entries representing over 580 proteins (including isoforms) from 216 species. A number of the proteins are isoforms (same gene, slightly different sequence) or homologs (related genes) found in multiple species or within a species. The database includes 370 food allergens and gliadins (proteins responsible for celiac disease) and 620 aeroallergens, as well as 51 proteins that are contact allergens, 92 venom and salivary proteins, and 55 proteins from parasites. Certainty of the identification of any protein as an allergen would require a biological challenge with subjects who experience allergies to the source of the protein, but for practical and ethical reasons that is rarely done. Instead, most were identified as allergens primarily by in vitro IgE binding with sera from individuals who are either specifically allergic to the source of the gene or generally allergic. However, studies have shown that in many cases, clear and specific IgE binding does not reflect biological activity. Presumably, this is because only one IgE epitope is bound, which is often the case for glycosylated proteins as some individual's IgE binds a single N-linked glycan (van Ree, 2002). As more allergens have been identified, comparisons of sequence, function, and structure indicate that many important allergens are clustered into a few structural families that are related through evolutionary divergence (Aalberse, 2000; Breiteneder and Mills, 2005; Jenkins et al., 2005). Yet many proteins identified as allergens do not fit into the major families, and many proteins in the families have not been identified as allergens. Therefore, a broad and efficient sequence comparison to all allergens offers the best way to screen proteins newly introduced in GMOs.

DEVELOPMENT OF GMOs

Development of GM crops for more efficient food and feed production began in the 1980s, following the development of new methods to specifically modify the DNA in bacteria. To transfer new functional genes into eukaryotic organisms and obtain successful expression of proteins, the inserted genes needed to include a promoter, the protein-coding regions, and a terminator (Astwood et al., 1997). Recombinant DNA has been introduced into plant cells by projection of DNA-coated microscopic metal particles into cells with a gene gun (Vain et al., 1995) and by infection with a recombinant bacterial or viral vector, e.g., *Agrobacterium tumefaciens* (Pniewski and Kapusta, 2005) carrying genetic elements intended for insertion in the plant DNA. The resulting transformed plant cells are selected and cultured with appropriate hormones to differentiate into reproductively mature plants, which are bred into commercial seed stock. Similar technology has been used to produce a small selection of GM animals, including fish, cows, sheep, and swine (Rahman et al., 1998; Hofmann et al., 2003; Houdebine, 2004). In most cases, the introduced DNA encodes an expressible protein, although the DNA may instead introduce regulatory elements to modify the expression of endogenous genes. Alternatively, the DNA may encode an antisense element designed to suppress translation of transcribed RNA (Bhalla and Singh,

2004). In addition to changes intentionally introduced by the added DNA, the inserted DNA may create an unexpected fusion protein or may up- or down-regulate endogenous genes, some of which may encode allergens. Characterization of the molecular insert and proximate flanking sequence is used to identify any potential and expected protein product and is included in the safety assessments required by various government agencies prior to the introduction of GMOs into the environment (Codex Alimentarius Commission, 2003).

ALLERGENICITY ASSESSMENT STRATEGY

Based on the scientific and clinical understanding of allergy and allergens, the U.S. Food and Drug Administration (1992) recommended that the allergenicity assessment of GM crops focus on testing to ensure that the allergenicity of the GM variety is not greater than that of the traditionally produced varieties of the same crop. In 1996, recommendations were elaborated as a detailed assessment strategy with a decision tree by a panel of experts sponsored by the International Life Sciences Institute/International Food Biotechnology Committee (Metcalfe et al., 1996). As described, the assessment of each new GM crop should evaluate the known allergenicity of the source of the gene, to help design an appropriate testing strategy (Metcalfe et al., 1996). While the primary focus is on the introduced or added gene and any newly expressed protein, the possibility that the expression of endogenous allergens has been up-regulated by gene insertion is evaluated for crops that are major causes of allergy (Metcalfe et al., 1996). Some have questioned the allergenicity approach and have proposed alternative strategies or tests (Food and Agriculture Organization of the United Nations-World Health Organization [FAO/WHO], 2001; Spok et al., 2005); however, many of the suggestions have turned out to be nonpredictive or very impractical (Goodman et al., 2005). A recent reevaluation by scientists and regulators led to an apparent international consensus regarding the allergenicity assessment (Codex Alimentarius Commission, 2003; European Food Safety Authority, 2004), which follows most of the original recommendations by Metcalfe et al. (1996). However, the practical details of the methods for performing the important bioinformatics comparison and of defined criteria for the level of sequence identity needed to predict potential cross-reactivity have not been well defined. Practical details and further evaluation of the criteria are examined in this chapter.

ALLERGENIC SOURCE OF THE GENE

If the source of the gene is known to frequently cause allergies, protein-specific in vitro serum IgE tests and possibly clinical tests (SPTs and food challenges) would be performed with appropriately allergic subjects to ensure that the protein encoded by the transferred gene is not an allergen. The purpose is to evaluate whether individuals with a higher likelihood of having been sensitized to the protein have developed IgE specific to the protein. It is hard to estimate the total number of allergenic sources, but a few relatively comprehensive lists of allergens can be found on the internet (e.g., http://www.allergome.org/ or http://allergenonline.com/). These sources list between 210 and 800 allergenic species that would include >1,100 allergenic genes. It should be easy to identify a sufficient number of individuals allergic to house dust mites, common grass pollen, birch pollen, or peanuts who might be willing and able to provide blood samples that could be used for in vitro laboratory tests to perform a statistically valid assessment (Codex Alimentarius Commission,

2003). However, many allergenic sources have not been reported as a common cause of allergies (e.g., shaggy mane mushroom [*Coprinus comatus*]). In such cases few, if any, allergic individuals are likely to be available for testing, meaning that an insufficient number of individuals with specific allergies might be identified to provide predictive test results (Codex Alimentarius Commission, 2003). The remaining steps in the safety assessment (sequence comparison to known allergens with appropriate serum testing if warranted, pepsin digestion, abundance, and possibly evaluation of the stability of the protein during heating) would then carry more weight. In such cases, the aggregate risk of allergenicity may be considered low, particularly since complex eukaryotic organisms such as rice contain >50,000 genes (Tyagi et al., 2004), while only five unique rice allergens are listed in AllergenOnline.

BIOINFORMATICS ASSESSMENT TO EVALUATE POTENTIAL ALLERGENICITY

Arguably, the most predictive single step of the allergenicity assessment is the comparison of the amino acid sequence of the transferred protein to those of known allergens (Metcalfe et al., 1996; FAO/WHO, 2001; Codex Alimentarius Commission, 2003). This comparison should identify proteins that the developer may not have known were identified by others as allergens or proteins with sequence similarities (and therefore structural similarities) that raise suspicion of cross-reactivity for those sensitized to a similar protein (Goodman et al., 2005). Additionally, a protein with high sequence identity compared to an allergen may be cross-reactive, due to shared sequential or conformational IgE epitopes, which should be tested. As originally envisioned by Metcalfe et al. (1996), an allergen database would be constructed by searching public databases of GenBank and SwissProt with keyword selections to identify allergens. The resulting database would then be searched using FASTA or BLASTP to find identical matches of eight or more amino acid segments in sequences of significant overall similarity. While the logic of using an eight-amino-acid match was based on the length of known T-cell and B-cell epitopes, the argument was not that any eight-amino-acid match might be an epitope but rather that the sequences are sufficiently similar to indicate they may contain identical or cross-reactive epitopes. This criterion has been the subject of considerable debate and will be discussed in detail below. If the results indicate a significant match, in vitro IgE binding using sera from individual donors allergic to the matched allergenic protein or the source of the matched allergenic protein would be performed. If needed for confirmation, in vivo clinical challenges (SPTs and/or food challenges) would also be performed, assuming that an adequate consenting population is available (Metcalfe et al., 1996). If the results of the assessment indicate the protein is an allergen or likely to be cross-reactive, the protein should not be transferred into a food crop, or foods produced using GM crop material containing this protein would need to be labeled so the added allergen (by allergenic source) is readily identified by consumers.

ALLERGENIC PROTEIN DATABASES

The list of known allergens has grown since the first GM crop was evaluated, when just over 200 allergenic and celiac-inducing protein sequences in public protein databases were identified using keywords such as "allergen" to select putative allergens (Chapters 10 and 11) (Astwood et al., 1997). Compilation of the first internet-accessible allergen sequence list was described in 1998 (Gendel, 1998a). Similar methods were used to compile the AllergenOnline database maintained by the University of Nebraska (http://www.allergenonline.com), which as of October 2004 listed 1,191 sequences representing 619 allergenic proteins, including homologs and isoforms of allergens from >200 species of organisms. Additional allergen databases that may help in the assessment process may be found at http://fermi.utmb.edu/SDAP/index.html, http://allermatch.org/database.html, http://www.ifr.bbsrc.ac.uk/protall/, http://www.iit.edu/~sgendel/fa.htm, and http://allergome.org/. Various algorithms and strategies may be employed to effectively search for sequence matches that indicate significant similarities between the query protein and known allergens.

BIOINFORMATICS COMPARISON (AMINO ACID SEQUENCE IDENTITIES)

As noted above, the first proposed method of comparison was to search for identical matches of eight contiguous amino acids shared between the query sequence (GM protein) and any allergen aligned by either FASTA or BLASTP (Metcalfe et al., 1996). The first published evaluation of search methods compared global alignments (e.g., Needleman and Wunsch, 1970) to local alignments (e.g., FASTA or BLASTP), discussed the limitations of only using the eight contiguous amino acid match to identify potentially cross-reactive allergens, and considered the possibility that searches for matches of eight identical amino acids might miss cross-reactive proteins (Gendel, 1998b). Other recommendations have emphasized the FASTA or BLAST comparisons (FAO/WHO, 2001; Codex Alimentarius Commission, 2003). In practice, the commercial developers of GM crops (e.g., Monsanto Company) have used two independent methods to identify potentially cross-reactive matches (Goodman et al., 2002; Hileman et al., 2002). The first search is typically a FASTA search, looking for matches with significantly small expectation scores (the statistical E value [expectation value]) and high percent identities, as calculated by the publicly available FASTA algorithm for each sequence alignment (Pearson, 1996, 2000). The second method is typically an automated string-search function similar to the find function in word processing programs, to search each sequence of the database with all possible eight-amino-acid segments of the query protein.

Details of the search options using FASTA or BLAST were never explicitly stated in regulatory guidance documents or in most publications. Academic advisors to the FAO/WHO (2001) recommended conservative criteria to predict potential cross-reactivity as a match of at least 35% identity over any segment of at least 80 amino acids (using either local alignment algorithm FASTA or BLASTP). They also stated that any identical match of six contiguous amino acids might indicate a potential for cross-reactivity (FAO/WHO, 2001). A number of researchers tested the criteria for short matches and demonstrated that six amino acids and even eight contiguous amino acids are more likely to occur by chance than to represent matches of allergenic significance (Goodman et al., 2002; Hileman et al., 2002; Kleter and Peijnenburg, 2002; Stadler and Stadler, 2003). Therefore, the advice incorporated into the international regulatory guidance document for the allergenicity assessment of GM crops (Codex Alimentarius Commission, 2003) recommends performing searches to identify matches of ≥35% identity over 80 or more amino acids as the primary criteria and, if scientifically justified, to perform searches for short matching segments of 6 or 8 amino acids. While there are a number of descriptions of synthetic peptides of 10 amino acids of allergens that bind IgE from allergic patients with specificities determined by 5 or 6 amino acids (Banerjee et al., 1999; Beezhold et al., 1999; Rabjohn et al., 1999), a single shared IgE-binding epitope will not cause cross-reactions, as IgE-mediated allergic reactions require cross-linking of high-affinity IgE receptors on the surface of mast cells. Effective cross-linking requires at least two spatially distinct IgE-binding epitopes on one protein or strong linkage (e.g., disulfide bonds) of peptides, each having at least one IgE epitope (Bannon, 2002; Kane, 1986). We have not been able to identify any reports identifying proteins that share only short peptide matches but not 35% identity matches, which share IgE-mediated in vivo allergic cross-reactivity. Therefore, considering the lack of evidence for including a search for short matching peptides and the apparent high rate of false-positive matches found by random chance (Hileman et al., 2002; Stadler and Stadler, 2003), such searches for matches of short peptides (eight or fewer amino acids) are not warranted in regulatory studies.

We are interested in defining the practical operational criteria for FASTA (or BLAST) searches and in evaluating evidence that a criterion of 35% identity in an 80-amino-acid segment is predictive of cross-reactivity. The sequence alignment tools (FASTA and BLASTP) were designed to identify proteins that are likely homologs, related by evolutionary divergence. Proteins with significant statistical scores and high percentages of identity and/or similarity are expected to have similar three-dimensional structures and similar biological and biochemical activities. While these algorithms were not written to identify allergenic similarities, they turn out to be quite useful, as proteins sharing significantly similar sequences usually share structures and are therefore likely to contain similar epitopes. Both FASTA and BLASTP algorithms search for local regions of similarity to initiate an alignment, rather than attempting to force alignment from the ends of the sequence (as global alignments do). Therefore, they should be able to match segments that could represent relatively short segments that were shared in evolution through mechanisms such as exon shuffling. Both algorithms use scoring matrices to produce alignments that are either based on a percentage of acceptable mutation (PAM) or block substitution matrices (BLOSUM), with a variety of options of penalties and rewards based on amino acid substitutions (e.g., PAM 30, BLOSUM 62, and BLOSUM 50) within the alignment to define the scoring of similar amino acid pairs (Henikoff and Henikoff, 1992). Hydrophobicity, potential charge, and structurally significant amino acids (Cys, Tyr, and Pro) that force structural constraints are weighted differently in some scoring matrices. The default matrix used with the FASTA3 program (Pearson, 2000) available at http://allergenonline.com, is BLOSUM 50, with a word size of 2 (minimum match for initiation of the alignment). The gap penalty will also affect alignments and scores. Optimal scoring has been defined for BLOSUM matrices to maximize the identification of distant homologs with a gap existence penalty of −10 with an additional −2 for each amino acid position in the gap. The gap penalty was empirically determined for the PAM series, based on the chosen target evolutionary distance (Reese and Pearson, 2002). The default values for common protein-protein BLASTP on Entrez (National Center for Biotechnology Information, National Institutes of Health) is BLOSUM 62, with word size of 3 and a gap existence penalty of 10 + 1 for each amino acid position in the gap (http://www.ncbi.nlm.nih.gov/BLAST/).

Alignments are performed by entering a protein query sequence, for instance, the protein introduced into a GM crop, which is then optimally aligned with each sequence in the database using the scoring matrix criteria. If a search is carried out on www.allergenonline.com (version 5), the query sequence will be compared to all 1,191 sequences, but only those matching the relatively insignificant threshold criteria (an E value of 10) will be displayed in the output. If the Entrez database is searched with BLASTP, the sequences that are searched would include hundreds of thousands of sequences, although the search may be restricted by limiting the search to sequences associated with a keyword (e.g., *allergen*), in which case, approximately 2,000 sequences would be searched. One flaw in using the *allergen* delimiter in the public databases is that thousands of sequences entered by researchers at the National Institute of Allergy and Infectious Diseases would be searched, based solely on the institutional affiliation. Therefore, many matches could be irrelevant.

Interpretation of FASTA3 or BLASTP results should include evaluation of the E value. This informative statistic is a calculated value that reflects the degree of similarity of the query protein to its corresponding matches. Matches of identical amino acids within the alignments are scored as most significant (~0), while matches of similar amino acids (charge and size) are given less significant values (>0). The E value depends on the overall length of joined (gapped) local sequence alignments, the quality (percent identity and similarity) of the overlap, and the size of the database. The size of the E value is inversely related to similarity of two proteins, meaning a very low E value (e.g., 1e 30) indicates a high degree of identity and similarity between the query sequence and the matching sequence from the database, while a value of 1 or higher indicates the proteins are not likely to be related in evolution or structure. In general, for a database the size of AllergenOnline, which contains many clusters of related proteins as well as unrelated proteins, two sequences might be considered related in evolutionary terms (i.e., diverged from a common ancestor and sharing common three-dimensional structure) if the E value of the FASTA query was <0.02 (Pearson, 1996). However, a value of 0.02 does not mean that the overall structures are likely to be sufficiently similar for antibodies (e.g., IgE from an allergic individual) against one protein to recognize the other protein. If the goal of the comparison is to identify proteins that may share immunologic or allergic cross-reactivity, matches with E values of >1e 7 are not very likely to identify relevant matches, while matches with E value results of <1e 30 are much more likely to be cross-reactive for at least some allergic individuals (Goodman et al., 2002; Hileman, et al., 2002). Since E values depend to a great degree on the scoring matrix, the size of the database, and many other factors, they have not been used in published studies to evaluate potentially cross-reactive allergens. Therefore, interpretation of immunological significance based only on E values should be viewed with caution.

PERCENT IDENTITY AND SEGMENT LENGTH

The percent identity of aligned proteins has often been used to identify proteins that may share cross-reactivity. While Pearson (1996) observed that homology is often indicated between proteins sharing 25% identity over ≥200 amino acids, Aalberse (2000) has suggested that proteins with >70% identical primary amino acid sequences throughout the length of the protein are likely to be allergenically cross-reactive, while those with < 50% identity are unlikely to be cross-reactive. Pearson (1996) also noted that short matching segments of 20 to 40 amino acids with roughly 50% identity can occur by random chance or by conservation of functional motifs; furthermore, there is little evidence that such short stretches of shared identity lead to IgE-mediated allergic cross-reactivity (Aalberse, 2000). In addition, many IgE epitopes involve binding to amino acid residues that are separated in linear sequence but are adjacent in spatial arrangement, due to the folding of the protein. These binding sites are referred to as conformational epitopes. The probability of having two shared IgE epitopes, either sequential or conformational, on two proteins that are not highly similar over a major portion of the proteins (i.e., are true homologs) is quite unlikely. The more closely related the species and the higher the identity of the proteins, the greater the likelihood of similar folding, similar potential epitopes, and therefore allergenic cross-reactivity. Results of FASTA3 alignments identified during bioinformatics searches are presented below and are discussed relative to published data on the cross-reactivity of these specific proteins. It must be remembered that sequence comparisons cannot absolutely predict cross-reactivity or the presence of IgE epitopes. Instead, the purpose is to try and predict whether two sequences are so similar that they share IgE cross-reactivity and clinical cross-reactivity. Strategies used in some examples include an overall FASTA3 alignment. In some examples, a comparison of the overall FASTA3 results with those produced by sequentially searching with segments of 80 amino acids, overlapping by 79 amino acids, as well as results using an 8-amino-acid string search algorithm, are presented. The AllergenOnline database, version 5, was used for all searches.

Example 1

In Europe, a number of birch pollen-allergic individuals experience allergic reactions when eating hazelnuts. Symptoms typically described are consistent with oral allergy syndrome, primarily swelling of the labia and glottis. However, some individuals experience more serious tracheal or laryngeal swelling. Similar reactions are experienced by some birch pollen-allergic individuals when they consume cherries, apples, carrots, or celery tuber (celeriac). One of the major hazelnut allergens (Cor a 1.04) is an 18-kDa homolog of the major birch pollen allergen Bet v 1 (Luttkopf et al., 2001). The Bet v 1 family of allergenic proteins includes homologs in nuts from closely related tree species, fruits of trees in the Rosaceae family, vegetative tissues of some members of the Apiaceae family (carrot, celery, and parsley), and in one example a protein in soybeans, a member of the Fabiaceae (legumes). These proteins belong to the pathogenesis-related protein family 10 (PR-10). Expression of many members of this family is induced when the plant is infected or when stressed by environmental factors. Heating these foods before ingestion often reduces or eliminates the reactions, due to denaturation of the protein. Over 50 isoforms of Bet v 1 in *Betula* sp. have been identified, primarily by cDNA sequencing. Based on published data, the genome of the hazelnut tree (*Corylus avellana*) includes at least three and possibly four homologous genes that may have slightly different functions. Cor a 1.01 is expressed in pollen, Cor a 1.04 is expressed in the nut, and Cor a 1.02 and Cor a 1.03 are expressed in leaves (Luttkopf et al., 2001). There are isoforms of each of the four proteins that vary in sequence by a few amino acids. There is considerable divergence among the four genes, as Cor a 1.01 is only about 65% identical to Cor a 1.04, indicating considerable evolutionary distance. In fact, the nut protein (Cor a 1.04) is more similar to the birch pollen allergen Bet v 1 than to the hazelnut pollen allergen Cor a 1.01.

The full-length protein sequence of Cor a 1.0401 (gi 5726304) was entered as the query sequence, and FASTA3 was used to search for sequence matches on http://allergenonline .com. Since Cor a 1.04 is in the database, self-alignment shows 100% sequence identity over the 160 amino acids and a calculated E value of 1e 57. The next three closest alignments were to other isoforms of this nut protein, and all had a >96% identity. The next-closest alignment was to birch pollen Bet v 1-Sc3 (gi 534898), with an E value of 5.3e 47 and an identity score of 85% over 160 amino acids. Next was European chestnut with an E value of 3.3e 39 and an identity of 72.5%. One of the hazelnut leaf homologs aligned with 70% identity, along with the second birch pollen allergen (Bet v 1-Sc1) (Fig. 1A). As can be seen in Fig. 1A, the identical residues are spread across the protein alignment. There is one identical segment of 14 contiguous amino acids, one of 10 amino acids, and another of 8 amino acids. If one did not already know that birch pollen and hazelnuts were allergens and that both commonly elicit allergic responses or that the proteins share IgE binding as tested by in vitro immunoassays, including reciprocal inhibition tests, the very small E value and high percent identities would provide a clear signal of probable cross-reactivity. Experimental evidence (Luttkopft et al., 2001) indicates that the birch pollen extract Bet v 1 and the hazelnut protein Cor a 1.0401 are more strongly cross-reactive than the hazelnut pollen allergen and the nut allergen (Cor a 1.01). The amino acid sequence of Cor a 1.01 is only 65% identical to Cor a 1.0401 (Fig. 1B) and is clearly not as good a match as Bet v and Sc-3. The sequence identity data correlate well with the IgE-binding data from at least a limited set of birch and hazelnut allergic patients in that IgE specifically bound to Cor a 1.04 and Bet v 1 with approximately equal avidity while binding less strongly to Cor a 1.01, based on the amount of protein needed to inhibit IgE binding in direct Western blots (Luttkopf et al., 2001). Furthermore, Cor a 1.01 was much less potent as an inhibitor of IgE binding to Cor a 1.04 in an enzyme allergosorbent (EAST) inhibition assay than Cor a 1.04, birch pollen extract, or hazelnut extract (Luttkopf et al., 2001).

Example 2

Other studies of PR-10 (Bet v 1) homologs have provided further confirmation of the general observation that the greater the percent identity over long segments of the proteins, the greater the probability or strength of cross-reactivity. A review of some well-designed studies comparing diverse homologs of Bet v 1 may help in evaluating the reliability of the FAO/WHO (2001) criteria of 35% identity over 80 amino acids as a level that might differentiate cross-reactive proteins from non-cross-reactive proteins. In one study, recombinant proteins Bet v 1a from birch pollen (gi 1542857), Pru av 1 (formerly known as Pru a 1) from cherries (gi 1513216), Api g 1 from celery tuber (gi 2147642), and Mal d 1 from golden delicious apple (gi 747852) were tested using sera from birch pollen-allergic individuals who were also allergic to cherry, celery tuber, or both (Scheurer et al., 1999). Analyzing the sequences of these proteins by FASTA3 on allergenonline.com indicated that all four proteins were likely to be evolutionary homologs, based on E values of <1e 16. The percent identities (Fig. 2) over the intact protein lengths ranged from a high-level identity of 84.9% between Pru av 1 and Mal d 1 (cherry and apple) to a modest level of identity of 37.7% between Mal d 1 and Api g 1 (apple and celery). Any of the individual matches would be considered to indicate potential allergenic cross-reactivity, based on the FAO/WHO

A

```
>>gi|534898|emb|CAA54696.1| 1 Sc-3 Birch pollen [Betula pendula](160 aa)
initn: 645 init1: 645 opt: 898  Z-score: 931.6  bits: 178.6 E(): 5.3e-47
Smith-Waterman score: 898; 85.625% identity (95.000% similar) in 160 aa
overlap (1-160:1-159) Bet v 1-Sc3

                      10        20        30        40        50        60
Cor a 1.0401 MGVFCYEDEATSVIPPARLFKSFVLDADNLIPKVAPQHFTSAENLEGNGGPGTIKKITFA
             :::: ::::::::: :::::::::::::::::::::.. .::::.:::::::::::::::
Bet v 1 Sc-3 MGVFNYEDEATSVIAPARLFKSFVLDADNLIPKVAPENVSSAENIEGNGGPGTIKKITFP
                      10        20        30        40        50        60

                      70        80        90       100       110       120
Cor a 1.0401 EGNEFKYMKHKVEEIDHANFKYCYSIIEGGPLGHTLEKISYEIKMAAAPHGGGSILKITS
             ::..::::::.:.:::::::::::::::::::: ::::::::::..::: ::::::::::
Bet v 1 Sc-3 EGSHFKYMKHRVDEIDHANFKYCYSIIEGGPLGDTLEKISYEIKIVAAP-GGGSILKITS
                      70        80        90       100        110

                     130       140       150       160
Cor a 1.0401 KYHTKGNASINEEEIKAGKEKAAGLFKAVEAYLLAHPDAYC
             ::::::. :.:::::::::::.:::::::: ::.:::.::
Bet v 1 Sc-3 KYHTKGDISLNEEEIKAGKEKGAGLFKAVENYLVAHPNAYN
           120       130       140       150       160
```

B

```
>>gi|22684|emb|CAA50325.1| Cor a 1.01 pollen [Corylus avellana](160 aa)
initn: 689 init1: 505 opt: 677  Z-score: 705.9  bits: 136.9 E(): 2e-34
Smith-Waterman score: 677; 65.000% identity (86.875% similar) in 160 aa
overlap (1-160:1-159)  Pollen allergen Cor a 1.01

                      10        20        30        40        50        60
Cor a 1.0401 MGVFCYEDEATSVIPPARLFKSFVLDADNLIPKVAPQHFTSAENLEGNGGPGTIKKITFA
             :::: :: :.::::: ::::::.:::.:.:::::::: .::.::.::::::::::.:::.
Cor a 1.01   MGVFNYEAETTSVIPAARLFKSYVLDGDKLIPKVAPQAITSVENVEGNGGPGTIKNITFG
                      10        20        30        40        50        60

                      70        80        90       100       110       120
Cor a 1.0401 EGNEFKYMKHKVEEIDHANFKYCYSIIEGGPLGHTLEKISYEIKMAAAPHGGGSILKITS
             ::...::.:..:.:.:..:: : :..:::  ::  :::. .:.:..::: ::::::::.:
Cor a 1.01   EGSRYKYVKERVDEVDNTNFTYSYTVIEGDVLGDKLEKVCHELKIVAAP-GGGSILKISS
                      70        80        90       100        110

                     130       140       150       160
Cor a 1.0401 KYHTKGNASINEEEIKAGKEKAAGLFKAVEAYLLAHPDAYC
             :.:.::.  :: ::::..:: :  :..::::.::::   :
Cor a 1.01   KFHAKGDHEINAEEIKGAKEMAEKLLRAVETYLLAHSAEYN
           120       130       140       150       160
```

Figure 1. FASTA3 alignment of hazelnut (nut) allergen Cor a 1.0401 with birch pollen allergen Bet v 1 Sc-3 (85.6% identity) (A) and hazelnut pollen allergen Cor a 1.01 (65% identity) (B).

Percent identity over the full-length (~160 amino acids)

Percent identity over the most similar 80 amino acid window (amino acid positions)

Allergen GI #	Bet v 1 1542857	Mal d 1 747852	Pru av 1 1513216	Api g 1 2147642
Bet v 1 1542857		55.6	59.4	40
Mal d 1 747852	61.3 (1-80)		84.9	37.7
Pru av 1 1513216	62.5 (1-80)	88.7 (11-90)		40
Api g 1 2147642	45 (34-113)	41.3 (1-80)	43.8 (1-80)	

Figure 2. Percent identity of Bet v 1 homologs. Values above and to the right of the diagonal represent the percent identity of the complete amino acid sequence of each protein, compared to each of the homologs. Values below and to the left of the diagonal represent the percent identity of the best aligned 80 amino acid segment of the homologous proteins, compared to the corresponding segment of each Bet v 1 homolog (Scheurer et al., 1999).

recommendations that 35% identity over 80 amino acids be viewed as indicating potential cross-reactivity. However, the published IgE-binding data demonstrate that binding specificity is not fully defined by the simple relationship of percent identity over the full-length of the four proteins or a single simple experiment (Scheurer et al., 1999). Western blot assays were designed to compare patient IgE binding to each protein and were performed using recombinant cherry allergen (rPru av 1), cherry extract, apple extract, recombinant birch allergen (rBet v 1), and birch pollen extract as solid-phase targets. Inhibition of binding to rPru av 1 by rBet v 1, rPru av 1, or recombinant celery allergen (rApi g 1) was tested with a pool of sera from four patients allergic to birch pollen who also had oral allergy syndrome to cherry. Birch extract and rBet v 1 totally inhibited binding to birch proteins, apple extract, cherry protein (rPru av 1), and celery protein (rApi g 1). Soluble rPru av 1 inhibited binding to cherry extract and rPru av 1 and inhibited binding to apple extract (~90%) but did not inhibit binding to birch extract or rBet v 1. Soluble rApi g 1, which is between 37 and 40% identical to the other three proteins (Bet v 1, Mal d 1, and Pru av 1), was unable to inhibit binding to any of those proteins, which suggested that 35% identity may not be sufficient to cause cross-reactivity (at least in this case). It is interesting to note that birch pollen and soluble rBet v 1a totally inhibited IgE binding to birch, apple, and cherry extracts, as well as rPru av 1, while soluble rPru av 1 was unable to inhibit binding to solid-phase birch pollen or rBet v 1, although they share 60% identity. This result indicates that IgE from the individual sera has much greater affinity for birch pollen and rBet v 1 than for rPru av 1.

Additional IgE-binding studies (EAST inhibition) were performed using different pools of serum from individuals allergic to birch and either cherry, celery, or both. The EAST inhibition provided quantitative data demonstrating that IgE had high-affinity to Bet v 1 but markedly lower affinities to rPru av 1, rMal d 1, and rApi g 1. By EAST inhibition, the highest concentration (100 μg/ml) of soluble rMal d 1 was only able to inhibit 50% of the birch-cherry-allergic IgE binding to solid-phase rPru av1 at a concentration 15- to 100-fold higher then the amount of soluble rPru av 1 required to produce 50% inhibition. Yet the sequence of Mal d 1 was 85% identical to that of Pru av 1, indicating the few amino acids that differ must markedly influence IgE-binding specificity. Additionally, soluble rApi g 1 did not inhibit IgE binding to solid-phase rPru av 1, and soluble rPru av 1 did not inhibit IgE binding to solid-phase rApr g 1, showing that in this case 40% identity was insufficient to produce measurable IgE cross-reactivity. Since some of the patients are reported as having oral allergy symptoms to both cherry and celery, it is likely that their IgE binds to different epitopes on these Bet v 1 homologs or that other proteins in cherry and celery are responsible for those individuals' allergic responses through cross-reactions that are dependent on different sequence similarities. In contrast, between 1 and 10 ng of soluble rBet v 1 caused a 50% reduction in IgE binding to solid-phase rPru av 1 or rApi g 1, while between 1 and 5 100 μg of rPru av 1 or rApi g 1 was needed to obtain 50% inhibition to the same protein on the solid phase. This 100- to 1,000-fold difference in concentration needed for equivalent inhibition demonstrates that the patients' IgE had much greater affinity for Bet v 1 than for Pru av 1 or Api g 1. These results suggest that an overall amino acid sequence identity of 40% (rBet v 1 compared to rApi g 1) may be sufficient to cause some biologically meaningful IgE cross-reactivity. The mixed results suggest that the Codex Alimentarius Commission (2003) criterion of 35% identity over 80 amino acids may be close to a biological threshold, at least for these closely related homologs. It is important to consider, however, that in vitro IgE binding does not prove clinical reactivity and that insufficient data were supplied to understand whether the clinical reactions of the celery allergic patients was due to Api g 1 or could be due to profilin (Api g 4) or another unidentified allergen.

The 1,000-fold difference in IgE-binding efficiency between Bet v 1 and Api g 1, as well as the >100-fold-greater IgE-binding to Bet v 1 than to Pru av 1 would be expected to have some biological relevance. A bioassay was performed using a rat basophil leukemia cell line (RBL-2H3), loaded with IgE from mice that were vaccinated with rBet v 1 to evaluate the doses of recombinant proteins required to trigger β-hexosaminidase release in the presence of various IgE antibodies (Scheurer et al., 1999). As in the EAST inhibition assays, the rBet v 1 was approximately 1,000-fold-more efficient than rPru av 1, rMal d 1, or rApi g 1 at inducing mediator release, although the epitope specificity of mouse IgE does not necessarily mimic human IgE-epitope specificity. It is also not known whether other natural isoforms of these proteins in birch, cherries, and celery have much higher-affinity binding by IgE from allergic patient sera that are responsible for eliciting a clinical reaction. On face value, these results do suggest that the quantitative IgE-binding data correlate with the ability to induce mast cell degranulation.

The data from the elegant study by Scheurer et al. (1999) may be interpreted as indicating that 40% identity is roughly in the range of biological meaning for IgE cross-reactivity. However, an additional consideration is that identical amino acids are not necessarily evenly distributed over 155 to 160 amino acids in these proteins. In fact, a major IgE-binding epitope in the Bet v 1 family of allergens is in the P-loop region (Spangfort et al., 1997; Scheurer et al., 1999), an 11-amino-acid segment with an alignment of 55 to 80% identity for these Bet v 1 homologs (Fig. 3). If this is a major IgE epitope for the study subjects, the differences in amino acids in this region may be partly responsible for differences in IgE-binding characteristics of the proteins, especially differences between rApi g 1 and the other homologs, due to three-amino-acid substitutions and the presence of a gap in the optimal alignment (Neudecker et al., 2003). A more recent study used site-directed mutagenesis and a highly specific monoclonal antibody to identify a second major IgE-binding epitope in Pru av 1 at amino acid 29 (Fig. 3) (Wiche et al., 2005). Again, Api g 1 contains an amino acid substitution at the same position as the point mutation, which may be responsible for reduced IgE binding to the celery allergen Api g 1. This substitution is in the region of highest identity among the Bet v 1 homologs and indicates an additional potential reason why the celery allergen is not as effective at binding IgE from birch pollen-allergic patients as is Pru av 1. Evaluation of these differences between the specific homologs of Bet v 1 is critical when considering the potential cause of differences in the clinical and in vitro cross-reactivity of these four proteins. A number of studies have demonstrated that most or all of the epitopes on Bet v 1 homologs are conformational rather than sequential (or linear), and it has therefore proven extremely difficult to identify other specific epitopes on these proteins (Neudecker et al., 2003). Considerably less data are available for most cross-reactive proteins.

The results of the FASTA3 comparison of the full-length sequences of the four Bet v 1 homologs did not fully correlate with the binding efficiency differences determined by Scheurer et al. (1999) and Wiche et al., (2005); specifically, with 40% identity Bet v 1 was able to inhibit binding to Api g 1, but Pru av 1 was not. Therefore, we used the FAO/WHO (2001) recommendation of searching overlapping segments of 80 amino acids to see if this

```
                10                  20                  30                  40
M G V F N Y E T E T T S V I P A A R L F K A F I L D G D N L F P K V A P Q A I S   Bet v 1
. . . Y T F . N . F . . E . . P S . . . . . . V . . A . . . I . . I . . . . . . K   Mal d 1
. . . . T . . S . F . . E . . P P . . . . . . V . . A . . . V . . I . . . . . . K   Pru av 1
. . . Q T H V L . L . . S V S . E K I . Q G . V I . V . T V L . . A . . G . Y K     Api g 1

                50                  60                * 70                  80
S V E N I E G N G G P G T I K K I S F P E G F P F K Y V K D R V D E V D H T N F   Bet v 1
Q A . I L . . . . . . . . . . . . T . G . . S Q Y G . . . H . I . S I . E A S Y   Mal d 1
H S . I L . . D . . . . . . . . . T . G . . S Q Y G . . . H K I . S I . K E . Y   Pru av 1
. . . I K - . D . . . . . L . I . T L . D . G . I T T M T L . I . G . N K E A L   Api g 1
      "P-loop"
                                    100                 110                 120
K Y N Y S V I E G G P I G D T L E K I S N E I K I V A T P D G G S I L K I S N K   Bet v 1
S . S . T L . . . D A L T . . I . . . . Y . T . L . . C G S . - . T I . S I S H   Mal d 1
S . S . T L . . . D A L . . . . . . . . Y . T . L . . S . S . . . . I . S T S H   Pru av 1
T F D . . . . D . D I L L G F I . S . E . H V V L . P . A . . . . . C . T T A I   Api g 1

                130                 140                 150                 160
Y H T K G D H E V K A E Q V K A S K E M G E T L L R A V E S Y L L A H S D A Y N   Bet v 1
. . . . . N I . I . E . H . . V G . . K A H G . F K L I . . . . K D . P . . . .   Mal d 1
. . . . . N V . I . E . H . . . G . . K A S N . F K L I . T . . K G . P . . . .   Pru av 1
F . . . . . A V . P E . N I . Y A N . Q N T A . F K . L . A . . I . N             Api g 1
```

Figure 3. Two prominent IgE-binding epitopes of Pru av 1(P-loop, from Scheurer et al., 1999; *, from Wiche et al., 2005), superimposed on the aligned sequences of birch pollen allergen Bet v 1, apple allergen Mal d 1, cherry allergen Pru av 1, and celery tuber allergen Api g 1.

method might uncover differences in the distribution of amino acid substitutions that could be more predictive regarding the identification of potential cross-reactivity. Computer scripts were written to search the allergenonline.com database using a scanning window of 80 amino acids, with 79-amino-acid overlaps with the FASTA3 algorithm to perform alignments. Matches of segments containing the 80-amino-acid segment with the highest identity for each query sequence are shown in Fig. 2, lower left). These values can be compared to the overall identity values shown in Fig. 2, upper right. The highest-scoring 80-amino-acid segment identity from each of the Bet v 1 homologs correlated slightly better with the in vitro IgE-binding results. Data from Scheurer et al. (1999) indicated that the IgE of sera used in the in vitro assays bound with approximately 100-times-greater affinity to Bet v 1 than to Pru av 1. These sequences share 62.5% identity, based on the highest-scoring 80-amino-acid match. The IgE affinity to Bet v 1 was approximately 1,000 times greater than for Api g 1, and these sequences share only 45% identity. The IgE-binding data (Scheurer et al., 1999) also indicated that IgE epitopes on Pru av 1 and Api g 1 are unlikely to overlap, as neither protein was able to inhibit binding to the other and their identity was only 43.8%. However, the IgE of the study subjects appears to be high for Bet v 1 but low for both Pru av 1 and Api g 1. Therefore, it is not clear whether the 43% identity can be seen as a lower limit for cross-reactivity if the IgE was of high affinity for one of the two related sequences. Clearly, further studies are needed to evaluate the predictive power of 35% versus 40% versus 45% identity. The affinity of the IgE antibodies should be expected to be important. But the data do suggest that there is a boundary of significant in vitro IgE cross-reactivity of somewhere around 35 to 40% identity, if the entire sequences are compared and of around 40 to 45% identity if a sliding window approach using 80 amino acids identifies potentially cross-reactive proteins. It will be interesting to see if that relationship holds true for other families of homologous proteins. Of course, the most important measure of meaningful criteria would be in vivo evidence of clinical cross-reactivity, something that is much harder to demonstrate.

Example 3

It is important to evaluate allergen sequence identities unrelated to Bet v 1 homologs. Food produced from a number of legumes, including peanuts, soybeans, chickpeas, beans, and lupines, is responsible for many food-related allergies. However, the legume family (Fabaceae) is a large and diverse family with a number of subfamilies. As one might predict, species that are most closely related taxonomically are more commonly clinically cross-reactive. Most of the legumes that are commonly consumed belong in the taxonomic order Fabales, family Fabaceae (prior name Leguminosae), subfamily Papilionoideae, which contains the allergenic legumes. Members of this family of plants typically do not shed much pollen and are often either self pollinated or insect pollinated. Therefore, most allergic sensitization probably occurs through consumption of the seeds and/or surrounding tissue (e.g., peapods) and not via the airway. Peanuts are in one tribe (Aeschynomeneae) of the subfamily Papilionoideae. Soybeans, lima beans, green beans, and kidney beans are in the tribe Phaseoleae. Garden (green) peas, fava beans, and lentils are in the tribe Vicieae. The chickpea is in the tribe Cicereae, and lupin is in the tribe Genisteae. Cross-reactivity can reach across tribes of this subfamily, as some individuals are apparently sensitized to peanuts and have mild to severe cross-reactions to peas, beans, lupins, or soybeans

(Moneret-Vautrin et al., 1999; Sanchez-Monge et al., 2004); however, it is often not clear whether the clinical reactions are due to cross-reaction or cosensitization. The ability of an extract of one species to inhibit in vitro IgE binding to an extract of the other species is the usual method used to prove cross-reactivity. The assays should be designed to test inhibition at various concentrations of inhibitor, something that is often not practical with the limited serum volumes available for study. Interestingly, many people who are allergic to peanuts do not react to the ingestion of soybeans or other legumes, although IgE from their sera binds to proteins from a number of other legume species, including soybeans, with apparent specificity (Wensing et al., 2003; Ferreira et al., 2004). People often assume incorrectly that cashew nuts are legumes and are closely related to peanuts. They assume the nuts are likely to be cross-reactive, something that is rarely observed (Rance et al., 2003). Cashew is in the order Sapindales, family Anacardiaceae (sumac family), and is more closely related to mangos and pistachios. Cross-reactivity between cashews and pistachio nuts has been observed (Rance et al., 2003). There are also unique allergens in peanuts and soybeans that are bound by IgE in some patients allergic to both (Eigenmann et al., 1996). While allergies to soybeans are thought to be common in young children, symptoms in many young patients are often restricted to atopic dermatitis. Most of these patients will become tolerant to the ingestion of soybeans while they are still young. However, a few children and adults develop asthmatic responses when they ingest soybean products, and a small number may experience anaphylaxis.

We wanted to compare the sequence search methods using the overall (full-length) sequence alignment with FASTA to those using FASTA comparisons of each possible 80-amino-acid segment and the string-search function of all possible 8-amino-acid segments, using allergens that are well characterized and have data demonstrating probable cross-reactivity across taxa. We therefore chose sequences of the major peanut allergens Ara h 1 (gi 1168390), a vicillin; Ara h 2 (gi 31322017), an 11S globulin; and Ara h 3 (gi 3703107), a 2S albumin. While the clinical and serological cross-reactivity of these allergens is not as well studied as that of the Bet v 1 homologs, there are many studies of individuals who either experience cross-reactions or are cosensitized, as measured by SPTs or IgE binding to corresponding homologs in other legumes or even some species producing tree nuts (Xiang et al., 2002; Ferreira et al., 2005; Wensing et al., 2003). One study tested in vitro IgE cross-reactivity of peanut (subclass Rosidae, order Fabales), cashew (subclass Rosidae, order Sapindales), almond (subclass Rosidae, order Rosales), hazelnut (subclass Hamamelidae, order Fagales), and Brazil nut (subclass Dilleniidae, order Lecythidales) with sera from four individuals having allergies to multiple nuts (de Leon et al., 2003). The results indicated extracts of roasted almond, hazelnut, and raw Brazil nut could inhibit binding to peanut when approximately 30- to 1,000-times-higher concentrations of protein were used with some subject's sera but not with others. Extract of cashews did not inhibit binding of any of the sera to peanuts. Extracts of hazelnut and Brazil nut were similarly ineffective at inhibiting binding for some serum samples. These data demonstrates that it is unlikely the major legume allergens (probably vicillins, 2S albumins, and 11S globulins) are as cross-reactive across broad taxonomic categories as is the Bet v 1 (PR-10) family of proteins.

The bioinformatics results of searches with the three peanut allergen sequences are presented in Table 1. The 7S albumin of peanut (Ara h 1) matched homologs in lentils and soybeans with >50% identity in the overall FASTA alignment, and many 80-amino-acid segments of those legumes matched with >50% identity (Table 1). In addition, the 7S globulin homologs of four nonlegumes (cashew, sesame, walnut, and hazelnut) were at or below 35% identity in the overall alignment but showed between 44 and 55% identity in at least one 80-amino-acid segment, compared to Ara h 1. The clinical and in vitro serological data suggest that there may be weak cross-reactivity to the nonlegume homologs, but the data are not clear (Beyer et al., 2002; Comstock et al., 2004; Wolf et al., 2004). Identity matches between the peanut 2S albumin and 2S albumins of sesame and Brazil nuts indicated probable homology but low identity (E values of <0.04 but overall identities of <30% and no segments of >35% identity over any 80-amino-acid segment (Table 1). The clinical and in vitro data suggest that while some people are cosensitized to sesame, peanut, and Brazil nut, there is no good evidence of significant IgE cross-reactivity to their homologous proteins, suggesting that the individuals are likely to have IgE antibodies with biologically significant differences in specificity of binding. Identity matches of >50% in the overall FASTA alignment to the peanut 11S globulin (Ara h 3) were restricted to soybeans (*Glycine max* and its wild relative, *Glycine soja*), although statistically significant matches (E values of <1e–20) were found for homologs in hazelnut, cashew, buckwheat, and Brazil nut; the overall identity of cashew Ana o 2 was 44% and for the Brazil nut protein was 38% (Table 1). However, FASTA alignments from searches with each possible 80-amino-acid segment of Ara h 3 produced multiple matches with >35% identity to the cashew, Brazil nut, and buckwheat 11S globulins, with highest identities of 60% for cashew, 48% for Brazil nut, and 46% for buckwheat. Based on recommendations of the Codex Alimentarius Commission (2003), proteins with similarities equivalent to the cashew, Brazil nut, and buckwheat proteins would be considered likely to be cross-reactive and would require further testing if they were used in a GMO. But it must be emphasized that there are currently insufficient clinical and serological data to conclusively support cross-reactivity of the 11S proteins from cashew, buckwheat, and Brazil nut to peanut Ara h 3.

Interestingly, the results of the exact eight-amino-acid string-searches do not add information to searches of any of the three major peanut allergens that would implicate additional candidate proteins as being potentially cross-reactive. All proteins with an 8-amino-acid match also had matches of >35% identity over 80 amino acids.

As indicated by the limited set of bioinformatics searches discussed here with results compared to published data on clinical and serological cross-reactivity, it appears that an overall FASTA search with matches of >50% identity indicates the matched proteins should be considered potentially cross-reactive unless proven otherwise. For matches with <50% identity over the length, an additional search for alignments using a sliding window of 80 amino acids may to help to identify proteins that have restricted regions of higher-than-average identities that could represent cross-reactive structures. It is clear that matches of only 35% identity over segments of 80 amino acids are not likely to identify cross-reactive structures, except possibly in very limited groups of highly homologous proteins such as the *Bet v 1* family of proteins. In at least some cases, in vitro IgE cross-reactivity may not be relevant to biologically relevant clinical reactions. How should one gauge the relevance of data indicating 50% inhibition requiring 1,000-fold-higher concentrations of the cross-reactive protein compared to the sensitizing allergen?

In cases with overall protein identities of 35 to 50%, the identification of matching regions of 80 or more amino acids with 45 to 60% may represent a highly conserved structure

Table 1. Sequence matches to peanut allergens[a]

Allergen and match (gi)	Species	Overall FASTA3			80-mer (>35%), best % ID (no. of matches)	8-mer (no. of identical matches)
		E Value	Overlap (aa)	Identity (%)		
Ara h 1 (gi 1168390)						
Ara h 1 (1168390)	*Arachis hypogaea*	5.5e–194	614	100	100 (587)	607
Ara h 1 (1168391)	*A. hypogaea*	6.6e–183	629	94.8	100 (585)	474
Len c 1.0101 (29539109)	*Lens culinaris*	1.3e–67	424	53.2	64 (400)	4
β-Conglycinin (169929)	*Glycine max*	1.8e–50	681	42.1	59 (451)	2
Len c 1.0012 (29539111)	*L. culinaris*	1.4e–46	424	52.1	64 (397)	4
β-Conglycinin (18536)	*G. max*	9.5e–31	642	45.2	61 (469)	5
β-Conglycinin (256427)	*G. max*	1.4e–26	457	51.2	64 (428)	2
7S globulin (13183177)	*Sesamum indicum*	5e–21	561	33.2	44 (165)	
Vicillin (6580762)	*Juglans regia*	1.9e–20	625	35.0	55 (307)	
48-KD glycop (19338630)	*Corylus avellana*	3.4e–20	495	34.1	46 (195)	
β-Conglycinin (169927)	*G. max*	1.2e–15	244	51.6	61 (207)	1
Vicillin (21666498)	*Anacardium occidentale*	4.6e–15	599	28.9	48 (84)	
Vicilin (21914823)	*A. occidentale*	5.2e–15	599	28.7	48 (84)	
Ara h 7 (5931948)	*A. hypogaea*	0.04	154	27.3		
Glycinin (736002)	*Glycine soja*	0.4	261	24.5		
Ara h 2 (gi 31322017)						
Ara h 2 (31322017)	*A. hypogaea*	21.e–36	169	100	100 (129)	162
Ara h 2.02 (26245447)	*A. hypogaea*	6e–36	169	99.4	100 (129)	155
Ara h 2 (14347293)	*A. hypogaea*	2.9e–19	166	91.0	99 (129)	126
Conglutin (17225991)	*A. hypogaea*	1.6e–10	167	52.7	73 (105)	8
Ara h 6 (5923742)	*A. hypogaea*	1e–09	155	51.6	76 (95)	8
Ara h 7 (5931948)	*A. hypogaea*	0.00032	137	41.6	65 (65)	1
2S albumin (5381323)	*S. indicum*	0.021	164	24.4		
2S albumin (17713)	*Bertholletia excelsa*	0.032	138	28.3		

Ara h 3 (gi 3703107)						
Ara h 3 (3703107)	*A. hypogaea*	4.6e–168	507	100	100 (480)	500
Ara h 3 / 4 (21314465)	*A. hypogaea*	4e–153	518	90.9	100 (476)	283
Glycinin (5712199)	*A. hypogaea*	1.1e–151	507	91.3	100 (477)	288
Glycinin GY3 (18639)	*G. max*	2.2e–51	507	58.2	78 (458)	10
Glycinin G1 (18635)	*G. max*	7.9e–51	515	57.5	75 (454)	16
Glycinin A1 (18615)	*G. max*	2.2e–50	515	56.7	71 (454)	14
Glycinin A3 (736002)	*G. soja*	2.4e–50	525	41.1	55 (336)	
Glycinin A2 (18609)	*G. max*	7e–50	508	58.3	74 (452)	13
Glycinin G2 (18637)	*G. max*	7e–50	508	58.5	75 (452)	15
11S globulin (18479082)	*C. avellana*	1.3e–37	504	46.2	61 (408)	
Glycinin A3 (169969)	*G. max*	1.4e–34	525	39.8	53 (320)	
Ana 0 2 (25991543)	*A occidentale*	6.3e–29	492	44.3	60 (355)	4
Legumin-like (2317670)	*Fagopyrum esculentum*	5.8e–23	556	35.1	46 (266)	
Fag e 1 (4895075)	*F. esculentum*	1.8e–22	527	36.1	46(263)	
Legumin-like (2317674)	*F. esculentum*	9.3e–22	500	36	46 (290)	
Glycinin (18641)	*G. max*	1.7e–21	569	38.5	52 (365)	
Glycinin GY4 (806556)	*G. soja*	1.7e–21	569	38.8	54 (365)	
Glycinin (732706)	*G. max*	4.2e–21	568	38.9	54 (368)	
Glycinin (169971)	*G. max*	4.4e–21	222	40.5	49 (150)	
11S globulin (30313867)	*B. excelsa*	9.8e–20	508	38	48 (280)	
Allergen (22353013)	*Fagopyrum tataricum*	3.7e–18	176	39.8	48 (114)	
Globulin (6979766)	*Fagopyrum gracilipes*	9.2e–16	166	37.3	48 (476)	
Vicilin-like (6580762)	*J. regia*	0.0029	562	22.2		
Conglycinin (18536)	*G. max*	0.0042	376	21.8		

[a]Peanut allergens (Ara h 1 [gi 1168390], Ara h 2 [gi 31322017], and Ara h 3 [gi 3703107]) were tested as query sequences using FASTA3 of the full-length proteins. Additionally, every possible 80-amino-acid segment of these query proteins was tested by FASTA3 for matches of >35% identity. The query sequences were also tested for exact string matches to all possible 8-amino-acid sequences in the database. Searches were against all sequences in an allergen database (http://allergenonline.com, version 5.0). The most similar allergens are listed by name and gi number (a public database reference identifier listed on the National Center for Biotechnology Information Web site). Results from the overall FASTA3 alignment include the E-value score, the number of amino acids in the aligned segment, and the percent identical amino acids in the alignment. Results from the 80-amino-acid alignment include the highest percent identity matched in the best single 80-amino-acid alignment and the number of individual 80-amino-acid alignments scoring at 35% or higher identity. Results of the 8-mer string search are the number of exact eight-amino-acid matches between the query and aligned proteins. aa, amino acids, ID, identity.

(motif) that could act as an important cross-reactive IgE epitope. The allergenicity assessment and our understanding of allergenic cross-reactivity in general may be improved by additional studies of clinically responsive subjects with clear histories of allergic reactions to multiple related foods. Such studies would also require negative control subjects; both groups should be tested by blinded food challenges to verify the classification. The laboratory tests should be validated to quantitatively test inhibition of binding with a wide dilution range of purified proteins, and the studies should be performed to test avidity (or affinity) and quantification of specifically bound IgE.

ADDITIONAL CONSIDERATIONS

It is not unusual to find individuals with low or moderate levels of IgE as measured by in vitro assay or SPTs that appear specific to food proteins in broad plant taxonomic groups. Yet these individuals often do not have obvious allergic reactions to more than one or two species. In at least some cases, the IgE that binds to food proteins from corn or wheat binds more tightly to proteins found in pollen from botanically related species, e.g., timothy grass, fescue, or Bermuda grass (Jones et al., 1995). It is often quite difficult, however, to ascertain whether the IgE binding is simply cross-reactive and not clinically important or whether it is capable of triggering mast cell degranulation in vivo. In some cases, it appears that much of the cross-reactivity is due to IgE binding to polymannose N-linked glycan structures with specific substitutions of β-1,2-linked xylose and/or α-1,3-linked fucose (Vieths et al., 2002). In many cases, IgE binding to the glycan does not seem to induce histamine release, unless other epitopes are present on the allergen and are recognized by the patient's IgE. Comparison of amino acid sequences will obviously not help to identify potentially cross-reactive proteins if the glycan is the target, except to help rule out shared peptide-specific epitopes.

OTHER BIOINFORMATICS SEARCH STRATEGIES AND METHODS

Approaches other than simple FASTA (or BLAST) searches or string searches have been evaluated to improve the sequence comparison predictive value. One group used the FAO/WHO (2001) criteria for matches of six contiguous amino acids as a preliminary screening tool, followed by a literature survey for IgE-binding epitopes and a theoretical evaluation of antigenic sites using the Hopp and Woods algorithm of the allergen-matched specific segment to reduce the false-positive rate (Kleter and Peijnenburg, 2002). However, the predictive value of the combined method has not been evaluated. Antigenicity prediction algorithms have not proven highly predictive for antibody-binding epitopes (Van Regenmortel and Pellequer, 1994), although ongoing studies of the identification of antigenic epitopes of pathogenic viruses may lead to better predictions (Enshell-Seijffers, et al., 2003). A motif-based allergenicity prediction method has been proposed to compare sequences based on protein structure by classifying 779 known or putative allergenic sequences into 52 distinct motifs (Stadler and Stadler, 2003), based on sequence similarity instead of a short-sequence peptide match. A comparison was made between the two methods, using a randomly chosen dataset of protein sequences selected from the Swiss-Prot sequence database. They were screened using a combination of the six contiguous amino acid-matching approach plus FASTA (35% identity match over any 80-amino-acid segment) and identified 200 proteins that matched at least one allergen by either method, although there is no published evidence of allergenicity for 199 of the 200 proteins (Stadler and Stadler, 2003). The same dataset of proteins was evaluated by the motif identification method, with results showing that the motif method falsely identified 9 of 10 proteins as allergens (Stadler and Stadler, 2003). While the motif method had nearly a 10-fold-lower rate of false positives than the six-amino-acid match, a 90% false-positive rate is still quite high. Another refined sequence and antigenicity motif prediction algorithm has been developed for the prediction of IgE-binding sites, based on previously identified sequences and structures (Ivanciuc et al., 2003). However, neither of these motif prediction algorithms has been compared to results that would be found using only an overall FASTA algorithm or a sliding 80-amino-acid FASTA method. It is therefore not clear whether there is an improvement in prediction by these alternative methods. But since the algorithms of both (Stadler and Stadler, 2003; Ivanciuc et al., 2003) are based to a large extent on sequence similarity, it would be surprising if their results differed markedly from those obtained by a direct FASTA (or BLAST) comparison.

Another group has worked to refine the FASTA3 search comparison by evaluating a combination of the percent identity matrix value and the overlapping sequence match length as vectors using nearest-neighbor analysis in a supervised learning system (Zorset et al., 2002). Further analysis with this approach tested two scoring matrices, BLOSUM 50 and BLOSUM 80, to statistically evaluate apparent true- and false-positive matches (Soeria-Atmadja et al., 2004). Further evaluation of this modified method is necessary with data of cross-reactive and nonallergenic proteins to understand if it is more predictive than a simple FASTA identity score.

The goal of the bioinformatics result should remain the identification of proteins that may be cross-reactive and would require further evaluation by serum IgE tests. The bioinformatics alone is not sufficiently robust to predict allergenicity unless there is a 100% match with an experimentally identified allergen.

SERUM IgE-BINDING TESTS

Allergen-specific allergic donor serum-binding studies were used to determine that the gene from Brazil nuts, which was inserted into soybeans as one of the first potential GM products, produced an allergenic protein that is now known as Ber e 1 (Nordlee et al., 1996). In that case, a sequence comparison did not lead to prediction of possible allergenicity because it was the first 2S albumin to be clearly identified as an allergen. Further, the bioinformatics strategy had not yet been fully developed. Instead the developers knew that a few individuals were allergic to Brazil nut, the source of the gene, and tested IgE binding with sera from individuals allergic to the source. Yet today, if someone were interested in transferring a high-methionine 2S albumin from one source into another crop to improve the nutrient profile of the grain, a bioinformatics search would likely lead to the conclusion that the protein should be tested for IgE binding with serum from patients allergic to the source and with serum from patients allergic to the matched allergen. It is of interest that a sesame seed 2S albumin has been transferred into rice to improve the nutritional qualities (Lee et al., 2005). As far as we are aware, the protein sequence was not compared to known allergens, although the authors were aware of the allergenicity of Ber e

1 and the sunflower 2S albumin. Based on our bioinformatics review, this protein would be suspected to be cross-reactive. In fact, another research group has identified this protein as an allergen (Pastorello et al., 2001). Clearly, it is important to use bioinformatics early in the process of developing GM products so that one can anticipate potential health risks and perform appropriate serum tests or further clinical tests to demonstrate safety or to prove the protein is a hazard. There are additional cases where genes encoding proteins that may be sufficiently similar to known or putative allergens have been transformed into crops used for food and feed without the benefit of a bioinformatics comparison or with misapplication of the search strategies (e.g., Sinagawa-Garcia et al., 2004; Tada et al., 2003; Katsube et al., 1999; de Sousa-Majer et al., 2004). As far as we know, none of these potentially hazardous products have become commercially available or have been approved for general planting or production. It is hoped that developers of such products will take note of the issues and that regulators in the various countries will require appropriate additional testing.

CONCLUSIONS

It is still fair to say that our ability to predict the potential allergenicity of any protein is limited to knowledge as to whether and how humans have been exposed to that protein or to highly similar proteins and whether any have experienced allergic reactions as a result. There are no absolute characteristics or tests that have been identified that will predict with great certainty whether a specific protein will sensitize individual consumers. However, appropriate application of the current allergenicity assessment strategy for GM crops (Codex Alimentarius Commission, 2003) should greatly reduce the risk of introducing a known allergen or a protein that is likely to be cross-reactive, as was demonstrated in the analysis of the potential GM soybean that contained the major Brazil nut protein that later became known as the allergen Ber e 1 (Taylor and Hefle, 2001). The bioinformatics search strategy is one of the pivotal studies in assuring safety. The success of the search is dependent on both the quality of the database used for the comparison and the algorithms and scoring criteria used to judge potential cross-reactivity. The data available at this time demonstrate that application of well-described, well-understood, and freely available programs (FASTA or BLAST) are efficient and reliable in the identification of proteins that are likely to pose significant risk of cross-reactivity. It is important to use appropriate scoring matrices, gap and mismatch penalties, and criteria that are robust enough to identify potentially cross-reactive proteins without a high likelihood of false-positive predictions. It is important to understand that serum and clinical tests needed to verify positive sequence matches are difficult to perform, due to the lack of availability of appropriate sera for many allergens and technical issues in developing and validating specific IgE-binding assays that are reliable and sufficiently quantitative to demonstrate specificity and relative avidity. Future research may provide more predictive comparisons than the straightforward FASTA or BLAST comparisons, but in our view it is more important to focus on clearly defining laboratory measures of cross-reactivity that correlate with biologically meaningful allergenicity to improve the reliability of sequence (or three-dimensional structural) prediction programs. The studies reviewed here clearly demonstrate that we do not have sufficient data to suggest that identities as low as 35% in 80 amino acids are predictive of cross-reactivity. In many cases, the subjects are likely to have been sensitized by both proteins. For a number of truly cross-reactive proteins, it appears there are often regions of much higher identity that may well represent the relevant structure for IgE binding. Further, there is considerable uncertainty about the clinical relevance of IgE that binds to one protein with <1/1,000 of the apparent avidity to the probable sensitizing allergen. Additional studies evaluating the clinical predictive values of various sequence comparison strategies and scoring criteria are likely to help improve the reliability of the current GM safety assessment process. The evidence suggests that proteins that are <50% identical to a proven allergen over their entire sequence provide little risk of significant allergic cross-reactivity. Therefore, a regulatory criterion of <35% identity over any 80-amino-acid segment appears to be extremely conservative.

REFERENCES

Aalberse, R. C. 2000. Structural biology of allergens. *J. Allergy Clin. Immunol.* **106:**228–238.

Astwood, J. D., R. L. Fuchs, and P. B. Lavrik. 1997. Food biotechnology and genetic engineering, p. 65–92. *In* D. D. Metcalfe, H. A. Sampson, and R. A. Simon (ed.), *Food Allergy: Adverse Reactions to Foods and Food Additives*, 2nd ed. Blackwell Science, Inc., Cambridge, Mass.

Banerjee, B., P. A. Greenberger, J. N. Fink, and V. P. Kurup. 1999. Conformational and linear B-cell epitopes of Asp f 2, a major allergen of *Aspergillus fumigatus*, bind differently to immunoglobulin E antibody in the sera of allergic bronchopulmonary aspergillosis patients. *Infect. Immun.* **67:**2284–2291.

Bannon, G., A., R. E. Goodman, J. N. Leach, E. Rice, R. L. Fuchs, and J. D. Astwood. 2002. Digestive stability in the context of assessing the potential allergenicity of food proteins. *Comments Toxicol.* **8:**271–285.

Beezhold, D. H., V. L. Hickey, J. E. Slater, and G. L. Sussman. 1999. Human IgE-binding epitopes of the latex allergen Hev b 5. *J. Allergy Clin. Immunol.* **103:**1166–1172.

Bernhisel-Broadbent, J., H. M. Dintzis, R. Z. Dintzis, and H. A. Sampson. 1994. Allergenicity and antigenicity of chicken egg ovomucoid (Gal d III) compared with ovalbumin (Gal d I) in children with egg allergy and in mice. *J. Allergy Clin. Immunol.* **93:**1047–1059.

Beyer, K., L. Bardina, G. Grishinan, and H. A. Sampson. 2002. Identification of sesame seed allergens by 2-dimensional proteomics and Edman sequencing: seed storage proteins as common food allergens. *J. Allergy Clin. Immunol.* **110:**154–159.

Bhalla, P. L., and M. B. Singh. 2004. Knocking out expression of plant allergen genes. *Methods* **32:**340–345.

Bindslev-Jensen, C., B. K. Ballmer-Weber, U. Bengtsson, C. Blanco, C. Ebner, J. Hourihane, A. C. Knulst, D. A. Moneret-Vautrin, K. Nekam, B. Niggemann, M. Osterballe, C. Ortolani, J. Ring, C. Schnopp, and T. Werfel. 2004. Standardization of food challenges in patients with immediate reactions to foods—position paper from the European Academy of Allergology and Clinical Immunology. *Allergy* **59:**690–697.

Breiteneder, H., and E. N. C. Mills. 2005. Molecular properties of food allergens. *J. Allergy Clin. Immunol.* **115:**14–23.

Burks, A. W., and R. L. Fuchs. 1995. Assessment of the endogenous allergens in glyphosate-tolerant and commercial soybean varieties. *J. Allergy Clin. Immunol.* **96:**1008–1010.

Burks, A. W., and H. Sampson. 1993. Food allergies in children. *Curr. Probl. Pediatr.* **23:**230–252.

Codex Alimentarius Commission. 2003. *Alinorm 03/34: Joint FAO/WHO Food Standard Programme*, p. 47–60. Codex Alimentarius Commission, Twenty-Fifth Session, Rome, Italy.

Comstock, S. S., G. McGranahan, W. R. Peterson, and S. S. Teuber. 2004. Extensive in vitro cross-reactivity to seed storage proteins is present among walnut (*Juglans*) cultivars and species. *Clin. Exp. Allergy* **34:**1583–1590.

de Leon, M. P., I. N. Glaspole, A. C. Drew, J. M. Rolland, R. E. O'Hehir, and C. Suphioglu. 2003. Immunological analysis of allergenic cross-reactivity between peanut and tree nuts. *Clin. Exp. Allergy* **33:**1273–1280.

de Sousa-Majer, M., N. C. Turner, D. C. Hardie, R. L. Morton, B. Lamont, and T. J. V. Higgins. 2004. Response to water deficit and high temperature of transgenic peas (Pisum sativum L.) containing a seed-specific alpha-amylase inhibitor and the subsequent effects on pea weevil (Bruchus pisorum L.) survival. *J. Exp. Botany* **55:**497–505.

Eigenmann, P. A., A. W. Burks, G. A. Bannon, and H. A. Sampson. 1996. Identification of unique peanut and soy allergens in sera absorbed with cross-reacting antibodies. *J. Allergy Clin. Immunol.* **98:**969–978.

Enshell-Seijffers, D., D. Denisov, B. L. Groisman, L. Smelyanski, R. Mehuhas, G. Gross, G. Genisova, J. M. Gershoni. 2003. The mapping and reconstitution of conformational discontinuous B-cell epitope of HIV-1. *J. Mol. Biol.* **334:**87–101.

European Food Safety Authority. 2004. Guidance document of the Scientific Panel on Genetically Modified Organisms for the risk assessment of genetically modified plants and derived food and feed. *EFSA J.* **99:**1–94.

Ferreira, F., T. Hawranek, P. Gruber, N. Wopfner, and A. Mari. 2004. Allergic cross-reactivity: from gene to the clinic. *Allergy* **59:**343–367.

Food and Agriculture Organization of the United Nations-World Health Organization. 2001. *Evaluation of Allergenicity of Genetically Modified Foods.* Report of a Joint FAO/WHO Expert Consultation on Allergenicity of Foods Derived from Biotechnology. Food and Agriculture Organization of the United Nations (FAO), Rome, Italy.

Gendel, S. M. 1998a. Sequence databases for assessing the potential allergenicity of proteins used in transgenic foods. *Adv. Food Nutr. Res.* **42:**63–92.

Gendel, S. M. 1998b. The use of amino acid sequence alignments to assess potential allergenicity of proteins used in genetically modified foods. *Adv. Food Nutr. Res.* **42:**45–62.

Gilissen, L. J. W. J., S. T. H. P. Bolhaar, G. J. A. Rouwendal, C. I. Matos, M. Boone, J., F. A. Krens, L. Zuidmeer, A. van Leeuwen, J. Akkerdaas, K. Hoffman-Summergruber, A. C. Knulst, D. Bosch, W. E. van de Weg, and R. van Ree. 2005. Silencing the major apple allergen Mal d 1 using the RNA interference approach. *J. Allergy Clin. Immunol.* **115:**364–369.

Goodman, R. E., S. L. Hefle, S. L. Taylor, and R. van Ree. 2005. Assessing genetically modified crops to minimize the risk of increased food allergy: A review. *Int. Arch. Allergy Immunol.* **137:**153–166.

Goodman, R. E., A. Silvanovich, R. E. Hileman, G. A. Bannon , E. A. Rice, and J. D. Astwood. 2002. Bioinformatic methods for identifying known or potential allergens in the safety assessment of genetically modified crops. *Comments Toxicol.* **8:**251–269.

Hagan, N. D., N. Upadhyaya, L. M. Tabe, and T. J. V. Higgins. 2003. The redistribution of protein sulfur in transgenic rice expressing a gene for a foreign, sulphur-rich protein. *Plant J.* **34:**1–11.

Hales, B. J., A. Bosco, K. L. Mills, L. A. Hazell, R. Loh, P. G. Holt, and W. R. Thomas. 2004. Isoforms of the major peanut allergen Ara h 2: IgE binding in children with peanut allergy. *Int. Arch. Allergy Immunol.* **135:**101–107.

Hamilton, R. G., E. L. Peterson, and D. R. Ownby. 2002. Clinical and laboratory-based methods in the diagnosis of natural rubber latex allergy. *J. Allergy Clin. Immunol.* **110**(Suppl.)**:**S47–S56.

Herman, E. M., R. M. Helm, R. Jung, and A. J. Kinney. 2003. Genetic modification removes an immunodominant allergen from soybean. *Plant Physiol.* **132:**36–43.

Hileman, R. E., A. Silvanovich, R. E. Goodman, E. A. Rice, G. Holleschak, J. D. Astwood, and S. L. Hefle. 2002. Bioinformatic methods for allergenicity assessment using a comprehensive allergen database. *Int. Arch. Allergy Immunol.* **128:**280–291.

Hofmann, A., B. Kessler, S. Ewerling, M. Weppert, B. Vogg, H. Ludwig, M. Stojkovic, M. Boelhauve, G. Brem, E. Wolf, and A. Pfeifer. 2003. Efficient transgenesis in farm animals by lentiviral vectors. *EMBO Rep.* **4:**1054–1060.

Houdebine, L. M. 2004. Preparation of recombinant proteins in milk. *Methods Mol. Biol.* **267:**485–494.

Ivanciuc, O., C. H. Schein, and W. Braun. 2003. SDAP: database and computational tools for allergenic proteins. *Nucleic Acids Res.* **31:**359–362.

Jenkins, J. A., S. Griffiths-Jones, P. R. Shewry, H. Breiteneder, and E. N. C. Mills. 2005. Structural relatedness of plant food allergens with specific reference to cross-reactive allergens: an in silico analysis. *J. Allergy Clin. Immunol.* **115:**163–170.

Jones, S. M., C. F. Magnolfi, S. K. Cooke, and H. A. Sampson. 1995. Immunologic cross-reactivity among cereal grains and grasses in children with food hypersensitivity. *J. Allergy Clin. Immunol.* **96:**341–351.

Kane, P., J. Erickson, C. Fewtrell, B. Baird, and D. Holowka. 1986. Cross-linking of IgE-receptor complexes at the cell surface: synthesis and characterization of a long bivalent hapten that is capable of triggering mast cells and rat basophilic leukaemia cells. *Mol. Immunol.* **23:**783–790.

Kanny, G., D. A. Moneret-Vautrin, J. Flabbee, E. Beaudouin, M. Morisset, and F. Thevein. 2001. Population study of food allergy in France. *J. Allergy Clin. Immunol.* **108:**133–140.

Katsube, T., N. Kurisaka, M. Ogawa, N. Maruyama, R. Ohtsuka, S. Utsumi, and F. Takaiwa. 1999. Accumulation of soybean glycinin and its assembly with the glutelins of rice. *Plant Physiol.* **120:**1063–1073.

Kelly, J. D., J. J. Hlywka, and S. L. Hefle. 2000. Identification of sunflower seed IgE-binding proteins. *Int. Arch. Allergy Immunol.* **121:**19–24.

Kleter, G. A., and A. A. C. M. Peijnenburg. 2002. Screening of transgenic proteins expressed in genetically modified food crops for the presence of short amino acid sequences identical to potential, IgE-binding linear epitopes of allergens. *BMC Struct. Biol.* **2:**8.

Knippels, L. M., F. van Wijk, and A. H. Penninks. 2004. Food allergy: what do we learn from animal models? *Curr. Opin. Allergy Clin. Immunol.* 4:205–209.

Konig, A., A. Cockburn, R. W. Crevel, E. Debryune, R. Graftstroem, U. Hammerling, I. Kimber, I. Knudsen, H. A. Kuiper, A. A. Peijnenburg, A. H. Penninks, M. Poulsen, M. Schauzu, and J. M. Wal. 2004. Assessment of the safety of foods derived from genetically modified (GM) crops. *Food Chem. Toxicol.* **42:**1047–1088.

Koppelman, S. J., M. Wensing, M. Ertmann, A. C. Knulst, and E. F. Knol. 2004. Relevance of Ara h1, Ara h2, and Ara h3 in peanut-allergic patients, as determined by immunoglobulin E western blotting, basophil-histamine release and intracutaneous testing: Ara h2 is the most important peanut allergen. *Clin. Exp. Allergy* **34:**583–590.

Kushimoto, J., and T. Aoki. 1985. Masked type I wheat allergy. Relation to exercise-induced anaphylaxis. *Arch. Dermatol.* **121:**355–360.

Lee, T. T. T., M.-C. Chung, Y.-W. Kao, C.-S. Wang, L.-J. Chen, and J. T. C. Tzen. 2005. Specific expression of a sesame storage protein in transgenic rice bran. *J. Cereal Sci.* **41:**23–29.

Luttkopf, D., U. Muller, P. S. Skov, B. W. Ballmer-Weber, B. Wuthrich, K. Skamstrup Hansen, L. K. Poulsen, M. Kastner, D. Haustein, and S. Vieths. 2001. Comparison of four variants of a major allergen in hazelnut (*Corylus avellana*) Cor a 1.04 with the major hazel pollen allergen Cor a 1.01. *Mol. Immunol.* **38:**515–525.

Metcalfe, D. D., J. D. Astwood, R. Townsend, H. A. Sampson, S. L. Taylor, and R. L. Fuchs. 1996. Assessment of the allergenic potential of foods derived from genetically engineered crop plants. *Crit. Rev. Food Sci. Nutr.* **36(S):**165–186.

Molvig, L., L. M. Tabe, B. O. Eggum, A. Moore, S. Craig, D. Spencer, and T. J. V. Higgins. 1997. Enhanced methionine levels and increased nutritive value of seeds of transgenic lupins (Lupinus angustifolius L.) expressing a sunflower seed albumin gene. *Proc. Natl. Acad. Sci. USA* **94:**8393–8398.

Moneret-Vautrin, D.-A., L. Guerin, G. Kanny, J. Flabbee, S. Fremont, and M. Morisset. 1999. Cross-allergenicity of peanut and lupin: the risk of lupine allergy in patients allergic to peanuts. *J. Allergy Clin. Immunol.* **104:**883–888.

Morisset, M., D. A. Moneret-Vaughtrin, F. Maadi, S. Fremont, L. Guenard, A. Croizner, and G. Kanny. 2003. Prospective study of mustard allergy: first study with double-blind placebo controlled food challenge trials (24 cases). *Allergy* **58:**295–299.

National Academy of Sciences. 2004. *Safety of Genetically Engineered Foods: Approaches To Assessing Unintended Health Effects*. The National Academies Press, Washington, D.C.

Needleman, S. B., and C. D. Wunsch. 1970. A general method applicable to the search for similarities in the amino acid sequence of two proteins. *J. Mol. Biol.* **48:**443–453.

Nordlee, J. A., S. L. Taylor, J. A. Townsend, L. A. Thomas, and R. K. Bush. 1996. Identification of a Brazil-nut allergen in transgenic soybeans. *N. Engl. J. Med.* **334:**688–692.

Park, J. W., D. B. Kang, C. W. Kim, S. H. Ko, H. Y. Yum, K. E. Kim, C.-S. Hong, and K. Y. Lee. 2000. Identification and characterization of the major allergens of buckwheat. *Allergy* **55:**1035–1041.

Pastorello, E. A., E. Varin, L. Farioli, V. Pravettoni, C. Ortolani, C. Trambaioli, D. Fortunato, M. G. Fiuffrida, F. Rivolta, A. Robino, A. M. Calamari, L. Lacava, and A. Conti. 2001. The major allergen of sesame seeds (*Sesamum indicum*) is a 2S albumin. *J. Chromatogr. B Biomed. Sci. Appl.* **756:**85–93.

Pearson, W. R. 1996. Effective protein sequence comparison. *Methods Enzymol.* **266:**227–258.

Pearson, W. R. 2000. Flexible sequence similarity searching with the FASTA3 program package. *Methods Mol. Biol.* **132:**185–219.

Pniewski, T., and J. Kapusta. 2005. Efficiency of transformation of Polish cultivars of pea (*Pisum sativum* L.) with various regeneration capacity by using hypervirulent Agrobacterium tumefaciens strains. *J. Appl. Genet.* **46:**139–147.

Rabjohn, P., E. M. Helm, J. S. Stanley, C. M. West, H. A. Sampson, A. W. Burks, and G. A. Bannon. 1999. Molecular cloning and epitope analysis of the peanut allergen Ara h 3. *J. Clin. Investig.* **103:**535–542.

Rahman, M. A., R. Mak, H. Ayad, A. Smith, and N. Maclean. 1998. Expression of a novel piscine growth hormone gene results in growth enhancement in transgenic tilapia (*Oreochromis niloticus*). *Transgenic Res.* **7:** 357–369.

Rance, F., E. Bidat, T. Bourrier, and D. Sabouraud. 2003. Cashew allergy: observations of 42 children without associated peanut allergy. *Allergy* **58:**1311–1314.

Ricci, G., M. Capelli, R. Miniero, G. Menna, L. Zannarini, P. Dillon, and M. Masi. 2003. A comparison of different allergometric tests, skin prick tests, Pharmacia UniCAP and ADVIA Centaur, for diagnosis of allergic diseases in children. *Allergy* **58:**38–45.

Roehr, C. C., G. Edenharter, S. Reimann, I. Ehlers, M. Worm, T. Zuberfier, and B. Niggemann. 2004. Food allergy and non-allergic food hypersensitivity in children and adolescents. *Clin. Exp. Allergy* **34:** 1534–1541.

Sachs, M. I., R. T. Jones, and J. W. Yunginger. 1981. Isolation and partial characterization of a major peanut allergen. *J. Allergy Clin. Immunol.* **67:**27–34.

Sampson, H. A. 2003. Anaphylaxis and emergency treatment. *Pediatrics* **111:**1601–1608.

Sampson, H. A. 2004. Update on food allergy. *J. Allergy Clin. Immunol.* **113:**805–819.

Sanchez-Monge, R., G. Lopez-Torrejon, C. Y. Pascual, J. Varela, M. Martin-Esteban, and G. Salcedo. 2004. Vicilin and convicilin are potential major allergens from pea. *Clin. Exp. Allergy* **34:**1747–1753.

Schafer, T., B. Hoelscher, H. Adam, J. Ring, H. E. Wichmann, and J. Heinrich. 2003. Hay fever and predictive value of prick test and specific IgE antibodies: a prospective study in children. *Pediatr. Allergy Immunol.* **14:**120–129.

Scheurer, S., D. Y. Son, M. Boehm, F. Karamloo, S. Franke, A. Hoffman, D. Haustein, and S. Vieths. 1999. Cross-reactivity and epitope analysis of Pru a 1, the major cherry allergen. *Mol. Immunol.* **36:**155–167.

Shewry, P. R., A. S. Tatham, and N. G. Halford. 2001. Genetic modification and plant food allergens: risks and benefits. *J. Chromatogr. B Biomed. Sci. Appl.***756:**327–335.

Shreffler, W. G., K. Beyer, T. H. Chu, A. W. Burks, and H. A. Sampson. 2004. Microarray immunoassay: association of clinical history, in vitro IgE function, and heterogeneity of allergenic peanut epitopes. *J. Allergy Clin. Immunol.* **113:**776–782.

Sinagawa-Garcia, S. R., Q. Rascon-Cruz, A. Valdez-Ortiz, S. Medina-Godoy, A. Escobar-Gutierrez, and O. Paredes-Lopez. 2004. Safety assessment by in vitro digestibility and allergenicity of genetically modified maize with an amaranth 11S globulin. *J. Agric. Food Chem.* **52:**2709–2714.

Soeria-Atmadja, D., A. Zorzet, M. G. Gustafsson, and U. Hammerling. 2004. Statistical evaluation of local alignment features predicting allergenicity using supervised classification algorithms. *Int. Arch. Allergy Immunol.* **133:**101–112.

Spangfort, M. D., M. Gajhede, P. Osmark, F. M. Poulsen, J. N. Larsen, C. Schou, and H. Loewenstein. 1997. Three-dimensional structure and epitopes of Bet v 1. *Int. Arch. Allergy Immunol.* **113:**243–245.

Spok, A., H. Gaugitsch, S. Laffer, G. Pauli, H. Saito, H. Sampson, E. Sibanda, W. Thomas, M. van Hage, and R. Valenta. 2005. Suggestions for the assessment of the allergenic potential of genetically modified organisms. *Int. Arch. Allergy Immunol.* **137:**167–180.

Stadler, M. B., and B. M. Stadler. 2003. Allergenicity prediction by protein sequence. *FASEB J.* **17:**1141–1143.

Tabar, A. I., M. J. Alvarez, S. Echechipia, S. Acero, B. E. Garcia, and J. M. Olaguibel. 1996. Anaphylaxis from cow's milk casein. *Allergy* **51:**343–345.

Tada, Y, S. Utsumi, and F. Takaiwa. 2003. Foreign gene products can be enhanced by introduction into low storage protein mutants. *Plant Biotechnol. J.* **1:**411–422.

Taylor, S. L., and S. L. Hefle. 2001. Will genetically modified foods be allergenic? *J. Allergy Clin. Immunol.* **107:**765–771.

Tyagi, A. K., J. P. Khurana, P. Khurana, S. Raghuvanshi, A. Gaur, A. Kapur, V. Gupta, D. Kumar, V. Ravi, S. Vij, P. Hhurana, and S. Sharma. 2004. Structural and functional analysis of rice genome. *J. Genet.* **83:**79–99.

U.S. Food and Drug Administration. 1992. Statement of policy: foods derived from new plant varieties. *Federal Register* **57:**22984.

Vain, P., J. De Buyser, V. Bui Trang, R. Haicour, and Y. Henry. 1995. Foreign gene delivery into monocotyledonous species. *Biotechnol. Adv.* **13:**653–671.

van Ree, R. 2002. Carbohydrate epitopes and their relevance for the diagnosis and treatment of allergic disease. *Int. Arch. Allergy Immunol.* **129:**189–197.

Van Regenmortel, M. H., and J. L. Pellequer. 1994. Predicting antigenic determinants in proteins: looking for unidemensional solutions to a three-dimensional problem? *Pep Res* **7:**224–278.

Vieths, S., D. Luttkopf, J. Reindl, M. D. Anliker, B. Wuthrich, and B. K. Ballmer-Weber. 2002. Allergens in celery and zucchini. *Allergy* **57**(Suppl.)**72:**100–105.

Vieths, S., S. Scheurer, and B. Ballmer-Weber. 2002. Current understanding of cross-reactivity of food allergens and pollen. *Ann. N. Y. Acad. Sci.* **964:**47–68.

Wensing, M., A. C. Knulst, S. Piersma, F. O'Kane, E. Knol, and S. J. Koppelman. 2003. Patients with anaphylaxis to pea can have peanut allergy caused by cross-reactive IgE to vicilin (Ara h 1). *J. Allergy Clin. Immunol.* **111:**420–424.

Wiche, R., M. Gubesch, H. Konig, K. Fotisch, A. Hoffman, A. Wangorsch, S. Scheurer, and S. Vieths. 2005. Molecular basis of pollen-related food allergy: identification of a second cross-reactive IgE epitope on Pru av 1, the major cherry (*Prunus avium*) allergen. *Biochem. J.* **385:**319–327.

Williams, P. B., S. Ahlstedt, J. H. Barnes, L. Soderstrom, and J. Portnoy. 2003. Are our impressions of allergy test performances correct? *Ann. Allergy Asthma Immunol.* **91:**26–33.

Xiang,, P., T. A. Beardslee, M. G. Zeece, J. Markwell, and G. Sarath. 2002. Identification and analysis of a conserved immunoglobulin E-binding epitope in soybean G1a and G2a. *Arch. Biochem. Biophys* **408:**51–57.

Zorzet, A., M. Gustafsson, and U. Hammerling. 2002. Prediction of food protein allergenicity: a bioinformatics learning systems approach. *In Silico Biol.* **2:**525–534.

Food Allergy
Edited by S. J. Maleki et al.

Chapter 10

Bioinformatics for Predicting Allergenicity

Steven M. Gendel

The ability of a food or a food protein to cause allergic sensitization and to elicit an allergic reaction is the result of a complex set of interactions involving both the immune and digestive systems (Sampson, 2004; Poulsen, 2005). Factors such as the physiology and genetics of the affected individual and the structure of the proteins involved play critical roles in determining whether an allergic reaction will occur and how severe it will be. Because we do not understand how sensitization occurs or how the nature and severity of a reaction are regulated, it is not possible to directly assess the potential allergenicity of a protein. However, it may be possible to infer potential allergenicity by comparing a protein of interest to known allergenic proteins.

One major tool for conducting such a comparison is bioinformatic analysis (Gendel, 2002). Bioinformatics can be used to compare primary sequences, secondary and tertiary structures, functional classifications, and evolutionary relationships for entire proteins or for domains within proteins. The utility of these comparisons depends on the availability of both appropriate data sources and analytical tools. Just as chemical or biological analyses should use characterized reagents and validated methods, bioinformatic analyses should be conducted using characterized databases (reagents) and validated algorithms (methods). Although a number of allergy-related databases are available, they are very different in design and content and in the degree to which information characterizing the content of the database is made available to users. A similar diversity exists among the available analytical resources for assessing potential allergenicity, and no standards or procedures have been developed for validating these resources.

To illustrate the extent of the diversity among allergen-related bioinformatic resources, representative online allergen databases and analytical resources are described below. Based on the descriptive information available for each of these resources, principles of good database practice (GDP) can be developed that will maximize the utility of these resources.

Steven M. Gendel • Food and Drug Administration, 6502 S. Archer Rd., Summit-Argo, IL 60501.

ALLERGEN DATABASES

Each of the major repository protein sequence databases, such as GeneBank and Swiss-Prot, contains a large number of allergen sequences. However, these databases cannot be used directly to carry out allergenicity assessments for several reasons, including the fact that no one database contains all the relevant sequences (Gendel, 1998; Brusic et al., 2003). Therefore, several specialized allergen databases have been constructed and made available over the internet. In general, there are two broad classes of allergen databases, those that focus on molecular data and those that include molecular data as an adjunct to biomedical or clinical information. The molecular databases all include amino acid sequences and may include additional information on protein structure, function, family relationships, and the location(s) of allergenic epitopes (when known). Examples of this type of database are shown in Table 1.

The oldest of these databases is the Biotechnology Information for Food Safety Database (Gendel, 1998). This database was initially constructed to support allergenicity assessments for bioengineered foods. Therefore, it takes a broad approach to sequence inclusion and provides information for each entry indicating why it was included. Each entry is identified as to source, protein name or type, and allergen designation (if any) and is linked to the source databases (SwissProt, Protein Information Resource, and the National Center for Biotechnology Information [NCBI] Entrez protein database) through source accession numbers. One of the unique strengths of this database is the complete comparison of sequences among the three source databases. That is, accession numbers with identical sequences in each source database are listed as a single entry, regardless of any cross-reference information contained in the source annotation. Sequences that differ between the source databases are listed as separate entries, again regardless of any cross-reference information in the original source annotations. Each isoallergen is also listed separately. This database is structured in a way that allows identification of complete, nonredundant data sets for food and nonfood allergens. The online data are updated regularly.

The AllergenOnline database (Chapter 9), maintained by the Food Allergy Research and Resource Program of the University of Nebraska, also contains a broadly defined set of allergen sequences that are updated regularly (Hileman et al., 2002). Each entry is identified by source, protein name, and allergen designation (if available), and is linked through a gene identifier (gi) number to the NCBI Entrez protein database. This site was also the first to allow users to compare a sequence to the online database using FASTA (see below).

Table 1. Online molecular databases

Database name	URL
Biotechnology Information for Food Safety Database	http://www.iit.edu/~sgendel/fa.htm
Allergen Online (Farrp)	http://allergenonline.com
Central Science Laboratory Allergen Database	http://www.csl.gov.uk/allergen/
Structural Database of Allergenic Proteins	http://fermi.utmb.edu/SDAP/sdap_ver.html
ALLERDB (AllerPredict)	http://sdmc.i2r.a-star.edu.sg/Templar/DB/Allergen/
AllerMatch	http://www.allermatch.org/
Allergen Database for Food Safety	http://mpj-srs.nihs.go.jp/allergen/index.jsp?pagen=top

Table 2. Online allergy databases

Database name	URL
Allergome	http://www.allergome.org/
AllAllergy	http://allallergy.net/
InformAll	http://www.foodallergens.info/

The Structural Database of Allergenic Proteins, currently the most ambitious of the molecular databases (Ivanciuc et al., 2003), is described in detail in Chapter 11. This database also includes primary sequences, epitope sequences, structural information, and links to protein classification servers. The associated analytical capabilities are described below.

The Central Science Laboratory (CSL) allergen database contains a set of allergen sequences, although the criteria for inclusion are not specified. This database includes epitope sequences and a field for links to structural information, but this field is not populated in most of the database records. The AllerMatch site does not contain an independent database but uses the SwissProt Protein Knowledgebase for allergens (http://www.expasy.org/cgi-bin/lists?allergen.txt) and the International Union for Immunological Societies listing of allergen nomenclature (http://www.allergen.org/). The Allergen Database for Food Safety is a new site that is under construction. This site appears to be derived from a combination of the SwissProt Protein Knowledgebase, the International Union for Immunological Societies allergen list, and the Biotechnology Information for Food Safety Database database; it includes links to epitope sequences and structural information. No descriptive information is given on the contents of the ALLERDB database at the AllerPredict site.

Examples of databases that link allergen protein information to allergy-related clinical, biochemical, and epidemiological information are given in Table 2. The Allergome database is a transitional form (Mari and Riccioli, 2004), built around a listing of allergen molecules but also containing information on biological function, routes of exposure, epidemiology (prevalence), and diagnostics. This is the only database that lists allergenic foods for which no allergenic proteins have yet been identified.

The AllAllergy site is among the most comprehensive allergy information resources. It includes a database of allergenic foods with information on allergenic proteins, along with extensive information on (and links to) relevant literature, other organizations, meetings, and training programs. The major drawback of this site is that it is abridged from the commercial Allergy Advisor database, which is not available online.

The InformAll database has recently replaced and extended the PROTALL database. The current version of InformAll is restricted to information on plant food allergens. The information is divided into layers, with an introductory layer for a lay audience that leads to layers of clinical information and biochemical data. The database contains structured text and links to sequences, protein structures, taxonomies, and articles. An unusual feature of InformAll is the use of a panel of expert referees to review each entry, as would be done for a published article.

ALLERGENICITY ASSESSMENT SITES

Several Web sites provide tools for analyzing allergen sequences and for assessing potential allergenicity. Each of these also contains a site-specific allergen database that is

Table 3. Allergenicity assessment Web sites

Site	URL
Allergen Online (Farrp)	http://allergenonline.com
AllerPredict	http://sdmc.i2r.a-star.edu.sg/Templar/DB/Allergen/
AllerMatch	http://www.allermatch.org/
Structural Database of Allergenic Proteins	http://fermi.utmb.edu/SDAP/sdap_ver.html
WebAllergen	http://weballergen.bii.a-star.edu.sg/
Allergen Database for Food Safety	http://mpj-srs.nihs.go.jp/allergen/index.jsp?pagen=top

used as a query target. Most of the sites that implement allergenicity assessments do so on the basis of the criteria described in the Food and Agriculture Organization of the United Nations-World Health Organization (2001) expert consultation. Examples of such sites are listed in Table 3.

The AllergenOnline site allows users to query the entire database using a simple FASTA search or to conduct a search with overlapping sets of 80 amino acid subsequences (an 80-mer sliding-window search). There is no information on the site as to how the 80-mer search is implemented. AllerPredict uses BLAST to carry out database searches and can search for regions with either 6 or more contiguous identical amino acids or >35% identity over 80 amino acids. The site does not describe how either search is implemented. It appears that this site is still under active construction. The Allermatch site permits full FASTA searches of either or both of the constituent databases, as well as short exact or 80-mer sliding-window searches. Allermatch allows the user to specify the target word length for the exact matches or the percent identity cutoff for the 80-mer sliding-window search. The analytical capabilities of the SDAP database are described in Chapter 11 and in Ivanciuc et al. (2003). This site also allows complete FASTA comparisons, short exact-match searches with a user-defined word length, 80-mer sliding-window searches with a user-defined identity cutoff, and peptide similarity searches for peptides of up to 30 amino acids in length. This site also has extensive links to other bioinformatics tools, including structural analysis programs.

The WebAllergen site uses a different approach to allergenicity assessment. In this case, a query sequence is compared to a library of motifs previously obtained by wavelet analysis of an allergen sequence library. Users do not have direct access to the motif sequence library. The site can also use wavelet analysis to generate a set of motifs from a set of sequences supplied by the user. The Allergen Database for Food Safety combines all types of analysis, BLAST-based searches of the entire database, BLAST-based searches of epitope sequences only, motif-based searches, a sliding-window search, and a search for short exact matches.

ALLERGEN DATABASES AND GOOD DATABASE PRACTICES (GDPs)

Most allergen databases have been constructed as in-house support for individual research programs. It is often difficult, if not impossible, for external users to obtain the descriptive information needed for complete characterization of a database. Similarly, none of the analytical resources provide information on the algorithms or parameters used, other than those sites that use FASTA or BLAST (none of which provide access to all program parameters). The development and application of appropriate GDPs would make it possible for users to validate the analyses and treat the databases as characterized reagents.

The fundamental goal of such GDPs would be to provide the users with complete characterization of a data resource, either a database or analytical resource. For databases, complete characterization includes a description of the contents of the database, the criteria used to determine which information is included, and information on how the database is updated. The variety of allergen databases described above demonstrates that database developers have used different criteria to determine which sequences to include, but these criteria are not often stated clearly. For example, in some cases all sequences that are annotated as allergens are included, even if that annotation is based solely on sequence homology. In other cases, only those proteins that have been tested with multiple immune sera are considered allergens. In still other cases, the criteria for inclusion may be implicit in the data source used. For example, the SwissProt database treats isoallergens as sequence variants described in the annotation, where the sequences can be accessed by analysis programs. It is generally not clear whether allergen database developers using SwissProt as the primary or sole source are aware of this limitation, nor would this be clear to users of the allergen database. Other problems relate to incomplete, inconsistent, or inaccurate cross-references among the source databases, different treatments of redundant sequences in the allergen databases, and the loss of historical context as allergen databases are updated. Table 4 characterizes the databases listed in Table 1 in relation to several potential GDP criteria.

Characterization and validation of analytical resources are more difficult than for the databases. In this case, characterization includes a complete description of the algorithms used and of any parameters that are not controlled by the user. For example, few sites specify the scoring matrix used for sequence comparison, despite the fact that this has a significant effect on the result obtained when two sequences are aligned. Sites that provide motif-based searches generally do not specify or describe either the motif library or the comparison algorithm used. This makes it impossible for the user to evaluate the significance of results obtained using these resources, or to validate analyses that use these resources. Table 5 characterizes the allergen analysis resources listed in Table 3 in relation to potential GDP criteria.

CONCLUSIONS

Allergen sequence databases and the related analytical resources are important tools for assessing the safety of genetically modified foods, for improving our understanding of why some proteins are allergens, and as models for developing and applying GDPs. As discussed above, one of the most important of these principles is transparency, providing users with complete descriptions of these bioinformatics resources. The second important principle is that of separation between the data and the analytical resources. The bioinformatics resources described above often present the data and the analytical resources as an integrated unit. Enforcing a separation between the two makes it possible to implement the third principle, that of providing users with access to complete data sets. Allowing user access to the data sets is equivalent to the sharing of reagents, plasmids, or clones, which is common among laboratory researchers. Such access is also needed to allow research results to be replicated and for comparative studies. Transparency and data access are also

Table 4. Characterization of allergen databases

Database	No. of data groupings	Data access[a]	Inclusion criteria specified[b]	Citation for individual entries[c]	Source database(s)[d]	Links to source database(s)[e]	Structural data[f]	Epitope sequences	Citations for epitope sequences[g]	Update information
Biotechnology Information for Food Safety Database	3	Y	Pub.	Y	NCBI, PIR, SP	Acc.	N	N	N/A	Y
AllergenOnline (Farrp)	11	Y	Site	N	NCBI	gi	N	N	N/A	Y
Central Science Laboratory	4	N	N	P	NCBI	gi	PDB	Y	N	N
Structural Database of Allergenic Proteins	12	N	Pub.		NCBI, SP	gi (NCBI), Acc. (SP)	Pfam	Y	Y	Y
ALLERGDB	1	N	N	N	NCBI, SP	Acc.	Pfam	N	N/A	N
AllerMatch	1	N	?	N/A	SP	N	N	N	N/A	Y
Allergen Database for Food Safety	8	N	Partial	N	NCBI, UniProt	gi (NCBI), Acc. (UniProt)	PDB	Y	Y	Y

[a]Y, access to data for more than a single allergen at a time. The BIFS and AllergenOnline databases provide access to members of a group of allergens.
[b]Inclusion criteria described in detail either in a publication (Pub.) or on the site (Site). The criteria used by the Allergen Database for Food Safety are not complete, and the AllerMatch database is taken intact from other sources.
[c]P, partial, provided for some entries; N/A, not applicable because the database is taken from other sources.
[d]NCBI, National Center for Biotechnology Information Entrez protein database (http://www.ncbi.nlm.nih.gov/Database/); SP, SwissProt plus TrEMBL database (http://us.expasy.org/sprot/); PIR, Protein Information Resource database(http://pir.georgetown.edu/); UniProt database (http://www.pir.uniprot.org/).
[e]gi, gene identifier number; Acc., accession number.
[f]PDB, Protein Database (http://www.rcsb.org/pdb/); Pfam, Protein Families Database (http://www.sanger.ac.uk/Software/Pfam/).
[g]N/A, not applicable.

Table 5. Allergen analysis resources

Name	Database search			Sliding window analysis			Exact match analysis		Motif analysis		
	Program	Scoring matrix	Controllable parameter(s)	Algorithm	Scoring matrix	% Identity	Algorithm	Length	Algorithm	Motif library	Other
Allergen Online (Farrp)	FASTA3	BLOSUM 50	E value, no. of alignments	FASTA	?	>35	N/A	N/A	N/A	N/A	N/A
AllerPredict	BLAST	?	Specific BLAST program	?	?	>35	?	6	N/A	N/A	N/A
AllerMatch	FASTA	?	Target database	FASTA	?	User specified	?	User specified	N/A	N/A	N/A
Structural Database of Allergenic Proteins	FASTA	?	None	FASTA	?	User specified	?	User specified	N/A	N/A	N/A
WebAllergen	N/A	N/A	N/A	N/A	N/A	N/A	N/A	N/A	?	Site specific	N/A
Allergen Database for Food Safety	BLAST	Five choices	Target database, word size, E value, low complexity mask	FASTA	?	User specified	FastA	User specified	N/A	N/A	N/A

tied to the fourth principle, interoperability. That is, to the extent possible, each data resource should use data structures and terminology that permit simple data exchange. For example, a common set of descriptors should be developed to describe the criteria used to determine why a protein is included in an allergen database. Such descriptors, which could be considered to be the start of a full allergen-related ontology, could be used by researchers to filter the sequence databases at different levels of stringency in assembling data sets for analysis.

Eventually, it will be possible to combine the information in these allergen databases with the rapidly increasing knowledge of human genomics and proteomics to reach a much better understanding of how and why people develop allergies and perhaps to use this knowledge to develop treatments and cures.

Acknowledgment. This work was supported by Cooperative Agreement FD000431 between the U.S. Food and Drug Administration and the National Center for Food Safety and Technology.

REFERENCES

Brusic, V., M. Millot, N. Petrovsky, S. Gendel, O. Gigonzac, and S. Stelman. 2003. Allergen databases. *Allergy* **58:**1093–1100.

Food and Agriculture Organization of the United Nations/ World Health Organization. 2001. *Evaluation of Allergenicity of Genetically Modified Foods: Report of a Joint FAO/WHO Consultation on Food Derived from Biotechnology*. Food and Agriculture Organization of the United Nations, Rome, Italy.

Gendel, S. 1998. Sequence databases for assessing the potential allergenicity of proteins used in transgenic foods. *Adv. Food Nutr. Res.* **42:**63–92.

Gendel, S. 2002. Sequence analysis for assessing potential allergenicity. *Ann. N.Y. Acad. Sci.* **964:**87–98.

Hileman, R. A. Silvanovich, R. Goodman, E. Rice, G. Holleschak, J. Astwood, and S. Hefle. 2002. Bioinformatic methods for allergenicity assessment using a comprehensive allergen database. *Int. Arch. Allergy Immunol.* **128:**280–291.

Ivanciuc, O., C. Schein, and W. Braun. 2003. SDAP: database and computational tools for allergenic proteins. *Nucleic Acids Res.* **31:**359–362.

Mari, A., and D. Riccioli. 2004. The Allergome web site—a database of allergenic molecules. Aim, structure and data of a web-based resource. *J. Allergy Clin. Immunol.* **113:**S301.

Poulsen, L. 2005. In search of a new paradigm: mechanisms of sensitization and elicitation of food allergy. *Allergy* **60:**549–558.

Sampson, H. 2004. Update on food allergy. *J. Allergy Clin. Immunol.* **113:**805–819.

Food Allergy
Edited by S. J. Maleki et al.

Chapter 11

Structural Database of Allergenic Proteins (SDAP)

Catherine H. Schein, Ovidiu Ivanciuc, and Werner Braun

The Structural Database of Allergenic Proteins (SDAP; http://fermi.utmb.edu/SDAP/) brings together data from diverse sources on over 800 allergen sequences and their epitopes in a cross-referenced format. SDAP-Food Allergens is a separate directory of food allergens, linking them to all the allergens in SDAP. SDAP is user friendly and freely available on the Web to clinicians, patients, food scientists, and industrial engineers. Clinicians use SDAP primarily to determine food sources that might contain cross-reacting antigens. Regulators and industrial researchers can, in addition, use the molecular analysis capabilities the database provides to derive rules for distinguishing allergens from other food proteins. A rapid FASTA search of allergenic proteins in SDAP can be used to group allergens, aid in determining proper nomenclature of novel allergenic proteins, and make general predictions about a protein's overall similarity to known allergens. Special methods developed for SDAP, such as the property distance (PD) scale, automatically detect peptides in the ensemble of allergen sequences that have physicochemical properties similar to known immunoglobulin E (IgE) epitopes. Experimentally determined and modeled structures of allergenic proteins can then be used to compare the structure and surface exposure of these peptides to that of the initial epitope. This introduction to SDAP describes the use of these methods to compare the epitopes of food allergens and detect areas that could contribute to observed clinical cross-reactivities of common foods.

Anyone who has ever seen a food sensitive individual go into anaphylactic shock knows that food allergies can present with devastating symptoms. Food allergies, mostly against milk, eggs, peanuts, soy, or wheat affect up to 8% of infants and young children (Sampson, 2005; Sampson, 1999a; Sampson, 1999b). Those with high levels of IgE to an allergen will probably not outgrow their sensitivity (Fleischer et al., 2004; Sicherer, 2003) and will face a life of careful avoidance of all potential sources of eliciting

Catherine H. Schein, Ovidiu Ivanciuc, and Werner Braun • Sealy Center for Structural Biology and Biophysics, Department of Biochemistry and Molecular Biology, University of Texas Medical Branch 301 University Blvd., Galveston TX 77555–0857.

proteins that trigger their reactions. In addition, many food allergies, such as those to shellfish, nuts, and fruits, can develop later in life. One hypothesis is that these late-onset allergies may be the result of the individual being sensitized by exposure to pollens, insect dust, or other aeroallergens (Rabjohn et al., 1999b; Sampson, 1999a; Scheurer et al., 1999; Vanek-Krebitz et al., 1995). Comparing the structure of food allergens with allergenic proteins from these other sources can give valuable information about the sensitization process. This also means that any novel protein introduced into a food crop should have a low potential to be an allergen. This chapter will deal with database approaches to determine the allergenic potential of a test protein, based on similarity to known allergens.

As our knowledge of allergenic proteins has increased, many clinical observations about cross-reactive proteins could be accounted for at the molecular level (Breiteneder and Ebner, 2000; Jenkins et al., 2005). For example, major allergenic proteins in peanuts have been isolated, and peptides from their sequences that react with IgE have been identified (Burks et al., 1997; Rabjohn et al., 1999b; Shin et al., 1998). Similar allergenic proteins from other foods that may also elicit response in peanut allergic individuals, such as tree nuts (de Leon et al., 2003), soy (Eigenmann et al., 1996), and legumes (Lopez-Torrejon et al., 2003; Wensing et al., 2003), have also been isolated. There have also been extensive studies of the major allergenic proteins from milk (Natale et al., 2004; Pourpak et al., 2004; Wal, 2004): casein (Cocco et al., 2003; Elsayed et al., 2004; Natale et al., 2004), and lactoglobulin (Adel-Patient et al., 2001; Ehn et al., 2004; Jarvinen et al., 2001). Studies that have been carried out of allergenic proteins from egg (Mine and Rupa, 2004) include those of ovomucoid (Mine and Zhang, 2002; Mine et al., 2003; Mizumachi and Kurisaki, 2003) and lysozyme (Fremont et al., 1997). Studies of proteins from shrimp and related species include those of tropomyosins from shrimp (Ayuso et al., 2002; Reese et al., 2002; Samson et al., 2004), parvalbumin from fish (Swoboda et al., 2002a; Swoboda et al., 2002b; Van Do et al., 2005), as well as albumins (Moreno et al., 2005a; Moreno et al., 2005b; Palomares et al., 2005; Robotham et al., 2005) and glycinins (Beardslee et al., 2000; Helm et al., 2000; Rabjohn et al., 1999a) from legumes. In addition, many plant allergens are classified as pathogenesis response proteins (Asensio et al., 2004; Elbez et al., 2002; Hoffmann-Sommergruber, 2002; Midoro-Horiuti et al., 2001a).

The decision by the American Society for Microbiology to sponsor this book is evidence of the growing awareness of food allergies as an environmental hazard. It is also a tribute to the accumulated body of knowledge about the proteins and their epitopes that elicit severe IgE-mediated reactions (Breiteneder and Mills, 2005; Gendel, 2004; Glaspole et al., 2005; Jenkins et al., 2005). Until recently, much of this information was distributed in many different literature sources and could not be assessed from the Internet (Brusic, 2003). In recent years, several databases have been constructed that contain sequences and information about allergenic proteins (Table 1; see Chapter 10 for more details). This chapter will concentrate on the SDAP, which has been specifically designed to allow combined analysis of the sequence, structure, and epitopes of allergens. Special tools have also been designed to permit unbiased statistical analysis and comparison of the structural and physical chemical properties of IgE epitopes (Ivanciuc et al., 2002; Ivanciuc et al., 2003a). In this chapter, we will show how SDAP and the methods for sequence and peptide comparison incorporated therein can be used to predict cross-reactive allergens in foods and determine common properties of their IgE epitopes.

Table 1. Web sites with information about allergens

Web site	URL	Information available
IUIS (International Union of Immunological Societies	http://www.allergen.org	Lists official names, grouped by source, and GenBank accession numbers of allergens
SDAP (Structural Database of Allergenic Proteins) and SDAP-Food	http://fermi.utmb.edu/SDAP	Allergen sequences, on-site and cross-referenced by source and protein type; links to all major sequence and structural databases, IgE epitope collection, tools for sequence and epitope comparison, on-site information about experimental structures of allergens, and high-quality protein models
Protall	http://www.ifrn.bbsrc.ac.uk/protall/	Allergen names, links to detailed biochemical, structural, and clinical data
National Center for Food Safety and Technology	http://www.ijt.edu_sgendel/fa.htm	Lists official names of food allergens with links to GenBank
CSL (Central Science Laboratory, United Kingdom)	http://www.csl.gov.uk/allergen/index.htm	Lists official names of allergens with sequence links to GenBank
Farrp	http://allergenonline.com/asp/public/login.asp	Lists official names of allergens, sequence links to Genbank, and a FASTA search for related sequences
Allergen Database for Food Safety (ADFS)	http://mpj-srs.nihs.go.jp/allergen/index.jsp?pagen=top	Allergen sequences; implements WHO allergenicity rules using FASTA
ALLERGEN	http://sdmc.i2r.a-star.edu.sg/Templar/DB/Allergen/	Lists official names of allergens and a BLAST search; implements the WHO allergenicity rules
InformAll	http://foodallergens.ifr.ac.uk/	Biochemical information, mainly for food allergens, epitopes, sequences, links to literature
ALLERbase	http://www.dadamo.com/allerbase/allerbase.cgi	List of allergens with links to protein databases
WEBAllergen	http://weballergen.bii.a-star.edu.sg/	Predicts the potential allergenicity of proteins using motifs found by a wavelet algorithm
Allermatch	www.allermatch.org/	Implements WHO allergenicity rules using FASTA
Allergome	http://www.allergome.org	Lists the official names of allergens, and links to PubMed and sequence databases
Swiss-Prot	http://us.expasy.org/cgi-bin/lists?	Allergen.txt list of allergens with sequence data
AllAllergy	http://allallergy.net/	A portal to allergy information, useful for the general public
Modbase	http://alto.rockefeller.edu/modbase-cgi/index.cgi	Automatically generated models for Swiss-Prot and TrEMBL sequences

AN INTRODUCTION TO SDAP AND ITS TOOLS

Everyone Can Use SDAP

Although SDAP was developed for basic research on the nature of allergenic proteins and to allow regulatory agencies, food scientists, and engineers a way to determine if a novel protein has allergenic potential, no special training is needed to use the data at the site. Food-allergic patients may first wish to consult databases that summarize clinical cross-reactivities, such as AllAllergy (http://allallergy.net/). The major use of SDAP by clinicians is to determine food sources that could induce cross-reactions in sensitive individuals. SDAP sequence searches can help in preparing dietary recommendations for allergic patients with a known sensitivity, so that they can avoid exposure to other foods likely to trigger an allergic reaction.

Environmental engineers and those in regulatory roles can immediately determine the similarity of novel proteins to known allergens according to current guidelines, using the FASTA search (Pearson, 1990) as recommended by the Food and Agriculture Organization of the United Nations (FAO) and the World Health Organization (WHO) (World Health Organization, 2000, 2001; Codex Alimentarius Commission, 2003) and the European Food Safety Authority (2004). Food scientists will appreciate the ability to relate allergenic proteins to one another and the tools to distinguish areas of the sequence that are known IgE epitopes.

But most of all, SDAP is designed to be useful to those who wish to define molecular properties of allergenic proteins and relate them to one another. These researchers will find specially designed methods for comparing peptides to known epitopes. These easy-to-use tools can aid in designing new proteins or selecting isoforms for reduced allergenicity. As we show in the second half of this chapter, SDAP has allowed us to rapidly collect data pertinent to very basic questions about allergens, such as whether enzymatic activity is related to allergenicity. The methods in SDAP are continuously updated, to attempt to answer that overwhelming question: what makes a protein an allergen?

SDAP's Basic Structure

SDAP contains information on sequence, three-dimensional (3D) structures, and epitopes of known allergens from published literature and databases compiled on the Web (King et al., 1994). The best way to learn to use SDAP is to go to the Web site at http://fermi.utmb.edu/SDAP/ and consult the lists of allergens on the main search page. Select a food or a single allergen of interest (for example, select Ara h 1, a major peanut allergen) from the table by clicking on it, and go to the descriptive page by a second click. Other data incorporated into the SDAP tables (which are assembled in MySQL under Linux) include the allergen name (according to the International Union of Immunological Societies [IUIS] Web site listing at http://allergen.org/); scientific and common names for the species; general source of the allergens; allergen type; species; systematic name; brief description; sequence accession numbers from SwissProt, the Protein Information Resource (PIR), and the National Center for Biotechnology Information (NCBI); and, where available, the Protein Database (PDB) file name. These data can be directly accessed by clicking on the links. More recently, lists of Pfam (http://www.sanger.ac.uk/Software/Pfam/) membership and links to structural models have been added. Figure 1 is an overview of the information and methods incorporated into the cross-referenced (MySQL-Linux) lists of data. The in-house

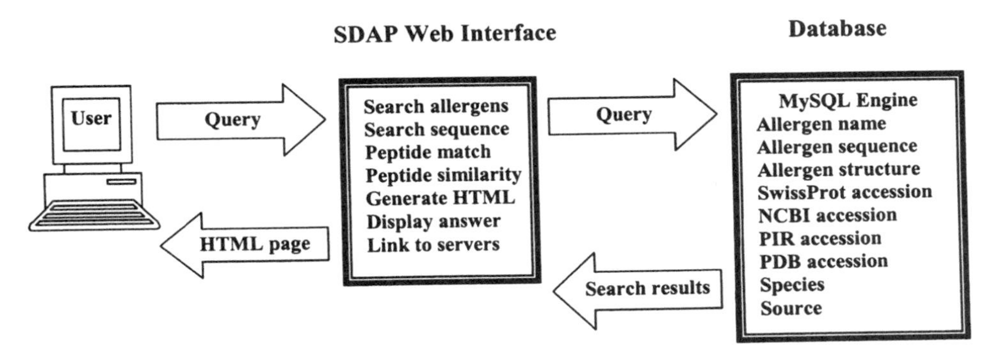

Figure 1. Basic structure of the SDAP database. The data are contained in cross-referenced lists. The user can send a query by pushing buttons at the site or go to specific search pages and type in a question.

bioinformatics methods permit almost-instantaneous FASTA and peptide similarity searches within SDAP and direct connections to much larger databases.

SDAP is also integrated with other bioinformatics servers, allowing the user to investigate structural similarity and neighbors with SCOP (Structural Classification of Proteins) (Conte et al., 2000), TOPS (Topological Representation of Protein Structure) (Gilbert et al., 1999), CATH (Class, Architecture, Topology, and Homologous superfamily) (Pearl et al., 2001), CE (Combinatorial Extension of the optimal path) (Shindyalov and Bourne, 1998), FSSP (Fold Classification Based on Structure-Structure Alignment of Proteins) (Holm and Sander, 1996), and VAST (Vector Alignment Search Tool) (Gibrat et al., 1996).

Epitope Lists for Allergenic Proteins

Among the unique features of SDAP are the lists of IgE-binding epitopes of allergenic proteins, assembled from the primary literature. The reader is cautioned that most of these sequence segments have been identified by in vitro binding to short peptides on solid phases and are assumed to represent epitopes that may be involved in eliciting allergic reactions. In a few cases, the biological importance of the epitopes has been tested, for example, by mutating these areas and showing that the IgE-binding capacity was thereby diminished (Bannon, 2001; Li, 2003; Rabjohn et al., 2002) or that the isolated peptides can interfere with IgE binding to the whole protein (Midoro-Horiuti et al., 2003; Midoro-Horiuti et al., 2006). Currently, IgE epitope information is available for the food allergens Ara h 1, Ara h 2, Ara h 3, Asp f 2, Gal d 1, and Pen i 1 and soybean glycinin G2, as well as for cross-reactive proteins from non-food sources such as latex (Hev b 1, Hev b 3, and Hev b 5), pollen (Par j 1, Par j 2, Jun a 1, Jun a 3, and Cry j 1), and other aerosol allergens.

Methods for Comparing Sequences in SDAP

SDAP has several different methods for comparing the sequences of allergens in its lists. These include standard database search methods such as FASTA (Pearson, 1990), which were designed to compare protein sequences rapidly in large databases. In addition, the PD method was developed in our group specifically to detect meaningful similarities in shorter sequences (Ivanciuc et al., 2002; Ivanciuc et al., 2003a).

FASTA for Comparing the Overall Sequences of Allergenic Proteins

The first step in determining the relationship between allergens is to compare their overall sequence. FASTA sequence comparisons can be run automatically from any sequence file in SDAP. This is a rapid method to determine related allergens and to assess the overall allergenic potential of a novel protein (see below). Table 2 is an example of a FASTA search result in SDAP for the food allergen Act c 1, from kiwifruit. Note that although the closest entries in SDAP for this cysteine protease are bromelain from pineapple (Ana c 2), papain from papaya (Car p 1), and thiol protease from soybean (Gly m 1), allergenic proteases from mites are also detected as near neighbors. Based on the results with these FASTA alignments, we expect the cross-reactivity between Act c 1 and other allergenic proteases to decrease as the similarity decreases.

Methods for Comparing Epitopes and Short-Sequence Segments

While FASTA can determine overall similarity of large proteins, it was not designed to compare short sequences, such as defined linear IgE epitopes. Further, comparison of sequences of linear epitopes is difficult with most scoring matrices, as what is probably important for IgE reactivity is the physicochemical properties of the exposed side chains and their spatial relationship to one another in the sequence, rather than only the absolute amino acid sequence.

Thus, two different tools were incorporated in SDAP to (i) determine short sequences identical to that of a known epitope or (ii) determine sequences that match the physicochemical properties of sequences of allergenic epitopes with those of all the proteins in the database.

SDAP Exact Search for Known IgE Epitopes

The WHO guidelines for predicting potential allergenicity (World Health Organization, 2001) specify that a protein might cross-react with an allergen if it contains an exact match of any peptide of six to eight amino acids with the allergen (Codex Alimentarius Commission, 2003). The SDAP peptide exact-match functionality allows the user to select a

Table 2. Output of an automatic FASTA search in SDAP starting from the file for allergen Act c 1 of kiwifruit (*Actinidia chinensis*)[a]

Allergen	Sequence	No. of amino acids	Bit score	E score
Act c 1	P00785	380	600.8	1.7e-173
Ana c 2	BAA21849	351	228.5	1.9e-61
Car p 1	AAB02650	345	217.2	4.6e-58
Gly m 1	AAB09252	379	189.1	1.5e-49
Eur m 1	P25780	321	126.1	1.1e-30
Der f 1	P16311	321	122.9	1.0e-29
Der p 1	P08176	320	115.1	2.3e-27
Blo t 1	AAK58415	221	113.8	4.0e-27

[a]This cysteine protease is most closely related to those in papaya and soybean but also to those from dust mites (last four allergens).

known IgE epitope and find other allergenic proteins that contain that exact sequence. Alternatively, the user can input a peptide whose sequence identity is then tested against all SDAP allergens. This simple function is useful in investigating allergen cross-reactivity, when a certain IgE epitope is suspected to be involved in clinically defined cross-sensitivities to several allergens. For example, an SDAP search for the Pen i 1 IgE epitope FLAEEADRK from shrimp tropomyosin (Shanti et al., 1993), reveals that the allergens Met e 1, Pan s 1, Cha f 1, and Hom a 1 all contain identical sequences (Ivanciuc et al., 2003a), consistent with the overall similarity of the proteins from shrimp, crab, and lobster (Leung et al., 1994; Leung et al., 1996; Leung et al., 1998a; Leung et al., 1998b).

Sequence Similarity Ranking in SDAP: the PD Scale

Because such exact sequence identity, even among known cross-reactive allergens, is rare, a method to compare sequences for their similarity with respect to physical chemical properties was developed. This detection is based on the descriptors E_1 to E_5 (Venkatarajan and Braun, 2001), which were determined by multidimensional scaling of 237 physical-chemical properties. Using these vectors, the distinct properties of the 20 naturally occurring amino acids can be numerically summarized as five values. Given a test peptide of length n (e.g., a known epitope), the PD tool (Ivanciuc et al., 2002; Ivanciuc et al., 2003a) converts it into a $5 \times n$ matrix of numbers that represents the physical chemical properties at each position. This matrix can then be compared to those for every other window of sequence length n in SDAP. For example, the side chains of serine and cysteine have some similarity in certain properties (e.g, size and hydrophilicity) but differ in others, such as their relative abundance in a database, their overall charge, and electron donor-acceptor status. The matrix allows a choice of amino acids at any position that is most consistent with all the properties of that residue in the query peptide. Peptides with identical sequences have a PD value of 0, and peptides with conservative substitutions of a few amino acids have a small PD value, typically in the range of 0 to 3. Peptides with a recognizable similarity in their physical chemical properties tend to have PD values of <10, and unrelated peptides have PD values of around 20, on average. In Table 3, we present several examples of allergen fragments similar to the Ara h 3 IgE epitope VTVRGGLRILSPDRK, which demonstrate the relationship between PD values and sequence similarity.

Mathematically formulated, to define the similarity of two sequences A and B of length n according to the E_1-E_5 descriptors, the property distance function PD can be calculated as follows (Ivanciuc et al., 2002; Ivanciuc et al., 2003a):

$$PD(A,B) = \frac{1}{n}\sum_{i=1}^{N}\left\{\sum_{j=1}^{5}\lambda_j[E_j(A_i) - E_j(B_i)]^2\right\}^{1/2}$$

where λ_j is the eigenvalue of the jth E component, $E_j(A_i)$ is the E_j value for the amino acid in the ith position from sequence A, and $E_j(B_i)$ is the E_j value for the amino acid in the ith position from sequence B.

Determining the Significance Level of PD Scores

The PD search is designed to compare protein regions with lengths comparable with those of published linear IgE epitopes. One of the first questions about such values is how significant a given match is, especially in a sizable database. Significance levels for the

Table 3. Sequences most closely related to epitope 3 of Ara h 3 from peanut by PD search[a]

No.	Allergen	PD	$z(PD_{min})$	$z(PD_{ave})$	Start residue	Matching region[b]	End residue
1	**Ara h 3**	**0.00**	**8.3189**	**9.2428**	**276**	**VTVRGGLRILSPDRK**	**290**
2	Ara h 4	2.94	6.3978	7.7709	299	VTVRGGLRILSPDGT	313
3	Gly m glycinin G2	6.32	4.1874	6.0774	253	VTVKGGLRVTAPAMR	267
4	**Gly m glycinin G1**	**6.68**	**3.9508**	**5.8961**	**256**	**VTVKGGLSVI**KPPTD*	**270**
5	Cor a 9	8.55	2.7250	4.9569	274	VKVEGRLQVVRPERS	288
6	Ber e 2	9.13	2.3485	4.6685	252	VRVEQGLKVIRPPRI	266
7	Asc s 1	10.09	1.7206	4.1874	235	SSLDTHLKWLSQEQK	249
8	Der p 10	10.86	1.2158	3.8007	191	VELEEELRVVGNNLK	205
9	Pan s 1	10.86	1.2158	3.8007	181	VELEEELRVVGNNLK	195
10	**Pen a 1**	**10.86**	**1.2158**	**3.8007**	**191**	**VELEEELRVVG**NNLK**	**205**
12	Hom a 1	10.86	1.2158	3.8007	191	VELEEELRVVGNNLK	205
16	Met e 1	10.86	1.2158	3.8007	181	VELEEELRVVGNNLK	195

[a]Known epitopes or parts of epitopes are in boldface type. The first column lists the order of sequences found. Epitopes have not been determined for the other three allergenic proteins. The sequence from Pen a 1 is included, as it is a defined epitope in a sequence region that is identical in many tropomyosins that are known food allergens. The average *PD* value for the best scoring sequence in the 828 full-length entries in SDAP was 12.72 (SD = 1.53); the average *PD* value for all 186,164 possible windows was 18.44 (SD = 1.99).

[b]*, the whole epitope is 253GAIVTVKGGLSVI265 (Beardslee et al., 2000); **, epitope of Pen a 1 (Ayuso et al., 2002).

sequence-similarity index PD have to be set high enough to detect all peptides that are similar to an IgE epitope but low enough to discriminate them from other regions in the ensemble of allergenic proteins that would match randomly. For the search, each area of all the sequences is individually matched, with a window for the sequence segment that moves progressively by one position. Thus, a 200-amino-acid protein would have 194 different sequence windows of 7 amino acids, and 191 for a 10-mer. All PD searches in SDAP are followed by two histograms (Fig. 2). In the first test, the lowest-scoring window is determined for the best-matching peptide in the 829 protein entries (at the time of this chapter) in SDAP. As shown in Fig. 2 for an Ara h 3 epitope, the lowest-scoring window clusters around 12 to 13 for this peptide, with a standard deviation (SD) of about 2, and very few scores below 10. According to this test, values below about 8 (mean value − [2 × SD]) would be significantly similar to the test peptide. However, there are many similar sequences and isoforms in SDAP, which tends to skew the statistics for peptides. As a better estimate of what a random match would be, a second histogram summarizes the scores for all ~190,000 windows of a given size in all the SDAP allergen sequences. The average values of this histogram range from 17 to 26, depending on the peptides. According to these statistics for a random match, peptides with PD values of <10 would be clear outliers. These data are summarized for the user by *z* scores (which indicate the quality of the match relative to the database random distribution), which are given with the PD value.

Using the PD distribution, the score $z(PD)$ for a given match is calculated as follows:

$$z(PD) = \frac{|\, PD_{min} - PD_{ave} \,|}{SD(PD)}$$

From the best-matching PD distribution (smallest PD value for each SDAP sequence), we obtain $z(PD_{min})$, while from the complete PD distribution, we obtain a $z(PD_{ave})$ value. Both

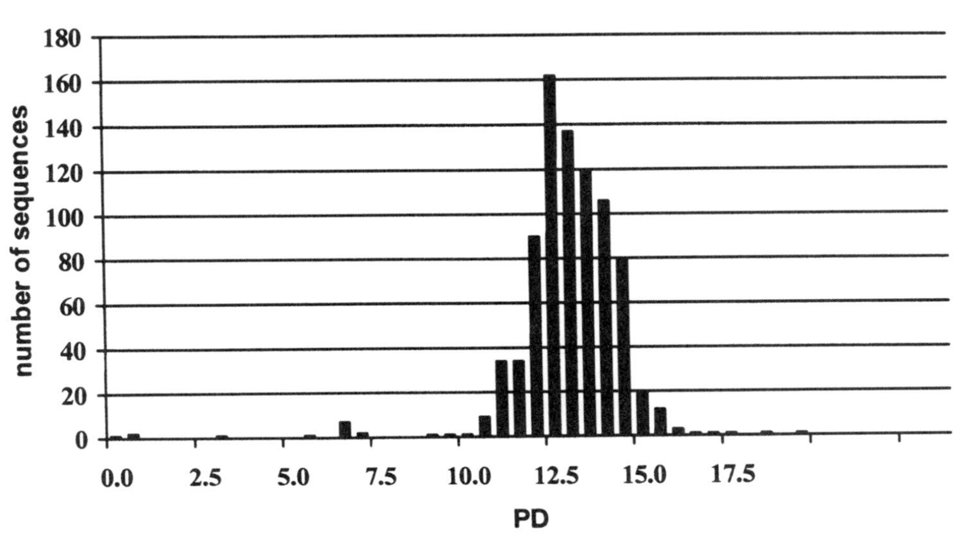

B

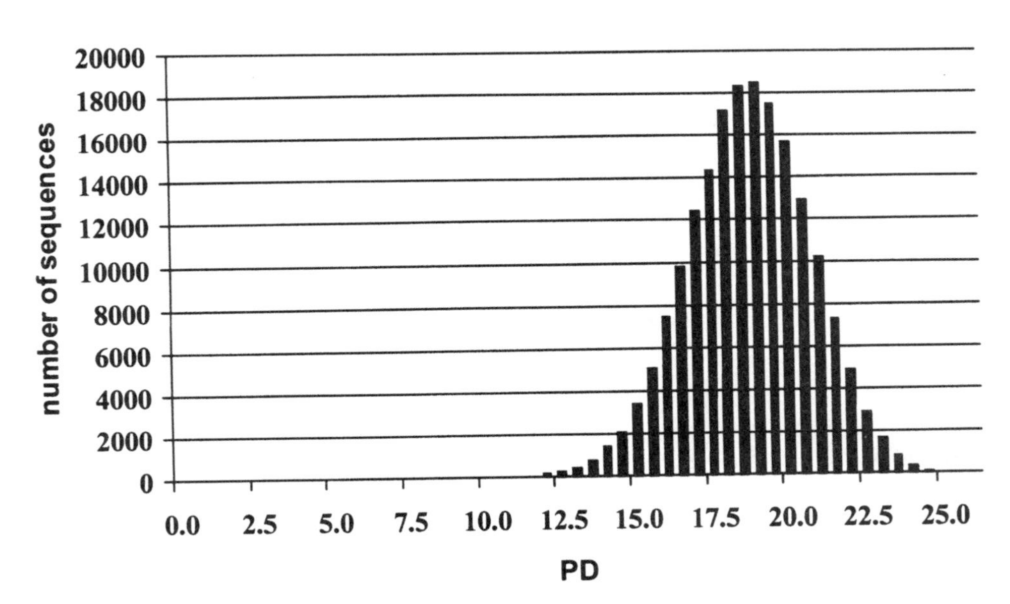

Figure 2. Histogram of PD values obtained for the Ara h 3 IgE epitope VTVRGGLRIL-SPDRK (Table 3) for the lowest-scoring window for each of the SDAP allergens (A) and the scores of all possible sequence windows in the SDAP allergens (B).

z scores are reported for each peptide compared to an epitope. The higher the z value, the more significant is the match between the two peptides. Note that because of the many similar sequences in SDAP, the $z(PD_{ave})$ score is a more realistic estimate of the random occurrence of a certain sequence similarity in an allergen database.

As another example of using the PD tool, several reports show that the G1 (Beardslee et al., 2000) and G2 (Helm et al., 2000) glycinins from soybean, which have not been submitted to the IUIS as allergens, have IgE-binding sites that are similar to those in known allergens such as peanut allergens. In SDAP, we have labeled these as Gly m glycinin G1 and G2, respectively, to allow their retrieval together with designated soybean allergens (for example, when the search term is "Gly m"). The 2 IgE epitopes identified for Gly m glycinin G1 (Beardslee et al., 2000) and the 11 epitopes for Gly m glycinin G2 (Helm et al., 2000) can be retrieved from SDAP. Consistent with the previously observed similarity (Beardslee et al., 2000), a PD search for the two Gly m glycinin G1 epitopes finds the known epitopes of Ara h 3, GNIFSGFTPEFLEQA ($PD = 3.32$, with z values of 5.8 and 8.32 to epitope 1 of the soybean G1) and VTVRGGLRILSPDRK ($PD = 2.92$, with z values of 5.12 and 6.87 to epitope 2 of soybean G1). This is further indication that the PD values indicate meaningful similarities between known epitopes.

While the PD search tool can provide very valuable information about potential IgE epitopes in homologous proteins, other criteria are needed to predict IgE reactivity of peptides from allergens. For example, IgE epitopes are often listed as the whole reactive peptide, which can be as long as 15 or more amino acids. The actual antibody-binding site probably comprises only a small portion of this, perhaps as little as three to six amino acids. Searching the protein sequence databases with such a small sequence will, on statistical grounds alone, generate far too many false positives to have this constitute a true test for allergenicity. On the other hand, limiting the search to an exact match of these amino acids will also ignore the fact that many amino acids can be altered within known epitope sequences without altering reactivity (Shin et al., 1998), while altering key amino acids completely eliminates IgE binding. As there is still a significant chance that a match with a PD value of <10 could occur by chance, it is essential to use other criteria to determine whether a given peptide is a possible match for a given epitope. Thus, additional classifiers that take position in the sequence into account are needed for any method to predict IgE-reactive sites on proteins.

Combining Sequence and Structural Information

Once similar sequences have been identified by PD values, the structural information in SDAP can be used to understand the characteristics of a given food allergen and identify other substances that resemble it in other foods or aerosol allergens (Mittag, 2004). SDAP allows direct access to the experimental structures (out of 586 SDAP allergens, 45 have known PDB structures). For those allergens where the structure has not yet been determined, most (we estimate >90%, based on results from the fold recognition server 3D-PSSM [http://www.sbg.bio.ic.ac.uk/~3dpssm/]) have sequences similar enough to other proteins of known structure that we can make reliable model structures. According to statistical studies, for large proteins (more than about 80 residues), if two proteins are >30% identical they will probably have the same fold (Abagyan and Batalov, 1997). Results in two rounds of the Critical Assessment of Techniques for Protein Structure Prediction (CASP4 and CASP5) competitions showed that we could prepare template-based models that were

Figure 3. Model structure of the vicilin allergen Ara h 1 (residues 172 to 586) from peanut, showing the internal dimer axis (i.e., the monomer has two equivalent areas of structure that have different sequences). Two epitopes on opposite sides of the protein (symmetric IgE epitopes) with side chains are shown, and the first and last amino acids are labeled. The template for homology modeling was canavalin from jack bean (PDB file 2CAV_A; resolution, 2 Å), which is 47% identical (162 of 346) with this area of Ara h 1.

very close to the crystal structures at even lower-percent identity (Ivanciuc et al., 2004; Tramontano et al., 2001).

We are thus in the process of using our MPACK modeling suite to prepare reliable models of the allergens in SDAP. We are concentrating first on molecular modeling of those allergenic proteins for which IgE epitope sites but no experimental structures are available (Greene and Thomas, 1992; Le Mao et al., 1992; Rabjohn et al., 1999b; Shin et al., 1998; Soman et al., 2000). An example of one of our models, for the Ara h 1 allergen of peanut, is shown in Fig. 3. Note that the monomer of Ara h 1 has two equivalent domains of structure, which differ greatly in their sequences. Epitopes, determined by peptide mapping (Burks et al., 1997), occur on both halves of the internal dimer axis, often in symmetric locations, as shown for one pair in the figure. Other symmetric epitope pairs can be identified based on their low PD value to one another (Schein et al., 2005).

Determining Surface-Exposed Residues with GETAREA

Assuming the undigested protein is allergenic, IgE epitope areas should have substantial surface exposure. A program developed in this group, GETAREA (http://www.scsb.utmb.edu/cgi-bin/get_a_form.tcl), rapidly determines the exact solvent exposure of all atoms of a protein structure. All that is required is a structure file in PDB format. The file is read through a browse window, and the user can select the option for per residue or per individual atom surface exposure, depending on the degree of complexity required. If the per residue option is chosen, GETAREA produces a table giving the percent solvent-exposed area of each residue in the structure, which is the ratio of the observed exposed area to the total area of the residue (as determined from the area in crystal structures for a given side chain). Results of residue surface exposure can be combined with alanine-scanning data

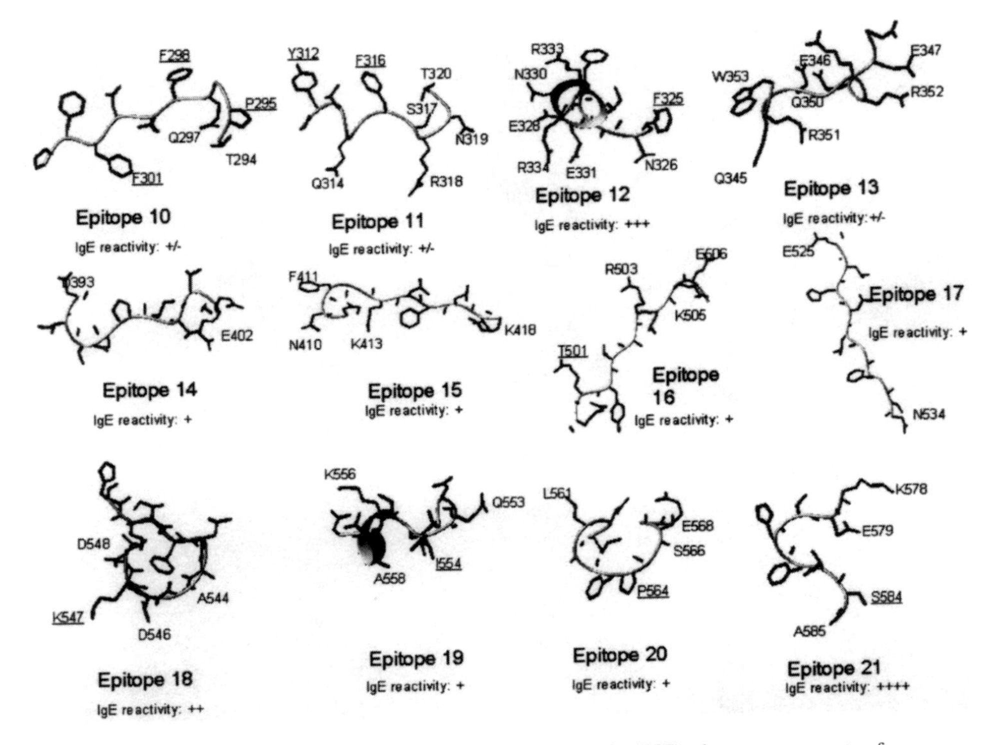

Figure 4. Isolated linear epitopes of Ara h 1 (Burks et al., 1997), shown as segments of the model shown in Fig. 3. The residues with significant (>30%) solvent exposure, according to GETAREA, are labeled; the surface-exposed residues known to play a role in reactivity of the epitope with IgE are underlined.

of a known epitope or PD values for a peptide to improve the prediction of sequences likely to be involved in antibody binding. Figure 4 shows the surface-exposed residues of epitopes of Ara h 1, as determined from linear peptide-mapping studies (Burks et al., 1997), extracted from the model structure of Fig. 3. This information, coupled with data on which residues are essential for binding according to peptide-scanning studies, can be used to predict antibody-binding sites within the peptide stretch.

Defining Motifs of Allergenic Proteins with MASIA

One other feature, residue conservation in related allergenic proteins of known clinical cross-reactivity, can also be used as a classifier for IgE epitopes. Several groups have sought to define motifs characteristic of allergenic proteins, based on amino acid conservation (Brusic et al., 2003; Li et al., 2004; Mills et al., 2002; Stadler and Stadler, 2003). In our work, we have chosen to concentrate on defining physical chemical properties of the epitopes of known allergens and also to use other methods to define motifs that may be common to allergens which are members of the same protein family. The underlying assumption is that for a group of cross-reactive allergenic proteins, the IgE epitopes are in positions common to all of them, and that the Web server MASIA (http://www.scsb.utmb

.edu/masia/masia.html) can quantify the extent of the surface-reactive group conservation. This method begins by aligning the sequences of known allergens that are related to one another, such as those in the tropomyosin or vicilin family. Once such an alignment is available, our automated method for sequence motif generation, MASIA (Zhu et al., 2000), can be used to determine PCP motifs based on the E_1-E_5 vectors (Venkatarajan and Braun, 2001) defined above for the PD search. MASIA finds sequence motifs in protein families by identifying regions with highly conserved physicochemical properties.

POSING QUESTIONS AND PROBLEM SOLVING WITH SDAP

SDAP was designed to allow researchers to answer scientific questions about the nature of allergens. Users can rapidly gather data on the structure, surface exposure, and residue conservation among related allergens for sequences identified as IgE epitopes. Below, we show how SDAP can be used to answer specific questions about allergenic proteins.

Using SDAP To Define Potential Sources of Cross-Reactions

Due to the severity of many food allergens, it is not prudent to subject an atopic individual to skin tests for all possible cross-reactive substances (Sicherer, 2001). Identification of the allergy spectrum is done by patient interview and then by identifying proteins that react with the IgE in patient sera in vitro. There are ongoing efforts to identify all possible proteins in foods that cause reactions in humans that bind patient IgE and to make tests for allergen sensitivity more specific (Eigenmann et al., 1996; Jenkins et al., 2005; Lehrer et al., 2002; Viquez, 2004; Wensing et al., 2003). Recent studies have explored doing even more fine tuning at this level, using peptide arrays (Shreffler et al., 2004). As this information becomes available, it will be incorporated into the SDAP tables. SDAP is particularly useful for the prediction of other possible cross-reactive food sources if there is specific information on the proteins in the food that the patient reacts to that are detected by IgE in the sera of the patient.

Many of the allergens identified in foods are related to known airway allergens. For example, Bet v 1 (the major allergen of birch pollen) and Mal d 1 (from apple) have 56% amino acid sequence identity and share IgE epitopes (Scheurer et al., 1999). This may account for why many patients allergic to pollen from birch, ragweed, and mugwort have unpleasant reactions when they eat apples (Vanek-Krebitz et al., 1995). On a positive note, similar proteins in other food sources might provide novel therapies (Pons, 2004). For this reason, searches started from SDAP-food entries will indicate any similarities to other allergens in the complete database. We hope to create lists of proteins in SDAP that have significant similarities, not just in their overall sequences (which can be determined by FASTA searches) but also in their experimentally determined IgE epitopes.

To test what PD values could indicate potential cross-reactivities between test peptides, we determined whether sequences with a low PD value (<8 to 10) to known epitopes were also IgE-reactive areas. For example, a PD search, starting from a known epitope of the peanut allergen Ara h 3 (Table 3) rapidly finds a known epitope sequence in Gly m glycinin G1 (PD value, 5.46) (Beardslee et al., 2000) and a similar sequence in Gly m glycinin G2 (Helm et al., 2000). Another known IgE epitope for the tropomyosin allergen of shrimp,

Pen a 1 (Ayuso et al., 2002), is also detected within the first 50 similar finds from SDAP. The Pen a 1 epitope sequence is identical in tropomyosins that are classified as allergenic from many other food sources. Other results with the peanut allergen epitopes (Schein et al., 2005) have identified epitopes with similar IgE reactivity and predicted structure (see below) for PD scores as high as 9.5 to 10.

We should at this point emphasize that the PD search is a computational way to define the sequence relationship between known IgE epitopes and other sequences in allergenic proteins. The correlation of PD values to meaningful IgE cross-reactivity and eventually to clinically relevant ones is ongoing. However, initial tests indicate that this is a rapid way to quantify local similarities in known allergens. To have more meaning, it is worthwhile to combine PD value search values with other descriptors.

The structure of epitopes and their location on the protein surface (solvent exposure) are other possible factors determining whether a given sequence will bind IgE or not. While structure is considered very important for airway allergens, there is still some discussion of whether it is a primary consideration in food allergens, which may be digested in the alimentary tract before presentation to immune regulator cells. There are three pieces of information that indicate structure must also play a role in food allergens. First, major IgE epitopes such as those in Ara h 1, where we have a dependable model, have considerable surface exposure (Fig. 3). Second, while sensitization may be a slow process, allergen-triggering reactions are often extremely fast. Anaphylaxis can occur rapidly after consumption of peanuts (Bock, 2001), for example. This suggests that the whole allergen, and not digested fragments, effectively induces the response (we note, however, that cooking and other preparation methods that encourage protein denaturation may affect the speed with which internal stretches of the protein are exposed to the mucosal membranes). Third, alteration of key amino acids at different points in the sequence of known epitopes completely eliminates IgE binding. If the binding site were primarily linear, as with T-cell epitope binding, one would expect more cooperation between individual side chains in establishing binding. It is thus obvious that certain key amino acids play more of a role in establishing an epitope, presumably by affecting the folding of the segment. We assume that food allergens have specific surface and physicochemical properties that enable them to first induce an IgE immune response and then react with IgE to effectively trigger cellular responses (Gounni et al., 1994; Minshall et al., 2000; Pawankar and Ra, 1998; Toru et al., 1998).

Predicting cross-reactivity based on these data is of course problematic. For example, soybean glycinin G1 shares IgE epitopes with Ara h 3, but the cross-reactivity is not clinically significant (Beardslee et al., 2000). True cross-reactivity probably depends on having several such sequences in the same allergenic protein (Schein et al., 2005).

Many questions about the nature of the IgE epitopes of allergens remain to be answered. Why, for example, do some individuals cross-react to homologous proteins in peanuts and tree nuts, while other individuals with strong allergies only react to one or another of the homologous proteins but not to others (Teuber and Beyer, 2004)? While single amino acid differences may be quite important in individual reactivity, a 3D view of the identified IgE-binding sites can provide missing information about the possible relationships between structure and sequence. If IgE-binding sequences of related proteins have similar properties, the proposed methods that combine PD values with structural clues will have predictive ability, if properly calibrated.

SDAP Can Aid in Determining Names for Newly Identified Allergens

Names of allergens are only official after approval by the IUIS. Submitted allergenic proteins are named by a complicated process, agreed to by the member societies. Allergens are named by abbreviating the Latin name of the species from which they were isolated (e.g., *Cryptomeria japonica* becomes Cry j), followed by a number that indicates the order in which they were identified (Cry j 1, a vicilin related to Jun a 1 from *Juniperus ashei* and similar allergens from other members of the Taxaceae). After the original rounds of naming, the IUIS nomenclature committee has tried, when possible, to maintain a structural or functional relationship across related taxa in the allergen numbering system. In an ideal case, the number would also be consistent with the protein class of the allergen. Thus, the pathogenesis-related 5 (PR5) allergens in cypress pollens, regardless of species, would be number 3 (Cry j 3, Jun a 3). However, the numbering is not always consistent, and the PR5 allergens from apple, cherry, and bell pepper are Mal d 2, Pru av 2, and Cap a 1, respectively. SDAP can aid at this stage in the nomenclature process by providing enough information to determine rapidly what other related allergens have similar sequences, and thus should have the same number,

Routine use of SDAP can prevent certain problems, such as those that may crop up when discoverers of a new allergen give it a name based on their understanding of how many other allergens have been previously isolated from the given biological source. Although this name may be changed by the IUIS, once a protein has been named in the literature, it is often difficult to obtain wide acceptance of a different designation (Schein, 2002). For example, an allergen identified in *Juniperus oxycedrus* was originally referred to as Jun o 2. When it was revealed that it was a different protein from the other cypress allergens named as type II, its name was changed officially to Jun o 4 (Weber, 2003). Still, most databases, including PIR, GenBank, and Swissprot, continue to identify this protein as Jun o 2.

Using SDAP to classify an allergen initially could aid in preventing such confusing situations. A FASTA search of SDAP (Table 4) rapidly clarifies sequence relationships between Jun o 2 and related pollen allergens. In SDAP, the protein is listed accurately, and the user is directed to the correct file name when Jun o 2 is typed in.

SDAP can help in classifying allergenic proteins accurately to reflect characteristics of their sequences. The other feature of allergen nomenclature illustrated by Table 4 is that allergens with very closely related sequences, such as those in the bottom half of the table, can have widely differing numbers. During our categorization of the SDAP entries, we noted that there was a lack of clarity in the nomenclature of food allergens that are generally categorized as seed storage proteins or albumins. Depending on their degree of identity to other proteins may categorize them as vicilins, proglycinins, 7S albumin, etc., which in turn can determine the name of the allergen. As it is unlikely that a radically different nomenclature scheme will be introduced anytime soon, database searches like this one will provide the best way to truly indicate which allergens are most related to one another.

Grouping Allergenic Proteins According to Major Sequence Families

As another aid to identifying potentially cross-reactive allergens, one can also identify proteins in SDAP that are significantly similar to one another according to their Pfam or enzyme classification. Pfam is a list of multiple-sequence alignments of related protein domains, classified in two ways. The Pfam-A database lists protein families that are grouped

Table 4. Results of two automatic FASTA searches in SDAP[a]

No.	Allergen	Sequence	Sequence length	Bit score	E score
1	Jun a 2	CAC05582	507	794.0	0.0e + 00
2	Cry j 2	P43212	514	579.2	9.5e – 167
4	Phl p 13	CAB42886	394	198.8	2.4e – 52
1	Jun o 2	O64943	165	229.0	2.7e – 62
2	Ole e 8	Q9M7R0	171	89.8	2.2e – 20
4	Syr v 3	P58171	81	62.5	1.7e – 12
5	Bra n 2	BAA09633	82	61.3	4.0e – 12
7	Bra r 2	Q39406	83	61.3	4.0e – 12
8	Aln g 4	O81701	85	61.3	4.1e – 12
9	Bet v 4	Q39419	85	60.9	5.4e – 12
11	Ole e 3	O81092	84	59.6	1.3e – 11
12	Phl p 7	O82040	78	58.8	2.2e – 11
15	Bra n 1	Q42470	79	52.8	1.4e – 09
16	Bra r 1	Q42470	79	52.8	1.4e – 09
17	Bet v 3	P43187	205	35.8	4.9e – 04
18	Gad c 1	P02622	113	34.9	4.9e – 04
19	Sal s 1	Q91482	109	33.6	1.1e – 03
20	Sco j 1	P59747	109	31.3	5.5e – 03

[a] The results of the searches rapidly reveal that while the Jun a 2 and Cry j 2 allergens are very close in sequence (i.e., have a very low E value) (top three rows), the allergen previously referred to as Jun o 2 (now called Jun o 4) is more closely related to other pollen allergens (bottom rows).

by their common function and sequence, using expert knowledge and experimental data. Most SDAP entries have now been classified to one of these groupings. Easy access to this Pfam classification for any allergen can be accessed from the List SDAP menu item on the SDAP Web site.

Is Allergenicity Related to a Catalytic or Specific Binding Activity of the Whole Protein?

Many allergenic proteins have enzymatic activities or closely resemble proteins with known catalytic function (Tables 5 and 6). While it has been suggested that this activity may be associated with their ability to induce an allergic response, this is a hard point to prove. For example, removing catalytic residues might leave vestiges of the active site that would be adequate for substrate binding, if not catalysis. The cross-referenced tables in SDAP allow one to rapidly summarize the previously attributed activities of allergenic proteins, based on their similarities to known proteins (Table 3). The table makes clear that the possible activities of allergens are even more diverse than previously reported and range from hydrolytic enzymes to proteins to those that serve a structural or storage role. Some allergens are toxic components of insect stings, such as hyaluronidase and phospholipases, while others, the pathogenesis-related proteins, are produced when plants are damaged or infected.

The data shown in Tables 5 and 6 indicate that many allergenic proteins serve some identifiable function in their host cells. However, the wide variety of possible functions suggests that the most important aspects that enable allergenic proteins to induce an allergic

Table 5. Allergenic molecules from similar enzymatic classes occur in many different sources

Class of allergen	No. known[a]	Example
Pectate lyase	4	Jun a 1
Cyclophilin	2	Bet v 5
Albumin	26	Fel d 2
2S albumin	9	Bra j 1
Vicilin	7	Ara h 1
Hyaluronidase	5	Api m 2
Antigen 5	17	Ves f 5
Lipid transfer protein	17	Zea M 14
Lipocalin	4	Bos d 2
Profilin	26	Bet v 2
Pathogenesis-related PR3	3	Hevein and prohevein
Pathogenesis-related PR5	4	Jun a 3
Pathogenesis-related PR10	13	Bet v 1
Pathogenesis-related PR14	5	Mal d 3
Tropomyosin	20	Der p 10
Glycinin	4	Ara h 3
Phospholipase	9	Api m 1

[a] From SDAP tables.

Table 6. SDAP food allergen classification according to protein type (from the IUIS database)

Protein	Allergen	Protein	Allergen
α-Amylase	Hor v 16	Isoflavone reductase	Pyr c 5
β-Amylase	Hor v 17	Lysozyme	Gal d 4
α-Lactalbumin	Bos d 4	Ovalbumin	Gal d 2
β-Lactoglobulin	Bos d 5	Ovomucoid	Gal d 1
2S albumin	Ber e 1, Bra j 1, Bra n 1, Jug r 1, Ric c 1, Ses i 1, Sin a 1	Lipid transfer protein	Jug r 3, Mal d 3, Pru ar 3, Pru av 3, Pru d 3, Pru p 3, Vit v 1, Zea m 14
Agglutinin	Tri a 18	Parvalbumin	Sal s 1
Caseins	Bos d 8	Patatin	Sola t 1
Conalbumin	Gal d 3	Prohevein	Bra r 2
Conglutin	Ara h 2, Ara h 6, Ara h 7	Secalin	Sec c 20
Cysteine	Act c 1	Profilin	Api g 4, Ara h 5, Gly m 3, Pru av 4, Pyr c 4
Protease			
Endochitinase	Pers a 1	Serum albumin	Bos d 6, Gal d 5
Gliadin	Tri a 19	Thaumatin	Mal d 2, Pru av 2
Glycinin	Ara h 3, Ara h 4	Tropomyosin	Met e 1, Pen a 1, Pen i 1, Tod p 1
Immunoglobulin	Bos d 7	Vicilin	Ara h 1, Jug n 2, Jug r 2

response must be related to their sequence or structure. This is also clear from the large number of allergens that have structural rather than catalytic roles in the cell. Apparently inert storage proteins, including vicilins, albumins, and glycinins such as Ara h 1-3, are some of the most potent allergens known in terms of their ability to induce a deadly anaphylactic shock. These facts imply that enzymatic function can play only a peripheral role in the ability of a given protein to provoke an allergic response. This again suggests that peripheral surface features, rather than conserved functional elements, are most likely to contribute to a protein's allergenicity. On one hand, this is a very fortunate result, as it suggests that we will not need to affect the functionality of proteins to alter their allergenicity (Akdis and Blaser, 2000). On the other hand, it suggests that the proposed similarities in IgE-binding sites will be subtle and probably undetectable by simple sequence comparison methods.

Using SDAP To Define the Allergenic Potential of Novel Proteins

For research purposes, we define an allergen in physical tests as a protein that reacts with IgE in sera from atopic individuals and induces a reaction in skin testing (Mari et al., 1996) or anaphylaxis in an animal model and/or stimulates basophilic degranulation by a variety of in vitro assays (Bond et al., 1991; Eigenmann et al., 1996; Ford et al., 1991; Moreno et al., 1995; Schramm et al., 1997; Shin et al., 1998; Spangfort et al., 1999; Sparholt et al., 1997). Some general characteristics of food allergens include size, relative abundance in the environment, and persistence in tissue (i.e., resistance to degradation). Recent concern over the possibility that genetically modified StarLink corn might be allergenic (FIFRA [Federal Insecticide, Fungicide, and Rodenticide Act] Scientific Advisory Panel, 2000) pointed out the limitations of these criteria in assessing the allergenic potential of new food products. An expert panel judged the Cry9C protein to have "moderate capacity to be an allergen," due to its size (between 10 and 70 kDa), resistance to acid and protease digestion, and limited results in animal model.

On the surface, prescreening novel proteins by sequence comparison with known allergens is straightforward, as the most significant clinically observable cross-reactivities are with allergens with >50% sequence identity (see Chapter 9 for more discussion of this point). There are several problems, the first of which is that all allergenic proteins should be identified and classified. This leads to obvious clinical questions, such as "how many patients should react to a protein before it is considered an allergen?" (a ranking system for risk assessment) and "what tests will be used to determine IgE reactivity?" Even after this is done, proteins are classified as allergens based on their ability to trigger responses in patients. Allergens may just be more-potent forms of other proteins with similar surface areas that may have been the true sensitizing antigens during development of the disease. Thus, the permissible degree of similarity of novel proteins to known triggers is not clear. The problem is made more difficult by the fact that some potent allergens can be rendered nonallergenic by selected point mutations (de Leon et al., 2003; Ferreira et al., 1998; Rabjohn et al., 2002; Scheurer et al., 1999); highly similar proteins, such as the glycinins of soybean and peanut (62% identity), provoke quite different degrees of response (Beardslee et al., 2000). It is clear that neither FASTA nor methods available on SDAP are going to accurately determine whether a protein is an allergen or is cross-reactive. Instead, either method provides some estimation that a protein may be cross-reactive. But in both cases, the decision points or criteria are estimated from additional evidence. In the final analysis, the predictions by either method must be tested experimentally, i.e., through specific testing with allergic subjects and controls by either in vitro or in vivo methods.

For these reasons, the methods for sequence comparison incorporated in SDAP were designed to permit a decision tree approach to assessing the allergenicity of novel proteins (Ivanciuc et al., 2002), based on that proposed by the International Life Sciences Institute/International Food Biotechnology Council (ILSI/IFBC) (Metcalfe et al., 1996). These guidelines have been adopted by the Food and Agriculture Organization and the World Health Organization (World Health Organization, 2000) and have been recently revised (World Health Organization, 2001). These say that any protein having an 80-amino-acid window of >35% identity to a known allergen or a stretch of six to eight residues that are identical, as determined by a FASTA search, might cross-react with the allergen. The allergens listed in SDAP were assembled from major sequence (SwissProt, PIR, and NCBI) and structure (PDB) databases, guided by the list of allergen names from the IUIS Web site. SDAP provides a platform to answer these bioinformatics questions, including the statistical significance of a given match, as discussed above.

Another approach to this problem is to determine whether a new protein contains sequences that are similar to known IgE epitopes. Standards proposed by the ILSI/IFBC require an exact match with a known allergen of at least eight contiguous amino acids (Metcalfe et al., 1996). However, examination of the diversity in sequences of proteins that induce cross-reactivity indicates that the requirement for absolute identity is probably too strenuous. Further, amino acid sequence alone may not be an adequate indicator of structural identity, since B-cell epitopes often require distinct conformations (Collins et al., 1996; Colombo et al., 1998; Donovan et al., 1994; van Neerven et al., 1998). We thus recommend an approach that combines sequence with structure and surface exposure, using the other tools in SDAP described in "An Introduction to SDAP and Its Tools," above.

For example, as the IgE epitopes have been determined for one PR5 allergen, we can take a sequence-structural approach to suggesting the location of epitopes in related allergens from food. Epitopes were defined for the Jun a 3 protein of mountain cedar pollen (Midoro-Horiuti et al., 2001b; Soman et al., 2000). Three food proteins, Mal d 2 from apple, Pru av 2 from cherry, and Cap a 1 from bell pepper, are also related by sequence to PR5 plant proteins that are expressed when plants are injured or infected. However, the epitopes of the three food proteins have not been characterized. The four proteins are all closely related in sequence (ca. 50% identical) to two other PR5 proteins of known structure, the pathogenesis-related protein 5d from tobacco and thaumatin from African berry (Soman et al., 2000). We thus prepared models of the three proteins, based on the tobacco protein 1AUN, and mapped the sequences with low PD values to a known epitope on each model (Fig. 5) (Ivanciuc et al., 2003b). As Fig. 5 shows, the predicted surface exposure for all four sequences is quite high. These data, coupled with the PD results for the four sequences, suggest strongly that this can be a shared epitope in these proteins. The tool allows the formation of rationally designed hypotheses that can then be tested using subjects with appropriate allergies. We are now engaged in such studies, in alliance with groups with access to patient sera.

Creating a Set of Nonallergenic Proteins

While it is important that all allergenic proteins are identified, it is equally important to define proteins that are not allergens. Homologs of the major allergens, such as nut vicilins and those related to Bet v 1, are present in many foods. Avoidance of every one of these proteins would lead to a very restricted diet for allergy sufferers, while even a careful patient might accidentally consume a food containing a potent trigger. We also need to provide guidelines for introducing novel proteins produced through agricultural biotechnology into foods. As more novel proteins are introduced into foods, medications, and

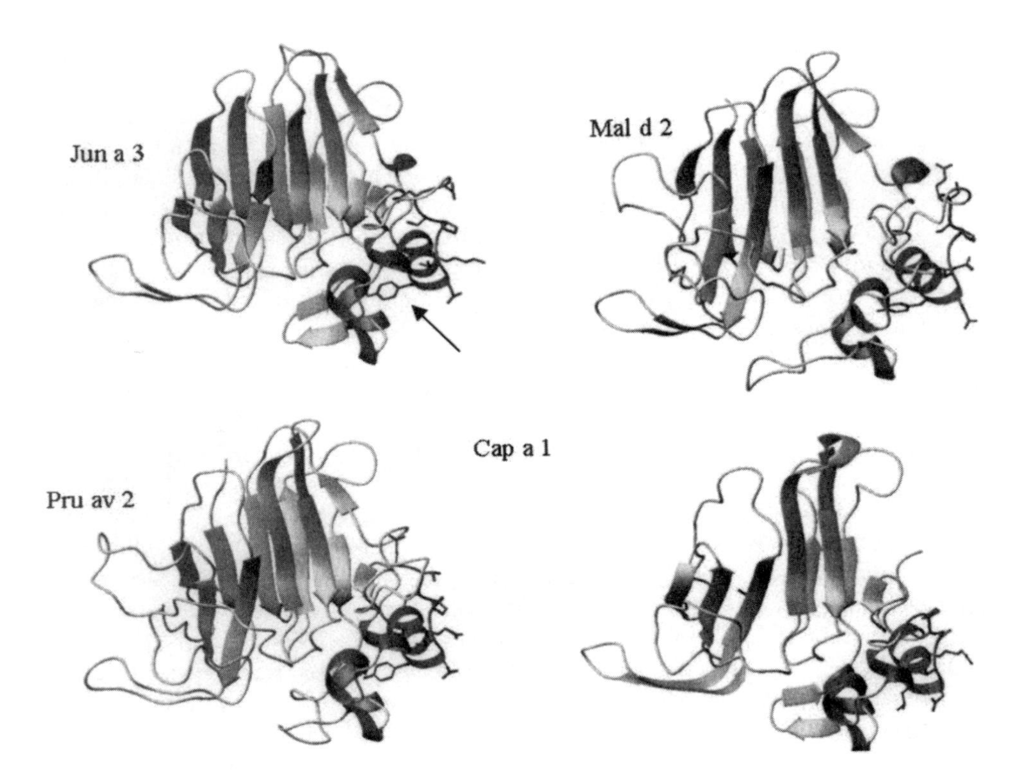

Figure 5. Use of structure, surface location, and sequence conservation as three ways to indicate potential epitopes of homologs of a protein where the IgE epitopes are known. The location of a known epitope of the cedar pollen allergen Jun a 3 is shown to be in a related position to PR5 allergenic proteins from three foods (Mal d 2 from apples, Pru av 2 from cherry, and Cap a 1 from bell pepper). The epitope (arrow), detected by mapping tryptic fragments of Jun a 3, is shown in black on all structures.

other products in our environment, distinguishing allergens from other proteins becomes more pressing (Gendel, 1998). We use FASTA searches primarily to group known allergens, but they may also be useful for discriminating allergens from nonallergens (Soeria-Atmadja et al., 2004).

Although bioinformatic methods have been suggested for discriminating nonallergenic proteins (Li et al., 2004; Stadler and Stadler, 2003), there is currently no set of proteins that have been shown, experimentally, not to bind IgE. We thus plan to construct a list of sequences and structures of proteins that do not react with IgE in conventional assays. These "generally recognized as nonallergenic" proteins (designated GRANAs) will play a major role in defining proteins that can be included in the foods of the 21st century. The list will not be simple to construct. Even essential cellular proteins can be allergens: two of the largest groups of allergens are the profilins and the tropomyosins, structural proteins involved in muscle function. Protein unfolding, degradation, and posttranslational modifications may also introduce potential epitopes. These all indicate how difficult it is for someone to define a protein as having no allergenic potential.

Perhaps defining nonallergens is more a matter of degree and definition. It is possible that nearly any protein of more than about 2,600 Da can be an allergen in the right context and for at least one individual. It seems that most humans are tolerant to self proteins, which would tend to reduce the probability that they would react to highly similar proteins. It is also possible that some of these self-proteins and similar proteins (mammalian albumins and tropomyosins) have physical characteristics such as digestibility and low thermal denaturation properties that make them less likely to act as major food allergens, while proteins from organisms adapted to very different environments are more stable (shrimp, etc.) and more likely to act as allergens.

CONCLUSIONS: FUTURE DEVELOPMENT OF SDAP FOR FOOD ALLERGENS

As discussed, similarities in sequence and structure of allergenic proteins can account for cross-reactivities among allergen sources (Eigenmann et al., 1996; Fedorov et al., 1997; Ipsen and Lowenstein, 1997; Lehrer et al., 2002; Leung et al., 1996; Scheurer et al., 1999; Sparholt et al., 1997), which can complicate the management of severely allergic patients (Bousquet et al., 1993; Lehrer et al., 2002; World Health Organization, 2001). The possibility that a novel protein could induce allergenic proteins is also a problem in introducing new foodstuffs and drugs (Bousquet et al., 1998; Huby et al., 2000; Ivanciuc et al., 2003b; Lehrer et al., 2002; Metcalfe et al., 1996; Platts-Mills et al., 1998; Sampson, 1999a; Sampson, 1999b; Sicherer, 2001; Smal et al., 1988; World Health Organization, 2001). For these reasons, it is vitally important to be able to distinguish allergenic from nonallergenic proteins. This chapter has outlined basic methodology to identify cross-reacting proteins in food sources and predict the potential allergenicity of novel proteins. As shown above in "An Introduction to SDAP and Its Tools," SDAP can be used quickly to gain a good deal of information about food allergens and to identify structural and sequence properties of food proteins that make them particularly effective in triggering a response. Recent identification of the sequence and structure of allergenic proteins from pollen and foods has revealed many similarities (Aalberse, 2000; Breiteneder and Ebner, 2000; Fedorov et al., 1997; Ipsen et al., 1997; Midoro-Horiuti et al., 2001a; Soman et al., 2000; Spangfort et al., 1999), which might offer a structural explanation for their allergenicity and cross-reactivity. SDAP and its tools were designed to help in determining the allergenic potential of a given sequence.

In the last analysis, the primary role of SDAP, rather than listing proteins to avoid, will be to allow us to make more-intelligent choices about which new proteins we can safely include in our foods and how to alter existing proteins to lower the possibility of evoking allergic responses. While many believe that the overriding policy must be to play it safe and reject all genetically modified products, we should recognize that many of these new proteins can enhance the quality of our foods by increasing the nutritional value and/or shelf life. These include tomatoes with improved ripening characteristics that provide enhanced flavor, canola oil enriched with high levels of oleic acid and a monounsaturated fatty acid, golden-rice with enhanced levels of vitamin A, and virus-resistant squash and papaya. Controlling the ripening of produce can eventually lead to reductions in the consumption of insecticides and chemical stabilizers, such as salts, benzoates and nitrates, which can also be health risks. We will thus continue to build on the SDAP database, adding new information regarding sequence, 3D structure, IgE epitopes of food allergens, and improved methods to discriminate allergens from the bulk of edible proteins.

Acknowledgments. This work was supported by grants from the Food and Drug Administration (FD-U-002249-01), an Advanced Technology Program grant, Texas Higher Education Coordinating Board (004952-0036-2003), and the Sealy Center for Structural Biology.

The structural depictions and surface exposure analysis of Fig. 4 were done by Farrar Elfstrom while she was a Bromberg Scholar in this group.

REFERENCES

Aalberse, R. C. 2000. Structural biology of allergens. *J. Allergy Clin. Immunol.* **106:**228–238.

Abagyan, R. A., and S. Batalov. 1997. Do aligned sequences share the same fold? *J. Mol. Biol.* **273:**355–368.

Adel-Patient, K., C. Creminon, D. Boquet, J. M. Wal, and J. M. Chatel. 2001. Genetic immunisation with bovine beta-lactoglobulin cDNA induces a preventive and persistent inhibition of specific anti-BLG IgE response in mice. *Int. Arch. Allergy Immunol.* **126:**59–67.

Akdis, C. A., and K. Blaser. 2000. Regulation of specific immune responses by chemical and structural modifications of allergens. *Int. Arch. Allergy. Immunol.* **121:**261–269.

Asensio, T., J. F. Crespo, R. Sanchez-Monge, G. Lopez-Torrejon, M. L. Somoza, J. Rodriguez, and G. Salcedo. 2004. Novel plant pathogenesis-related protein family involved in food allergy. *J. Allergy Clin. Immunol.* **114:**896–899.

Ayuso, R., S. B. Lehrer, and G. Reese. 2002. Identification of continuous, allergenic regions of the major shrimp allergen Pen a 1 (tropomyosin). *Int. Arch. Allergy Immunol.* **127:**27–37.

Bannon, G., G. Cockrell, C. Connaughton, C. M. West, R. Helm, J. S. Stanley, N. King, P. Rabjohn, H. A. Sampson, and A. W. Burks. 2001. Engineering, characterization and in vitro efficacy of the major peanut allergens for use in immunotherapy. *Int. Arch. Allergy Immunol.* **124:**70–72.

Beardslee, T. A., M. G. Zeece, G. Sarath, and J. P. Markwell. 2000a. Soybean glycinin G1 acidic chain shares IgE epitopes with peanut allergen Ara h 3. *Int. Arch. Allergy Immunol.* **123:**299–307.

Bock, S., A. Munoz-Furlong, and H. A. Sampson. 2001. Fatalities due to anaphylactic reactions to foods. *J. Allergy Clin. Immunol.* **107:**191–193.

Bond, J. F., R. D. Garman, K. M. Keating, T. J. Briner, T. Rafnar, and D. G. Klapper. 1991. Multiple Amb a I allergens demonstrate specific reactivity with IgE and T cells from ragweed-allergic patients. *J. Immunol.* **146:**3380–3385.

Bousquet, J., J. Knani, A. Hejjaoui, R. Ferrando, P. Cour, H. Dhivert, and F.-B. Michel. 1993. Heterogeneity of atopy. 1. Clinical and immunological characteristics of patients allergic to cypress pollen. *Allergy* **48:** 183–188.

Bousquet, J., R. Lockey, H. J. Malling, E. Alvarez-Cuesta, G. W. Canonica, M. D. Chapman, P. J. Creticos, J. M. Dayer, S. R. Durham, P. Demoly, R. J. Goldstein, T. Ishikawa, K. Ito, D. Kraft, P. H. Lambert, H. Lowenstein, U. Muller, P. S. Norman, R. E. Reisman, R. Valenta, E. Valovirta, and H. Yssel. 1998. Allergen immunotherapy: therapeutic vaccines for allergic diseases. *Ann. Allergy Asthma Immunol.* **81:**401–405.

Breiteneder, H., and C. Ebner. 2000. Molecular and biochemical classification of plant-derived food allergens. *J. Allergy Clin. Immunol.* **106:**27-36.

Breiteneder, H., and E. N. C. Mills. 2005. Molecular properties of food allergens. *J. Allergy Clin. Immunol.* **115:**14–23.

Brusic, V., and N. Petrovsky. 2003. Bioinformatics for characterisation of allergens, allergenicity and allergic crossreactivity. *Trends Immunol.* **24:**225–228.

Brusic, V., M. Millot, N. Petrovsky, S. M. Gendel, O. Gigonzac, and S. J. Stelman. 2003. Allergen databases. *Allergy* **58:**1093–1100.

Burks, A. W., D. Shin, G. Cockrell, J. S. Stanley, R. M. Helm, and G. A. Bannon. 1997. Mapping and mutational analysis of the IgE-binding epitopes on Ara h 1, a legume vicilin protein and a major allergen in peanut hypersensitivity. *Eur. J. Biochem.* **245:**334–339.

Cocco, R. R., K. M. Jarvinen, H. A. Sampson, and K. Beyer. 2003. Mutational analysis of major, sequential IgE-binding epitopes in α (s1)-casein, a major cow's milk allergen. *J. Allergy Clin. Immunol.* **112:**433–437.

Codex Alimentarius Commission. 2003. *Report of the Fourth Session of the Codex Ad Hoc Intergovernmental Task Force on Foods Derived from Biotechnology.* ALINORM 03/34A. World Health Organization, Yokohama, Japan. ftp://ftp.fao.org/docrep/fao/meeting/006/y9220e.pdf.

Collins, S. P., G. Ball, E. Vonarx, C. Hosking, M. Shelton, D. Hill, and M. E. H. Howden. 1996. Absence of continuous epitopes in the house dust mite major allergens Der p I from *Dermatophagoides pteronyssinus* and Der f I from *Dermatophagoides farinae*. *Clin. Exp. Allergy* **26:**36–42.

Colombo, P., D. Kennedy, T. Ramsdale, M. A. Costa, G. Duro, V. Izzo, S. Salvadori, R. Guerrini, R. Cocchiara, M. G. Mirisola, S. Wood, and D. Geraci. 1998. Identification of an immunodominant IgE epitope of the Parietaria judaica major allergen. *J. Immunol.* **160:**2780–2785.

Conte, L. L., B. Ailey, T. J. P. Hubbard, S. E. Brenner, A. G. Murzin, and C. Chothia. 2000. SCOP: a structural classification of proteins database. *Nucleic Acids Res.* **28:**257–259.

de Leon, M. P., I. N. Glaspole, A. C. Drew, J. M. Rolland, R. E. O'Hehir, and C. Suphioglu. 2003. Immunological analysis of allergenic cross-reactivity between peanut and tree nuts. *Clin. Exp. Allergy* **33:**1273–1280.

Donovan, G. R., M. D. Street, B. A. Baldo, D. Alewood, P. Alewood, and S. Sutherland. 1994. Identification of an IgE-binding determinant of the major allergen Myr p I from the venom of the Australian jumper and Myrmecia pilosula. *Biochim. Biophys. Acta* **1204:**48–52.

Ehn, B. M., B. Ekstrand, U. Bengtsson, and S. Ahlstedt. 2004. Modification of IgE binding during heat processing of the cow's milk allergen beta-lactoglobulin. *J. Agric. Food Chem.* **52:**1398–1403.

Eigenmann, P. A., A. W. Burks, G. A. Bannon, and H. A. Sampson. 1996. Identification of unique peanut and soy allergens in sera adsorbed with cross-reacting antibodies. *J. Allergy Clin. Immunol.* **98:**969–978.

Elbez, M., C. Kevers, S. Hamdi, M. Rideau, and G. Petit-Paly. 2002. The plant pathogenesis-related PR-10 proteins. *Acta Botanica Gallica* **149:**415-444.

Elsayed, S., D. J. Hill, and T. V. Do. 2004. Evaluation of the allergenicity and antigenicity of bovine-milk alpha s1-casein using extensively purified synthetic peptides. *Scand. J. Immunol.* **60:**486–493.

European Food Safety Authority. 2004. *Guidance Document of the GMO Panel for the Risk Assessment of Genetically Modified Plants and Derived Food and Feed.* http://www.efsa.eu.int/science/gmo/gmo_guidance/660_en.html.

Fedorov, A. A., T. Ball, N. M. Mahoney, R. Valenta, and S. C. Almo. 1997. The molecular basis for allergen cross-reactivity: crystal structure and IgE epitope mapping of birch pollen profilin. *Structure* **5:**33–45.

Ferreira, F., C. Ebner, B. Kramer, G. Casari, P. Briza, R. Grimm, B. Jahn-Schmid, H. Breiteneder, D. Kraft, M. Breitenbach, H. J. Rheinberger, and O. Scheiner. 1998. Modulation of IgE reactivity of allergens by site-directed mutagenesis: potential use of hypoallergenic variants for immunotherapy. *FASEB J.* **12:**231–242.

FIFRA Scientific Advisory Panel. 2000. *Meeting Report: A Set of Scientific Issues Being Considered by the Environmental Protection Agency Regarding: Assessment of Additional Scientific Information Concerning StarLink Corn.* SAP report no. 2000-06. http://www.epa.gov/scipoly/sap/2000/november/one.pdf.

Fleischer, D., M. Conover-Walker, L. Christie, A. Burks, and R. Wood. 2004. Peanut allergy: recurrence and its management. *J. Allergy Clin. Immunol.* **114:**1195–1201.

Ford, S., B. Baldo, R. Panzani, and D. Bass. 1991. Cypress (Cupressus sempervirens) pollen allergens: identification by protein blotting and improved detection of specific IgE antibodies. *Int. Arch. Allergy Appl. Immunol.* **95:**178–183.

Fremont, S., G. Kanny, J. P. Nicolas, and D. A. MoneretVautrin. 1997. Prevalence of lysozyme sensitization in an egg-allergic population. *Allergy* **52:**224–228.

Gendel, S. M. 1998. Sequence databases for assessing the potential allergenicity of proteins used in transgenic foods. *Adv. Food Nutr. Res.* **42:**63-92.

Gendel, S. M. 2004. Bioinformatics and food allergens. *J. AOAC Int.* **87:**1417–1422.

Gibrat, J. F., T. Madej, and S. H. Bryant. 1996. Surprising similarities in structure comparison. *Curr. Opin. Struct. Biol.* **6:**377–385.

Gilbert, D., D. Westhead, N. Nagano, and J. Thornton. 1999. Motif-based searching in TOPS protein topology databases. *Bioinformatics* **15:**317–326.

Glaspole, I. N., M. P. de Leon, J. M. Rolland, and R. E. O'Hehir. 2005. Characterization of the T-cell epitopes of a major peanut allergen, Ara h 2. *Allergy* **60:**35–40.

Gounni, A. S., B. Lamkhioued, E. Delaporte, A. Dubost, J. P. Kinet, A. Capron, and M. Capron. 1994. The high-affinity IgE receptor on eosinophils: from allergy to parasites or from parasites to allergy? *J. Allergy Clin. Immunol.* **94:**1214–1216.

Greene, W. K., and W. R. Thomas. 1992. IgE binding structures of the major house dust mite allergen Der p I. *Mol. Immunol.* **29:**257–262.

Helm, R. M., G. Cockrell, C. Connaughton, H. A. Sampson, G. A. Bannon, V. Beilinson, D. Livingstone, N. C. Nielsen, and A. W. Burks. 2000. A soybean G2 glycinin allergen - 1. Identification and characterization. *Int. Arch. Allergy Immunol.* **123:**205–212.

Hoffmann-Sommergruber, K. 2002. Pathogenesis-related (PR)-proteins identified as allergens. *Biochem. Soc. Trans.* **30:**930–935.

Holm, L., and C. Sander. 1996. Mapping the protein universe. *Science* **273:**595–602.

Huby, R. D. J., R. J. Dearman, and I. Kimber. 2000. Why are some proteins allergens? *Toxicol. Sci.* **55:**235–246.

Ipsen, H., and H. Lowenstein. 1997. Basic features of crossreactivity in tree and grass pollen allergy. *Clin. Rev. Allergy Immunol.* **15:**389–396.

Ivanciuc, O., C. H. Schein, and W. Braun. 2002. Data mining of sequences and 3D structures of allergenic proteins. *Bioinformatics* **18:**1358–1364.

Ivanciuc, O., C. H. Schein, and W. Braun. 2003a. SDAP: database and computational tools for allergenic proteins. *Nucleic Acids Res.* **31:**359–362.

Ivanciuc, O., V. Mathura, T. Midoro-Horiuti, W. Braun, R. M. Goldblum, and C. H. Schein. 2003b. Detecting potential IgE-reactive sites on food proteins using a sequence and structure database, SDAP-Food. *J. Agric. Food Chem.* **51:**4830–4837.

Ivanciuc, O., N. Oezguen, V. Mathura, C. H. Schein, Y. Xu, and W. Braun. 2004. Using property based sequence motifs and 3D modeling to determine structure and functional regions in CASP5 targets. *Curr. Med. Chem.* **11:**583–593.

Jarvinen, K. M., P. Chatchatee, L. Bardina, K. Beyer, and H. A. Sampson. 2001. IgE and IgG binding epitopes on alpha-lactalbumin and beta-lactoglobulin in cow's milk allergy. *Int. Arch. Allergy Immunol.* **126:** 111–118.

Jenkins, J., S. Griffiths-Jones, P. Shewry, H. Breiteneder, and E. N. Clare Mills. 2005. Structural relatedness of plant food allergens with specific reference to cross-reactive allergens: an in silico analysis. *J. Allergy Clin. Immunol.* **115:**163–170.

King, T. P., D. Hoffman, H. Lowenstein, D. G. Marsh, T. A. E. Platts-Mills, W. Thomas, et al. 1994. Allergen nomenclature. *Int. Arch. Allergy Immunol.* **105:**224–233.

Lehrer, S. I., R. Ayuso, and G. Reese. 2002. Current understanding of food allergens. *Ann. N. Y. Acad. Sci.* **964:**69–85.

Le Mao, J., A. Weyer, J. C. Mazie, S. Rouyre, F. Marchand, A. Le Gall, and B. David. 1992. Identification of allergenic epitopes on Der f I, a major allergen of Dermatophagoides farinae, using monoclonal antibodies. *Mol. Immunol.* **29:**205–211.

Leung, P. S. C., K. H. Chu, W. K. Chow, A. Ansari, C. I. Bandea, H. S. Kwan, S. M. Nagy, and M. E. Gershwin. 1994. Cloning, expression, and primary structure of Metapenaeus ensis tropomyosin, the major heat-stable shrimp allergen. *J. Allergy Clin. Immunol.* **94:**882–890.

Leung, P. S. C., W. K. Chow, S. Duffey, H. S. Kwan, M. E. Gershwin, and K. H. Chu. 1996. IgE reactivity against a cross-reactive allergen in crustacea and mollusca: evidence for tropomyosin as the common allergen. *J. Allergy Clin. Immunol.* **98:**954–961.

Leung, P. S. C., Y. C. Chen, M. E. Gershwin, H. Wong, H. S. Kwan, and K. H. Chu. 1998a. Identification and molecular characterization of Charybdis feriatus tropomyosin, the major crab allergen. *J. Allergy Clin. Immunol.* **102:**847–852.

Leung, P. S. C., Y. C. Chen, D. L. Mykles, W. K. Chow, C. P. Li, and K. H. Chu. 1998b. Molecular identification of the lobster muscle protein tropomyosin as a seafood allergen. *Mol. Mar. Biol. Biotechnol.***7:**12–20.

Li, K. B., P. Issac, and A. Krishnan. 2004. Predicting allergenic proteins using wavelet transform. *Bioinformatics* **20:**2572–2578.

Li, X., K. Srivastava, J. W. Huleatt, K. Bottomly, A. W. Burks, and H. A. Sampson. 2003. Engineered recombinant peanut protein and heat-killed Listeria monocytogenes coadministration protects against peanut-induced anaphylaxis in a murine model. *J. Immunol.* **170:**3289–3295.

Lopez-Torrejon, G., G. Salcedo, M. Martin-Esteban, A. Diaz-Perales, C. Y. Pascual, and R. Sanchez-Monge. 2003. Len c 1, a major allergen and vicilin from lentil seeds: protein isolation and cDNA cloning. *J. Allergy Clin. Immunol.* **112:**1208–1215.

Mari, A., G. Di Felice, C. Afferni, B. Barletta, R. Tinghino, F. Sallusto, and C. Pini. 1996. Assessment of skin prick test and serum specific IgE detection in the diagnosis of Cupressaceae pollinosis. *J. Allergy Clin. Immunol.* **98:**21–31.

Metcalfe, D. D., J. D. Astwood, R. Townsend, H. A. Sampson, S. L. Taylor, and R. L. Fuchs. 1996. Assessment of allergenic potential of foods derived from genetically engineered crop plants. *Crit. Rev. Food Sci. Nutr.* **36:**S165–S186.

Midoro-Horiuti, T., E. G. Brooks, and R. M. Goldblum. 2001a. Pathogenesis-related proteins of plants as allergens. *Ann. Allergy Asthma Immunol.* **87:**261–271.

Midoro-Horiuti, T., R. M. Goldblum, and E. G. Brooks. 2001b. Identification of mutations in the genes for the pollen allergens of eastern red cedar (Juniperus virginiana). *Clin. Exp. Allergy* **31:**771–778.

Midoro-Horiuti, T., V. S. Mathura, C. H. Schein, W. Braun, S. Yu, M. Watanabe, J. C. Lee, E. G. Brooks, and R. M. Goldblum. 2003. Major linear IgE epitopes of mountain cedar pollen allergen Jun a 1 map to the pectate lyase catalytic site. *Mol. Immunol.* **40:**555–562.

Midoro-Horiuti, T., C. Schein, V. Mathura, W. Braun, E. Czerwinski, A. Togawa, Y. Kondo, T. Oka, M. Watanabe, and R. Goldblum. 2006. Structural basis for epitope sharing between group 1 allergens of cedar pollen. *Mol. Immunol.* **43:**509–518.

Mills, E. N., J. Jenkins, N. Marigheto, P. S. Belton, A. P. Gunning, and V. J. Morris. 2002. Allergens of the cupin superfamily. *Biochem. Soc. Trans.* **30:**925–929.

Mine, Y., and J. W. Zhang. 2002. Identification and fine mapping of IgG and IgE epitopes in ovomucoid. *Biochem. Biophys. Res. Commun.* **292:**1070–1074.

Mine, Y., and P. Rupa. 2004. Immunological and biochemical properties of egg allergens. *Worlds Poult. Sci. J.***60:**321–330.

Mine, Y., E. Sasaki, and J. W. Zhang. 2003. Reduction of antigenicity and allergenicity of genetically modified egg white allergen, ovomucoid third domain. *Biochem. Biophys. Res. Commun.* **302:**133–137.

Minshall, E., J. Chakir, M. Laviolette, S. Molet, Z. Zhu, R. Olivenstein, J. A. Elias, and O. Hamid. 2000. IL-11 expression is increased in severe asthma: association with epithelial cells and eosinophils. *J. Allergy Clin. Immunol.* **105:**232–238.

Mittag, D., J. Akkerdaas, B. K. Ballmer-Weber, L. Vogel, M. Wensing, W. M. Becker, S. J. Koppelman, A. C. Knulst, A. Helbling, S. L. Hefle, R. Van Ree, and S. Vieths. 2004. Ara h 8, a Bet v 1-homologous allergen from peanut, is a major allergen in patients with combined birch pollen and peanut allergy. *J. Allergy Clin. Immunol.* **114:**1410–1417.

Mizumachi, K., and J. Kurisaki. 2003. Localization of T cell epitope regions of chicken ovomucoid recognized by mice. *Biosci. Biotechnol. Biochem.* **67:**712–719.

Moreno, F., M. Blanca, C. Mayorga, S. Terrados, M. Moya, E. Perez, R. Suau, J. M. Vega, J. Garcia, and A. Mirando. 1995. Studies of the specificities of IgE antibodies found in sera from subjects with allergic reactions to penicillins. *Int. Arch. Allergy Immunol.* **108:**74–81.

Moreno, F. J., B. M. Maldonado, N. Wellner, and E. N. C. Mills. 2005a. Thermostability and in vitro digestibility of a purified major allergen 2S albumin (Ses i 1) from white sesame seeds (Sesamum indicum L.). *Biochim. Biophys. Acta* **1752:**142–153.

Moreno, F. J., F. A. Mellon, M. S. J. Wickham, A. R. Bottrill, and E. N. C. Mills. 2005b. Stability of the major allergen Brazil nut 2S albumin (Ber e 1) to physiologically relevant in vitro gastrointestinal digestion. *FEBS J.* **272:**341–352.

Natale, M., C. Bisson, G. Monti, A. Peltran, L. P. Garoffo, S. Valentini, C. Fabris, E. Bertino, A. Coscia, and A. Conti. 2004. Cow's milk allergens identification by two-dimensional immunoblotting and mass spectrometry. *Mol. Nutr. Food Res.* **48:**363–369.

Palomares, O., J. Cuesta-Herranz, R. Rodriiguez, and M. Villalba. 2005. A recombinant precursor of the mustard allergen Sin a 1 retains the biochemical and immunological features of the heterodimeric native protein. *Int. Arch. Allergy Immunol.* **137:**18–26.

Pawankar, R., and C. Ra. 1998. IgE-Fc εRI-mast cell axis in the allergic cycle. *Clin. Exp. Allergy* **28:**6–14.

Pearl, F. M. G., N. Martin, J. E. Bray, D. W. A. Buchan, A. P. Harrison, D. Lee, G. A. Reeves, A. J. Shepherd, I. Sillitoe, A. E. Todd, J. M. Thornton, and C. A. Orengo. 2001. A rapid classification protocol for the CATH Domain Database to support structural genomics. *Nucleic Acids Res.* **29:**223–227.

Pearson, W. 1990. Rapid and sensitive sequence comparison with FASTP and FASTA. *Methods Enzymol.* **183:**63–98.

Platts-Mills, T., G. Mueller, and L. Wheatley. 1998. Future directions for allergen immunotherapy. *J. Allergy Clin. Immunol.* **102:**335–343.

Pons, L., U. Ponnappan, R. A. Hall, P. Simpson, G. Cockrell, C. M. West, H. A. Sampson, R. M. Helm, and A. W. Burks. 2004. Soy immunotherapy for peanut-allergic mice: modulation of the peanut-allergic response. *J. Allergy Clin. Immunol.* **114:**915–921.

Pourpak, Z., A. Mostafaie, Z. Hasan, G. A. Kardar, and M. Mahmoudi. 2004. A laboratory method for purification of major cow's milk allergens. *J. Immunoassay Immunochem.* **25:**385–397.

Rabjohn, P., A. W. Burks, H. A. Sampson, and G. A. Bannon. 1999a. Mutational analysis of the IgE-binding epitopes of the peanut allergen, Ara h 3: a member of the glycinin family of seed-storage proteins. *J. Allergy Clin. Immunol.* **103:**S101.

Rabjohn, P., E. M. Helm, J. S. Stanley, C. M. West, H. A. Sampson, A. W. Burks, and G. A. Bannon. 1999b. Molecular cloning and epitope analysis of the peanut allergen Ara h 3. *J. Clin. Investig.* **103:**535–542.

Rabjohn, P., C. West, C. Connaughton, H. Sampson, R. Helm, A. Burks, and G. Bannon. 2002. Modification of peanut allergen Ara h 3: effects on IgE binding and T cell stimulation. *Int. Arch. Allergy Immunol.* **128:**15–23.

Reese, G., R. Ayuso, S. M. Leong-Kee, M. Plante, and S. B. Lehrer. 2002. Epitope mapping and mutational substitution analysis of the major shrimp allergen Pen a 1 (tropomyosin). *J. Allergy Clin. Immunol.* **109:** S307–S307.

Robotham, J. M., F. Wang, V. Seamon, S. S. Teuber, S. K. Sathe, H. A. Sampson, K. Beyer, M. Seavy, and K. H. Roux. 2005. Ana o 3, an important cashew nut (Anacardium occidentale L.) allergen of the 2S albumin family. *J. Allergy Clin. Immunol.* **115:**1284–1290.

Sampson, H. 2005. Food allergy: when mucosal immunity goes wrong. *J. Allergy Clin. Immunol.* **115:**139–141.

Sampson, H. A. 1999a. Food allergy. Part 2: diagnosis and management. *J. Allergy Clin. Immunol.* **103:** 981–989.

Sampson, H. A. 1999b. Food allergy. Part 1: immunopathogenesis and clinical disorders. *J. Allergy Clin. Immunol.* **103:**717–728.

Samson, K. T. R., F. H. Chen, K. Miura, Y. Odajima, Y. Iikura, M. N. Rivas, K. Minoguchi, and M. Adachi. 2004. IgE binding to raw and boiled shrimp proteins in atopic and nonatopic patients with adverse reactions to shrimp. *Int. Arch. Allergy Immunol.* **133:**225–232.

Schein, C. H. 2002. The shape of the messenger: using protein structural information to design novel cytokine-based therapeutics. *Curr. Pharm. Des.* **8:**213–230.

Schein, C. H., O. Ivanciuc, and W. Braun. 2005. Common physical-chemical properties correlate with similar structure of the IgE epitopes of peanut allergens. *J. Agric. Food Chem.* **53:**8752–8759.

Scheurer, S., D. Y. Son, M. Boehm, F. Karamloo, S. Franke, A. Hoffmann, D. Haustein, and S. Vieths. 1999. Cross-reactivity and epitope analysis of Pru a 1, the major cherry allergen. *Mol. Immunol.* **36:**155–167.

Schramm, G., A. Bufe, A. Petersen, H. Haas, M. Schlaak, and W. M. Becker. 1997. Mapping of IgE-binding epitopes on the recombinant major group I allergen of velvet grass pollen, rHol 1 1. *J. Allergy Clin. Immunol.* **99:**781–787.

Shanti, K. N., B. M. Martin, S. Nagpal, D. D. Metcalfe, and P. V. S. Rao. 1993. Identification of tropomyosin as the major shrimp allergen and characterization of its Ige-binding epitopes. *J. Immunol.* **151:**5354–5363.

Shin, D. S., C. M. Compadre, S. J. Maleki, R. A. Kopper, H. Sampson, S. K. Huang, A. W. Burks, and G. A. Bannon. 1998. Biochemical and structural analysis of the IgE binding sites on Ara h 1, an abundant and highly allergenic peanut protein. *J. Biol. Chem.* **273:**13753–13759.

Shindyalov, I. N., and P. E. Bourne. 1998. Protein structure alignment by incremental combinatorial extension (CE) of the optimal path. *Prot. Eng.* **11:**739–747.

Shreffler, W. G., K. Beyer, T. H. Chu, A. W. Burks, and H. A. Sampson. 2004. Microarray immunoassay: association of clinical history, in vitro IgE function, and heterogeneity of allergenic peanut epitopes. *J. Allergy Clin. Immunol.* **113:**776–782.

Sicherer, S., A. Munoz-Furlong, and H. A. Sampson. 2003. Prevalence of peanut and tree nut allergy in the United States determined by means of a random digit dial telephone survey: a 5-year follow-up study. *J. Allergy Clin. Immunol.* **112:**1203–1207.

Sicherer, S. H. 2001. Clinical implications of cross-reactive food allergens. *J. Allergy Clin. Immunol.* **108:** 881–890.

Smal, M. A., B. A. Baldo, and D. G. Harle. 1988. Drugs as allergens. The molecular basis of IgE binding to trimethoprim. *Allergy* **43:**184–191.

Soeria-Atmadja, D., A. Zorzet, M. G. Gustafsson, and U. Hammerling. 2004. Statistical evaluation of local alignment features predicting allergenicity using supervised classification algorithms. *Int. Arch. Allergy Immunol.* **133:**101–112.

Soman, K. V., T. Midoro-Horiuti, J. C. Ferreon, R. M. Goldblum, E. G. Brooks, A. Kurosky, W. Braun, and C. H. Schein. 2000. Homology modeling and characterization of IgE epitopes of mountain cedar allergen Jun a 3. *Biophys. J.* **79:**1601–1609.

Spangfort, M. D., O. Mirza, J. Holm, J. N. Larsen, H. Ipsen, and H. Lowenstein. 1999. The structure of major birch pollen allergens—epitopes, reactivity and cross-reactivity. *Allergy* **50:**23–26.

Sparholt, S. H., J. N. Larsen, H. Ipsen, C. Schou, and R. J. van Neerven. 1997. Crossreactivity and T-cell epitope specificity of Bet v 1-specific T cells suggest the involvement of multiple isoallergens in sensitization to birch pollen. *Clin. Exp. Allergy* **27:**932–941.

Stadler, M. B., and B. M. Stadler. 2003. Allergenicity prediction by protein sequence. *FASEB J.* **17:**1141–1143.

Swoboda, I., A. Bugajska-Schretter, R. Valenta, and S. Spitzauer. 2002a. Recombinant fish parvalbumins: candidates for diagnosis and treatment of fish allergy. *Allergy* **57:**94–96.

Swoboda, I., A. Bugajska-Schretter, P. Verdino, W. Keller, W. R. Sperr, P. Valent, R. Valenta, and S. Spitzauer. 2002b. Recombinant carp parvalbumin, the major cross-reactive fish allergen: a tool for diagnosis and therapy of fish allergy. *J. Immunol.* **168**:4576–4584.

Teuber, S. S., and K. Beyer. 2004. Peanut, tree nut and seed allergies. *Curr. Opin. Allergy Clin. Immunol.* **4:**201–203.

Toru, H., R. Pawankar, C. Ra, J. Yata, and T. Nakahata. 1998. Human mast cells produce IL-13 by high-affinity IgE receptor cross-linking: enhanced IL-13 production by IL-4-primed human mast cells. *J. Allergy Clin. Immunol.* **102:**491–502.

Tramontano, A., R. Leplae, and V. Morea. 2001. Analysis and assessment of comparative modeling predictions in CASP4. *Proteins* **45:**22–38.

Van Do, T., I. Hordvik, C. Endresen, and S. Elsayed. 2005. Characterization of parvalbumin, the major allergen in Alaska pollack, and comparison with codfish allergen M. *Mol. Immunol.* **42:**345–353.

Vanek-Krebitz, M., K. Hoffmann-Sommergruber, M. L. D. Machado, M. Susani, C. Ebner, D. Kraft, O. Scheiner, and H. Breiteneder. 1995. Cloning and sequencing of Mal d 1, the major allergen from apple (*Malus domestica*), and its immunological relationship to Bet-V-1, the major birch pollen allergen. *Biochem. Biophy. Res. Commun.* **214:**538–551.

van Neerven, R. J., S. H. Sparholt, C. Schou, and J. N. Larsen. 1998. Preserved epitope-specific T cell activation by recombinant Bet v 1-MBP fusion proteins. *Clin. Exp. Allergy* **28:**423–433.

Venkatarajan, M. S., and W. Braun. 2001. New quantitative descriptors of amino acids based on multidimensional scaling of a large number of physical-chemical properties. *J. Mol. Mod.* **7:**445–453.

Viquez, O. M., K. N. Konan, and H. W. Dodo. 2004. Genomic organization of peanut allergen gene, Ara h 3. *Mol. Immunol.* **41:**1235–1240.

Wal, J. M. 2004. Bovine milk allergenicity. *Ann. Allergy Asthma Immunol.* **93:**S2–S11.

Weber, R. W. 2003. Patterns of pollen cross-reactivity. *Curr. Rev. Allergy Clin. Immunol.* **112:**229–239.

Wensing, M., A. C. Knulst, S. Piersma, F. O'Kane, E. F. Knol, and S. J. Koppelman. 2003. Patients with anaphylaxis to pea can have peanut allergy caused by cross-reactive IgE to vicilin (Ara h 1). *J. Allergy Clin. Immunol.* **111:**420–424.

World Health Organization. 2000. *Safety Aspects of Genetically Modified Foods of Plant Origin.* Report of a joint FAO/WHO expert consultation. World Health Organization, Geneva, Switzerland.

World Health Organization. 2001. *Evaluation of Allergenicity of Genetically Modified Foods.* Report of a joint FAO/WHO expert consultation. World Health Organization, Geneva, Switzerland.

Zhu, H., C. H. Schein, and W. Braun. 2000. MASIA: recognition of common patterns and properties in multiple aligned protein sequences. *Bioinformatics* **16:**950–951.

Section V

REDUCING ALLERGENICITY OF THE FOOD SUPPLY

Food Allergy
Edited by S. J. Maleki et al.

Chapter 12

Hypoallergenic Foods beyond Infant Formulas

Peggy Ozias-Akins, Maria Laura Ramos, and Ye Chu

The most common allergenic foods are milk, eggs, shellfish, fish, peanuts, and tree nuts (Chapter 3). Although milk allergy is observed in 2.5% of children, it is much less prevalent in adults (0.3%). On the contrary, only a small percentage of the 0.8% of children who exhibit peanut allergy will outgrow it (Sampson, 2004). The persistence of food allergies into adulthood thus requires lifelong diligence to avoid certain foods, particularly when a severe allergic reaction such as anaphylaxis can be anticipated. Avoidance is complicated, however, by the widespread use of some food products, such as soy protein in processed foods or in facilities where processing streams may overlap. Although soybean is not considered to be one of the most common or most dangerous of allergenic foods, numerous soybean proteins have been identified as having IgE immunoreactivity, the immune system response characteristic of allergenic foods (Table 1; see Chapter 3) (www.allergome.org).

Since the immune system reacts largely to the protein component of allergenic foods, the idea has emerged to eliminate or alter the protein antigens in the consumed or inhaled plant parts. Before such a strategy can be realized, however, one must consider the role of these plant-derived proteins in plant growth and development. This review is written primarily from the perspective of manipulating the plant rather than the human immune system.

ROLE OF PLANT-DERIVED ANTIGENIC PROTEINS IN PLANT GROWTH AND DEVELOPMENT

A thorough review of plant and seed proteins that are allergenic has recently been published (Mills et al., 2004). Food allergens most frequently are storage proteins found in seeds (Boulter and Croy, 1997). The seed comprises (usually) a seed coat, an embryo (an incipient plant), and a reservoir of food for the embryo (endosperm). The endosperm is the predominant site of storage protein accumulation in cereals such as wheat and rice, whereas the endosperm is largely consumed by the developing embryo in other plant families; storage protein subsequently accumulates in the seed leaves (cotyledons). The

Peggy Ozias-Akins, Maria Laura Ramos, and Ye Chu • Department of Horticulture, The University of Georgia Tifton Campus, Tifton, GA 31793–0748.

seeds of many legumes (e.g., peanut, soybean, and pea) consist almost entirely of the two cotyledons, the bulk of the embryo. This pattern of storage protein accumulation in cotyledons also is observed with nuts such as walnuts. The botanical definition of a nut is a hard, dry, indehiscent fruit with one seed (www.mobot.org). The edible portion of a nut is actually the seed of a plant. Although not all of the following are botanically nuts, Brazil nuts, cashews, peanuts (a legume), and walnuts are colloquially referred to as nuts; each accumulates storage proteins, as well as oils, in their cotyledons. Allergenic proteins have been identified in each of these species (Table 1). Many of these allergenic proteins are abundant in seeds and actually fall into the storage protein category.

Shewry and Casey (1999) have identified three broad functional groups of seed proteins: (i) storage proteins, whose primary function is to store reserves of carbon, nitrogen,

Table 1. Allergenic proteins

Protein type and class	Species	Allergen	Accession no.
Storage			
2S albumin	*Anacardium occidentale* (cashew)	Ana o 3	AY081853
	Arachis hypogaea (peanut); groundnut	Ara h 2 (conglutin)	AY117434
	A. hypogaea (peanut); groundnut	Ara h 6 (conglutin)	AF092846
	A. hypogaea (peanut); groundnut	Ara h 7 (conglutin)	AF091737
	Bertholletia excelsa (Brazil nut)	Ber e 1	S14947
	Helianthus annuus (sunflower)	Hel a 2S albumin	X56686
	Sinapis alba (white mustard)	Sin a 1	S01791
7S globulin (vicilin)	*A. occidentale* (cashew)	Ana o 1	
	A. hypogaea (peanut); groundnut	Ara h 1 (conarachin)	L34402
	Corylus avellana (hazelnut)	Cor a 11	
	Glycine max (soybean)	Gly m Bd28K	AB046874
	G. max (soybean)	Gly m 1 (β-conglycinin)	P24337
	Pisum sativum (pea)	Pis s 1	AJ626897
11S globulin[a] (legumin; glycinin)	*A. occidentale* (cashew)	Ana o 2	AF453947
	A. hypogaea (peanut); groundnut	Ara h 3,4 (arachin)	AF093541
	B. excelsa (Brazil nut)	Ber e 2	AY221641
	G. max (soybean)	Gly m Glycinin G1	M36686
Oleosin[b]	*A. hypogaea* (peanut); groundnut	Ara h Oleosin	AF325917
Metabolic			
Profilin[c]	*A. hypogaea* (peanut); groundnut	Ara h 5	AF059619
	G. max (soybean)	Gly m 3	AJ223981
	Hevea brasiliensis (rubber tree)	Hev b 8	Y15042
	Malus domestica (apple)	Mal d 4	AF129428
	Zea mays (corn, maize)	Zea m 12	O22655
Protective			
Agglutinin[d]	*A. hypogaea* (peanut); groundnut	Ara h Agglutinin	S42352
	G. max (soybean)	Gly m Lectin	K00821
	Triticum aestivum (wheat)	Tri a 18	
Protease inhibitor[e]	*A. hypogaea* (peanut); groundnut	Ara h 2 (AAI domain)	AY117434
	A. hypogaea (peanut); groundnut	Ara h 3 (AAI domain)	AF093541
	G. max (soybean)	Gly m TI	AF128268
Protease[f]	*Carica papaya* (papaya)	Car p 3 (papain)	
	G. max (soybean)	Gly m Bd30K (P34)	J05560
	Lolium perenne (ryegrass)	Lol p 11	A54002

(*continued*)

Table 1. *(continued)*

Protein type and class	Species	Allergen	Accession no.
Lipid transfer protein[g]	*A. hypogaea* (peanut); groundnut	Ara h LTP	
	C. avellana (hazelnut)	Cor a 8	AF329829
	Hordeum vulgare (barley)	Hor v LTP	X05168
	Pyrus communis	Pyr c 3	AF221503
	Zea mays (corn, maize)	Zea m 14	P19656
Other pathogenesis[h] related	*A. hypogaea* (peanut); groundnut	Ara h 8	AY328088
	Betula verrucosa (birch)	Bet v 1 (ribonuclease)	AJ002106
	G. max (soybean)	Gly m 4	X60043
	Malus domestica (apple)	Mal d 1	X83672
	Hevea brasiliensis (rubber tree)	Hev b 11 (chitinase)	AJ238579

[a] InterPro accession no. IPR006044; "Plant seed storage proteins, whose principal function appears to be the major nitrogen source for the developing plant, can be classified, on the basis of their structure, into different families. 11-S are non-glycosylated proteins which form hexameric structures [*1, 2*]. Each of the subunits in the hexamer is itself composed of an acidic and a basic chain derived from a single precursor and linked by a disulphide bond." The quotations in this and the following footnotes are from Mulder et al., 2005.

[b] InterPro accession no. IPR000136; "Oleosins are the proteinaceous components of plants' lipid storage bodies called oil bodies. Oil bodies are small droplets (0.2 to 1.5 μm in diameter) containing mostly triacylglycerol that are surrounded by a phospholipid/oleosin annulus. Oleosins may have a structural role in stabilizing the lipid body during dessication of the seed, by preventing coalescence of the oil. They may also provide recognition signals for specific lipase anchorage in lipolysis during seedling growth."

[c] InterPro accession no. IPR005455; "Profilin is a small eukaryotic protein that binds to monomeric actin (G-actin) in a 1:1 ratio, thus preventing the polymerization of actin into filaments (F-actin). It can also in certain circumstance promote actin polymerization."

[d] InterPro accession no. IPR008985; "Lectins and glucanases exhibit the common property of reversibly binding to specific complex carbohydrates. The lectins/glucanases are a diverse group of proteins found in a wide range of species from prokaryotes to humans . . . Members of this family are diverse, and include the lectins: legume lectins, cereal lectins, viral lectins, and animal lectins. Plant lectins function in the storage and transport of carbohydrates in seeds, the binding of nitrogen-fixing bacteria to root hairs, the inhibition of fungal growth or insect feeding, and in hormonally regulated plant growth."

[e] InterPro accession no. IPR003612; "Plant lipid transfer/seed storage/trypsin-alpha amylase inhibitor. This domain is found in several proteins, including plant lipid transfer protein, seed storage protein and trypsin-alpha amylase inhibitor. The domain forms a four-helical bundle with an internal cavity."

[f] InterPro accession no. IPR000668; "The papain family has a wide variety of activities, including broad-range (papain) and narrow-range endo-peptidases, aminopeptidases, dipeptidyl peptidases and enzymes with both exo- and endo-peptidase activity. Members of the papain family are widespread, found in baculovirus, eubacteria, yeast, and practically all protozoa, plants and mammals."

[g] InterPro accession no. IPR000528; "Plant cells contain proteins, called lipid transfer proteins (LTP) which transfer phospholipids, glycolipids, fatty acids and sterols from liposomes or microsomes to mitochondria. These proteins, whose subcellular location is not yet known, could play a major role in membrane biogenesis by conveying phospholipids such as waxes or cutin from their site of biosynthesis to membranes unable to form these lipids."

[h] InterPro accession no. IPR000916; "Bet v 1-type allergens are related to pathogenesis-related proteins."

and sulfur to support germination and early seedling growth; (ii) structural and metabolic proteins that are required for normal growth and development; and (iii) protective proteins that play a role in desiccation tolerance or in resistance to pathogens and insect pests. These three functional categories are not mutually exclusive. For example, note that in Table 1, Ara h 2 and Ara h 3 can be found in the protease inhibitor biological function category, as well as in the storage protein category. Antigenic storage proteins themselves can be split into two broad categories, the 2S albumins and the globulins (7S globulins or vicilins and 11S globulins or legumins) (Shewry and Casey, 1999). The albumins are defined as being water soluble (in the absence or presence of salt) and coagulated by heat. Albumins comprise a significant proportion of the water-soluble fraction, ranging, for example, from 20% in peanut or 30% in Brazil nut to 62% in sunflower (Youle and Huang, 1981).

Major allergens in peanut, Ara h 2, and Brazil nut, Ber e 1, are 2S albumins. In addition to their role as seed storage proteins, 2S albumins have been shown to have secondary activity as enzyme inhibitors (e.g., α-amylase–trypsin), antifungal proteins, sweet proteins (compare mabilinin), and allergens (Shewry and Pandya, 1999). Seed globulins have recently been reviewed (Casey, 1999). Globulins are soluble in a dilute salt solution but not in water. The nomenclature for globulins often is derived from the group of plants where a particular globulin was first described. For example, phaseolin is a 7S globulin from *Phaseolus vulgaris* and cruciferin is a 12S globulin from the family Cruciferae (Brassicaceae). Globulins are multimeric proteins that are encoded by gene families. Both 7S and 11S globulins were derived from a common ancestral protein (Shewry et al., 1995) and are members of the cupin superfamily (Breitender and Radauer, 2004).

Most major allergens, defined as recognition by serum immunoglobulin E (IgE) from >50% of allergic individuals, are seed storage proteins. Some seed storage, structural-metabolic, and protective proteins also can be minor allergens. An example is profilin, the InterPro description of which can be found at http://www.ebi.ac.uk/interpro/IEntry?ac=IPR005455. The immune system can be exposed to profilin through contact (with latex), inhalation (of pollen), or ingestion (of seeds) because profilin is present in virtually all eukaryotic cells. Profilin is a protein catalogued under the biological process of actin polymerization and/or depolymerization (www.geneontology.org). Actin is a component of all eukaryotic cells, playing an essential role in cell divison, shape, and motility. Profilin binds to actin; consequently, like actin, this interacting companion protein also plays an essential role in cell biology. Even though profilins are known allergens, removing all profilin from a cell would have deleterious consequences. Not all proteins within a particular functional category are necessarily allergenic. Functional diversification of multigene families such as profilin has been found in complex organisms such as flowering plants (Kandasamy et al., 2002). Diversification can be due to changes in gene regulation or altered activity of different isoforms (Kovar et al., 2001). Isoform variation could result in altered antigenicity.

Allergens that fall into the protective protein category are numerous. Seventeen families of pathogenesis-related proteins, those proteins that are synthesized in the early events of the plant defense response, have been listed by Mills et al. (2004), but they can be divided into only four broad categories: (i) enzyme inhibition (e.g., protease inhibitors), (ii) hydrolytic enzymes (e.g., chitinases), (iii) membrane interactive (e.g., lipid transfer proteins), and (iv) carbohydrate binding (e.g., agglutinins). Some members of all four of these categories have been determined to be allergenic, which can be found at the Allergome Web site (www.allergome.org). This site is described in more detail in Chapter 2. Nine subfamilies of protease inhibitors (pathogenesis-related family 6 [PR-6]) (Mills et al., 2004) have been described by Valueva and Mosolov (1999), and the majority of these subfamilies contain certain members that are known allergens (www.allergome.org). Considerable effort is being placed on the development of improved plant cultivars that are more resistant to pests and diseases. Given the known allergenicity of many plant-protective proteins, the goal of developing hypoallergenic foods, i.e., reducing or eliminating food allergens, may sometimes be at odds with the goal of enhanced pest resistance. It also becomes clear that there is a certain level of risk in introducing a gene for a known allergen into a crop plant where a similar allergen has not been identified. A well-known example is the transfer of a 2S albumin gene from Brazil nut into sunflower. The 2S albumin in the Brazil nut is known to be allergenic; therefore, it is not surprising that the same protein, when expressed in sunflower, was also allergenic (Nordlee et al., 1996). The strategy to improve the amino acid balance in sunflower by expression of an allergenic 2S albumin was not pursued for commercialization.

METHODS FOR DOWN-REGULATING ALLERGEN GENE EXPRESSION

One must carefully evaluate approaches to eliminate allergenic proteins from a plant, given that many such proteins have a function in the plant or contribute to the balanced nutrition that can be obtained from a plant. Examples of elimination of specific proteins exist where the result was not obviously deleterious. In cases where deleterious effects are observed, a protein replacement strategy could be considered.

Two methods potentially could be used to biologically remove an allergen from a plant (i.e., preharvest, not postharvest). One would be to knockdown gene expression; the second approach would be to knockout gene expression. A knockdown approach is most feasible through the mechanism of RNA silencing, which can function either posttranscriptionally or transcriptionally. Transcriptional gene silencing due to methylation of promoter sequences can in some cases be homology dependent but RNA initiated, with certain elements in common with posttranscriptional gene silencing (Mette et al., 2000). RNA-mediated posttranscriptional gene silencing acts through a similar pathway in both plants and animals. This phenomenon is now more frequently referred to as RNA interference (RNAi) or RNA silencing. A second method to remove an allergen from a plant would be to knock out a gene through mutation or to discover a natural variant of a particular food crop plant that is missing a functional allergen gene. The relative merits of each of these approaches will be considered below.

RNA silencing has recently received extraordinary attention from scientists. Numerous recent reviews focus on RNA silencing in plants (Baulcombe, 2004), utility of the method for genetic improvement of crop plants (Kusaba, 2004), a historical account of RNA silencing from the plant perspective (Matzke and Matzke, 2004), RNA silencing across kingdoms (Meister and Tuschl, 2004; Mello and Conte, 2004; Susi et al., 2004), and RNA silencing in genome regulation (Matzke and Birchler, 2005). Although this phenomenon was first described for plants 15 years ago (Napoli et al., 1990; van der Krol et al., 1990), intensive investigation of the mechanisms involved progressed rapidly once double-stranded RNA (dsRNA) was shown to be the trigger for RNA silencing in *Caenorhabditis elegans* (Fire et al., 1998), and small RNAs were correlated with homology-dependent silencing in plants (Hamilton and Baulcombe, 1999). The similarities in mechanisms among plants, fungi, and animals argue for an ancient process that has been retained and elaborated into slightly different pathways among different organisms. For example, RNA-mediated silencing of retrotransposons, mobile elements within plant genomes that transpose via an RNA intermediate, is now considered to be a major mechanism for protecting plant genomes from rampant invasion by these type 1 mobile elements (Waterhouse et al., 2001). RNA silencing also can allow a plant to escape from viral infection; in mammals, viral suppression through RNAi does not appear to be a widespread defense mechanism (Baulcombe, 2004). In both plants and animals, however, RNA silencing has been shown to be an essential process for regulating certain developmental genes (Carrington and Ambros, 2003).

RNAi is a posttranscriptional control mechanism that is highly conserved among eukaryotes and is thought to have evolved as a defense mechanism, at least in plants, against viral genomes. This mechanism was first discovered in plants but is now being applied to drug design and therapeutics for viral infections and possibly cancer. The mechanisms and applications of RNAi have been summarized in several recent reviews (Baulcombe, 2004, Meister and Tuschl et al., 2004, Susi et al., 2004), and the main events are shown in Fig. 1. All of these functions depend on the presence of dsRNA. dsRNA is recognized, bound, and cleaved by an enzyme (Dicer or Dicer like) with a dsRNA-binding domain and RNAse III-type endonuclease activity. The products of Dicer activity are 21- to 26-nucleotide (26-nt) short interfering RNAs (siRNAs). The siRNAs can potentially originate from any transcript that is recognized by the host cell as aberrant, foreign, or in excess. These transcripts can be derived from endogenous genes, transgenes, or viral genes. These siRNAs form a ribonucleoprotein complex with Argonaut proteins called the RNA-induced silencing complex (RISC) or the RNA-induced transcriptional silencing complex. RISC and the RNA-induced transcriptional silencing complex recruit long single-stranded mRNAs that have homology to the siRNA in the complex and target them for degradation, which is effected by a slicer activity. siRNAs also may pair with their complementary mRNAs and serve as primers for RNA synthesis by RNA-dependent RNA polymerase (RdRP) (at least in plants), or RdRP activity can be primer independent. RdRP is required for transitive RNA silencing, i.e., the production of secondary siRNAs either upstream or downstream of the primary siRNA sequence. dsRNA resulting from RdRP activity can be cleaved by Dicer, and the resulting siRNAs can enter the same RISC-associated cleavage process, perpetuating the sequence-dependent degradation of homologous RNA molecules. Another type of short RNA (miRNA) is associated with regulation of developmental processes and is encoded by specific genes in the eukaryotic host. Target genes of miRNAs in plants have perfectly complementary nucleotide sequences and enter the RISC-associated RNA degradation pathway (Tang, 2005). Since miRNAs have not been shown to be involved in allergen gene expression, they are not considered relevant to this review. siRNAs, however, particularly those resulting from transgenes, are key components to effecting a knockdown of allergen gene expression. A key feature of RNA silencing in plants and worms is the systemic spread of a silencing signal. In plants, siRNAs of different lengths are proposed to participate differentially in short-and long-range signaling through the phloem tissue. Recently, a single-stranded RNA-binding protein that selectively attaches to 25-nt RNAs in the phloem and is an integral part of long-range signaling has been identified (Yoo et al., 2004). Silencing could operate systemically with expression of a silencing gene in one part of the plant and actual silencing of an endogenous gene expressed in a remote plant part.

To design a transgene (the synthetic gene that will be introduced by transformation) that has a high probability of silencing an allergen gene, the gene of interest should be characterized at the nucleotide sequence level from both the genome and the transcriptome. Allergen genes, particularly the globulin seed storage protein genes, may be members of multigene families. Therefore, a single cDNA sequence may not contain enough sequence information for the most effective construct development. Members of multigene families may have diverged significantly in coding sequence and/or in regulatory domains. An example is the conglutin (2S albumin) gene family in peanut. This gene family is represented by a major allergen, Ara h 2, but other members previously identified are

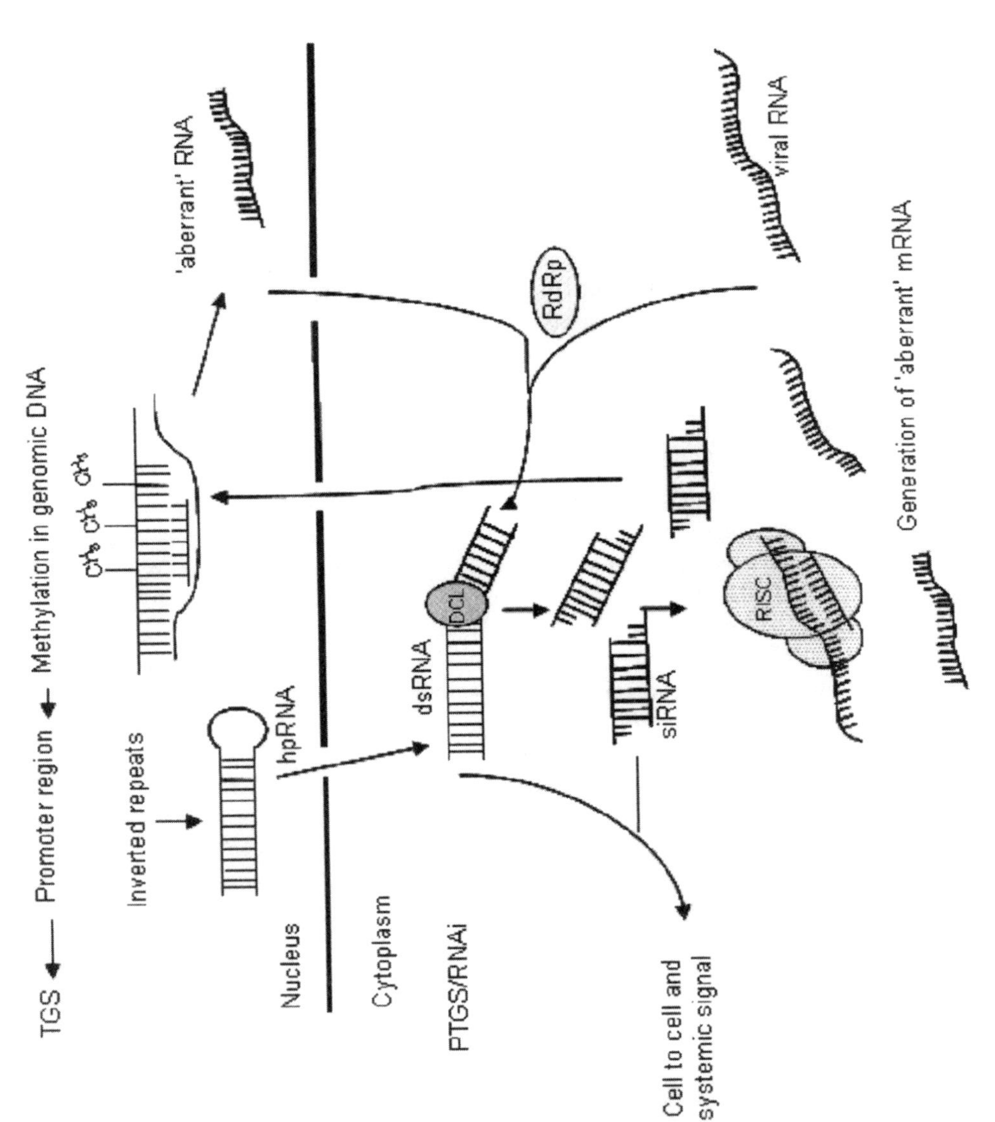

Figure 1. Basic steps in RNA silencing pathway.

considered to be minor allergens (Ara h 6 and Ara h 7) (Kleber-Janke et al., 1999). Two isoforms of Ara h 2 are likely to be expressed in every cultivated peanut genotype because peanut is a tetraploid species with A and B genomes that have been proposed to originate from the diploid progenitors *Arachis duranensis* (A genome) and *Arachis ipaensis* (B genome) (Kochert et al., 1996). Only two isoforms of Ara h 2 are predicted from the nucleotide sequences expressed in the peanut cultivar Georgia Green (Ramos et al., 2006). A different situation is encountered with Ara h 6, where only two isoforms are predicted, but these isoforms originate from three genes, one from the A genome and two from the B genome (Ramos et al., 2006). An alignment of the nucleotide sequences from the coding regions of *ara h 2* and *ara h 6* from Georgia Green is shown in Fig. 2, and a dendrogram showing the relationship of *Arachis hypogaea* gene family members with each other and other legume conglutins is shown in Fig. 3. If both members of this gene family in peanut were targeted for gene silencing, one would need to select a region that maximizes nucleotide identity among all candidates to have the greatest potential for success. It is possible, however, that certain genes or regions of a gene are less susceptible to silencing than others (Kerschen et al., 2004), but this susceptibility can only be empirically determined at this time.

ALLERGEN GENE DOWN-REGULATION USING RNA SILENCING

RNA silencing of an allergen gene can be accomplished with the complete mRNA sequence by expressing it in the antisense orientation, as was done 15 years ago for a fruit-ripening (polygalacturonase) gene in tomato (Smith et al., 1990). This antisense approach is less effective than the directed siRNA approach, although both probably function through the sequence-specific RNA degradation pathway described above (Di Serio et al., 2001). One of the first attempts to silence an allergen gene used antisense technology (Tada et al., 1996). Rice seeds contain 14- to 16-kDa storage proteins of the α-amylase–trypsin inhibitor type, and these proteins were determined to be major rice allergens (Ory s aA/TI) (www.allergome.org) (Izumi et al., 1992). Sequencing of multiple cDNA clones indicated that the proteins were encoded by a multigene family. The cDNA for one of the 16-kDa allergen genes was cloned in an antisense direction behind a promoter that regulated seed-specific expression and then introduced into rice protoplasts by electroporation. Antibiotic selection (a constitutively expressed hygromycin resistance gene provided antibiotic resistance), plant regeneration, and DNA analysis resulted in the identification of 120 transformed plants, of which 11 were fertile and could set seed. In several fertile, transgenic lines, the detection of 14- to 16-kDa seed proteins by a monoclonal antibody was markedly reduced compared with the wild type. Both sense and antisense transcripts were detected upon Northern blot analysis, but the level of sense transcripts was reduced in the antisense-expressing plants. With a monoclonal antibody and purified 16-kDa protein in a competitive enzyme-linked immunosorbent assay, it was determined that antisense expressing lines contained only ~20% of the allergenic protein compared with the wild type. In subsequent work, the same group explored the expression of gene family members with different sequence identities (Tada et al., 2003). Multiple cDNA sequences had shown that the 14- to 16-kDa group of proteins are derived from a gene family with a nucleotide sequence identity that ranged from 79 to 88% when aligned with the rice allergen (RA17) gene used for the antisense construct (Tada et al., 2003). After immunoblots

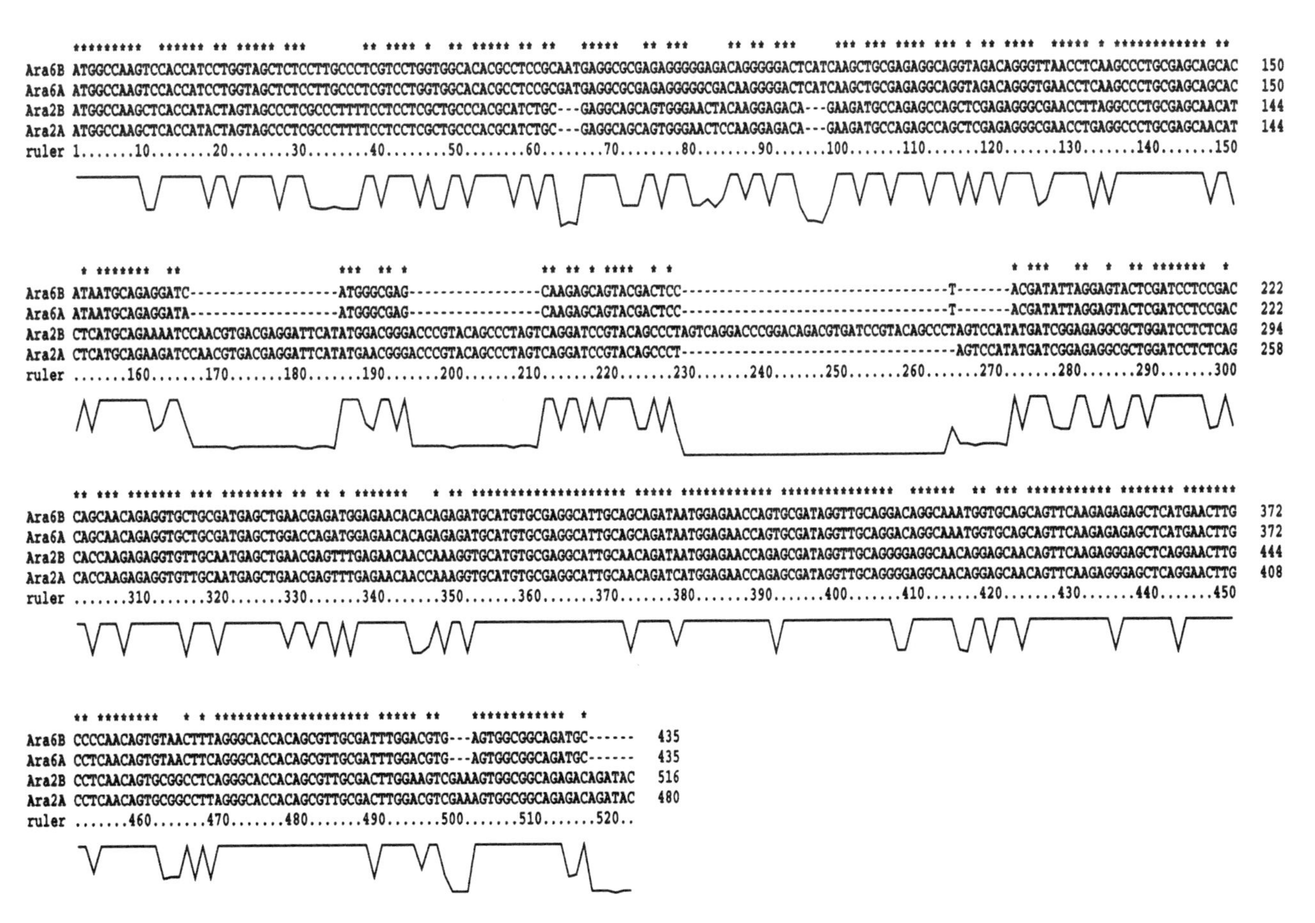

Figure 2. Alignment of A- and B-genome sequences from Georgia Green (the variety of peanut most commonly marketed in the United States) of *ara h 2* and *ara h 6*. Only coding regions are shown.

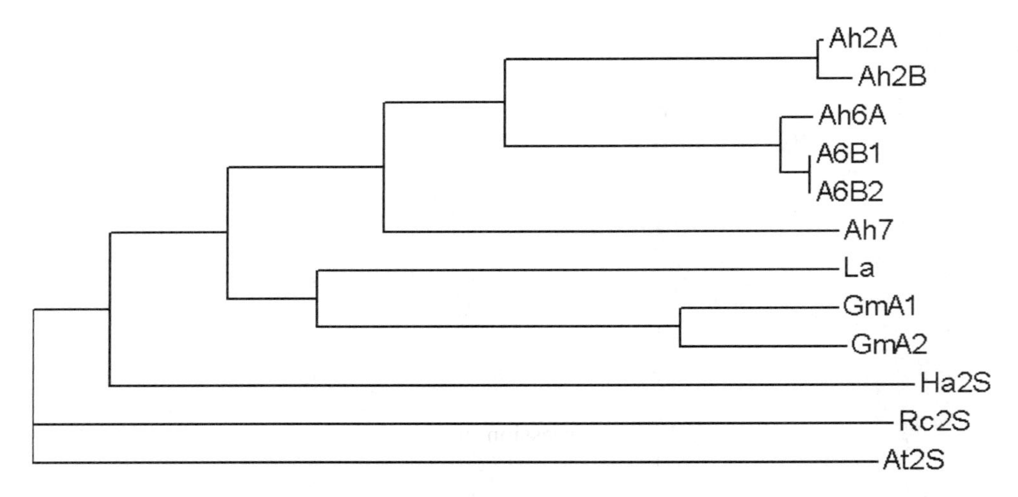

Figure 3. Phylogenetic tree based on nucleotide sequence alignments of the genes shown in Fig. 2 plus another conglutin gene from *A. hypogaea* (*ara h 7*) and genes from two related genera in the legume family. La, *Lupinus albus*; Gm, *Glycine max*. Related genes outside the legumes were from *Helianthus annuus* (Ha), *Ricinus communis* (Rc), and *Arabidopsis thaliana* (At).

of two-dimensional protein gels were examined, expression of certain members of the multigene family was selectively reduced, including the protein corresponding to the gene used to develop the antisense construct. The proteins in the 14- to 16-kDa family that were not detectably reduced were thought to be those that had the least homology to the transgene (79 to 80%).

An antisense strategy also has been used to reduce the expression of a 2S albumin seed storage protein in *Arabidopsis thaliana*, albeit one that is not known to be an allergen because *Arabidopsis* is not a food crop (Goossens et al., 1999). This experiment was designed to compare the expression of a heterologous seed protein, arcelin 5 (Arc5) from common bean (*Phaseolus vulgaris*), in transgenic *Arabidopsis* seed when endogenous seed protein was expressed at either normal or suppressed levels. Suppression was achieved by insertion of the antisense 2S2 albumin gene from *Arabidopsis* under the control of its native regulatory sequences. When expression of the 2S albumin was not suppressed, the Arc content of transgenic seeds ranged from 1 to 15% (on average, around 5%). When a knockdown of 2S albumin expression was achieved, the Arc content ranged from 7 to 24%. These experiments show that the relative contributions of individual seed storage proteins to total seed protein can be adjusted to compensate for the loss of a single protein while normal protein levels are retained. As in the Tada et al. study (1996), gene expression was not totally suppressed by the antisense RNA transcripts. 2S albumin from *A. thaliana* is also a gene family (Krebbers et al., 1988), and it was speculated by Goossens et al. (1996) that the level of suppression was sequence dependent and that genes for certain isoforms were not sufficiently homologous to respond to the antisense transcript.

The multisubunit soybean protein, β-conglycinin, has been suppressed by antisense expression of a 5′ untranslated leader from an α subunit (Kinney et al., 2001). The α and α′ subunits are derived from a small gene family (Harada et al., 1989), are 90% similar in their mRNA sequence, but have only 70% homology to the β subunit. α/α′ subunit transcript abundance was greatly reduced in the antisense-expressing transgenic soybean lines with a concurrent reduction in α/α'-subunit proteins. β subunit protein expression was similar in control and transgenic lines. The total seed protein content, which did not change significantly, had an elevated proportion of glycinin that apparently compensated for the reduction in β-conglycinin. Surprisingly, a substantial amount of the glycinin remained as the unprocessed proglycinin. A third type of allergenic soybean seed storage protein, the papain-like P34 (Gly m Bd 30 K), could be detected as unprocessed precursor protein, in addition to processed protein. The unprocessed proteins were located to protein bodies, unlike the typical localization of mature proteins in protein storage vacuoles. In subsequent work from the same group, the immunodominant soybean allergen P34 was silenced by sense suppression (Herman et al., 2003). Sense suppression or cosuppression was the first observation of what is now known to be RNA silencing (Napoli et al., 1990), although sense strand initiation of RNAi is not as efficacious as antisense or inverted-repeat initiation. For P34 suppression, transgenic soybean lines were recovered after microprojectile bombardment of somatic embryos and selection for antibiotic (hygromycin) resistance. The P34-suppressed phenotype was analyzed for immunoreactivity of seed protein extracts to monoclonal and serum IgE antibodies. Both immunoassays indicated that P34 was dramatically reduced or absent. Based on proteomic analysis, seed protein profiles of the transgenic line were only altered for the level of P34 compared with a nontransgenic control. Although there were no apparent phenotypic or developmental abnormalities in P34-suppressed soybean seeds, there is speculation that P34 plays a role in pathogen resistance. This function can now be tested with P34-deficient soybean lines. A review of P34 properties and silencing has recently been published (Herman, 2005).

In addition to allergenic seed proteins, allergens produced in pollen have been silenced using antisense technology. An example is the knockdown of Lol p 5 expression in ryegrass (*Lolium perenne*). Lol p 5 is a PR-10–ribonuclease-type allergen that is less abundant than Lol p 1, an expansin, but is recognized by IgE from >90% of allergic individuals. This major allergen succumbed to almost complete knockdown by an antisense strategy (Bhalla et al., 1999). The antisense Lol p 5 coding sequence was inserted under the control of a rice pollen-specific promoter. Microprojectile bombardment of embryogenic callus, followed by selection on kanamycin-Geneticin, resulted in the recovery of transgenic plants that were subsequently assayed for allergen production. An immunoassay was conducted against pollen proteins from a transgenic and a control plant with monoclonal antibody to Lol p 1, monoclonal antibody to Lol p 5, and IgE antibodies from a pollen-allergic individual. Although there was no difference in the amount of Lol p 1 detected, Lol p 5 was dramatically reduced and in fact was undetectable with the monoclonal antibody. Based on pollen viability and functional assays, Lol p 5 did not appear to play an essential role in pollen growth and development.

Another pollen allergen, Bet v 1, is also a member of the PR-10 group of pathogenesis-related proteins. Sensitization to Bet v 1 can occur through inhalation of birch pollen; subsequently, cross-reactivity to similar allergens in certain fruits, vegetables, and nuts can elicit an allergic reaction. The apple is one of the primary fruits to which this cross-reactivity occurs. Although the allergic reaction usually is mild, fruit avoidance by allergic individuals deprives them of the nutritional benefit of many fruits from the Rosaceae family in addition to apple, such as peach, cherry, and pear. Recently, Gilissen et al. (2005) silenced the immunodominant and Bet v 1-related apple allergen, Mal d 1, by RNAi. In this example of allergen silencing, they used a silencing construct design that was shown to be more effective than the antisense strategy, i.e., an intron-containing inverted repeat that resulted in an intron-spliced hairpin RNA (Smith et al., 2000). Since RNAi is initiated by dsRNA, inverted repeat constructs were designed to take advantage of this property. Using Mal d 1 DNA sequence information, one set of PCR primers was designed to amplify a DNA fragment encompassing the 5′ untranslated region (5′UTR-exon 1-intron 1-exon 2 (partial) (= 516 bp) region and a second set of PCR primers amplified only the 5’UTR-exon 1 (= 276 bp) region. The inverted 5'UTR-exon 1 repeat separated by the Mal d 1 intron was expressed under the control of an enhanced cauliflower mosaic virus 35S promoter. Since Mal d 1 is expressed in leaves as well as fruit tissues, it was not necessary to grow transgenic plants to reproductive maturity prior to analysis for allergen knockdown. *Agrobacterium*-mediated transformation of leaf disks, followed by shoot regeneration on selective (kanamycin) medium, resulted in nine putative transformed plants that were subjected to further analysis. A skin prick test was conducted with extracts from six plants plus controls and a significantly reduced response was measured for three of the six plants. Immunoblots of protein extract from all transgenic plantlets showed reduced or no Mal d 1 protein as detected by either a monoclonal antibody to Bet v 1 or by IgE from apple- and birch pollen-allergic individuals.

RNA silencing can target some sequences more efficiently than others (Kerschen et al., 2004). Although sense suppression of P34 appears to be relatively complete and genetically stable, evidence from silencing of other genes strongly suggests that constructs designed to produce hairpin RNAs would result in a higher frequency of transgenic lines with complete posttranscriptional silencing (Chuang and Meyerowitz, 2000; Smith et al., 2000). Posttranscriptional gene silencing can be reversed by viral suppressors of silencing (Moissiard and Voinnet, 2004), which could be of concern for hypoallergenic genotypes that rely on RNA silencing for their hypoallergenicity. Viral suppressors of silencing evolved as a counterdefense to plants and target a variety of steps in the silencing pathway. Transcriptional silencing due to DNA methylation could be more stable than posttranscriptional silencing that requires the production of short RNAs, since maintenance methylation of a promoter sequence should continue even in the absence of the initiating dsRNA silencing signal (Mette et al., 2000). A major advantage of RNA silencing over mutational methods (described below) for eliminating allergens is the ability to target multiple members of a gene family, as long as they have sufficient sequence homology to the inverted repeat region cloned into a transformation vector (Table 2) (Lawrence and Pikaard, 2003). The threshold of nucleotide identity for silencing of a gene family member may be dependent in part on the length of the homologous fragment. Silencing of viral sequences could be observed with as few as 60 nt of identical sequence but was much more efficient with 490 nt (Sijen et al., 1996). Similarly, silencing could be observed with a fragment of the green fluorescent protein gene (*gfp*) as small as 23 nt, although 28 to 818 nt was far more effective (Thomas et al., 2001). In this case, a 27-nt region of homology with *gfp* that had a single nucleotide mismatch was not able to initiate silencing.

MUTATION STRATEGIES FOR DOWN-REGULATING ALLERGEN GENES

For the lowest level of risk that might arise from reactivation of a suppressed gene, a mutation strategy is preferred. The major disadvantage of a mutation strategy is again

Table 2. Methods for eliminating allergens

Method	Advantage(s)	Disadvantages
RNAi knockdown	Can target multiple genes with one construct, provided that genes have sufficient sequence homology	Requires genetic engineering (plant transformation)
		Potentially unstable or reversible
		Efficiency of silencing can vary among lines
Mutational knockout	Does not require genetic engineering (plant transformation)	Recovery of multiple mutated alleles probably would be necessary to find a true knockout
	Stable phenotype with low probability of reversion	Only one targeted gene is likely to be mutated in each line

related to the expression of many allergens from multiple members of a gene family. For a mutation strategy to succeed, a knockout of each expressed gene in the gene family is necessary. The probability of achieving multiple knockouts simultaneously in the same individual plant would be low. An alternative would be to knock out one allergen gene at a time and then combine the loss-of-function mutations by conventional crossing. Methods of screening would not be as simple as an enzyme-linked immunosorbent assay for allergen reduction because it is unlikely that knocking out expression of one gene family member would result in a significant reduction in allergen content. If each protein product of a gene family member is biochemically distinct in terms of size, charge, or immunoreactivity, it may be possible to screen by protein two-dimensional gel methods, but such screening would be costly and time consuming. This type of screening has been conducted for the P34 allergen of soybean in numerous accessions from the soybean core collection plus related *Glycine* species (Yaklich et al., 1999). P34 was found in all of the samples analyzed. A third method of screening for individual gene knockouts would be through the reverse genetics method of targeting induced local lesions in genomes (TILLING) (McCallum et al., 2000). It will be described below and is summarized in Fig. 4.

Mutations in DNA can be induced by chemical mutagens, by ionizing radiation, or by biological mechanisms (transposons or T-DNA insertions). Ionizing radiation such as gamma rays or fast neutrons has a greater tendency to cause deletions and chromosomal rearrangements than chemical mutagenesis (Lightner and Caspar, 1998). TILLING, a chemical-based mutagenesis method, most frequently has been accomplished with mutant populations generated with ethylmethane sulfonate, an alkylating agent that alkylates guanine, allowing it to pair with thymine rather than cytosine, followed by a PCR-based method of screening for these mutations. In 98 to 99% of the ethylmethane sulfonate-generated mutants examined, the mutation was a G/C-to-A/T transition (Greene et al., 2003), as would be expected if the mutagenesis was due to alkylation of guanine. Mutation detection in TILLING requires prior knowledge of gene sequence because the TILLING method searches directly for changes in nucleotide sequence. TILLING requires that PCR primers be designed to amplify across part or all of the coding region of the gene of interest.

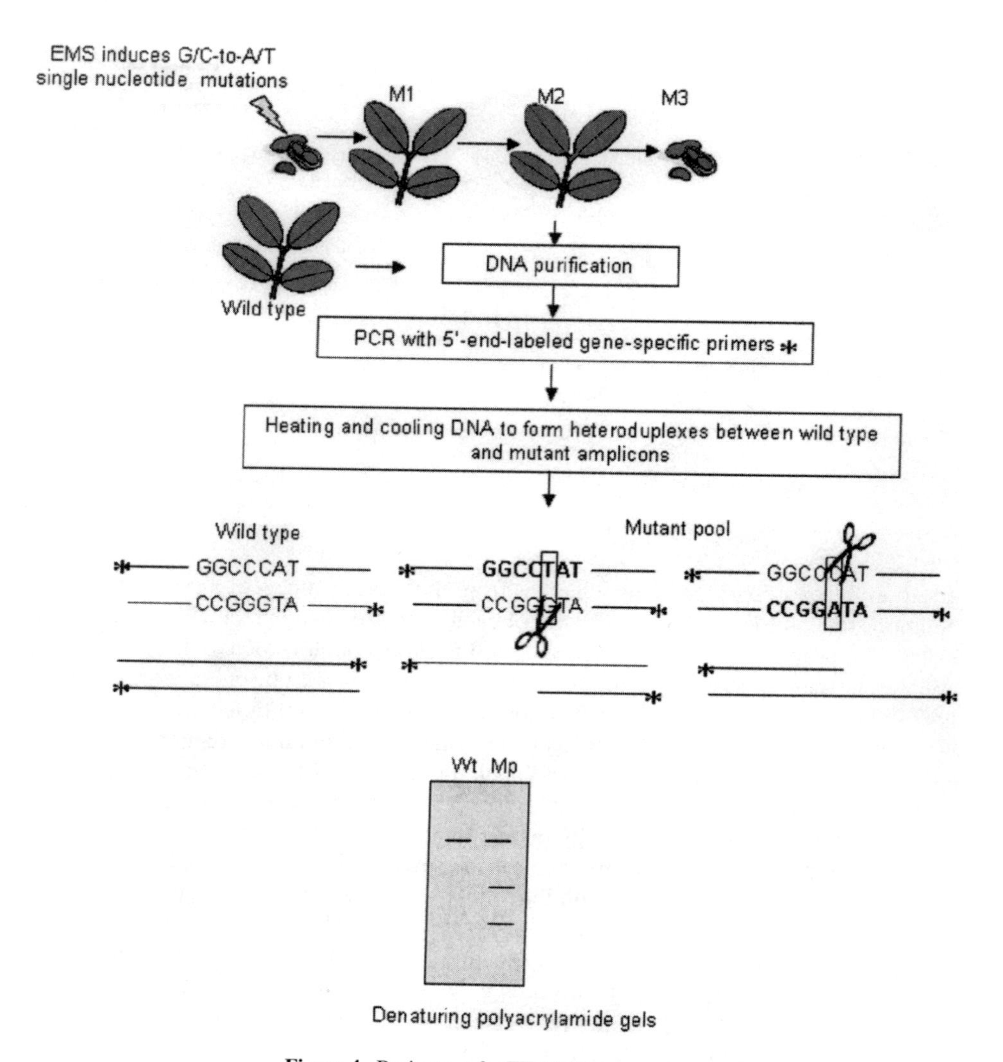

Figure 4. Basic steps for TILLING in peanut.

A computer program, CODDLE, facilitates selection of primers from regions of the gene that have the greatest probability of yielding missense or nonsense mutations (Henikoff and Comai, 2003). These PCR primers should specifically amplify a 1.0- to 1.5-kb piece of DNA. Although this seems to be a trivial task, specific amplification of a single gene is not always straightforward, particularly in crop plants, which are often polyploid. We have begun to generate a TILLING population for peanut, but there is very little gene sequence available for this crop plant (Luo et al., 2005). Furthermore, peanut is tetraploid as described above, and we have shown that the *ara h 2* and *ara h 6* genes from the two subgenomes of peanut are highly similar. To optimize the potential for success with TILLING for *ara h 2* mutants, we need to design gene-specific primers that will amplify *ara h*

2 from only the A or B genome. cDNA sequence alone has not provided sufficient nucleotide differences to accomplish gene-specific amplification across the coding region. In this example, it has been necessary to sequence upstream regulatory regions to identify segments with sufficient nucleotide polymorphism to allow gene-specific primer design (Fig. 5) (Ramos et al., 2006). The use of upstream sequence also allows the entire coding region of *ara h 2*, a short (<1 kb) intronless gene, to be amplified.

Once a gene of interest is specifically amplified, TILLING can begin by collection of tissue from M2 plants (the M1 generation is grown from mutagenized seeds and the M2 is progeny from the M1) and extraction of DNA. Equal amounts of DNA from at least two plants should be pooled so that PCR with the gene-specific primers will amplify one mutant and one wild-type allele. The mixture of PCR products is allowed to reanneal and will contain mutant-mutant, mutant–wild-type, and wild-type–wild-type double-stranded DNA molecules. The mutant-mutant and wild-type–wild-type molecules will be perfect matches, but the mutant–wild-type molecules will contain one or more mismatches. The mismatched nucleotides can be recognized and cleaved by single-strand specific nucleases (Till et al., 2004a). Since each primer is labeled with a fluorescent tag, the smaller cleavage products can be detected by polyacrylamide gel electrophoresis (McCallum et al., 2000). The high-throughput capacity of TILLING was first demonstrated with *Arabidopsis*, where the *Arabidopsis* TILLING Project discovered >1,000 mutations in 100 genes, an average of 10 mutations per gene in a population of ~3,000 M2 plants (Till et al., 2003). The maize (*Zea mays*) TILLING project achieved a mutation frequency of ~50% of that found with *Arabidopsis*, allowing the discovery of 17 mutations after 11 genes in 750 plants were screened (Till et al., 2004b). The highest mutation frequency reported in a TILLING population to date has been with wheat (*Triticum aestivum*), where an average of 1 mutation every 24,000 bp was identified (Slade et al., 2005). Wheat is probably able to tolerate a higher mutation load because it is hexaploid; and so for every gene knockout in the A genome, there are potentially two other functional genes in the B and D genomes. Because of polyploidy, TILLING with wheat is complicated by the need to design gene-specific primer pairs. It was experimentally determined that gene-specific primers were much more efficient for detecting mutations than primers that amplified from more than one locus.

The best allergen gene candidates for TILLING will be those for which there is only a single copy in the genome. Such genes will be rare because, as already mentioned, most

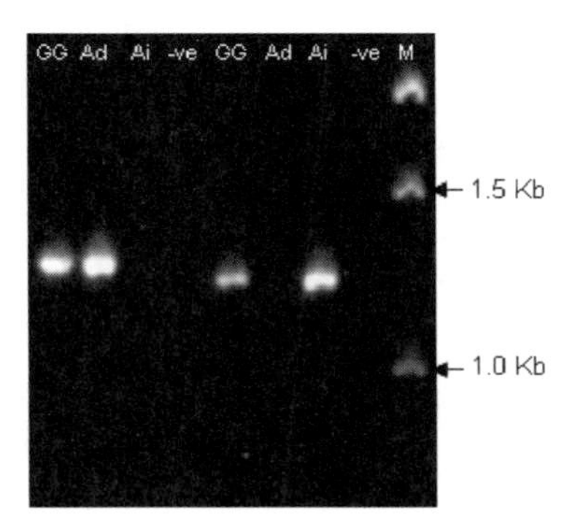

Figure 5. Gene (allele)-specific PCR of the A-genome (lanes 1 and 2 from the left) and B-genome (lanes 5 and 7 from the left) copies of *ara h 2*. The upstream PCR primers 815 and 816 are specific for each gene. The downstream primer 371 is common to both genes and is positioned in the 3′UTR. These primer sets should be suitable for TILLING. GG, Georgia Green; Ad, *A. duraensis*; Ai, *A. ipaensis*; -ve, negative control; M, marker.

crop plants are polyploid and will have at least one copy per subgenome. Our data support the presence of only two copies of *ara h 2* in peanut, one from the A genome and one from the B genome, both of which are expressed and predicted to give rise to two isoforms per peanut genotype. There could be other isoforms discovered within the species, due to genetic variability at the nucleotide level that would result in an amino acid substitution in the coding sequence. Such natural variability is not likely to have a large impact on reducing allergenicity of *ara h 2* because 10 linear epitopes recognized by IgE have been identified in this protein (Stanley et al., 1997). A spontaneous missense mutation could, however, affect the stability of a protein; algorithms are available that can predict the severity of such a mutation (Ng and Henikoff, 2001). It is also possible that a spontaneous nonsense mutation could be discovered in the gene pool of peanut. Such deleterious missense or nonsense mutations could be sought by applying the method of EcoTILLING (Comai et al., 2004), which is simply TILLING for natural variation across different genotypes rather than induced mutants of a single genotype. In EcoTILLING, one genome is used as a reference in all reactions that contain DNA from the reference and potentially variant genomes.

If independent deleterious mutations were identified in each member of an allergen gene family, could these mutations be combined by crossing? Combining the two mutant genes might be difficult to achieve in cases where two genes are closely linked but very easy to achieve when the two genes are on separate chromosomes. Two scenarios from our work might be to combine an A-genome *ara h 2* knockout with a B-genome *ara h 2* knockout or to combine an *ara h 2* knockout with an *ara h 6* knockout. The first result should be relatively simple to achieve, since chromosomes from the A and B genomes of peanut rarely pair. Therefore, the two mutant *ara h 2* genes would follow Mendel's law of independent assortment. On the other hand, we recently determined that the B-genome copy of *ara h 2* is physically linked with one B-genome copy of *ara h 6* at a distance of ~8,000 bp (Ramos et al., 2006). Positional cloning of genes requires that genetic recombination between markers that are tightly linked with a locus of interest be recovered. Such recombinants are rare and are usually found only after screening thousands of progeny. The physical distances that typically separate the closest flanking, genetically recombining markers are tens of thousands of base pairs. Recombination between these B-genome copies of *ara h 2* and *ara h 6* would probably be a rare event. Since *ara h 2* and *ara h 6* (as well as *ara h 7*) are related members of the conglutin family, would knocking out *ara h 2* by mutation significantly decrease the allergenicity of peanut? Ara h 6 currently is classified as a minor allergen because it is recognized by the serum IgE of 38% of peanut allergic individuals (Kleber-Janke et al., 1999). We also have some evidence at the RNA level that expression of *ara h 6* in the seed is less than that of *ara h 2* (Ramos et al., 2006). Questions that remain to be addressed are as follows. (i) Would Ara h 6 become immunodominant in the absence of Ara h 2? (ii) Would levels of Ara h 6 increase significantly in the absence of Ara h 2? Such questions can be addressed through experimentation with the molecular tools now available to facilitate such research.

The mutation approach to genetically eliminating seed storage proteins has been accomplished and the consequences studied with two legumes, common bean and soybean. In common bean, Arc is a seed protein in the α-amylase–trypsin inhibitor category that provides resistance to certain insects (Hartweck et al., 1997). In most bean genotypes, the globulin or vicillin, phaseolin, rather than Arc, is the most abundant seed protein, comprising up to 60% of total protein. Genetic mutants that were null for phaseolin were used to develop near-isogenic lines that were either phaseolin null or expressed phaseolin and also contained variant alleles of Arc and phytohemagglutinin (Hartweck and Osborn, 1997). The absence of phaseolin resulted in compensation with other seed proteins so that total protein amounts varied little among the isogenic lines. One allelic variant of Arc (Arc1) provided the greatest resistance to bruchid pests (Hartweck et al., 1997). Although this was the first study to use a seed storage protein mutant to reap the benefit of compensation by an insect resistance protein, previous work to alter bean seed storage protein composition due to mutation was reviewed.

In addition to common bean, the seed storage protein composition of soybean has also been altered by spontaneous or induced mutations. In the most extreme example, both glycinin and β-conglycin were eliminated through the combination of three glycinin-null mutations, each of which affected one of the three glycinin subunit groups (I, IIa, and IIb), a null mutant that removed α and α′ subunits of β-conglycin, and a dominant allele of the *Scg* (suppressor of β-conglycin) gene from a wild soybean that suppresses all β-conglycin expression (Takahashi et al., 2003). These mutant soybean seeds had a nitrogen content similar to that of wild-type seed, but the level of nitrogen attributable to free amino acids was significantly higher. The seed protein profile also showed an elevation in proportion of nonglycinin proteins, one of which was P34, the immunodominant soybean allergen. Therefore, there are possible scenarios where allergen content could be inadvertently increased after manipulation of other seed storage proteins.

HOW REALISTIC IS A STRATEGY FOR ALLERGEN GENE REPLACEMENT?

Seed storage proteins have physical properties that not only contribute to packaging and stability of the protein in the plant but also affect processed food characteristics dependent on solubility and emulsification (Casey, 1999). A strategy that has been proposed as one route to reduce food allergenicity is one of protein replacement. Replacement of one seed storage protein by another would require a parallel knockout or knockdown of the endogenous gene(s). If RNA-mediated knockdown were attempted, the DNA sequence chosen to effect the knockdown would need to have low sequence similarity with a replacement gene. Regions of low sequence similarity are more likely to occur in 5′- and 3′UTRs; such regions have been shown to be effective for gene silencing and could be engineered to be dissimilar in a replacement gene versus an endogenous gene. A high-throughput gene silencing strategy has used the terminator from the nopaline synthase (*nos*) gene as an inverted repeat fused with any endogenous gene to transitively silence the respective endogenous gene (Brummell et al., 2003). The *nos* terminator is more commonly fused to coding sequences in constructs used for plant transformation than used for silencing. For gene replacement, its more traditional use could be exploited by fusing the protein-coding region of an allergen gene with any seed-specific promoter sequence and the *nos* terminator to create a synthetic gene. The native gene potentially could be silenced with its own 3′UTR (or 5′UTR), whose sequence would differ from the *nos* terminator (3′UTR) or the engineered 5′UTR in the synthetic gene. Although silencing and/or replacement probably would be feasible from the seed development perspective, the product might have altered food qualities. Some amino acid changes are certain to alter conformational properties of proteins and such conformational changes may be undesirable.

Proteins that are structurally stable and abundant and can associate with membranes may be allergenic (Mills et al., 2004). It is not yet possible to predict the allergenicity of any novel protein or food, although progress has been made (Chapter 4) (Oehlschlager et al., 2001). For immunotherapy purposes, considerable work has been carried out on altering IgE recognition of pollen or food antigens while retaining their ability to stimulate T cells (Bannon et al., 2001). For example, the pollen allergen Phlp 5 b has been altered by site-directed mutagenesis; the variants were shown to have significantly reduced, but not abolished, T-cell reactivity (Schramm et al., 2003). The ability to stimulate T-cell proliferation was retained by two mutants of Bet v 1 (birch pollen allergen), although binding to serum IgE was significantly reduced (Holm et al., 2004). A similar approach was taken with Ara h 2, where 10 linear IgE-binding epitopes have been identified (Stanley et al., 1997). Mutation of IgE epitopes 3, 4, 6, and 7 within this protein and analysis of mutant and wild-type proteins produced from overexpression vectors showed that binding to serum IgE from peanut-allergic individuals was significantly reduced for the mutant protein in 12 of 16 samples tested (Burks et al., 1999). T-cell lines proliferated equally well upon stimulation with either mutant or wild-type allergen. Surprisingly, one IgE sample showed increased binding to the mutant protein. Ara h 1 from peanut is a larger and more complex protein than Ara h 2 because its quaternary structure consists of three monomeric subunits predicted to associate along hydrophobic regions at the alpha-helical ends of each molecule. It has been shown that linear epitopes cluster at these contact points that are necessary for trimer formation (Bannon et al., 1999; Maleki et al., 2000). Modification of these epitopes may interfere with subunit association and thereby alter protein structure, turnover, packaging, or function. While a replacement strategy may be feasible for some allergen proteins, it may be difficult for others, particularly where epitopes lie in regions critical for protein structure. Furthermore, the likelihood of de novo sensitization by a modified allergen would need to be extensively tested in animal models prior to replacement of a native allergen.

CONCLUSIONS

The term hypoallergenic should only be used in marketing when there is little likelihood that a food will cause an allergic reaction. Controversy has arisen over the use of this term in marketing baby food formulas, since virtually all formulas contain proteins that are potentially allergenic. The term hypoallergenic, as used in this review, refers to the significant reduction or elimination of individual known allergens from foods that would be consumed throughout life, recognizing that additional allergenic proteins may remain in a food product but that any reduction may prove beneficial to human health. Many allergenic foods are derived from the fruits and seeds of plants, and the allergenic proteins usually perform essential functions in these plant parts. The functions are often redundant, however, and the potential exists to compensate for the loss of one allergenic protein by replacement by a less or nonallergenic protein. This capacity opens opportunities for allergen reduction with the tools of biotechnology to knock down or knock out allergen gene expression. The clinical significance of such an approach will only become apparent after many years of testing.

REFERENCES

Bannon, G. A., G. Cockrell, C. Connaughton, C. M. West, R. Helm, J. S. Stanley, N. King, P. Rabjohn, H. A. Sampson, and A. W. Burks. 2001. Engineering, characterization and in vitro efficacy of the major peanut allergens for use in immunotherapy. *Int. Arch. Allergy Immunol.* **124:**70–72.

Bannon, G., D. Shin, S. Maleki, and R. Kopper. 1999. Tertiary structure and biophysical properties of a major peanut allergen, Implications for the production of a hypoallergenic protein. *Int. Arch. Allergy Immunol.* **118:**315–316.

Baulcombe, D. 2004. RNA silencing in plants. *Nature* **431:**356–363.

Bhalla, P. L., I. Swoboda, and M. B. Singh. 1999. Antisense-mediated silencing of a gene encoding a major ryegrass pollen allergen. *Proc. Natl. Acad. Sci. USA* **96:**11676–11680.

Boulter, D., and R. R. D. Croy. 1997. The structure and biosynthesis of legume seed storage proteins: a biological solution to the storage of nitrogen in seeds. *Adv. Biol. Res.* **27:**1–84.

Breitender, H., and C. Radauer. 2004. A classification of plant food allergens. *J. Allergy Clin. Immunol.* **113:** 821–830.

Brummell, D. A., P. J. Balint-Kurti, M. H. Harpster, J. M. Palys, P. W. Oeller, and N. Gutterson. 2003. Inverted repeat of a heterologous 3′-untranslated region for high-efficiency, high-throughput gene silencing. *Plant J.* **33:**793–800.

Burks, A., N. King, and G. Bannon. 1999. Modification of a major peanut allergen leads to loss of IgE binding. *Int. Arch. Allergy Immunol.* **118:**313–314.

Carrington, J. C., and V. Ambros. 2003. Role of microRNAs in plant and animal development. *Science* **301:** 336–338.

Casey, R. 1999. Distribution and some properties of seed globulins, p. 159–169. *In* P. R. Shewry and R. Casey (ed.), *Seed Proteins*. Kluwer, Dordrecht, The Netherlands.

Chuang, C. F., and E. M. Meyerowitz. 2000. Specific and heritable genetic interference by double-stranded RNA in *Arabidopsis thaliana*. *Proc. Natl. Acad. Sci. USA* **97:**4985–4990.

Comai, L., K. Young, B. J. Till, S. H. Reynolds, E. A. Greene, C. A. Codomo, L. C. Enns, J. E. Johnson, C. Burtner, A. R. Odden, and S. Henikoff. 2004. Efficient discovery of DNA polymorphisms in natural populations by EcoTILLING. *Plant J.* **37:**778–786.

Di Serio, F., H. Scob, A. Iglesias, C. Tarina, and E. Bouldoires. 2001. Sense- and antisense-mediated gene silencing in tobacco is inhibited by the same viral suppressors and is associated with accumulation of small RNAs. *Proc. Natl. Acad. Sci. USA* **98:**6506–6510.

Fire, A., S. Xu, M. K. Montgomery, S. A. Kostas, S. E. Driver, and C. C. Mello. 1998. Potent and specific genetic interference by double-stranded RNA in *Caenorhabditis elegans*. *Nature* **391:**806–811.

Gilissen, L. J. W. J., S. T. H. P. Bolhaar, C. I. Matos, G. J. A. Rouwendal, M. J. Boone, F. A. Krens, Zuidmeer L., A. van Leeuwen, J. Akkerdaas, K. Hoffmann-Sommergruber, A. C. Knulst, D. Bosch, W. E. van de Weg, and R. van Ree. 2005. Silencing the major apple allergen Mal d 1 by using the RNA interference approach. *J. Allergy Clin. Immunol.* **115:**364–369.

Goossens, A., M. V. Montague, and G. Angenon. 1999. Co-introduction of an antisense gene for an endogenous seed storage protein can increase expression of a transgene in *Arabidopsis thaliana* seeds. *FEBS Lett.* **456:**160–164.

Greene, E. A., C. A. Codomo, N. E. Taylor, J. G. Henikoff, B. J. Till, S. H. Reynolds, L. C. Enns, C. Burtner, J. E. Johnson, A. R. Odden, L. Comai, and S. Henikoff. 2003. Spectrum of chemically induced mutations from a large-scale reverse-genetic screen in *Arabidopsis*. *Genetics* **164:**731–740.

Hamilton, A. J., and D. C. Baulcombe. 1999. A species of small antinsene RNA in posttranscriptionl gene silencing in plants. *Science* **286:**950–952.

Harada, J. J., S. J. Barker, and R. B. Goldberg. 1989. Soybean β-conglycinin genes are clustered in several DNA regions and are regulated by transcriptional and posttranscriptional processes. *Plant Cell* **1:** 415–425.

Hartweck, L. M., and T. C. Osborn. 1997. Altering protein composition by genetically removing phaseolin from common bean seeds containing arcelin or phytohemagglutinin. *Theor. Appl. Genet.* **95:**1012–1017.

Hartweck, L. M., C. Cardona, and T. C. Osborn. 1997. Brucid resistance of common bean lines having an altered seed protein-composition. *Theor Appl Genet* **95:**1018–1023.

Henikoff, S., and L. Comai. 2003. Single-nucleotide mutations for plant functional genomics. *Annu. Rev. Plant Biol.* **54:**375–401.

Herman, E. 2005. Soybean allergenicity and suppression of the immunodominant allergen. *Crop Sci.* **45:**462–467.

Herman, E. M., R. M. Helm, R. Jung, and A. Kinney. 2003. Genetic modification removes an immunodominant allergen from soybean. *Plant Physiol.* **132:**36–43.

Holm, J., M. Gajhede, M. Ferreras, A. Henriksen, H. Ipsen, J. N. Larsen, L. Lund, H. Jacobi, A. Millner, P. A. Wurtzen, and M. D. Spangfort. 2004. Allergy vaccine engineering: epitope modulation of recombinant Bet v 1 reduces IgE binding but retains protein folding pattern for induction of protective blocking-antibody responses. *J. Immunol.* **173:**5258–5267.

Izumi, H., T. Adachi, N. Fujii, T. Matsuda, R. Nakamura, K. Tanaka, and Y. Urisu Kurosawa. 1992. Nucleotide sequence of a cDNA clone encoding a major allergenic protein in rice seeds. Homology of the deduced amino acid sequence with members of alpha-amylase/trypsin inhibitor family. *FEBS Lett.* **302:**213–216.

Kandasamy, M. K., E. C. McKinney, and R. B. Meagher. 2002. Plant profilin isovariants are distinctly regulated in vegetative and reproductive tissues. *Cell Motil. Cytoskeleton* **52:**22–32.

Kerschen, A., C. A. Napoli, R. A. Jorgensen, and A. E. Muller. 2004. Effectiveness of RNA interference in transgenic plants. *FEBS Lett.* **566:**223–228.

Kinney, A. J., R. Jung, and E. M. Herman. 2001. Cosuppression of the α subunits of β-conglycinin in transgenic soybean seeds induces the formation of endoplasmic reticulum-derived protein bodies. *Plant Cell* **13:**1165–1178.

Kleber-Janke, T., R. Crameri, U. Appenzeller, M. Schlaak, and W.-M. Becker. 1999. Selective cloning of peanut allergens including profilin and 2S albumins, by phage display technology. *Int. Arch. Allergy Immunol.* **119:**265–274.

Kochert, G., H. Stalker, M. Gimenes, L. Galgaro, C. Lopes, and K. Moore. 1996. RFLP and cytogenetic evidence on the origin and evolution of allotetraploid domesticated peanut, *Arachis hypogaea* (Leguminosae). *Am. J. Bot.* **83:**1282–1291.

Kovar, D. R., B. K. Drobak, D. A. Collings, and C. J. Staiger. 2001. The characterization of ligand-specific maize (*Zea mays*) profilin mutants. *Biochem. J.* **358:**49–57.

Krebbers, E., L. Herdies, A. DeClercq, J. Seurinch, J. Leemans, J. Van Damme, M. Segura, G. Gheysen, M. Van Montague, and J. Vanderkerckhove. 1988. Determination of the processing sites of an *Arabidopsis* 2S albumin and characterization of the complete gene family. *Plant Physiol.* **87:**859–866.

Kusaba, M. 2004. RNA interference in crop plants. *Curr. Opin. Biotechnol.* **15:**139–143.

Lawrence, R. J., and C. S. Pikaard. 2003. Transgene-induced RNA interference: a strategy for overcoming gene redundancy in polyploids to generate loss-of-function mutations. *Plant J.* **36:**114-–121.

Lightner, J., and T. Caspar. 1998. Seed mutagenesis of *Arabidopsis*, p. 91–103. *In* J. Martinez-Zapater and J. Salinas (ed.), *Arabidopsis Protocols*. Humana Press, Totowa, N.J.

Luo, M., P. Dang, B. Z. Guo, G. He, C. C. Holbrook, M. G. Bausher, and R. D. Lee. 2005. Generation of expressed sequence tags (ESTs) for gene discovery and marker development in cultivated peanut. *Crop Sci.* **45:**346–353.

Maleki, S., R. Kopper, D. Shin, C. Park, C. Compadre, H. Sampson, A. Burks, and G. Bannon. 2000. Structure of the major peanut allergen Ara h 1 may protect IgE-binding epitopes from degradation. *J. Immunol.* **164:**5844–5849.

Matzke, M. A., and A. J. M. Matzke. 2004. Planting the seeds of a new paradigm. *PLoS Biol.* **2:**582–586.

Matzke, M., and J. Birchler. 2005. RNAi-mediated pathways in the nucleus. *Nat. Rev. Genet.* **6:**24–35.

McCallum, C., L. Comai, E. Greene, and S. Henikoff. 2000. Targeting induced local lesions in genomes (TILLING) for plant functional genomics. *Plant Physiol.* **123:**439–442.

Meister, G., and T. Tuschl. 2004. Mechanisms of gene silencing by double-stranded RNA. *Nature* **431:** 343–349.

Mello, C. C., and D. Conte. 2004. Revealing the world of RNA interference. *Nature* **431:**338–342.

Mette, M. F., W. Aufsatz, J. van der Winden, M. A. Matzke, and A. J. M. Matzke. 2000. Transcriptional silencing and promoter methylation triggered by double-stranded RNA. *EMBO J.* **19:**5194–5201.

Mills, E. N. C., J. A. Jenkins, M. J. C. Alcocer, and P. R. Shewry. 2004. Structural, biological, and evolutionary relationships of plant food allergens sensitizing via the gastrointestinal tract. *Crit. Rev. Food Sci. Nutr.* **44:** 379–407.

Moissiard, G., and O. Voinnet. 2004. Viral suppression of RNA silencing in plants. *Mol. Plant Pathol.* **5:**71–82.

Mulder, N. J., R. Apweiler, T. K. Attwood, A. Bairoch, A. Bateman, D. Binns, P. Bradley, P. Bork, P. Bucher, L. Cerutti, R. Copley, E. Courcelle, U. Das, R. Durbin, W. Fleischmann, J. Gough, D. Haft, N. Harte, N. Hulo, D. Kahn, A. Kanapin, M. Krestyaninova, D. Lonsdale, R. Lopez, I. Letunic,

M. Madera, J. Maslen, J. McDowall, A. Mitchell, A. N. Nikolskaya, S. Orchard, M. Pagni, C. P. Ponting, E. Quevillon, J. Selengut, C. J. Sigrist, V. Silventoinen, D. J. Studholme, R. Vaughan, and C. H. Wu. 2005. InterPro, progress and status in 2005. *Nucleic Acids Res.* **33:**D201–205.

Napoli, C., C. Lemieux, and R. Jorgensen. 1990. Introduction of a chimeric chalcone synthase gene into petunia results in reversible co-suppression of homologous genes in trans. *Plant Cell* **2:**279–289.

Ng, P., and S. Henikoff. 2001. Predicting deleterious amino acid substitutions. *Genome Res.* **11:**863–874.

Nordlee, J. A., S. L. Taylor, J. A. Townsend, L. A. Thomas, and R. K. Bush. 1996. Identification of a Brazil-nut allergen in transgenic soybeans. *N. Engl. J. Med.* **334:**688–692.

Oehlschlager, S., P. Reece, A. Brown, E. Hughson, H. Hird, J. Chisholm, H. Atkinson, C. Meredith, R. Pumphrey, P. Wilson, and J. Sunderland. 2001. Food allergy—towards predictive testing for novel foods. *Food Add. Contam.* **18:**1099–1107.

Ramos, M. L., G. Fleming, Y. Chu, Y. Akiyama, M. Gallo, and P. Ozias-Akins. 2006. Chromosomal and phylogenetic context for conglutin genes in *Arachis* based on genomic sequence. *Mol. Gen. Genomics* DOI 10.1007/s00438-006-0114-z.

Sampson, H. A. 2004. Update on food allergy. *J. Allergy Clin. Immunol.* **113:**805-819.

Schramm, G., H. Kahlert, R. Suck, B. Weber, H. Stuwe, W. Muller, A. Bufe, W. Becker, U. Lepp, M. Schlaak, L. Jager, O. Cromwell, and H. Fiebig. 2003. Variants of allergen Phlp 5 b and reduction of anaphylactogenic potential. *Rev. Fr. Allergol. Immunol. Clin.* **43:**56–58.

Shewry, P. R., and R. Casey. 1999. Seed proteins, p. 1–10. *In* P. R. Shewry and R. Casey (ed.), *Seed Proteins*. Kluwer, Dordrecht, The Netherlands.

Shewry, P. R., and M. J. Pandya. 1999. The 2S albumin storage proteins, p. 563–586. *In* P. R. Shewry and R. Casey (ed.), *Seed Proteins*. Kluwer, Dordrecht, The Netherlands.

Shewry, P., J. Napier, and A. Tatham. 1995. Seed storage proteins: structures and biosynthesis. *Plant Cell* **7:**945–956.

Sijen, T., J. Wellink, J. B. Hiriart, and A. Van Kammen. 1996. RNA-mediated virus resistance: role of repeated transgenes and delineation of targeted regions. *Plant Cell* **8:**2277–2294.

Slade, A. J., S. I. Fuerstenberg, D. Loeffler, M. N. Steine, and D. Facciotti. 2005. A reverse genetic, nontransgenic approach to wheat crop improvement by TILLING. *Nat. Biotechnol.* **23:**75–81.

Smith, C. J., C. F. Watson, P. C. Morris, C. R. Bird, G. B. Seymour, J. E. Gray, C. Arr, G. A. Tucker, W. Schuch, S. Harding, et al. 1990. Inheritance and effect on ripening of antisense polygalacturonase genes in transgenic tomatoes. *Plant Mol. Biol.* **14:**369–379.

Smith, N. A., S. P. Singh, M.-B. Wang, P. A. Stoutjesdijk, A. G. Green, and P. M. Waterhouse. 2000. Total silencing by intron-spliced hairpin RNAs. *Nature* **407:**319–320.

Stanley, J. S., N. King, A. W. Burks, S. K. Huang, H. Sampson, G. Cockrell, R. M. Helm, C. M. West, and G. A. Bannon. 1997. Identification and mutational analysis of the immunodominant IgE binding epitopes of the major peanut allergen Ara h 2. *Arch. Biochem. Biophys.* **342:**244–253.

Susi, P., M. Hohkuri, T. Wahlroos, and N. J. Kilby. 2004. Characteristics of RNA silencing in plants: similarities and differences across kingdoms. *Plant Mol. Biol.* **54:**157–174.

Tada, Y., H. Akagi, T. Fujimura, and T. Matsuda. 2003. Effect of an antisense sequence on rice allergen genes comprising a multigene family. *Breed. Sci.* **53:**61–67.

Tada, Y., M. Nakase, T. Adachi, R. Nakamura, H. Shimada, M. Takahashi, T. Fujimura, and T. Matsuda. 1996. Reduction of 14–16 kDa allergenic proteins in transgenic rice plants by antisense gene. *FEBS Lett.* **391:**341–345.

Takahashi, M., Y. Uematsu, K. Kashiwaba, K. Yagasaki, M. Hajika, R. Matsunaga, K. Komatsu, and M. Ishimoto. 2003. Accumulation of high levels of free amino acids in soybean seeds through integration of mutations conferring seed protein deficiency. *Planta* **217:**577–586.

Tang, G. 2005. siRNA and miRNA: an insight into RISCs. *Trends Biochem. Sci.* **30:**106–114.

Thomas, C. L., L. Jones, D. C. Baulcombe, and A. J. Maule. 2001. Size constraints for targeting post-transcriptional gene silencing and for RNA-directed methylation in *Nicotiana benthamiana* using a potato virus X vector. *Plant J.* **25:**417–425.

Till, B. J., S. H. Reynolds, E. A. Greene, C. A. Codomo, L. C. Enns, J. E. Johnson, C. Burtner, A. R. Odden, K. Young, N. E. Taylor, J. G. Henikoff, L. Comai, and S. Henikoff. 2003. Large-scale discovery of induced point mutations with high-throughput TILLING. *Genome Res.* **13:**524–530.

Till, B. J., C. Burtner, L. Comai, and S. Henikoff. 2004a. Mismatch cleavage by single-strand specific nucleases. *Nucleic Acids Res.* **32:**2632–2641.

Till, B. J., S. H. Reynolds, C. Weil, N. Springer, C. Burtner, K. Young, E. Bowers, C. A. Codomo, L. C. Enns, A. R. Odden, E. A. Greene, L. Comai, and S. Henikoff. 2004b. Discovery of induced point mutations in maize genes by TILLING. *BMC Plant Biol.* **4:**12.

Valueva, T. A., and V. V. Mosolov. 1999. Protein inhibitors of proteinases in seeds. 1. Classification, distribution, structure, and properties. *Russ. J. Plant Physiol.* **46:**307–321.

van der Krol, A., L. Mur, P. de Lange, J. Mol, and A. Stuitje. 1990. Inhibition of flower pigmentation by antisense CHS genes: promoter and minimal sequence requirements for the antisense effect. *Plant Mol. Biol.* **14:**457–466.

Waterhouse, P. M., M.-B. Wang, and T. Lough. 2001. Gene silencing as an adaptive defence against viruses. *Nature* **411:**834–842.

Yaklich, R. W., R. M. Helm, G. Cockrell, and E. M. Herman. 1999. Analysis of the distribution of the major soybean seed allergens in a core collection of *Glycine max* accessions. *Crop Sci.* **39:**1444–1447.

Yoo, B. C., F. Kragler, E. Varkonyi-Gasic, V. Haywood, S. Archer-Evans, Y. M. Lee, T. J. Lough, and W. J. Lucas. 2004. A systemic small RNA signaling system in plants. *Plant Cell* **16:**1979–2000.

Youle, R., and A. Huang. 1981. Occurrence of low molecular weight and high cysteine containing albumin storage proteins in oilseeds of diverse species. *Am. J. Bot.* **68:**44–48.

Food Allergy
Edited by S. J. Maleki et al.

Chapter 13

The Effects of Processing Methods on Allergenic Properties of Food Proteins

Soheila J. Maleki and Shridhar K. Sathe

Food allergies are increasing in developed countries and, some evidence indicates, in less-studied, developing countries (Perr et al., 2004; Pereira et al., 2002; Visitsunthorn et al., 2002; Wang and Li, 2005; Pitche et al., 1996; Hill et al., 1999). With the globalization of markets and trade, many new foods or foods processed in novel ways are introduced into our daily diet. Most allergenic proteins, particularly plant allergens, appear to be members of a few families of proteins and share common characteristics that may render them allergenic (Jenkins et al., 2004; Radauer and Breiteneder, 2006), although not all of these properties are well defined. The food source and the different methods of storing and processing foods complicate analysis of the structural and functional properties of allergens that contribute to enhanced or reduced allergenicity (for a review, see Maleki et al., 2004). Furthermore, foods are often processed in combination with other food ingredients that can result in biochemical interactions among the different proteins and ingredients in that food. These biochemical interactions may result in the generation of new (or cryptic) proteins and peptides and novel fusions of sugars or other compounds and proteins. And when the genetic variation among individuals and differential allergic responses by sensitive individuals to allergic proteins in foods are considered, the issue of what makes a protein an allergen or a more potent allergen becomes highly complex.

To elicit an immunoglobulin E (IgE)-mediated allergenic response, the allergen must cross-link IgE molecules on the surface of mediator release cells. Patient IgE may recognize linear stretches of amino acid sequence (linear epitope) or a three-dimensional structural motif (conformational epitope) on an allergen. Understanding the IgE-allergen interaction at the molecular level is therefore important in designing ways to reduce or eliminate allergenicity. However, most allergens, before or after processing, remain to be characterized at the molecular level.

In this chapter, we will attempt to describe some of the structural and molecular com-

Soheila J. Maleki • UAMS (USDA-ARS), Department of Microbiology and Immunology, 4301 W. Markham St., Slot 511, Little Rock, AR 72205. ***Shridhar K. Sathe*** • Department of Nutrition, Food & Exercise Sciences, 402 Sandels Building, Florida State University, Tallahassee, FL 32306-1493.

ponents that contribute to allergenicity of proteins and foods and the processes involved in altering the allergenicity of food proteins. Also, we will address possible options for the reduction of allergenicity in foods.

ALLERGEN STRUCTURE

Food allergens have been identified as proteins or, mostly, as glycoproteins. The protein amino acid sequence is referred to as the primary protein structure and dictates the overall structure of the protein, such as secondary structure formation (α helices, loops, and βsheets). The interaction of the secondary structural elements with each other in a three-dimensional space describes the tertiary structure of a protein molecule, and the binding of an individual protein molecule with itself or other proteins is referred to as the quaternary structure. Each of these structural components can contribute to the allergenic properties of various proteins, and the relative contribution of each structural feature may vary considerably between different allergic proteins. The structure of an allergic protein will dictate characteristics such as digestibility and absorption upon ingestion, stability to heat and other processing treatments, functional properties, and recognition by the immune system (antibodies, T cells, antigen-presenting cells, etc.). The multitude of biochemical reactions that occur during processing of foods may have a variety of effects on primary, secondary, tertiary, and quaternary proteins or peptide structures that could alter allergenic properties of proteins.

When a food is ingested, it encounters chewing and mixing with α-amylase, an enzyme known to digest carbohydrates including those from the surface of proteins. As food travels through the gut and intestines, it is subjected to enzymes such as pepsin (Kopper et al., 2004) in the stomach and trypsin and chymotrypsin in the intestine. Food proteins are digested into peptides and amino acids and absorbed into the bloodstream after passing from the lumen of the intestine. Intestinal enzymes such as carboxypeptidase, aminopepsidase, dipeptidase, and tripeptidase within enterocyte brush border membranes represent the intermediate and final stages of protein hydrolysis in the intestine (Kushak and Winter, 1999; Gray and Cooper, 1971; Kim et al., 1974) and may play an important role in determining allergenicity of proteins. In fact, Felix et al. (2002) showed that the immunodominant T-cell epitopes of gliadin are particularly resistant to digestion by dipeptidyl peptidase and carboxypeptidase and likely pass intact into the bloodstream. The catalytic and regulatory activities of these enzymes are affected by carbohydrates, lipids, and other nutrients within a food (Kushak and Winter, 1999). Following absorption into the bloodstream, food proteins are often internalized and processed by the internal enzymes (i.e., cathepsin D) within antigen-presenting cells. These cells go on to present processed protein fragments to immune cells, such as T-helper cells and B cells leading to the production of protein-specific IgE (sensitization phase). The IgE-mediated allergic experience (allergic phase) is due to the cross-linking of bound IgE antibodies on the surface of mast cells and other mediator-releasing cells, which discharge chemical mediators upon IgE-allergen interactions. For the cross-linking of IgE molecules to take place, the absorbed allergen fragments must be large enough to contain multiple (at least two) IgE-binding sites. The minimum length of a peptide fragment necessary for this cross-linking is thought to be approximately 8 to 10 amino acids, longer than the typical peptide digestion product. For this reason, resistance to digestion is thought to be an important characteristic of allergenic proteins. In our findings, the absorbed peptides and proteins can be as large as (or larger than) 20 to 25 kDa (S. J. Maleki, unpublished data), capable of cross-linking adjacent IgE molecules, supporting structural configurations, and influencing digestibility and recognition by the immune system.

Allergens are characteristically resistant to digestive enzymes (Astwood et al., 1996), although a plausible mechanism for this observation was not described until a few years ago. It was shown that the quaternary structure of an allergic protein, Ara h 1, might dictate digestive enzyme accessibility and sites for digestion (Maleki et al., 2000a) and possibly define the immunodominant epitopes of a molecule. The quaternary structure of the allergic molecule was shown to protect the IgE-binding sites from enzymatic digestion. Thus, understanding the structure of allergens can be very important in describing and predicting allergenicity, processing-induced alterations, and their effect on allergenicity.

FUNCTIONAL PROPERTIES

Three dominating plant food allergen groups, the prolamin, cupin, and pathogenesis-related (PR) proteins, have been identified. The members of the prolamin superfamily encompass allergic 2S albumins, nonspecific lipid transfer proteins (LTPs), and cereal α-amylase–trypsin inhibitors. These proteins are structurally homologous, highly heat stable, and resistant to digestion with enzymes. The cupin superfamily proteins include the 7S and 11S globulin seed storage proteins from legumes and tree nuts. This group is capable of forming higher-order structures and is relatively heat stable.

PR proteins can inhibit the growth of pathogenic microorganisms such as bacteria, fungi, and viruses and interfere with feeding of pests (i.e., insects) on the plant. So far, 17 classes of PR have been identified and include chitinases, proteinase inhibitors, thionins, endoproteinase, peroxidase (POD), and LTPs (for a review, see Mills et al., 2004).

The functional properties of most allergic proteins from these plants have been described, properties that may or may not contribute to their allergenicity. One exception is the prolamine-cupin family of plant proteins, where the only described function is to serve as nutrients for development and growth of the embryo. For example, the 11S, 7S, and 2S albumin seed storage proteins are members of different plant families. Each has diverse as structures and functions within the seed and can present as allergen sources in foods (Mills et al., 2004). While some of these proteins may not have an active function in the plant, they may have properties that enhance their allergenicity in humans. Some have shown pepsin resistance (for a review, see Mills et al., 2004), which is possibly due to their structural aggregation, as previously discussed for Ara h 1. In one case, Ara h 3, a glycinin and major allergen from peanut, was specifically shown to act as a trypsin inhibitor (Dodo et al., 2004). Also, some 2S albumins and α-amylase and trypsin inhibitors of the prolamin cereal family have been shown to be inhibitors of human digestive enzymes. Inhibition of human digestive enzymes is a property that likely contributes to the allergenic properties of these proteins. Prolonged access of intact protein or peptides to the gastrointestinal immune system may enhance absorption through the gut mucosa. Maleki et al. (2003) specifically showed that an enzyme inhibitory molecule such as Ara h 2, the most potent peanut allergen (Koppelman et al., 2004), which acts as a trypsin inhibitor, not only protects itself from digestion in vitro but may also protect other allergens from digestion, possibly causing a synergistic enhancement of allergenic potential.

THE SPECIFIC REACTIONS IN PROCESSING THAT INFLUENCE ALLERGENIC PROPERTIES

Thermal processing such as roasting, curing, and various types of cooking can cause multiple, nonenzymatic, biochemical reactions to occur in foods (Shahidi et al., 1998). Therefore, it is important to understand which, if any, of these reactions influence the allergenicity of known allergens or other proteins within a food matrix. For example, it is known that heating of foods can cause a caramelization or browning effect in which proteins are decorated with a variety of sugar modifications and thus altered. This important phenomenon is known as the Maillard reaction (Maillard, 1912a and 1912b); it occurs during processing or browning of foods and is important in the development of flavor and color in many processed foods. In studies of peanut allergens, it has been shown that roasted peanut extracts bind serum IgE from allergic individuals at significantly higher levels than do raw peanut extracts. The observed effect was attributed, at least in part, to the Maillard reaction (Maleki et al., 2000b; Chung and Champagne, 2001; Chung et al., 2003). These studies demonstrate that peanut protein extracts Ara h 1 and Ara h 2 (two of the major allergens) from roasted peanuts are less soluble, less digestible, and have higher levels of IgE binding than raw peanuts and that the allergens undergo structural and functional alterations that enhance their allergenic properties. Furthermore, a comparison of the sensitization potency of raw and roasted peanuts demonstrated that mice are more likely to become sensitized to roasted than to raw peanuts (K. Yamaki and S. J. Maleki, unpublished data).

Although their numbers are rapidly increasing, relatively few studies have been performed to determine the molecular mechanisms behind the alterations in allergenicity due to processing. Interestingly, the mechanisms thought to influence the allergenic potential of different allergens under diverse processing conditions vary. These structural mechanisms include oligomerization (altering quaternary structure), reduction (altering tertiary structure), denaturation, and modification (possibly altering secondary structure) through reaction with other molecules (sugars, phosphor lipids, peptides, chemicals, etc). The majority of peanut (Maleki et al., 2001), almond, cashew nut, and walnut proteins (Su et al., 2004; Venkatachalam et al., 2002) and other food proteins, such as egg and milk (Moreno et al., 2005c), modified via the Maillard reaction or heat treatment are typically less soluble and sometimes bind higher levels of IgE. Reversible Ara h 1 trimers purified from raw peanuts and heated in the presence of reducing sugars develop irreversible, covalently cross-linked oligomeric complexes. The Ara h 1 oligomer complex was shown to be less soluble and more resistant to digestive enzymes than an Ara h 1 monomer or reversible trimer (Maleki et al., 2000b). In theory, survival of digestion-resistant fragments of Ara h 1 will be more likely to contain more than one intact IgE-binding site (Maleki et al., 2000b). Additionally, it is highly possible that the IgE-binding epitopes identified in proteins obtained from raw sources do not present the same epitopes as proteins from processed sources. Proteins such as LTP and 2S albumins, which have α-helical structures held together by multiple disulfide bonds, are highly resistant to heat and digestion in their nonaggregate forms (Moreno et al., 2005a, b). Ara h 2, with protein domain homology to trypsin inhibitors, was purified from both raw and roasted peanuts, and the properties of the two proteins were compared (Maleki et al., 2003). Trypsin inhibitory activity of Ara h 2 purified from roasted peanuts was approximately fourfold higher than that of Ara h 2 from raw peanuts. Ara h 2 is a protein that consists mostly of α helices that can form four disulfide linkages thought to be important in structural stabilization of the molecule. Partial reduction (reduction of some but not all disulfide bonds) of the protein, which had very little impact on the secondary structure, is thought to contribute to the difference in trypsin inhibitory function. At the molecular level, the Maillard reaction can cause loss, disruption, or rearrangement of sulfhydryl groups; therefore, the disulfide bonds of Ara h 2 may be affected during roasting. Interestingly, roasted Ara h 2 was more resistant to trypsin digestion than its native counterpart from raw peanuts; it also protected Ara h 1 from digestion with trypsin. This protective effect of one allergen on another is a good example of allergen intensification upon processing, as well as synergistic enhancement of allergenic properties of different proteins in the same food matrix. Grubber et al. (2005) assessed the specific molecular influence of the Maillard reaction on the allergenicity of peptides and intact recombinant Ara h 2. In this study, recombinant Ara h 2 subjected to the Maillard reaction was shown to have increased serum-specific IgE binding, confirming previous studies. However, when synthetic immunodominant epitopes or peptides containing lysine residues were analyzed, it was shown that the lysines, even if modified, did not contribute to enhanced IgE binding as previously speculated (Maleki et al., 2003). The immunodominant epitope 7, which does not contain lysine or arginine residues, showed a higher level of IgE binding following the Maillard reaction. This study demonstrated that nonbasic amino acids might play a key role in enhanced IgE binding to Ara h 2 via the Maillard reaction. Gruber et al. also demonstrated the complexity of assessing the effects of processing on allergenicity by showing that that the Maillard reaction decreased the IgE-binding capacity of peanut lectin and some of the Ara h 2 peptides. It would be very interesting to investigate allergenicity of Ara h 2 polypeptides after selective mutation or deletion of the amino acids critical to IgE binding to determine if they are the same as those modified by the Maillard reaction. Similar investigations could also target the amino acids modified via the Maillard reaction in the intact recombinant form of Ara h 2, as opposed to using peptides, to determine if the IgE-binding characteristics are similar.

Fruit LTPs from the Rosaceae family are highly homologous and immunologically cross-reactive. Sancho et al. (2005) investigated the effects of thermal processing in the presence and absence of glucose on Mal d 3 purified from apple peel. They found that mild heat treatment (90°C; 20 min) with or without sugar had no effect on structure or IgE reactivity. However, when subjected to more severe heat treatment (100°C; 2 h), the IgE reactivity and biological activity (histamine release) of the Mal d 3 was decreased significantly in the absence of sugar. When Mal d 3 was heated in the presence of sugar, the decrease in biological activity and IgE reactivity was much less significant. Therefore, glycation was shown to have a protective effect on the structure and allergenicity of this protein. This finding is particularly significant in the case of fruits, which contain high levels of sugar. Although, allergens can be studied in purified form, this observation suggests that they must be studied within the food matrix (van Wijk et al., 2005) to obtain a more realistic understanding of the effects of processing on a particular allergen. Studies clearly show that other ingredients within a processed food can also influence allegenicity (Kato et al., 2000; Kato et al., 2001; Gruber et al., 2004; Ehn et al., 2004; Mouecoucou et al., 2004; Moreno et al., 2005c; van Wijk et al., 2005).

CLINICAL AND BIOLOGICAL STUDIES

Heat-processed celery, maize, and roasted hazelnuts were assessed by in vitro IgE-binding assays and in vivo by double-blind, placebo-controlled food challenge for allergenicity (Ballmer-Weber et al., 2002; Pastorello et al., 2003; Skamstrup et al., 2003). Ballmer-Weber et al. (2002) found that all patients reacted to raw celery, some reacted only to raw celery, but no patients were found that reacted only to cooked celery, which argues against the formation of neoallergens in cooked celery. Meanwhile, results from mediator release experiments using rat basophilic leukemia (RBL) cells argued for either the formation of neoallergens or cryptic epitopes in cooked celery or the formation of novel IgE molecules in the rats sensitized with cooked celery. Comparison of the clinical and in vitro data did not allow the authors to predict which patients would be reactive to cooked celery. RBL cells loaded with mouse serum containing IgE against raw celery resulted in mediator release to Api g 1, a major celery allergen, but no mediator release was obtained when serum from mice sensitized to cooked celery was used. This implies that the allergenicity of Api g 1 is eliminated or highly reduced with cooking and that other allergenic moieties are responsible for mediator release. These data suggest either the formation of neoallergens in cooked celery or the formation of novel IgE molecules in the rats sensitized with cooked celery that do not recognize raw celery allergens. Interestingly, this finding was confirmed by enzyme allergosorbent testing with cooked celery as an inhibitor and using serum from an individual that exclusively recognized Api g 1. Pastorello et al. (2003) compared the allergenicity of maize proteins in raw and thermally processed forms and suggested that the LTP was heat resistant and most likely did not contain conformational epitopes. While this study makes an important contribution demonstrating that cooking does not eliminate the allergenicity of maize, the reduced solubility of maize allergens should be considered as a contributing factor in addition to enhanced stability. A reduction in the solubility of higher-molecular-weight proteins (possibly due to aggregation, denaturation, modification, or degradation) can be seen when the protein profiles of the soluble portion of raw and time-dependent, heated maize are compared (Pastorello et al. 2003).

Skamstrup et al. (2003) showed that studies with roasted hazelnuts had yielded results that differed from those with peanut, maize, and celery. Responses to IgE binding, RBL mediator release, skin prick test, and double-blind, placebo-controlled food challenge to hazelnut were all reduced by roasting. However, it is important to note that the study by Skamstrup et al. (2003) was carried out with a population that reacted to birch pollen and the pollen-related allergen, Cor a 1.04, which is homologous to and exhibits cross-reactivity with Bet v 1. Some of the patients also recognized recombinant Cor a 2, a profilin, but none of the individuals reacted to Cor a 8, an LTP, or the seed storage proteins (e.g., Cor a 9). Previous reports indicated that heat treatment reduces the reactivity of Cor a 1 with both birch pollen- and hazelnut-specific IgE. Although these results are not surprising, they demonstrate that cooking processes or treatments can eliminate or reduce the allergenicity of a particular food for at least a certain population. This observation also shows that studies with homologous, cross-reactive antigens can be useful in determining epitope or allergen inactivation due to treatments such as heat. Despite these findings, approximately 30% of the population tested still recognized and had a mild reaction to roasted hazelnut. One possibility is that some of the Cor a 1 remained partially intact in the insoluble fraction and was not detected by IgE-binding assays in which only the soluble fraction of the defatted hazelnut was used. The formation of a neoallergen (for the individuals tested) or reactivity of IgE to a known allergen other than Cor a 1 in the roasted hazelnut cannot be discounted. Unfortunately, the in vitro assays could not predict which patients would react to an oral challenge with roasted nuts. Considering that the Maillard reaction is known to enhance IgE binding and occurs during hazelnut roasting, it would be interesting to repeat these studies with patients with sera directed against the non-pollen-related hazelnut allergens, particularly the seed storage proteins. While the specific effects of processing and the Maillard reaction on the allergenicity of different foods are not fully understood, it is also intriguing that a major hazelnut allergen may be mostly degraded or denatured during the roasting process and rendered less allergenic. This suggests that roasting may increase or decrease allergenicity, depending on the particular allergic protein or perhaps even the IgE-binding capacity of a particular allergic population. These studies clearly demonstrate the complexity of the effects of processing on allergenicity.

REDUCING ALLERGENICITY

Since there are no medical treatments currently available that can cure allergies, the best option for a consumer suffering from food allergies remains complete avoidance of the offending agent. However, for various reasons, such avoidance may not always be possible. Among the several ways available to help protect sensitive consumers from unwanted exposure to an allergen or allergenic food are (i) accurate and proper food labeling; (ii) robust, sensitive, and specific detection methods for the targeted allergenic food; (iii) physical removal of an allergen; and (iv) reduction or elimination of an allergen through processing.

With the increased worldwide accessibility to modern processing technologies and the rapid advances in the ability to manipulate genes, more options may become available in the future to address the problem of food allergens. These methods include (i) physical removal of the targeted allergen; (ii) food processing, including allergen modification by various chemical or biochemical treatments; and (iii) genetic manipulation, a topic explored in Chapter 12. With the exception of physical removal to reduce or eliminate biopotency of an allergenic protein, one must alter the structure of the targeted allergenic protein or mask, delete, or alter the IgE-binding epitopes. Any loss in secondary or tertiary structure, commonly referred to as protein denaturation, is likely to alter the conformational epitopes, thereby potentially influencing an allergen's biological activity. Protein denaturation is, however, typically not expected to modify the allergenicity of linear epitopes, unless the conformational epitopes shield the linear epitope and thereby make the linear epitope inaccessible to IgE binding. It is therefore useful to know whether the immunodominant epitopes in the targeted food allergen are linear or conformational. To alter immunodominant linear epitopes, one may use (i) chemical or enzymatic hydrolysis, (ii) chemical modification of a critical amino acid residue(s), (iii) substitution of critical amino acid residues (genetic mutations), or (iv) a combination of these methods.

Physical Removal

The physical removal of an allergen from a food source can be effective when the allergen is present in a discrete location within the food, when an allergen has specific physical and biochemical properties distinct from the rest of the components in the target food (food system), or when the allergen is preferentially soluble or degradable.

The Allergen Is Present in Discrete Locations within the Food

For example, the peach (*Prunus persica*) is a recognized allergenic food. The major peach allergen, Pru p 1, is typically located in the epicarp of the peach (Lloonart et al., 1992) and its homologs are found in many other fruits (Martinez et al., 1997). Since Pru p 1 is a basic protein (pI, >9) and because peach pulp is typically acidic (pH 3.7 to 3.8), Pru p 1 is solubilized when peaches are processed to prepare peach juice, nectar, jam, or peach segments canned in syrup (Pastorello et al., 1999). Brenna et al. (2000) reported that lye peeling of peaches and two-stage ultrafiltration of peach juice (the molecular mass cutoff of the membrane was 6 to 8 kDa for the first stage and 5 kDa for the second) were effective in substantially removing the Pru p 1, as demonstrated by immunoblotting experiments. It should be noted here that since lye has a high pH (usually >10), some of the reduction might be attributed to protein denaturation. However, Brenna and coworkers (2000) also demonstrated the stability of Pru p 1 in peach homogenates prepared from peaches exposed to lye peeling or peach nectar exposed to heat treatment (10 and 30 min at 121°C). Taken together, these results suggest that Pru p 1 allergenicity may be primarily due to linear epitopes.

The Allergen Has Specific Physical and Biochemical Properties Distinct from the Rest of the Components in the Target Food (Food System)

By taking advantage of the tendency of β-lactoglobulin to reversibly polymerize in fluid milk, Chiancone and Gattoni (1993) demonstrated that β-lactoglobulin could be effectively removed from fluid milk to produce hypoallergenic milk. The authors demonstrated that β-lactoglobulin could be removed from milk without the use of any buffers by using β-lactoglobulin-coupled Sepharose 4B. The advantage of this method is that it does not remove any of the major milk caseins, thereby preserving the nutritional quality of fluid milk. Small-molecular-weight polypeptides and proteins can be effectively removed from fluid foods by reverse osmosis or ultrafiltration. Obviously, such methods will not be practical for solid and semisolid foods. When the targeted allergen constitute the bulk of the targeted food or is an important component of food quality, physical removal of the allergen may not be economically or physically practical.

The Allergen Is Soluble or Degradable

Although not an example using food, removal of allergen from latex gloves illustrates this principle. Kawahara et al. (2004) were able to remove allergenic protein from natural rubber by incubating rubber latex with urea in the presence of a surfactant. Under optimized conditions of temperature, pH, and incubation time, the nitrogen content of the rubber was decreased from 0.38 to 0.02% (by weight). Further treatment with proteolytic enzyme in the presence of a surfactant, followed by incubation in areas, decreased the allergen content to 0.7 mg/ml, which was a lower level than the one found in commercially manufactured gloves. Simple washing with solvents, such as water or aqueous alcohol, may be effective in substantially or completely removing allergens that may be present on the surface of food. For example, small amounts of airborne allergens (such as pollens, aerosolized foods, fine particles from grinding and milling operations, and particles present in aerosols and foams) may contaminate certain foods. Such situations are likely to arise where multiple food ingredients are handled in the same facility (vicinity) or when food preparation or manufacturing equipment is shared. Chen and Eggleston (2001) reported that an aqueous 5.25% solution of sodium hypochlorite (Clorox) was effective in eliminating allergenicity of Mus m 1, Fel d 1, Bla g 1, and Der p 1 as assessed by in vitro assays (Western blotting and protein determination).

Allergen Modification

Allergen modification by various processing methods has been attempted in different foods. Depending on the food and the molecular properties of the allergen, mixed results have been obtained. For example, major allergens in almond, cashew nut, and walnut have been reported to be stable towards various food processing treatments including gamma irradiation alone (up to 25 kGy) or when followed by thermal treatments such as blanching, auto claving, dry roasting, microwave heating, or frying (Su et al., 2004). On the other hand, commercial kiwifruit processing (steam cooking at 100°C for 5 min and homogenization), has recently been reported to eliminate kiwifruit sensitivity in certain kiwifruit-allergic children (Fiocchi et al., 2004).

Some of the other compounds that may influence the allergenicity of certain foods during processing include phytic acid, lipid oxidation intermediates, peroxidase (POD), polyphenol oxidase (PPO), and lysinoalanine (LAL).

Phytic acid has been shown to inhibit digestive enzymes such as proteases, α-amylases, and lipase (O'Dell and De Boland, 1976; Knuckles, 1987; and Knuckles, 1988). Phytic acid binds to proteins (Cheryan, 1980) and, as a result, affects protein digestibility and bioavailability.

Lipid oxidation intermediates such as malondialdehyde are formed during processing (Chung and Champagne, 2001), and allergens modified by fatty acids and malondialdehyde may be less allergenic than the unmodified proteins (Akita and Nakai, 1990; Matsuda et al., 1993).

POD is a heme-containing enzyme catalyzing the oxidation of a variety of phenolic compounds. The POD-generated *o*-quinones can react with another phenolic, amino, or sulfhydryl group to form a cross-linked structure. A study (Chung et al., 2003) indicated that POD was able to cross-link major peanut allergens. The resultant cross-link was thought to mask the IgE-binding sites.

PPO is a copper-containing enzyme involved in the enzymatic browning of many fruits and vegetables (Seo et al., 2003). The tyrosine residues in the proteins can be oxidized with PPO to the corresponding *o*-quinone and may then condense with each other or with the amino and sulfhydryl groups of other proteins to form protein cross-links or polymers (Matheis and Whitaker, 1984).

LAL is a product formed under heat and or alkaline conditions in food proteins from a reaction between lysine and dehydroalanine residues (Finot, 1983). The latter residue is generated by β-elimination of a cysteine or serine residue (Friedman and Pearce, 1989). Various levels of LAL have been detected in heat-processed food proteins, including soybean, casein, peanut, meat, and ultra-heat-treated milk (Savoie et al., 1991; Dehn-Muller et al., 1991; Nishino and Uchida, 1995; Davies, 1993; Pellegrino et al., 1996). LAL affects the availability of lysine residues, which is thought to change the allergenic properties of proteins (Hussein et al., 1995; Swart et al., 1996).

PROCESSING AND GEOGRAPHIC LOCATION

Many reasons have been suggested for the rise in prevalence of food allergies, particularly in developed countries. These suggestions include increased consumption or exposure to allergic foods, environmental and genetic factors, drugs, exposure by unusual routes such as the skin (Lack et al., 2003; Sampson et al., 2002), exposure through natural routes such as breast milk (Vadas et al., 2001), too much vaccination and hygiene (for a review, see Teixeira et al, 2005), consumption of novel and diversified foods through globalization of trade, the advancement of food technology, and the development of novel processing methods. For example, while other legumes such as lentils, soy, peas, and beans have proteins that are highly homologous (or similar) in sequence and in structure to peanut allergens, they rarely cause reactions as severe as peanut. Also, allergies to other legumes are often outgrown. It is possible that different forms in which these products are processed and consumed may influence the allergenic properties of these foods. Age, dose, and time of exposure may be responsible for the heightened allergenicity of particular proteins rather than the inherent allergenicity of the proteins in that food. Different food processing methods may influence the age, dose, and time of exposure. For example, in the Mediterranean area, one of the main plant foods causing allergic reactions is lentil, whereas the lentil is not a major allergen-causing legume in the United States (Lopez-Torrejon et al., 2003). Soy plays a similar allergic role in Japan, but it is not considered a major allergic food in the United States (Cordle, 2004; Bruno et al., 1997). Also, buckwheat, which is not considered a major food allergen in North America and Europe, causes deadly allergic responses in Japan (Takahashi et al., 1998). While it is possible, proteins that exhibit IgE cross-reactivity are not necessarily clinically allergenic (for a review, see Sicherer, 2001, and Ferreira et al., 2004).

Indeed, cross-reactivity among legumes has been documented to some extent but is rarely clinically significant (Sicherer, 2001). For example, patients with allergies to pea are most often not reactive to peanut, but three patients with anaphylactic reactions to pea have recently been documented to have a comparatively mild clinical cross-reactivity with the peanut allergen Ara h 1 (Wensing et al., 2003). One possibility is that some legumes such as peas, beans, and soy are processed so that an infant can consume large quantities at an early age and become tolerized. Studies have shown that exposure to small intermittent doses of a food favors sensitization, while exposure to large, regular doses favors tolerization. Interestingly, in less-industrialized countries (such as countries in Africa), peanuts are processed in such a manner that infants and children consume these products at a much earlier age and in larger quantities. In South Africa, a semiliquid, fermented, peanut product is often used as a weaning food for infants. However, the incidence of peanut allergy in these countries is very low. Due the magnitude of processes that different foods undergo before reaching consumers and the limited number of studies in this area, the relationship between food antigenicity and processing is poorly understood.

CONCLUSIONS

Ongoing studies indicate the following. (i) Future analyses of the allergenicity of foods should be carried out on foods in the form in which they are ingested. (ii) Clinicians should use extracts from foods in the form in which they are ingested for more-accurate diagnostic tests. (iii) Analysis of the allergenic potential of genetically modified foods containing recombinant proteins should be first processed by typical methods (i.e., harvesting, storage, manufacturing, and cooking) prior to being assessed for allergenic properties. (iv) A multifaceted and integrated research approach that would include attempts to reduce allergenicity via processing or genetic engineering in conjunction with the development of immunotherapeutic and computational prediction tools would allow more rapid advances towards solving or treating allergic disease than any single independent approach.

REFERENCES

Aba-Alkhail, B. A., and F. M. El-Gamal. 2000. Prevalence of food allergy in asthmatic patients. *Saudi Med. J.* **21:**81–87.

Akita, E. M., and S. Nakai. 1990. Lipophilization of beta-lactoglobulin: effect on allergenicity and digestibility. *J. Food Sci.* **55:**718–723.

Annan, W. D., and W. Manson. 1981.The production of lysinoalanine and related substances during processing of proteins. *Food Chem.* 6:255–261.

Astwood, J. D., J. N. Leach, and R. L. Fuchs. 1996. Stability of food allergens to digestion in vitro. *Nat. Biotechnol.* **10:**1269–1273.

Ballmer-Weber, B. K., A. Hoffmann, B. Wuthrich, D. Luttkopf, C. Pompei, A. Wangorsch, M. Kastner,and S. Vieths. 2002. Influence of food processing on the allergenicity of celery: DBPCFC with celery spice, cooked celery in patients with celery allergy. *Allergy* **57:**228–235.

Brenna, O., C. Pompei, C. Ortolani, V. Pravettoni, L. Farioli, and E. Pastorello. 2000. Technological processes to decrease the allergenicity of peach juice and nectar. *J. Agric. Food Chem.* **48:**493–497.

Bruno, G., P. G. Giampietro, M. J. Del Guercio, P. Gallia, L. Giovannini, C. Lovati, P. Paolucci, L. Quaglio, E. Zoratto, and L. Businco. 1997. Soy allergy is not common in atopic children: a multicenter study. *Pediatr. Allergy Immunol.* **8:**190–193.

Chen, P., and P. A. Eggleston. 2001. Allergenic proteins are fragmented in low concentrations of sodium hypochlorite. *Clin. Exp. Allergy* **31:**1086–1093.

Cheryan, M. 1980. Phytic acid interactions in food systems. *CRC Crit. Rev. Food Sci. Nutr.* **3:**297–335.

Chiancone, E., and M. Gattoni. 1993. Selective removal of β-lactoglobulin directly from cow's milk and preparation of hypoallergenic formulas: a bioaffinity method. *Biotechnol. Appl. Biochem.* **18:**1–8.

Chung, S. Y., and E. T. Champagne. 2001. Association of end-product adducts with increased IgE binding of roasted peanuts. *J. Agric. Food Chem.* **49:**3911–3916.

Chung, S. Y., C. L. Butts, S. J. Maleki, and E. T. Champagne. 2003. Linking peanut allergenicity to the processes of maturation, curing and roasting. *J. Agric. Food Chem.* **51:**4273–4277.

Chung, S. Y., S. J. Maleki, and E. T. Champagne. 2004. Allergenic properties of roasted peanut allergens may be reduced by peroxidase. *J. Agric. Food Chem.* **52:**4541–4545.

Cordle, C. T. 2004. Soy protein allergy: incidence and relative severity. *J. Nutr.* **134:**1213S–1219S.

Davies, R. L. 1993. D-Lysine, allisoleusine and lysinoalanine in supplementary proteins with different lysine availabilities. *J. Sci. Food Agric.* **61:**151–154.

Dehn-Muller, B., B. Muller, and H. F. Erbersdobler. 1991. Studies on protein damage in UHT milk. *Milchwissenschaft* **46:**431–434.

Dodo, H., O. M. Viquez, S. J. Maleki, and K. N. Konan. 2004. cDNA cloning of a putative peanut trypsin inhibitor with homology to peanut allergens Ara h 3 and Ara h 4. *J. Agric. Food Chem.* **52:**1404–1409.

Ehn, B. M., B. Ekstrand, U. Bengtsson, and S. Ahlstedt. 2004. Modification of IgE binding during heat processing of the cow's milk allergen beta-lactoglobulin. *J. Agric. Food Chem.* **52:**1398–1403.

Ferreira, F., T. Hawranek, P. Gruber, N. Wopfner, and A. Mari. 2004. Allergic cross-reactivity: from gene to the clinic. *Allergy* **59:**243–267.

Finot, P. A. 1983. Lysinoalanine in food proteins. *Nutr. Abstr. Rev. Clin. Nutr. A* **53:**67–80.

Fiocchi, A., P. Restani, L. Bernardo, A. Martelli, C. Ballabio, E. D'Auria, and E. Riva. 2004. Tolerance of heat-treated kiwi by children with kiwifruit allergy. *Pediatr. Allergy Immunol.* **15:**454–458.

Friedman, M., and K. N. Pearce. 1989. Copper (II) and cobalt (II) affinities of LL- and LD-lysinoalanine diastereomers: implication for food safety and nutrition. *J. Agric. Food Chem.* **37:**123–127.

Gray, G. M., and H. L. Cooper. 1971. Protein digestion and absorption. *Gastroenterology* **61:**535–544.

Gruber, P., S. Vieths, A. Wangorsch, J. Nerkamp, and T. Hofmann. 2004. Maillard reaction and enzymatic browning affect the allergenicity of Pru av 1, the major allergen from cherry (Prunus avium). *J. Agric. Food Chem.* **52:**4002–4007.

Gruber, P., W. M. Becker, and T. Hofmann. 2005. Influence of the Maillard reaction on the allergenicity of rAra h 2, a recombinant major allergen from peanut (Arachis hypogaea), its major epitopes, and peanut agglutinin. *J. Agric. Food Chem.* **53:**2289–2296.

Hansen, K. S., B. K. Ballmer-Weber, D. Luttkopf, P. S. Skov, B. Wuthrich, C. Bindslev-Jensen, S. Vieths, and L. K. Poulsen. 2003. Roasted hazelnuts-allergic activity evaluated by double-blind placebo-controlled food challenge. *Allergy* **58:**132–138.

Hill, D. J., C. S. Hosking, and R. G. Heine. 1999. Clinical spectrum of food allergy in children in Australia and south-east Asia: identification and targets for treatment. *Ann. Med.* **31:**272–281.

Hussein, S., E. Gelencser, and G. Hajos. 1995. Reduction of allergenicity and increasing the biological value of buffalo's milk proteins by enzymatic modification. *J. Food Biochem.* **19:**239–252.

Jenkins, J. A., S. Griffiths-Jones, P. R. Shewry, H. Breiteneder, and E. N. C. Mills. 2004. Structural relatedness of plant food allergens with specific reference to cross-reactive allergens: an in silico analysis. *J. Allergy Clin. Immunol.* **115:**163–170.

Kato, Y., H. Watanabe, and T. Matsuda. 1999. Ovomucoid rendered insoluble by heating with wheat gluten but not with milk casein. *Biosci. Biotechnol. Biochem.* **64:**198–201.

Kato, Y., E. Oozawa, and T. Matsuda. 2001. Decrease in antigenic and allergenic potentials of ovomucoid by heating in the presence of wheat flour: dependence on wheat variety and intermolecular disulfide bridges. *J. Agric. Food Chem.* **49:**3661–3665.

Kawahara, S., W. Klinklai, H. Kuroda, and Y. Isono. 2004. Removal of proteins from natural rubber with urea. *Polymers for Advanced Technologies* **15:**181–184.

Kim, Y. S., J. A. Nicholson, and K. J. Curtis. 1974. Intestinal peptide hydrolases: peptide and amino acid absorption. *Med. Clin. North Am.* **58:**1397–1412.

Knuckles, B. E. 1988. Effect of phytate and other myo-inositol phosphate esters on lipase activity. *J. Food Sci.* **53:**250.

Knuckles, B. E., and A. A. Betschart. 1987. Effect of phytate and other myo-inositol phosphate esters on α-amylase digestion of starch. *J. Food. Sci.* **52:**719.

Koppelman, S. J., M. Wensing, M. Ertmann, A. C. Knulst, and E. F. Knol. 2004. Relevance of Ara h1, Ara h2 and Ara h3 in peanut-allergic patients, as determined by immunoglobulin E Western blotting, basophil-histamine release and intracutaneous testing: Ara h2 is the most important peanut allergen. *Clin. Exp. Allergy* **34:**583–590.

Koppelman, S. J., C. A. D. M. Bruijnzeel-Koomen, M. Hessing, H. H. de Jongh. 1999. Heat induced conformational changes of Ara h 1, a major peanut allergen, do not affect its allergenic properties. *J. Biol. Chem.* **274:**4770–4777.

Kopper, R. A., N. J. Odum, M. Sen, R. M. Helm, S.J. Stanley, and A. W. Burks. 2004. Peanut protein allergens: gastric digestion is carried out exclusively by pepsin. *J. Allergy Clin. Immunol.* **114:**14–618.

Kushak, R. I., and H. S. Winter. 1999. Regulation of intestinal peptidases by nutrients in human fetuses and children. *Comp. Biochem. Physiol. A Mol. Integr. Physiol.***24:**191–198.

Lack, G., D. Fox, K. Northstone, J. Golding, and the Avon Longitudinal Study of Parents and Children Study Team. 2003. Factors associated with the development of peanut allergy in childhood. *N. Engl. J. Med.* **348:**977–985.

Lleoonart, R., A. Cisteró, J. Carrelra, A. Batista, and J. Moscoso del Prado. 1992. Food allergy: identification of the major IgE binding component of peach (Prunus persica). *Ann. Allergy* **69:**128–130.

Lopez-Torrejon, G., G. Salcedo, M. Martin-Estevan, A. Diaz-Perales, C. Y. Paschual, and R. Sanchez-Monge. 2003. Len c 1, a major allergen and vicilin from lentil seeds: protein isolation and cDNA cloning. *J. Allergy Clin. Immunol.* **112:**1208–1215.

Magni, C., C. Ballabio, P. Restani, E. Sironi, A. Scarafoni, C. Poiesi, and M. Duranti. 2005. Two-dimensional electrophoresis and Western-blotting analyses with anti Ara h 3 basic subunit IgG evidence the cross-reacting polypeptides of Arachis hypogaea, Glycine max, and Lupinus albus seed proteomes. *J. Agric. Food Chem.* **53:**2275–2281.

Maillard, L. C. 1912a. Action des acides amines sur les sucres: formation des melanoidines par voie methodique. *C. R. Acad. Sci.* **154:**66–68.

Maillard, L. C. 1912b. Formation d'humus et de combustibles mineraux sans intervention de l'oxygiene atmospherique, de microorganismes, de hautes temperatures, ou des fortes pressions. *C. R. Acad. Sci.* **155:**1554–1558.

Maleki, S. J. 2004. Food processing: effects on allergenicity. *Curr. Opin. Allergy Clin. Immunol.* **3:**241–245.

Maleki, S. J., R. A. Kopper, D. S. Shin, S. J. Stanley, H. Sampson, A. W. Burks, and G. A. Bannon. 2000a. Structure of the major peanut allergen Ara h 1 may protect IgE-binding epitopes form degradation. *J. Immunol.* **164:**5844–5849.

Maleki, S. J., S. Y. Chung, E. T. Champagne, and J. P. Raufman. 2000b. The effects of processing on the allergenic properties of peanut proteins. *J. Allergy Clin. Immnol.* **106:**763–768.

Maleki, S. J., S. Y. Chung, E. T. Champagne, and K. G. Khalifah. 2001. Allergic and biophysical properties of peanut proteins before and after roasting. *Food Allergy Intolerance* **2:**211–221

Maleki, S. J., O. Viquez, T Jacks, H. Dodo, E. T. Champagne, S-Y. Chung, and S. Laundry. 2003. The major peanut allergen, Ara h 2, functions as a trypsin inhibitor and roasting enhances this function. *J. Allergy Clin. Immunol.* **112:**190–195.

Martinez, A., M. Fernandez-Rivas, and R. Palacios. 1997. Improvement of fruit allergenic extracts for immunoblotting experiments. *Allergy* **52:**155–161.

Matheis, G., and J. Whitaker. 1984. Modification of proteins by polyphenol oxidase and peroxidase and their products. *J. Food Biochem.* **8:**137–162.

Matsuda, T., T. Ishii, K. Yamamoto, and R. Nakamura. 1993. Studies on allergenicity of soybean protein modified with oxidized lipid. *Rep. Soy Protein Res. Comm. Jpn.* **14:**21–27.

Mills, E. N., J. A. Jenkins, M. J. Alcocer, and P. R. Shewry. 2004. Related structural, biological, and evolutionary relationships of plant food allergens sensitizing via the gastrointestinal tract. *Crit. Rev. Food Sci. Nutr.* **44:**379–407.

Moreno, F. J., B. M. Maldonado, N. Wellner, and E. N. Mills. 2005a. Thermostability and in vitro digestibility of a purified major allergen 2S albumin (Ses i 1) from white sesame seeds (Sesamum indicum L.). *Biochim. Biophys. Acta* **1752:**142–153.

Moreno, F. J., F. A. Mellon, M. S. Wickham, A. R. Bottrill, and E. N. Mills. 2005b. Stability of the major allergen Brazil nut 2S albumin (Ber e 1) to physiologically relevant in vitro gastrointestinal digestion. *FEBS J.* **272:**341–352.

Moreno, F. J., A. R. Mackie, and E. N. Mills. 2005c. Phospholipid interactions protect the milk allergen α-lactalbumin from proteolysis during in vitro digestion. *J. Agric. Food Chem.* **53:**9810–9816.

Mouecoucou, J., S. Fremont, C. Sanchez, C. Villaume, and L. Mejean. 2004. In vitro allergenicity of peanut after hydrolysis in the presence of polysaccharides. *Clin. Exp. Allergy* **34:**1429–1437.

Nishino, N., and S. Uchida. 1995. Formation of lysinoalanine following alkaline processing of soya bean meal in relation to the degradability of protein in the rumen. *J. Sci. Food Agric.* **68:**59–64.

O'Dell, B. L., and A. De Boland. 1976. Complexation of phytate with proteins and cations in corn germ and oilseed meals. *J. Agric. Food Chem.* **24:**804.

Pastorello, E. A., L. Farioli, V. Pravettoni, C. Ortolani, M. Ispano, M. Monza, C. Baroglio, E. Scibola, R. Ansaloni, C. Incorvaia, and A. Conti. 1999. The major allergen of peach (Prunus persica) is a lipid transfer protein. *J. Allergy Clin. Immunol.* **103:**520–526.

Pastorello, E. A., C. Pompei, V. Pravettoni, L. Farioli, A. M. Calamari, J. Scibilia, A. M. Robino, A. Conti, S. Iametti, D. Fortunato, S. Bonomi, and C. Ortolani. 2003. Lipid-transfer protein is the major maize allergen maintaining IgE-binding activity after cooking at 100 degrees C, as demonstrated in anaphylactic patients and patients with positive double-blind, placebo-controlled food challenge results. *J. Allergy Clin. Immunol.* **112:**775–783.

Pellegrino, L., P. Resmini, I. De Noni, and F. Masotti. 1996. Sensitive determination of lysinoalanine for distinguishing natural from imitation Mozzarella cheese. *J. Dairy Sci.* **79:**725–734.

Pereira, M. J., M. T. Belver, C. Y. Pascual, and M. Esteban. 2002. The allergenic significance of legumes. *Allergol. Immunopathol.* (Madrid) **30:**346–353.

Perr, H. A. 2004. Novel foods to treat food allergy and gastrointestinal infection. *Curr. Gastroenterol. Rep.* **6:**254–260.

Pitche, P., A. Bahounde, K. Agbo, and K. Tchangai-Walla. 1996. Etiology of isolated pruritus in dermatology consultations at Lome (Togo). *Sante* **6:**17–19.

Radauer, C., and H. Breiteneder. 2006. Pollen allergens are restricted to few protein families and show distinct patterns of species distribution. *J. Allergy Clin. Immunol.* **117:**141–147.

Sampson, H. A. 2002. Clinical practice. Peanut allergy. *N. Engl. J. Med.* **346:**1294–1299.

Sancho, A. I., N. M. Rigby, L. Zuidmeer, R. Asero, G. Mistrello, S. Amato, E. Gonzalez-Mancebo, M. Fernandez-Rivas, R. van Ree, and E. N. Mills. 2005. The effect of thermal processing on the IgE reactivity of the non-specific lipid transfer protein from apple, Mal d 3. *Allergy* **10:**1262–1268.

Sathe, S. K., H. H. Kshirsagar, and K. H. Roux. 2005a. Advances in seed protein research: a perspective on seed allergens. *J. Food Sci.* **70:**R93–R120.

Sathe, S. K., S. S. Teuber, and K. H. Roux. 2005b. Effects of food processing on the stability of food allergen. *Biotech. Adv.* **23:**423–429.

Savoie, L., G. Parent, and I. Galibois. 1991. Effects of alkali treatment on the in vitro digestibility of proteins and the release of amino acids. *J. Sci. Food Agric.* **56:**363–372.

Seo, S. Y., V. K. Sharma, and N. Sharma. 2003. Mushroom tyrosinase: recent prospects. *J. Agric. Food Chem.* **51:**2837–2853.

Shahidi, F., and T.-C. Ho. 1998. *Process-Induced Chemical Changes in Foods*, 2nd ed. Plenum Press, New York, N.Y.

Sicherer, S. H. 2001.Clinical implications of cross-reactive food allergens. *J. Allergy Clin. Immunol.* **108:**881–890.

Su, M., M. Venkatachalam, S. S. Teuber, K. H. Roux, and S. K. Sathe. 2004. Impact of γ-irradiation and thermal processing on the antigenicity of almond, cashew nut and walnut proteins. *J. Sci. Food Agric.* **84:**1119–1125.

Swart, P. J., M. E. Kuipers, C. Smit, R. Pauwels, M. P. deBethune, E. de Clercq, D. K. F. Meijer, and J. G. Huisman. 1996. Antiviral effects of milk proteins: acylation results in polyanionic compounds with potent activity against human immunodeficiency virus types 1 and 2 in vitro. *AIDS Res. Hum. Retrovir.* **12:**769–775.

Takahashi, Y., S. Ichikawa, Y. Aihara, and S. Yokota. 1998. Buckwheat allergy in 90,000 school children in Yokohama. *Arerugi* **47:**26–33.

Teixeira, M. Z. 2005. The hygiene hypothesis revisited. *Homeopathy* **94:**248–251.

Ugolev, A. M., N. M. Mityushova, V. V. Egorova, I. K. Gozite, and G. G. Koltushkina. 1979. Catalytic and regulatory properties of the triton and trypsin forms of brush border hydrolases *Gut* **20:**737–742.

Vadas, P., Y. Wai, A. W. Burks, and B. Perelman. 2001. Detection of peanut allergens in breast milk of lactating women. *JAMA* **285:**1746–1748.

vanWijk, F., S. Nierkens, I. Hassing, M. Feijen, S. J. Koppelman, G. A. de Jong, R. Pieters, and L. M. Knippels. 2005. The effect of the food matrix on in vivo immune responses to purified peanut allergens. *Toxicol Sci.* **86:**333–341.

Venkatachalam, M., S. S. Teuber, K. H. Roux, and S. K. Sathe. 2002. Effects of roasting, blanching, autoclaving, and microwave heating on antigenicity of almond (*Prunus dulcis* L.) proteins. *J. Agric. Food Chem.* **50:**3544–3548.

Visitsunthorn, N., A. Tiranathanakul, R. Netrakul, and P. Vichyanond. 2002. Evaluation of consistency between local and imported seafood allergen extracts. *J. Med. Assoc. Thai.* **85**(Suppl. 2)**:**S593–S598.

Wang, N. R., and H. Q. Li. 2005. Prognoses of food allergy in infancy. *Zhonghua Er Ke Za Zhi* **43:**777–781.

INDEX